AN INTRODUCTION TO CLASSICAL MECHANICS

AN INTRODUCTION TO CLASSICAL MECHANICS

[For BSc (Hons) and MSc Students in Physics]

Dr Binoy Bhattacharyya PhD (Moscow), MIE
Formerly Sr Scientific Officer, Structural Engineering
Research Centre, Roorkee and Chennai
Formerly Superintending Engineer (Civil)
Damodar Valley Corporation, Kolkata

New Central Book Agency (P) Ltd
LONDON
HYDERABAD ERNAKULAM BHUBANESWAR
NEW DELHI KOLKATA PUNE GUWAHATI

NCBA

RECD OFFICE
8/1 Chintamoni Das Lane, Kolkata 700 009, India
email: ncbapvtltd@eth.net

OVERSEAS
NCBA (UK) Ltd, 149 Park Avenue North, Northampton, NN3 2HY, UK
email: ncbauk@yahoo.com

EXPORT
NCBA Exports Pvt. Ltd, 212 Shahpur Jat, New Delhi 110 049
email: ncbaexports@ncbapvtltd.com

BRANCHES

2/27 Ansari Road, 1st Floor
Daryaganj, **New Delhi** 110 002
email: ncbadel@ncbapvtltd.com

Shop Nos. 3 & 4, Vinayak Towers
681/B Budhwar Peth, Appa Balwant Chowk
Pune 411 002
email: ncbapune@ncbapvtltd.com

Shop No. 15, Bharati Towers, Ground Floor
Forest Park
Bhubaneswar 751 001
email: ncbabub@ncbapvtltd.com

House No. 3-1-315, 1st Floor
Nimboliadda, Kachiguda, **Hyderabad** 500 027
email: ncbahydb@ncbapvtltd.com

GSS Shopping Complex, 1st Floor
Opposite College Post Office, Convent Road
Ernakulam 682 035
0484-4021054/53

Radhamayee Complex (Second Floor)
Opp. SBI, Panbazar, Jaswanta Road
Guwahati 781 001
email: ncbaguwahati@ncbapvtltd.com

AN INTRODUCTION TO CLASSICAL MECHANICS • Dr Binoy Bhattacharyya

First Published: January 2011
With Additions: January 2012
Thoroughly Revised and Enlarged Second Edition: July 2016

PUBLISHER AND TYPESETTER
New Central Book Agency (P) Ltd
8/1 Chintamoni Das Lane, Kolkata 700 009

PRINTER
New Central Book Agency (P) Ltd
Web-Offset Division, Dhulagarh, Sankrail, Howrah

PROOF CHECKING
Dipan Roy

PROJECT TEAM
Subir Pal and Pradip Biswas

ISBN: 978 81 7381 646 8

Price ₹ 295.00

To

my wife

Ashoka

Contents

Preface to the Second Edition

The warm acceptance of my book "An Introduction to Classical Mechanics" by the students and readers is really gratifying to me. I am, therefore, very much pleased to present this second edition of the book.

In this edition, additional worked-out problems have been added at the end of almost every chapter. The concept of artificial satellites have been included in Chapter-3. The chapters on "Anharmonic Oscillations" and "Canonical Transformations and Brackets" have been enlarged.

I sincerely believe that this book will be of considerable benefit to the students and readers for years to come.

I am thankful to my son, Dr. Anirban Bhattacharyya, Ph.D (Ohio), for his valuable suggestions. Thanks are also due to Mr. Amitabha Sen, Managing Director, New Central Book Agency (P) Ltd, for his keen interest in bringing this second edition to light. The proof-checking job performed by Sri Dipan Roy is commendable.

May, 2016 Dr Binoy Bhattacharyya
Tollygunge
Kolkata 700 040

Preface to the First Edition

I have observed that textbooks on Classical Mechanics have been written mostly for the benefits of the Postgraduate students. Hence the students at the Undergraduate level do not feel pretty comfortable with them. This book has been written to serve the purpose of both the Honours and Postgraduate students. It is a humble attempt to make the students familiar with the subject right from the beginning—gradually and smoothly—in the simplest possible manner.

In addition to the knowledge of elementary calculus and differential equations, basic knowledge of vector analysis and matrix operations are needed to understand this book. Appendix A for Vector Analysis and Appendix B for Rudiments of Matrices have, therefore, been provided at the end of the text.

I have also discussed the Special Theory of Relativity including relevant advanced topics like Minkowski space.

I would feel myself honoured if even a single student finds this book useful.

I would like to thank Sri P. K. Chowdhury for his excellent proof reading job. Special thanks go to Mr Amitabha Sen, Director, New Central Book Agency (P) Ltd, and his competent production team for taking active interest in bringing this book to light.

Constructive criticisms and valuable suggestions from every corner will be gratefully accepted.

Dr Binoy Bhattacharyya

January, 2011
Tollygunge
Kolkata 700 040

CHAPTER 1

Mechanics of a Single Particle

1.1 Introduction

The position of a particle moving in space at different instants of time is described with respect to a chosen coordinate system called ***reference frame***. If a particle under no external force moves in a straight line with constant velocity (or remains at rest), then the coordinate system used is called ***inertial frame***. Thus, ***an inertial frame is the one in which the law of inertia holds good***. The inertial frames are also called ***Galilean frames*** of reference in the name of the famous scientist Galileo Galilei who first stated the law of inertia. Newton's laws of motion are valid in inertial frames only.

1.2 Velocity and Acceleration

Let the position of a particle at any instant of time be specified by its position vector $\mathbf{r}$, the magnitude and direction of which are functions of time. Let, in time Δt, the position vector of the particle change to $\mathbf{r} + \Delta\mathbf{r}$; then the velocity of the particle is given by

$$\mathbf{v} = \lim_{\Delta t \to 0} \frac{(\mathbf{r} + \Delta\mathbf{r}) - \mathbf{r}}{\Delta t} = \lim_{\Delta t \to 0} \frac{\Delta\mathbf{r}}{\Delta t} = \frac{d\mathbf{r}}{dt} \tag{1.1}$$

Similarly, if the velocity vector changes from $\mathbf{v}$ to $\mathbf{v} + \Delta\mathbf{v}$ in time Δt, we obtain acceleration of the particle

$$\mathbf{a} = \lim_{\Delta t \to 0} \frac{(\mathbf{v} + \Delta\mathbf{v}) - \mathbf{v}}{\Delta t} = \lim_{\Delta t \to 0} \frac{\Delta\mathbf{v}}{\Delta t} = \frac{d\mathbf{v}}{dt} \tag{1.2}$$

In Cartesian coordinate system, we write

$$\mathbf{r} = x\mathbf{i} + y\mathbf{j} + z\mathbf{k} \tag{1.3}$$

where $\mathbf{i}, \mathbf{j}, \mathbf{k}$ are unit vectors along x-, y- and z-axes. These unit vectors are constants both in magnitude and direction.

Therefore, the velocity of the particle is

$$\mathbf{v} = \frac{d\mathbf{r}}{dt} = \frac{dx}{dt}\mathbf{i} + \frac{dy}{dt}\mathbf{j} + \frac{dz}{dt}\mathbf{k} \tag{1.4}$$

The components of the velocity vector along coordinate axes are

$$v_x = \dot{x};\ v_y = \dot{y};\ v_z = \dot{z} \tag{1.5}$$

where we have denoted the time derivative by a dot.

Similarly, the acceleration of the particle is

$$\mathbf{a} = \frac{d\mathbf{v}}{dt} = \ddot{x}\mathbf{i} + \ddot{y}\mathbf{j} + \ddot{z}\mathbf{k} \tag{1.6}$$

so that the components of the acceleration along the coordinate axes are

$$a_x = \ddot{x};\ a_y = \ddot{y};\ a_z = \ddot{z}. \tag{1.7}$$

1.3 Derivative of a Vector

Consider a vector $\mathbf{r}$ referenced to a fixed coordinate system. If in a time Δt, the vector changes to $\mathbf{r} + \Delta\mathbf{r}$, then its derivative w.r.t. time is defined as

$$\frac{d\mathbf{r}}{dt} = \lim_{\Delta t \to 0} \frac{(\mathbf{r} + \Delta\mathbf{r}) - \mathbf{r}}{\Delta t} = \lim_{\Delta t \to 0} \frac{\Delta\mathbf{r}}{\Delta t} \tag{1.8}$$

which gives the total derivative.

Consider now a vector $\mathbf{r}$ referenced to a rotating coordinate system. If $\mathbf{i_r}$ be a unit vector along r, then

$$\mathbf{r} = r\mathbf{i_r} \tag{1.9}$$

so that

$$\frac{d\mathbf{r}}{dt} = \frac{dr}{dt}\mathbf{i_r} + r\frac{d}{dt}(\mathbf{i_r}) \tag{1.10}$$

Now consider the unit vector $\mathbf{i_r}$ rotating about an axis ON as shown in FIG. 1.1. In a time Δt, the end point of the vector will undergo a change equal to

$$(i_r \sin\phi)\Delta\theta \tag{1.11}$$

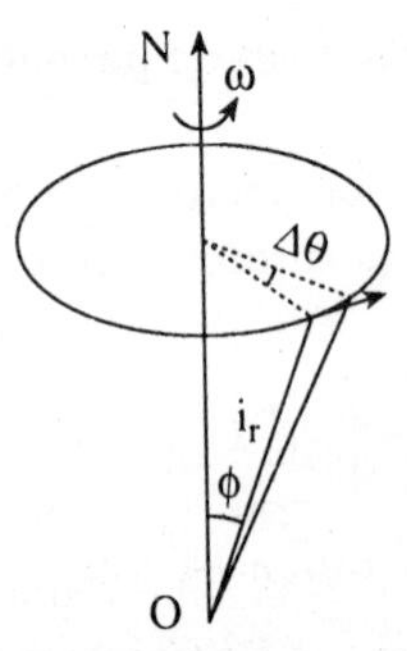

FIG. 1.1

along the tangent to the circle.

Hence, its rate of change along the tangent is

$$\lim_{\Delta t \to 0} (i_r \sin\phi)\frac{\Delta\theta}{\Delta t} = \omega i_r \sin\phi \tag{1.12}$$

where ω is the angular velocity. Hence, we write:

$$\frac{d}{dt}(\mathbf{i_r}) = \omega i_r \sin\phi = \boldsymbol{\omega} \times \mathbf{i_r} \tag{1.13}$$

Thus, the total derivative of the rotating vector $\mathbf{r}$ is

$$\frac{d\mathbf{r}}{dt} = \frac{d\mathbf{r}}{dt} + \boldsymbol{\omega} \times \mathbf{r} \tag{1.14}$$

Thus, we find that the total derivative of a rotating vector $\mathbf{r}$ consists of two parts: (1) the first part denotes the rate of change when the unit vector is fixed and (2) the second part arises due to the rotation of the unit vector.

1.4 Radial and Transverse Components

Consider a particle moving along a curve fixed in a stationary XY-plane (FIG. 1.2). Let $\mathbf{i_r}$ and $\mathbf{i_\theta}$ be unit vectors along radial and transverse directions. These unit vectors are not constants in direction. The position vector of the particle at any point P is given by

$$\mathbf{r} = r\mathbf{i_r} \tag{1.15}$$

Then the velocity of the particle is

$$\frac{d\mathbf{r}}{dt} = \frac{dr}{dt}\mathbf{i_r} + r\frac{d}{dt}(\mathbf{i_r}) \tag{1.16}$$

Now, the unit vector $\mathbf{i_r}$ is rotating with an angular velocity $\dot{\boldsymbol{\theta}}$ about an axis through the origin perpendicular to the XY-plane. Hence, its derivative by previous article is the cross product of the vectors $\dot{\boldsymbol{\theta}}$ and $\mathbf{i_r}$. That is

$$\frac{d}{dt}(\mathbf{i_r}) = \dot{\boldsymbol{\theta}} \times \mathbf{i_r} = \dot{\theta}\mathbf{i_\theta} \tag{1.17}$$

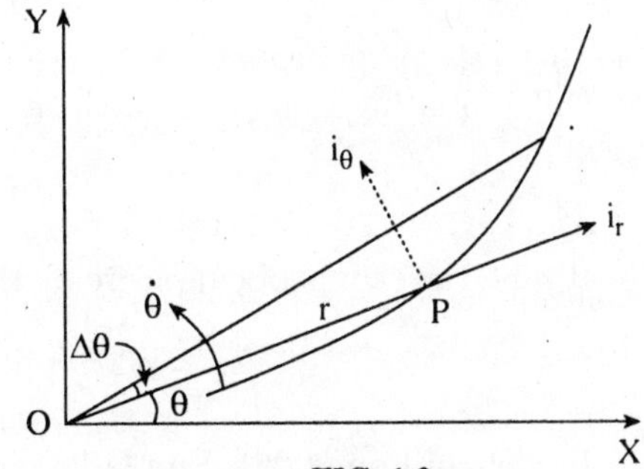

FIG. 1.2

Eqn (1.16) is then written as

$$\frac{d\mathbf{r}}{dt} = \frac{dr}{dt}\mathbf{i_r} + r\dot{\theta}\mathbf{i_\theta}, \text{ i.e., } \mathbf{v} = \dot{r}\mathbf{i_r} + r\dot{\theta}\mathbf{i_\theta} \tag{1.18}$$

which gives the velocity vector. Thus, the radial and transverse components of velocity are

$$v_r = \dot{r}; \; v_\theta = r\dot{\theta}. \tag{1.19}$$

The acceleration of the particle is determined by differentiating Eqn (1.18). Thus,

$$\mathbf{a} = \ddot{\mathbf{r}} = \ddot{r}\mathbf{i_r} + (r\ddot{\theta} + \dot{r}\dot{\theta})\mathbf{i_\theta} + r\dot{\theta}\frac{d}{dt}(\mathbf{i_\theta}) \tag{1.20}$$

As before

$$\frac{d}{dt}(\mathbf{i_\theta}) = \dot{\boldsymbol{\theta}} \times \mathbf{i_\theta} = -\dot{\theta}\mathbf{i_r} \tag{1.21}$$

Thus, Eqn (1.20) is written as

$$\mathbf{a} = \ddot{r}\mathbf{i_r} + (r\ddot{\theta} + \dot{r}\dot{\theta})\mathbf{i_\theta} - r\dot{\theta}^2\mathbf{i_r}, \text{ i.e., } \mathbf{a} = (\ddot{r} - r\dot{\theta}^2)\mathbf{i_r} + (r\ddot{\theta} + \dot{r}\dot{\theta})\mathbf{i_\theta} \tag{1.22}$$

Eqn (1.22) determines the acceleration of the particle. The radial and transverse components of acceleration are given by

$$\left.\begin{aligned} a_r &= \ddot{r} - r\dot{\theta}^2 \\ a_\theta &= r\ddot{\theta} + 2\dot{r}\dot{\theta} \end{aligned}\right\} \tag{1.23}$$

1.5 Tangential and Normal Components

We consider, as before, the motion of a particle along a curve fixed in XY-plane (FIG. 1.3). Let $\mathbf{i_t}$ and $\mathbf{i_n}$ be unit vectors along the tangential and normal directions to the curve. These unit vectors constantly change their directions along with the motion of the particle. At any instant of time, let s be the arc length of the point P where the particle is situated. Then the velocity of the particle is

$$\mathbf{v} = \lim_{\Delta t \to 0} \frac{\Delta \mathbf{r}}{\Delta t} = \lim_{\Delta t \to 0} \frac{\Delta \mathbf{r}}{\Delta s}\frac{\Delta s}{\Delta t}. \tag{1.24}$$

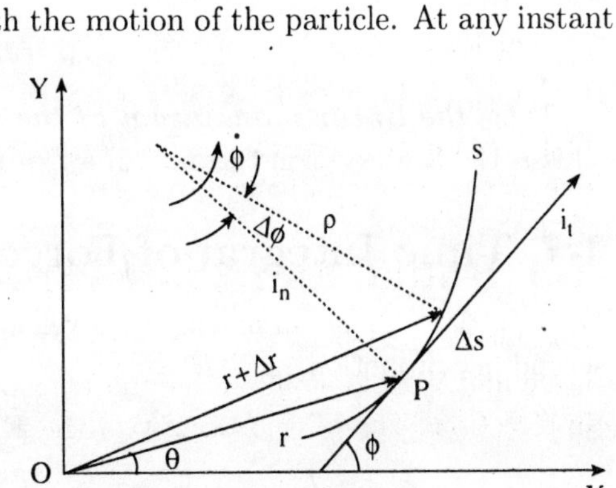

FIG. 1.3

Now

$$\lim_{\Delta t \to 0} \frac{\Delta \mathbf{r}}{\Delta s} = \mathbf{i_t} \tag{1.25}$$

which is the unit vector along the tangent to the curve. Hence, Eqn (1.24) reduces to

$$\mathbf{v} = \frac{ds}{dt}\mathbf{i_t} = \dot{s}\mathbf{i_t} \tag{1.26}$$

which determines the velocity along tangential direction. The two normals drawn at points $\mathbf{r}$ and $\mathbf{r} + \Delta\mathbf{r}$ meet at a point making a small angle $\Delta\phi$ between them. These normals will be along the corresponding radii of curvatures. Hence, $\dot{\phi}$ is the angular rate of the radius of curvature. Then we can write:

$$\frac{d}{dt}(\mathbf{i_t}) = \dot{\boldsymbol{\phi}} \times \mathbf{i_t} = \dot{\phi}\mathbf{i_n} \tag{1.27}$$

Differentiating Eqn (1.26) with the use of Eqn (1.27) we obtain the acceleration

$$\mathbf{a} = \frac{d\mathbf{v}}{dt} = \ddot{s}\mathbf{i_t} + \dot{s}\dot{\phi}\mathbf{i_n} \tag{1.28}$$

If ρ be the radius of curvature at the point,

$$\Delta s = \rho\Delta\phi \left(\text{in the limit } \rho = \frac{ds}{d\phi}\right), \text{ i.e., } \dot{s} = \rho\dot{\phi} \tag{1.29}$$

In view of Eqn (1.29), Eqn (1.28) reduces to

$$\mathbf{a} = \ddot{s}\mathbf{i_t} + \frac{\dot{s}^2}{\rho}\mathbf{i_n} \tag{1.30}$$

$$= \ddot{s}\mathbf{i_t} + \rho\dot{\phi}^2\mathbf{i_n} \tag{1.31}$$

Thus, the components of acceleration along tangential and normal directions are

$$a_t = \ddot{s} = \frac{dv}{dt}; \; a_n = \frac{\dot{s}^2}{\rho} = \frac{v^2}{\rho} \tag{1.32}$$

1.6 Conservation of Linear Momentum

Consider a particle of mass m moving under the action of an external force $\mathbf{F}$. Then by Newton's second law of motion

$$\mathbf{F} = \frac{d\mathbf{p}}{dt} = \frac{d}{dt}(m\mathbf{v}) \tag{1.33}$$

where $\mathbf{v}$ is the velocity and $\mathbf{p}$ is the linear momentum of the particle.

Hence, if the external force $\mathbf{F}$ acting on the particle is zero, then

$$\frac{d\mathbf{p}}{dt} = \frac{d}{dt}(m\mathbf{v}) = 0 \tag{1.34}$$

which yields

$$\mathbf{p} = m\mathbf{v} = \text{constant} \tag{1.35}$$

Thus, ***the linear momentum of the particle under no external force is conserved***. This is the conservation theorem of a free particle.

1.7 Time Integral of Force

For a particle of mass m moving with velocity $\mathbf{v}$ under an external force $\mathbf{F}$, we have by Newton's second law of motion

$$\mathbf{F} = \frac{d}{dt}(m\mathbf{v}) \tag{1.36}$$

Hence, we get

$$\int_{t_1}^{t_2} \mathbf{F}dt = \int_{v_1}^{v_2} d(m\mathbf{v}) \tag{1.37}$$

The vector $\mathbf{F}dt$ is called ***impulse of the force*** $\mathbf{F}$ acting in time dt. Thus, Eqn (1.37) yields the impulse of the force acting during the time from t_1 to t_2 and its value is

$$\int_{t_1}^{t_2} \mathbf{F}dt = m\mathbf{v_2} - m\mathbf{v_1} \tag{1.38}$$

which is equal to the change in momentum. If the force is constant, Eqn (1.38) reduces to

$$\mathbf{F}(t_2 - t_1) = m\mathbf{v_2} - m\mathbf{v_1} \tag{1.39}$$

The impulsive force is a large force acting for a very short time so that the product is a finite value.

1.8 Conservation of Angular Momentum

The angular momentum $\mathbf{J}$ of a particle of mass m about a point O is defined as (FIG. 1.4)

$$\mathbf{J} = \mathbf{r} \times \mathbf{p} \tag{1.40}$$

If the force acting on the particle be $\mathbf{F}$, then the moment of the force or torque about the origin is defined as

$$\mathbf{N} = \mathbf{r} \times \mathbf{F} \tag{1.41}$$

Differentiating Eqn (1.40) w.r.t. time,

$$\frac{d\mathbf{J}}{dt} = \frac{d}{dt}(\mathbf{r} \times \mathbf{p}) = \frac{d\mathbf{r}}{dt} \times \mathbf{p} + \mathbf{r} \times \frac{d\mathbf{p}}{dt} \tag{1.42}$$

But

$$\frac{d\mathbf{r}}{dt} \times \mathbf{p} = \mathbf{v} \times m\mathbf{v} = 0 \tag{1.43}$$

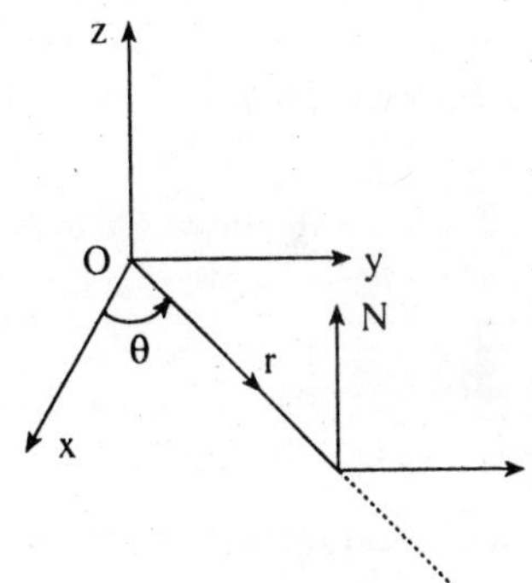

FIG. 1.4

Hence, from Eqns (1.42) and (1.43), we get

$$\frac{d\mathbf{J}}{dt} = \mathbf{r} \times \frac{d\mathbf{p}}{dt} = \mathbf{r} \times \mathbf{F} \tag{1.44}$$

Therefore,

$$\mathbf{N} = \mathbf{r} \times \mathbf{F} = \frac{d\mathbf{J}}{dt} \tag{1.45}$$

Thus, ***the torque acting on the particle is equal to the rate of change of the angular momentum.***

If $\mathbf{N} = 0$, then $\frac{d\mathbf{J}}{dt} = 0$, i.e., $\mathbf{J}$ = constant.

Therefore, ***in absence of external torque, the angular momentum of the particle is constant.*** This is the ***conservation theorem*** of angular momentum of a particle.

1.9 Work and Kinetic Energy

Work done by an external force $\mathbf{F}$ upon a particle in displacing it from point 1 to point 2 is defined as

$$W = \int_1^2 \mathbf{F} \cdot d\mathbf{r} \tag{1.46}$$

Integral on the RHS is a line integral of force.

Now, if mass of the particle is constant,

$$\mathbf{F}\cdot d\mathbf{r} = m\frac{d\mathbf{v}}{dt}\cdot\frac{d\mathbf{r}}{dt}dt = m\frac{d\mathbf{v}}{dt}\cdot\mathbf{v}dt = m\frac{d}{dt}\left(\frac{1}{2}\mathbf{v}\cdot\mathbf{v}\right) = d\left(\frac{1}{2}mv^2\right) \tag{1.47}$$

Therefore, the work done is

$$W_{12} = \int_1^2 \mathbf{F}\cdot d\mathbf{r} = \int_1^2 d\left(\frac{1}{2}mv^2\right) = \frac{1}{2}mv_2^2 - \frac{1}{2}mv_1^2 \tag{1.48}$$

The kinetic energy T of a particle of mass m moving with velocity v is the scalar quantity

$$T = \frac{1}{2}mv^2 \tag{1.49}$$

Then Eqn (1.48) is written as

$$W_{12} = \int_1^2 \mathbf{F}\cdot d\mathbf{r} = T_2 - T_1 \tag{1.50}$$

where T_1, T_2 are the kinetic energies of the particle at points 1 and 2 respectively. Thus, ***the work done by the force acting on the particle is equal to the change in the kinetic energy***. This is known as ***work-energy theorem.***

If $W_{12} > 0$, the work is done by the force on the particle. In this case, $T_2 > T_1$, that is, the kinetic energy of the particle has increased.

If $W_{12} < 0$, the work is done by the particle against the force. In this case, $T_2 < T_1$, that is, the kinetic energy of the particle has decreased.

1.10 Conservative Force

The force $\mathbf{F}$ is said to be conservative if the work done by it in moving a particle from point 1 to point 2 along any possible path is the same. Thus, for a conservative force $\mathbf{F}$, shown in FIG. 1.5,

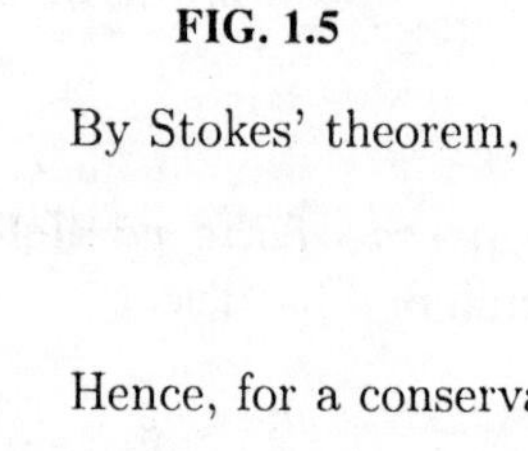

FIG. 1.5

$$\int_1^2 \mathbf{F}\cdot d\mathbf{r} \text{ along curve } A = \int_1^2 \mathbf{F}\cdot d\mathbf{r} \text{ along curve } B,$$

which yields

$$\oint \mathbf{F}\cdot d\mathbf{r} = 0 \tag{1.51}$$

Thus, if the force is conservative, the work done along a closed path is zero.

By Stokes' theorem,

$$\oint \mathbf{F}\cdot d\mathbf{r} = \iint_S \nabla\times\mathbf{F}\cdot d\mathbf{S} \tag{1.52}$$

Hence, for a conservative force-field,

$$\iint_S \nabla\times\mathbf{F}\cdot d\mathbf{S} = 0 \tag{1.53}$$

Since the surface element $dS \neq 0$, we obtain

$$\nabla\times\mathbf{F} = 0 \tag{1.54}$$

This is the necessary and sufficient condition for a force $\mathbf{F}$ to be conservative.

1.11 Potential Energy

If the force **F** is conservative, $\nabla \times \mathbf{F} = 0$, which indicates that the force **F** can be expressed as

$$\mathbf{F} = -\text{grad } V \tag{1.55}$$

since $\nabla \times \nabla V = 0$. The scalar point function V is called the ***potential energy*** of the particle at that point.

Thus, we write:

$$\mathbf{F} = -\left(\mathbf{i}\frac{\partial V}{\partial x} + \mathbf{j}\frac{\partial V}{\partial y} + \mathbf{k}\frac{\partial V}{\partial z}\right) \tag{1.56}$$

Then the work done

$$W_{12} = \int_1^2 \mathbf{F}\cdot d\mathbf{r} = -\int_1^2 \left(\mathbf{i}\frac{\partial V}{\partial x} + \mathbf{j}\frac{\partial V}{\partial y} + \mathbf{k}\frac{\partial V}{\partial z}\right)\cdot(\mathbf{i}dx + \mathbf{j}dy + \mathbf{k}dz)$$

$$= -\int_1^2 \left(\frac{\partial V}{\partial x}dx + \frac{\partial V}{\partial y}dy + \frac{\partial V}{\partial z}dz\right) = -\int_1^2 dV = V_1 - V_2 \tag{1.57}$$

That is, the work done by the force in displacing the particle from point 1 to point 2 is equal to the change in potential energy of the particle.

• Discussion

Eqn (1.55) does not change if any constant C is added to the potential energy, i.e., we can also write:

$$\mathbf{F} = -\text{grad }(V + C) \tag{1.58}$$

Hence, Eqn (1.55) does not define the potential energy uniquely. We can determine only the differences in potential energies and therefore, the reference level of the potential function V can be chosen at convenience.

1.12 Law of Conservation of Energy

From Eqns (1.50) and (1.57), we have the work done by a force in moving a particle from point 1 to point 2 as $W_{12} = T_2 - T_1 = V_1 - V_2$. Hence, we get

$$T_1 + V_1 = T_2 + V_2 = \text{const.} = E \tag{1.59}$$

Thus, ***the sum of potential and kinetic energies of a particle at every point in a conservative force-field is constant.***

This is known as the ***law of conservation of energy***. Thus, we write:

$$T + V = E \tag{1.60}$$

in a conservative force-field.

EXAMPLE 1.1 *The trajectory of a particle of mass m moving on a plane is given by*

$$x = x_0 \cos\omega_1 t \text{ and } y = y_0 \sin\omega_2 t.$$

(a) *Find the force* **F**, *torque* **N** *about the origin acting on the particle. Find also the angular momentum* **J** *about the origin.*

(b) *Find the expressions for potential energy and kinetic energy of the particle at any instant of time. Is the total energy conserved?* **[C.U. (HONS) 2003]**

Solution. The position vector of the particle at any instant of time is

$$\mathbf{r} = \mathbf{i}x + \mathbf{j}y = \mathbf{i}x_0 \cos\omega_1 t + \mathbf{j}y_0 \sin\omega_2 t \tag{1}$$

$$\therefore\ \dot{\mathbf{r}} = -\mathbf{i}\omega_1 x_0 \sin\omega_1 t + \mathbf{j}\omega_2 y_0 \cos\omega_2 t \tag{2}$$

$$\ddot{\mathbf{r}} = -\mathbf{i}\omega_1^2 x_0 \cos\omega_1 t - \mathbf{j}\omega_2^2 y_0 \sin\omega_2 t \tag{3}$$

$$= -\mathbf{i}\omega_1^2 x - \mathbf{j}\omega_2^2 y \tag{4}$$

Hence, the force is

$$\mathbf{F} = m\ddot{\mathbf{r}} = -m(\mathbf{i}\omega_1^2 x + \mathbf{j}\omega_2^2 y) \tag{5}$$

Torque $\mathbf{N}$ is

$$\begin{aligned}\mathbf{N} = \mathbf{F} \times \mathbf{r} &= -m(\mathbf{i}\omega_1^2 x + \mathbf{j}\omega_2^2 y) \times (\mathbf{i}x + \mathbf{j}y)\\ &= -m(\mathbf{k}\omega_1^2 xy - \mathbf{k}\omega_2^2 xy)\\ &= -m(\omega_1^2 - \omega_2^2)xy\mathbf{k}\end{aligned} \tag{6}$$

Momentum of the particle is

$$\mathbf{p} = m\mathbf{v} = m\dot{\mathbf{r}} = m(-\mathbf{i}\omega_1 x_0 \sin\omega_1 t + \mathbf{j}\omega_2 y_0 \cos\omega_2 t) \tag{7}$$

Hence, the angular momentum about the origin is

$$\begin{aligned}\mathbf{J} = \mathbf{r} \times \mathbf{p} &= (\mathbf{i}x_0 \cos\omega_1 t + \mathbf{j}y_0 \sin\omega_2 t) \times m(-\mathbf{i}\omega_1 x_0 \sin\omega_1 t + \mathbf{j}\omega_2 y_0 \cos\omega_2 t)\\ &= \mathbf{k}m x_0 y_0(\omega_2 \cos\omega_1 t \cos\omega_2 t + \omega_1 \sin\omega_1 t \sin\omega_2 t)\end{aligned}$$

From Eqn (2), the velocity components are $v_x = -\omega_1 x_0 \sin\omega_1 t$ and $v_y = \omega_2 y_0 \cos\omega_2 t$. Hence,

$$v^2 = v_x^2 + v_y^2 = \omega_1^2 x_0^2 \sin^2\omega_1 t + \omega_2^2 y_0^2 \cos^2\omega_2 t$$

Therefore, the kinetic energy is

$$T = \frac{1}{2}mv^2 = \frac{1}{2}m\left(\omega_1^2 x_0^2 \sin^2\omega_1 t + \omega_2^2 y_0^2 \cos^2\omega_2 t\right) \tag{8}$$

Now

$$\nabla \times \mathbf{F} = \begin{vmatrix} \mathbf{i} & \mathbf{j} & \mathbf{k} \\ \dfrac{\partial}{\partial x} & \dfrac{\partial}{\partial y} & \dfrac{\partial}{\partial z} \\ -m\omega_1^2 x & -m\omega_2^2 y & 0 \end{vmatrix} = 0 \tag{9}$$

Hence, the force $\mathbf{F}$ is conservative. If V is the potential energy, then

$$\mathbf{F} = -m\left(\mathbf{i}\omega_1^2 x + \mathbf{j}\omega_2^2 y\right) = -\left(\mathbf{i}\frac{\partial V}{\partial x} + \mathbf{j}\frac{\partial V}{\partial y}\right)$$

Hence,

$$\left.\begin{aligned}\frac{\partial V}{\partial x} &= m\omega_1^2 x\\ \frac{\partial V}{\partial y} &= m\omega_2^2 y\end{aligned}\right\}$$

Integrating these equations,

$$V = \frac{1}{2}m(\omega_1^2 x^2 + \omega_2^2 y^2) = \frac{1}{2}m(\omega_1^2 x_0^2 \cos^2\omega_1 t + \omega_2^2 y_0^2 \sin^2\omega_2 t) \tag{11}$$

The total energy, from Eqns (8) and (11), is

$$E = T + V = \frac{1}{2}m\left(\omega_1^2 x_0^2 + \omega_2^2 y_0^2\right)$$

which is a constant. Thus, the total energy is conserved.

1.13 Equation of Motion

Considering the particle of constant mass m, the equation of its motion is in general of the following type:

$$m\ddot{\mathbf{r}} = \mathbf{F}(\mathbf{r}, \dot{\mathbf{r}}, t) \tag{1.61}$$

An analytical solution of this equation of motion for types of problems of interest are not possible.

We therefore intend to discuss simple problems under the following types of forces:

(1) The force is constant, i.e., $\mathbf{F} = m\mathbf{a}, \mathbf{a} =$ constant
(2) The force is velocity-dependent, i.e., $\mathbf{F} = \mathbf{F}(\mathbf{v})$
(3) The force is position-dependent, i.e., $\mathbf{F} = \mathbf{F}(\mathbf{r})$
(4) The force is time-dependent, i.e., $\mathbf{F} = \mathbf{F}(t)$.

In all the cases, the equation of motion is supplemented with some conditions called initial conditions. These conditions provide the position and velocity of the particle at some initial time $t = t_0$.

1.14 Motion under a Constant Force

The best example of a constant force-field is the gravity near the earth's surface where the value of the acceleration due to gravity g is taken to be a constant.

We consider the motion of a projectile of mass m thrown from the ground level with velocity u at an angle α with the horizon. Then the scalar equations of motion are (FIG. 1.6)

$$\ddot{z} = -g \tag{1.62}$$

$$\ddot{x} = 0 \tag{1.63}$$

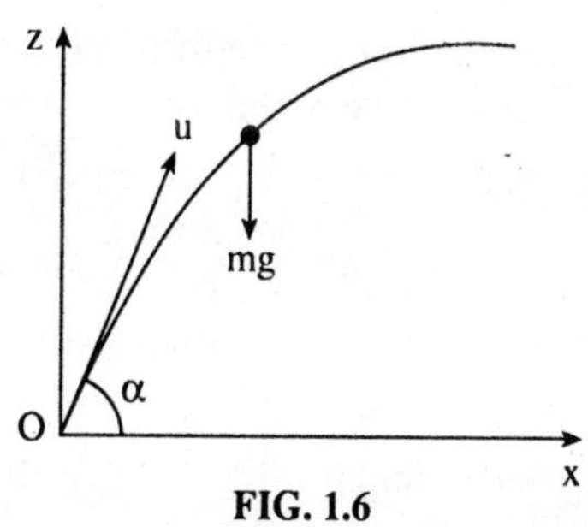

FIG. 1.6

Integrating once w.r.t. time and using the initial condition,

$$\left.\begin{aligned} \dot{x} &= u\cos\alpha \\ \dot{z} &= u\sin\alpha \end{aligned}\right\} \tag{1.64}$$

at $t = 0$, we find

$$\dot{z} = -gt + u\sin\alpha \tag{1.65}$$

$$\dot{x} = u\cos\alpha \tag{1.66}$$

Integrating again and noting that $x = 0$ and $z = 0$ at $t = 0$, we get

$$z = ut\sin\alpha - \frac{1}{2}gt^2 \tag{1.67}$$

$$x = ut\cos\alpha \tag{1.68}$$

The total time of flight T is obtained by putting $z = 0$ in Eqn (1.67). Thus,

$$T = \frac{2u}{g}\sin\alpha \tag{1.69}$$

The horizontal distance covered during time T gives the range of the projectile. Thus, from Eqn (1.68) using the value of T from Eqn (1.69), we find

$$R = \frac{2u^2}{g}\sin\alpha\cos\alpha = \frac{u^2}{g}\sin 2\alpha \tag{1.70}$$

The path of the projectile can be obtained by eliminating t between Eqn (1.67) and Eqn (1.68). Thus, we find

$$z = x\tan\alpha - \frac{gx^2}{2u^2\cos^2\alpha} \tag{1.71}$$

which represents a parabola.

1.15 Motion under a Velocity-Dependent Force

The situation occurs when a particle moves in a viscous medium which offers resistance to the motion. The resisting force, known as ***drag force***, is a function of velocity. The magnitude of this drag force as first shown by Newton is given by

$$F_D = \frac{1}{2}\rho u^2 A C_D \tag{1.72}$$

where ρ is the density of the medium, A is the area of the body projected on a plane normal to the direction of velocity u, and C_D is a coefficient called ***drag coefficient***.

Consider the motion of a particle of mass m falling under gravity, the resisting force being

$$F = \mu m u^2 \tag{1.73}$$

and the initial condition being $u = 0$ at $t = 0$. The equation of motion of the particle is (here z-axis is taken vertically downwards)

$$m\frac{du}{dt} = mg - \mu m u^2 \tag{1.74}$$

Integrating we get

$$\int_0^u \frac{du}{(g/\mu) - u^2} = \mu\int_0^t dt$$

which yields

$$u = \sqrt{\frac{g}{\mu}}\tanh\left(\sqrt{\frac{g}{\mu}}\,t\right) \tag{1.75}$$

This determines the velocity of the particle at any instant of time.

As $t \to \infty$, $\tanh\left(\sqrt{\frac{g}{\mu}}\,t\right) \to 1$.

Hence, the terminal velocity is $\sqrt{\frac{g}{\mu}}$.

Since $u = \dot{z}$, we integrate Eqn (1.75) once and obtain

$$z = \frac{1}{\mu}\ln\left[\cosh\left(\sqrt{\frac{g}{\mu}}\,t\right)\right] \tag{1.76}$$

which determines the height of the particle at any instant of time.

Again writing

$$\frac{du}{dt} = \frac{du}{dz}\frac{dz}{dt} = u\frac{du}{dz} = \frac{1}{2}\frac{d}{dz}(u^2)$$

we recast Eqn (1.74) in the form:

$$\frac{d}{dz}(u^2) + \mu u^2 = g \tag{1.77}$$

This upon integration yields

$$u^2 = \frac{g}{\mu}\left[1 - \exp(-2\mu z)\right] \tag{1.78}$$

EXAMPLE 1.2 *A canon has its maximum range given by* R_{max}. *Prove that when the maximum range is actually realised,* (i) *the height reached in such a case is* $\frac{1}{4}R_{\text{max}}$ *and* (ii) *the total time of flight is* $2\sqrt{R_{\text{max}}/2g}$. **[C.U. (HONS) 1995]**

Solution. The horizontal range of a projectile is given by

$$R = \frac{u^2}{g}\sin 2\alpha \tag{1}$$

Its maximum value occurs at $\sin 2\alpha = 1$, i.e., $\alpha = \pi/4$. Hence,

$$R_{\text{max}} = \frac{u^2}{g} \tag{2}$$

From this relation, we get

$$u = \sqrt{gR_{\text{max}}} \tag{3}$$

The total time of flight is, by Eqn (1.69),

$$T = \frac{2u}{g}\sin\alpha = \frac{2}{g}\sqrt{gR_{\text{max}}}\frac{1}{\sqrt{2}} = 2\sqrt{\frac{R_{\text{max}}}{2g}} \quad \textbf{(Proved)} \tag{4}$$

The time to reach the maximum height is

$$t = \frac{T}{2} = \sqrt{\frac{R_{\text{max}}}{2g}} \tag{5}$$

By Eqn (1.67), the maximum height reached is

$$H = ut\sin\alpha - \frac{1}{2}gt^2 \tag{6}$$

Using Eqns (3) and (5) for $\alpha = \pi/4$

$$H = \sqrt{gR_{\text{max}}}\sqrt{\frac{R_{\text{max}}}{2g}}\cdot\frac{1}{\sqrt{2}} - \frac{1}{2}g\frac{R_{\text{max}}}{g} = \frac{1}{4}R_{\text{max}} \quad \textbf{(Proved)}$$

EXAMPLE 1.3 *A boat of mass m was travelling with velocity u when the boatman stopped rowing. Assuming the retarding force due to water be* $a\exp(bv)$ *at velocity v, find the time required by the boat to stop. Assume the water to be stationary.* **[C.U. (HONS) 2005]**

Solution. When rowing is stopped, the only force acting on the boat is the retarding force due to water. Hence, the equation of motion of the boat is

$$m\frac{dv}{dt} = -ae^{bv}, \ (a, b \text{ constants}) \tag{1}$$

$$\text{i.e., } \frac{dv}{e^{bv}} = -\frac{a}{m}dt$$

Integrating, we get

$$\int e^{-bv}dv = -\frac{a}{m}\int dt + C, \text{ i.e., } -\frac{1}{b}e^{-bv} = -\frac{a}{m}t + C \tag{2}$$

Initially when $t = 0$, $v = u$. Putting these values in Eqn (2), we get

$$-\frac{1}{b}e^{-bu} = C \tag{3}$$

From Eqns (2) and (3),

$$\frac{1}{b}\left(e^{-bu} - e^{-bv}\right) = -\frac{a}{m}t \text{ or, } t = \frac{m}{ab}\left(e^{-bv} - e^{-bu}\right) \tag{4}$$

When the boat stops, $v = 0$. Hence, the corresponding time is

$$t = \frac{m}{ab}\left(1 - e^{-bu}\right).$$

1.16 Motion under a Position-Dependent Force

Consider the motion of a particle of constant mass m in one dimension. The external force being position-dependent, its equation of motion is

$$m\frac{dv}{dt} = F(x) \tag{1.79}$$

Since,

$$\frac{dv}{dt} = \frac{dv}{dx}\frac{dx}{dt} = v\frac{dv}{dx} = \frac{1}{2}\frac{d}{dx}(v^2)$$

we write Eqn (1.79) in the form:

$$d(v^2) = \frac{2}{m}F(x)dx.$$

Integrating, we find

$$v^2 = u^2 + \frac{2}{m}\int_0^x F(x)dx \tag{1.80}$$

where we have assumed that the velocity of the particle is u at $x = 0$.

If we consider two points at $x = x_1$ and $x = x_2$, work done in moving a particle between these two points back and forth is

$$\int_{x_1}^{x_2} F(x)dx + \int_{x_2}^{x_1} F(x)dx = 0 \tag{1.81}$$

Hence, the force is conservative. The force $F(x)$ being arbitrary, we conclude that:

All one-dimensional position-dependent forces are conservative.

We can, therefore, find a potential function $V(x)$ such that

$$V(x) = V(0) - \int_0^x F(x)dx \tag{1.82}$$

Then the total energy of the particle is

$$E = T + V = \frac{1}{2}mv^2 + V(x) \tag{1.83}$$

This equation yields

$$v = \frac{dx}{dt} = \sqrt{\frac{2}{m}[E - V(x)]} \tag{1.84}$$

Integrating Eqn (1.84), we find

$$t = \sqrt{\frac{m}{2}} \int_0^x \frac{dx}{\sqrt{E - V(x)}} \tag{1.85}$$

EXAMPLE 1.4 *A particle of mass m moves along a straight line under a repulsive force $F = \mu x$. Describe its motion.*

Solution. Let the repulsive force be measured from a fixed point O taken as origin.

The equation of motion is

$$m\frac{dv}{dt} = \mu x \tag{1}$$

Writing $\dfrac{dv}{dt} = \dfrac{1}{2}\dfrac{d}{dx}(v^2)$, Eqn (1) becomes $d(v^2) = \dfrac{2\mu}{m}x\,dx$, which upon integration yields

$$v^2 = \frac{\mu}{m}x^2 + C \tag{2}$$

We assume that, $v = 0$ at $x = a$. Then from Eqn (2), we find

$$C = -\frac{\mu}{m}a^2 \tag{3}$$

Hence,

$$v^2 = \frac{\mu}{m}(x^2 - a^2) \tag{4}$$

$$\therefore\ v = \frac{dx}{dt} = \sqrt{\frac{\mu}{m}(x^2 - a^2)} \tag{5}$$

We have taken the positive sign since the velocity is away from the origin. Eqn (5) shows that the velocity increases continually. Integrating Eqn (5),

$$\sqrt{\frac{\mu}{m}}\,t = \int \frac{dx}{\sqrt{x^2 - a^2}} \quad \text{or,} \quad \sqrt{\frac{\mu}{m}}\,t = \log\left[x + \sqrt{x^2 - a^2}\right] + B \tag{6}$$

Since $x = a$ at $t = 0$, we have

$$0 = \log(a) + B \tag{7}$$

Hence,

$$\sqrt{\frac{\mu}{m}}\,t = \log\frac{x + \sqrt{x^2 - a^2}}{a} \quad \text{or,} \quad x + \sqrt{x^2 - a^2} = a\exp\left(\sqrt{\frac{\mu}{m}}\,t\right) \tag{8}$$

Now,

$$x - \sqrt{x^2 - a^2} = \frac{a^2}{x + \sqrt{x^2 - a^2}}$$

Hence,

$$x - \sqrt{x^2 - a^2} = a\exp\left(-\sqrt{\frac{\mu}{m}}\,t\right) \tag{9}$$

Adding Eqns (8) and (9), we get

$$x = \frac{a}{2}\left[\exp\left(\sqrt{\frac{\mu}{m}}\,t\right) + \exp\left(-\sqrt{\frac{\mu}{m}}\,t\right)\right] \tag{10}$$

Eqn (10) shows that as t increases, the distance of the particle from O continually increases.

1.17 Motion under a Time-Dependent Force

Let the force acting on a particle of mass m be given as $\mathbf{F}(t)$. Then the equation of its motion is

$$m\frac{d\mathbf{v}}{dt} = \mathbf{F}(t) \tag{1.86}$$

We assume that the initial conditions given are

$$\left.\begin{aligned} \mathbf{v} = \mathbf{v_0} \text{ at } t = t_0 \\ \text{and } \mathbf{r} = \mathbf{r_0} \text{ at } t = t_0 \end{aligned}\right\} \tag{1.87}$$

Integrating Eqn (1.86) once w.r.t. time we find, after using the initial condition,

$$\mathbf{v} - \mathbf{v_0} = \int_{t_0}^{t} \frac{\mathbf{F}(t)}{m} dt \tag{1.88}$$

A further integration yields the position vector in the form:

$$\mathbf{r} - \mathbf{r_0} = \int_{t_0}^{t} \left[\mathbf{v_0} + \int_{t_0}^{t} \frac{\mathbf{F}(t)}{m} dt\right] dt \tag{1.89}$$

Time-dependent force of the type

$$\mathbf{F} = \mathbf{F_0} \sin \omega t \tag{1.90}$$

is experienced by charged particles in the fields of alternating currents. If we consider one-dimensional motion under the force given by Eqn (1.90), then we can find using Eqns (1.88) and (1.89)

$$\mathbf{v} = \mathbf{v_0} + \frac{\mathbf{F_0}}{m\omega}(1 - \cos \omega t) \tag{1.91}$$

$$\text{and } \mathbf{x} = \mathbf{x_0} + \left(\mathbf{v_0} + \frac{\mathbf{F_0}}{m\omega}\right) t - \frac{\mathbf{F_0}}{m\omega^2} \sin \omega t \tag{1.92}$$

1.18 Cylindrical Polar Coordinates

The coordinates in the cylindrical polar system are (r, θ, z). The Cartesian coordinates corresponding to these cylindrical coordinates are (FIG. 1.7)

$$\left.\begin{aligned} x &= r\cos\theta \\ y &= r\sin\theta \\ z &= z \end{aligned}\right\} \tag{1.93}$$

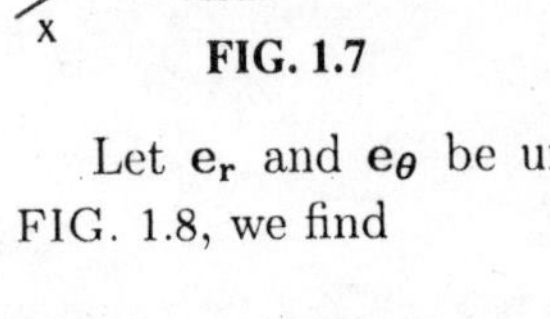

FIG. 1.7

Thus, $x^2 + y^2 = r^2 =$ constant describes a family of right circular cylinder with z-axis as its own axis.

The position vector of any point is

$$\mathbf{r} = \mathbf{i}r\cos\theta + \mathbf{j}r\sin\theta + \mathbf{k}z \tag{1.94}$$

Let $\mathbf{e_r}$ and $\mathbf{e_\theta}$ be unit vectors along the increasing r and θ directions. Then, from the FIG. 1.8, we find

$$\mathbf{e_r} = \mathbf{i}\cos\theta + \mathbf{j}\sin\theta \tag{1.95}$$

$$\mathbf{e_\theta} = -\mathbf{i}\sin\theta + \mathbf{j}\cos\theta \tag{1.96}$$

Hence,

$$\frac{d}{d\theta}\mathbf{e_r} = \mathbf{e_\theta}, \ \frac{d}{d\theta}\mathbf{e_\theta} = -\mathbf{e_r} \tag{1.97}$$

The radius vector in cylindrical coordinates is

$$\mathbf{r} = r\mathbf{e_r} + z\mathbf{k} \tag{1.98}$$

Differentiating this with respect to t, we find

$$\mathbf{v} = \dot{\mathbf{r}} = r\dot{\mathbf{e}}_\mathbf{r} + \dot{r}\mathbf{e_r} + \dot{z}\mathbf{k} \tag{1.99}$$

Using Eqn (1.97), we find

$$\begin{aligned}\mathbf{v} &= r\frac{d\mathbf{e_r}}{d\theta}\frac{d\theta}{dt} + \dot{r}\mathbf{e_r} + \dot{z}\mathbf{k} \\ &= \dot{r}\mathbf{e_r} + r\dot{\theta}\mathbf{e}_\theta + \dot{z}\mathbf{k}\end{aligned} \tag{1.100}$$

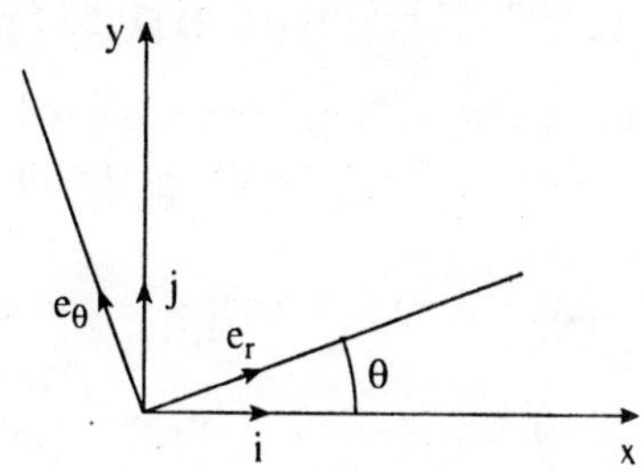

FIG. 1.8

Similarly, for the acceleration vector, we find

$$\begin{aligned}\mathbf{a} = \dot{\mathbf{v}} &= (\ddot{r}\mathbf{e_r} + \dot{r}\dot{\mathbf{e}}_\mathbf{r}) + \left(r\dot{\theta}\dot{\mathbf{e}}_\theta + r\ddot{\theta}\mathbf{e}_\theta + \dot{r}\dot{\theta}\mathbf{e}_\theta\right) + \ddot{z}\mathbf{k} \\ &= \left(\ddot{r}\mathbf{e_r} + \dot{r}\dot{\theta}\mathbf{e}_\theta\right) + \left(-r\dot{\theta}^2\mathbf{e_r} + r\ddot{\theta}\mathbf{e}_\theta + \dot{r}\dot{\theta}\mathbf{e}_\theta\right) + \ddot{z}\mathbf{k} \\ &= \left(\ddot{r} - r\dot{\theta}^2\right)\mathbf{e_r} + \left(r\ddot{\theta} + 2\dot{r}\dot{\theta}\right)\mathbf{e}_\theta + \ddot{z}\mathbf{k}\end{aligned} \tag{1.101}$$

1.19 Spherical Polar Coordinates

The coordinates in this system are (r, θ, ϕ). The Cartesian coordinates of any point are related to the spherical polar coordinates in the form:

$$\left.\begin{aligned} x &= r\sin\theta\cos\phi \\ y &= r\sin\theta\sin\phi \\ z &= r\cos\theta \end{aligned}\right\} \tag{1.102}$$

Let $\mathbf{e_r}, \mathbf{e}_\theta$ and $\mathbf{e}_\phi$ be unit vectors along the spherical polar coordinates (FIG. 1.9). Using Eqns (1.95) and (1.96), we find

$$\begin{aligned}\mathbf{e_r} &= \boldsymbol{\rho}\sin\theta + \mathbf{k}\cos\theta \\ &= (\mathbf{i}\cos\phi + \mathbf{j}\sin\phi)\sin\theta + \mathbf{k}\cos\theta \\ &= \mathbf{i}\cos\phi\sin\theta + \mathbf{j}\sin\phi\sin\theta + \mathbf{k}\cos\theta\end{aligned} \tag{1.103}$$

$$\begin{aligned}\mathbf{e}_\theta &= \boldsymbol{\rho}\cos\theta - \mathbf{k}\sin\theta \\ &= \mathbf{i}\cos\phi\cos\theta + \mathbf{j}\sin\phi\cos\theta - \mathbf{k}\sin\theta\end{aligned} \tag{1.104}$$

$$\mathbf{e}_\phi = -\mathbf{i}\sin\phi + \mathbf{j}\cos\phi. \tag{1.105}$$

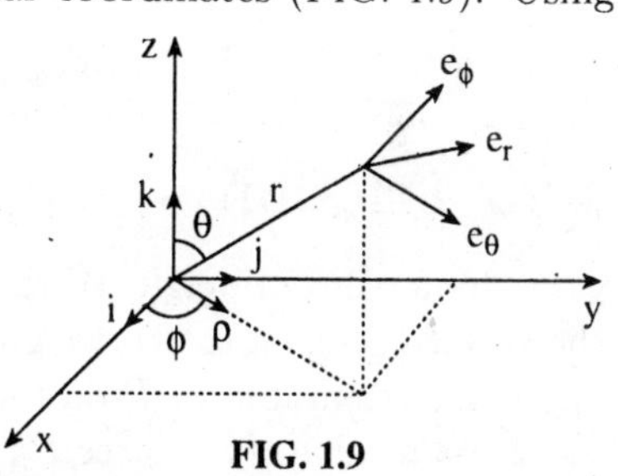

FIG. 1.9

The derivatives of unit vectors are obtained by direct differentiation. Thus,

$$\frac{\partial}{\partial\theta}\mathbf{e_r} = \mathbf{i}\cos\phi\cos\theta + \mathbf{j}\sin\phi\cos\theta - \mathbf{k}\sin\theta = \mathbf{e}_\theta \tag{1.106}$$

$$\frac{\partial}{\partial\phi}\mathbf{e_r} = -\mathbf{i}\sin\phi\sin\theta + \mathbf{j}\cos\phi\sin\theta = \mathbf{e}_\phi\sin\theta \tag{1.107}$$

$$\frac{\partial}{\partial\theta}\mathbf{e}_\phi = 0$$

$$\frac{\partial}{\partial\phi}\mathbf{e}_\phi = -\mathbf{i}\cos\phi - \mathbf{j}\sin\phi = -\mathbf{e_r}\sin\theta - \mathbf{e}_\theta\cos\theta \tag{1.108}$$

$$\frac{\partial}{\partial\phi}\mathbf{e}_\theta = -\mathbf{i}\sin\phi\cos\theta + \mathbf{j}\cos\phi\cos\theta = \mathbf{e}_\phi\cos\theta \tag{1.109}$$

$$\frac{\partial}{\partial\theta}\mathbf{e}_\theta = \mathbf{i}\cos\phi\sin\theta - \mathbf{j}\sin\phi\sin\theta - \mathbf{k}\cos\theta = -\mathbf{e_r} \tag{1.110}$$

The radius vector in spherical coordinates is

$$\mathbf{r} = r\mathbf{e_r} \tag{1.111}$$

Hence, the velocity of the particle is

$$\mathbf{v} = \dot{\mathbf{r}} = r\dot{\mathbf{e}}_\mathbf{r} + \dot{r}\mathbf{e_r} \tag{1.112}$$

Now

$$\dot{\mathbf{e}}_\mathbf{r} = \left(\frac{\partial}{\partial\theta}\mathbf{e_r}\right)\frac{d\theta}{dt} + \left(\frac{\partial}{\partial\phi}\mathbf{e_r}\right)\frac{d\phi}{dt} = \mathbf{e}_\theta\dot{\theta} + \mathbf{e}_\phi \sin\theta\dot{\phi} \tag{1.113}$$

Thus,

$$\mathbf{v} = \dot{r}\mathbf{e_r} + r\dot{\theta}\mathbf{e}_\theta + r\dot{\phi}\sin\theta\mathbf{e}_\phi \tag{1.114}$$

Similarly, the acceleration of the particle is

$$\begin{aligned}\mathbf{a} = \dot{\mathbf{v}} &= (\ddot{r}\mathbf{e_r} + \dot{r}\dot{\mathbf{e}}_\mathbf{r}) + \left(\dot{r}\dot{\theta}\mathbf{e}_\theta + r\ddot{\theta}\mathbf{e}_\theta + r\dot{\theta}\dot{\mathbf{e}}_\theta\right)\\ &+ \left(\dot{r}\dot{\phi}\sin\theta\mathbf{e}_\phi + r\ddot{\phi}\sin\theta\mathbf{e}_\phi + r\dot{\phi}\dot{\theta}\cos\theta\mathbf{e}_\phi + r\dot{\phi}\sin\theta\dot{\mathbf{e}}_\phi\right)\end{aligned} \tag{1.115}$$

Now

$$\frac{d}{dt}\mathbf{e}_\theta = \left(\frac{\partial}{\partial\theta}\mathbf{e}_\theta\right)\frac{d\theta}{dt} + \left(\frac{\partial}{\partial\phi}\mathbf{e}_\theta\right)\frac{d\phi}{dt} = -\mathbf{e_r}\dot{\theta} + \mathbf{e}_\phi\cos\theta\dot{\phi} \tag{1.116}$$

$$\frac{d}{dt}\mathbf{e}_\phi = \left(\frac{\partial}{\partial\phi}\mathbf{e}_\phi\right)\frac{d\phi}{dt} = (-\mathbf{e_r}\sin\theta - \mathbf{e}_\theta\cos\theta)\,\dot{\phi} \tag{1.117}$$

Substituting these values in Eqn (1.115), we get after algebraic simplification

$$\begin{aligned}\mathbf{a} &= \left(\ddot{r} - r\dot{\theta}^2 - r\dot{\phi}^2\sin^2\theta\right)\mathbf{e_r} + \left(r\ddot{\theta} + 2\dot{r}\dot{\theta} - r\dot{\phi}^2\sin\theta\cos\theta\right)\mathbf{e}_\theta\\ &+ \left(r\ddot{\phi}\sin\theta + 2\dot{r}\dot{\phi}\sin\theta + 2r\dot{\theta}\dot{\phi}\cos\theta\right)\mathbf{e}_\phi\end{aligned} \tag{1.118}$$

1.20 Friction

Consider a block of weight W resting on a horizontal ***rough*** surface. The normal reaction N on the surface is equal to the weight in magnitude but opposite in direction (FIG. 1.10). Now we apply a horizontal force P on the block. It will be experienced that the block will not move up to a certain value of the force P_c; it will start slipping beyond that particular value of P_c. Hence, there must be a horizontal reaction force R keeping the block in equilibrium. This reaction force is called the ***frictional force***. This frictional force is found experimentally to depend on the nature of the contacting surfaces only. It is also found that the limiting horizontal force P_c is proportional to the normal reaction N. Hence, we write:

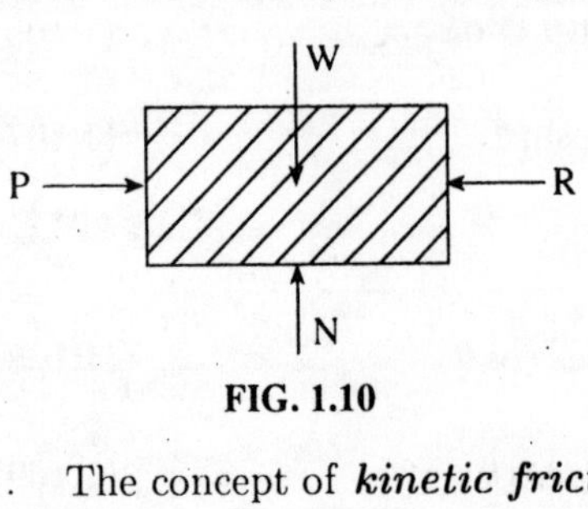

FIG. 1.10

$$P_c = \mu_s N, \tag{1.119}$$

where μ_s is called the ***static friction coefficient***. The angle of friction ε is defined as

$$\tan\varepsilon = \mu_s \text{ or, } \varepsilon = \tan^{-1}\mu_s \tag{1.120}$$

The concept of ***kinetic friction*** is introduced in the same manner. If a body moves along a rough surface, the frictional force is found to oppose the motion and to be proportional to the normal force.

Thus, as shown in FIG. 1.11,

$$f = \mu N \tag{1.121}$$

where μ is the ***coefficient of kinetic friction.***

The angle of kinetic friction is, as before, defined as

$$\tan \varepsilon = \mu \text{ or, } \varepsilon = \tan^{-1} \mu \tag{1.122}$$

For a given pair of contacting surfaces, it is always found that

$$\mu_s > \mu. \tag{1.123}$$

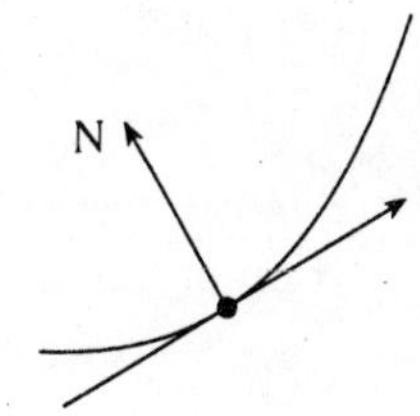

FIG. 1.11

1.21 Constrained Motion

The motion of a particle along some specified curve or surface is called ***constrained motion.*** The curve or surface along which the particle is constrained to move is called a ***constraint.*** In order to study the constrained motion, we use the tangential and normal components of velocity and acceleration discussed in Section 1.5.

Let us discuss the motion of a particle of mass m along a rough inclined surface having a constant coefficient of friction μ (FIG. 1.12). The forces acting on the particle are:

(1) the weight mg vertically downwards,
(2) the normal reaction $N = mg \cos \alpha$,
(3) the frictional force f directed up the incline.

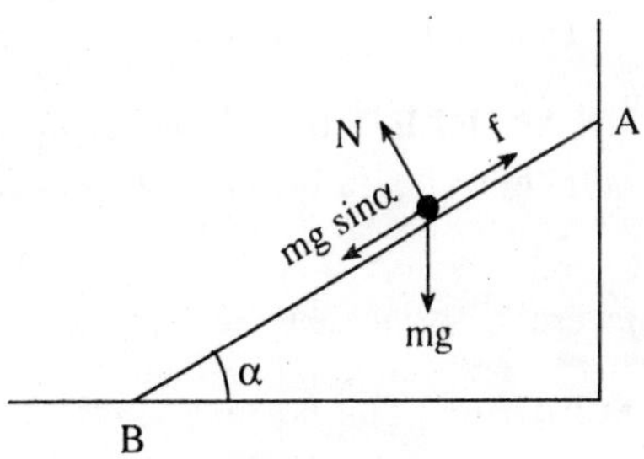

FIG. 1.12

We measure the distance traversed s from the top of the plane. Then the equation of motion is

$$m\ddot{s} = mg \sin \alpha - \mu mg \cos \alpha, \text{ i.e., } \ddot{s} = g(\sin \alpha - \mu \cos \alpha) \tag{1.124}$$

Eqn (1.124) shows that for the motion of the particle,

$$\sin \alpha - \mu \cos \alpha > 0, \text{ i.e., } \tan \alpha > \mu \tag{1.125}$$

Integrating Eqn (1.124) once w.r.t. time and assuming the initial condition that at $t = 0$, the velocity $v = 0$, we find

$$v = \dot{s} = g(\sin \alpha - \mu \cos \alpha)t \tag{1.126}$$

Integrating once more and assuming that at $t = 0$, $s = 0$, we find

$$s = \frac{1}{2} g(\sin \alpha - \mu \cos \alpha)t^2 \tag{1.127}$$

EXAMPLE 1.5 *A bead of mass m slides down a smooth circular wire in a vertical plane starting from rest at the end of a horizontal diameter. Find its angular velocity. Also show that the maximum normal reaction on the wire is* $3\,mg$.

Solution. Let θ be the angle of inclination at the centre of the particle at any instant. If y be the vertical fall, then (FIG. 1.13)

$$y = a \sin \theta \tag{1}$$

If v be the velocity at that instant, then from energy conservation, we have

$$v^2 = 2gy = 2ag \sin \theta \text{ or, } a^2\omega^2 = 2ag \sin \theta.$$

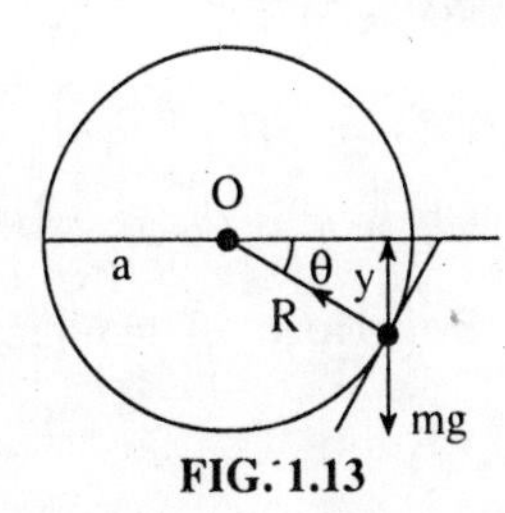

FIG. 1.13

Hence,

$$\omega = \sqrt{\frac{2g}{a}\sin\theta} \tag{2}$$

Eqn (2) determines the angular velocity. To determine the normal reaction, we write the equation of motion along the normal. Thus,

$$\frac{mv^2}{a} = R - mg\sin\theta$$

$$\therefore\ R = mg\sin\theta + \frac{mv^2}{a} \tag{3}$$

Substituting the value of v^2 from Eqn (2) into Eqn (3), we get

$$R = mg\sin\theta + m(2g\sin\theta) = 3mg\sin\theta$$

Maximum value of R occurs at $\theta = \pi/2$, when its value is $3mg$.

It may be noted that the velocity of the particle is also maximum at $\theta = \pi/2$.

EXAMPLE 1.6 *A block of mass m slides down a rough plane of incline θ from rest. Show that the velocity of the block at any point is*

$$v = \sqrt{2gh(1-\mu\cot\theta)},$$

where h is the vertical displacement of the block from initial position.

Solution. The normal force on the plane

$$N = mg\cos\theta \tag{1}$$

Hence the friction force

$$f = \mu N = \mu mg\cos\theta \tag{2}$$

Let s be the displacement of the block along the plane after time t. Then

$$s = \frac{h}{\sin\theta}. \tag{3}$$

Taking zero potential line at this position, the principle of energy gives

$$\left(\frac{1}{2}mv^2 - 0\right) - mgh = -f\cdot s$$

$$\text{i.e.,}\ \frac{1}{2}mv^2 = mgh - \mu mg\cos\theta\cdot\frac{h}{\sin\theta}$$

$$\therefore\ v = \sqrt{2gh(1-\mu\cot\theta)}$$

EXAMPLE 1.7 *A particle of mass m is thrown upward with velocity v_0. If air-resistance is proportional to the velocity of the particle, show that the maximum height attained by it is given by*

$$h = \frac{v_0}{k} + \frac{g}{k^2}\ln\left(\frac{g}{g+kv_0}\right),$$

where k is the resistive force constraint and g is the acceleration due to gravity.

Solution. The equation of motion of the particle is

$$m\frac{d^2x}{dt^2} = -mg - mkv,\ \text{i.e.,}\ v\frac{dv}{dx} = -(g+kv) \tag{1}$$

Hence,

$$\frac{vdv}{g+kv} = -gdx, \text{ i.e., } \frac{1}{k}\left(1 - \frac{g}{g+kv}\right) dv = -gdx \tag{2}$$

Integrating,

$$v - g\frac{1}{k}\ln(g+kv) = -kx + c. \tag{3}$$

The initial conditions are

$$x = 0, \text{ when } v = v_0 \tag{4}$$

Using this in (3),

$$v_0 - \frac{g}{k}\ln(g+kv_0) = C \tag{5}$$

Hence, from (3) and (5), we get

$$v - \frac{g}{k}\ln(g+kg) = -kx + v_0 - \frac{g}{k}\ln(g+kv_0) \tag{6}$$

When x is maximum $= h$, velocity $v = 0$.

Hence, from (6) we get

$$0 - \frac{g}{k}\ln g = -kh + v_0 - \frac{g}{k}\ln(g+kv_0)$$

$$\therefore h = \frac{v_0}{k} + \frac{g}{k^2}\ln\left(\frac{g}{g+kv_0}\right)$$

EXAMPLE 1.8 *A particle moves under the force-field*

$$\mathbf{F} = a(\mathbf{i}\sin\omega t + \mathbf{j}\cos\omega t)$$

If the particle be initially at rest, prove that at an instant t, the work done on the particle is

$$\frac{a^2}{m\omega^2}(1 - \cos\omega t)$$

Solution. The equation of motion is

$$m\frac{d\mathbf{v}}{dt} = \mathbf{F}(t) = a(\mathbf{i}\sin\omega t + \mathbf{j}\cos\omega t) \tag{1}$$

Integrating w.r.t. time, we find

$$\mathbf{v} - \mathbf{v_0} = \int_0^t \frac{\mathbf{F}(t)}{m}dt = \int_0^t \frac{a}{m}(\mathbf{i}\sin\omega t + \mathbf{j}\cos\omega t)dt$$

$$= \frac{a}{\omega m}[(-\mathbf{i}\cos\omega t + \mathbf{j}\sin\omega t) + \mathbf{i}] \tag{2}$$

Initially, the particle is at rest. Hence, $\mathbf{v_0} = 0$. Then

$$\mathbf{v} = \frac{a}{m\omega}(-\mathbf{i}\cos\omega t + \mathbf{j}\sin\omega t) + \mathbf{i} \tag{3}$$

Work done on the particle at any instant is

$$W = \int \mathbf{F}\cdot d\mathbf{r} = \int m\frac{d\mathbf{v}}{dt}\cdot\frac{d\mathbf{r}}{dt}dt = \int \frac{1}{2}m\frac{d}{dt}(\mathbf{v}\cdot\mathbf{v})$$

$$= \int \frac{1}{2}md(\mathbf{v}\cdot\mathbf{v}) = \frac{1}{2}m(\mathbf{v}\cdot\mathbf{v}) \tag{4}$$

Inserting **v** from (3),

$$W = \frac{1}{2}m\left(\frac{a}{m\omega}\right)^2 (-\mathbf{i}\cos\omega t + \mathbf{j}\sin\omega t + \mathbf{i}) \cdot (-\mathbf{i}\cos\omega t + \mathbf{j}\sin\omega t + \mathbf{i})$$

$$= \frac{a^2}{2m\omega^2}\left[(1-\cos\omega t)^2 + \sin^2\omega\right]$$

$$= \frac{a^2}{m\omega^2}(1-\cos\omega t) \quad \textbf{(Proved)}$$

EXAMPLE 1.9 *A particle of mass m is projected upwards at an angle α to the horizon. The air resistance acting upon it is λ times the instantaneous velocity. Discuss the motiion.*

Solution. We assume the motion in yz-plane. Let **v** and $\mathbf{v_0}$ be the instantaneous and initial velocity of the particle.

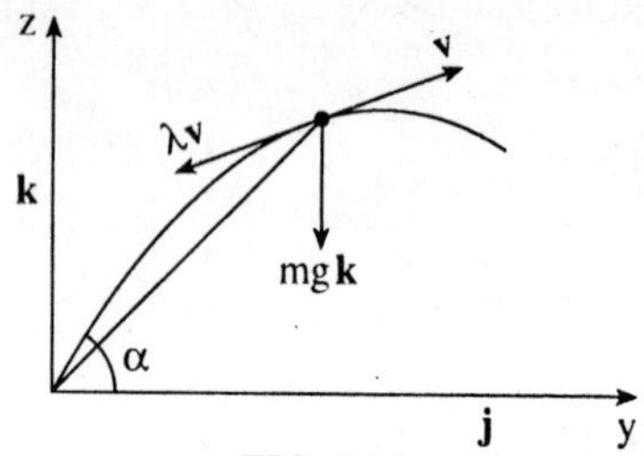

FIG. 1.14

The equation of motion is (FIG. 1.14)

$$m\frac{d\mathbf{v}}{dt} = -mg\mathbf{k} - \lambda\mathbf{v}$$

$$\text{i.e.,} \quad \frac{d\mathbf{v}}{dt} + \frac{\lambda}{m}\mathbf{v} = -g\mathbf{k} \tag{1}$$

Its integrating factor

$$= e^{\int \lambda/m dt} = e^{\lambda t/m} \tag{2}$$

Then Eqn (1) can be written as

$$\frac{d}{dt}\left\{e^{\lambda t/m}\mathbf{v}\right\} = (-g\mathbf{k})e^{\lambda t/m} \tag{3}$$

Integrating,

$$e^{\lambda t/m}\mathbf{v} = -\frac{mg}{\lambda}e^{\lambda t/m}\mathbf{k} + C \tag{4}$$

At $t = 0$, the velocity is

$$\mathbf{v_0} = \mathbf{j}v_0\cos\alpha + \mathbf{k}v_0\sin\alpha \tag{5}$$

Inserting this in Eqn (4), we have

$$C = \mathbf{j}v_0\cos\alpha + \mathbf{k}v_0\sin\alpha + \mathbf{k}\frac{mg}{\lambda} \tag{6}$$

From Eqns (4) and (6), we find

$$\mathbf{v} = (\mathbf{j}v_0\cos\alpha + \mathbf{k}v_0\sin\alpha)e^{-\lambda t/m} - \frac{mg}{\lambda}(1 - e^{-\lambda t/m})\mathbf{k} \tag{7}$$

Since $\mathbf{v} = d\mathbf{r}/dt$, we find after integrating Eqn (7),

$$\mathbf{r} = -\frac{m}{\lambda}(\mathbf{j}v_0\cos\alpha + \mathbf{k}v_0\sin\alpha)e^{-\lambda t/m} - \frac{mg}{\lambda}\left(t + \frac{m}{\lambda}e^{-\lambda t/m}\right)\mathbf{k} + D \tag{8}$$

At $t = 0$, $\mathbf{r} = 0$. Hence, from (8) we find

$$D = \frac{m}{\lambda}(\mathbf{j}v_0\cos\alpha + \mathbf{k}v_0\sin\alpha) + \frac{m^2 g}{\lambda^2}\mathbf{k} \tag{9}$$

From (8) and (9), we write,

$$\mathbf{r} = \frac{mv_0}{\lambda}(\mathbf{j}\cos\alpha + \mathbf{k}\sin\alpha)\left(1 - e^{-\lambda t/m}\right) - \frac{mg}{\lambda}\left(t + \frac{m}{\lambda}e^{-\lambda t/m} - \frac{m}{\lambda}\right)\mathbf{k} \tag{10}$$

At the highest point, the vertical component of the velocity is zero. If t^* be the time taken to reach this point, we get from (7),

$$0 = v_0 \sin\alpha e^{-\lambda t^*/m} - \frac{mg}{\lambda}\left(1 - e^{-\lambda t^*/m}\right)$$

$$\text{i.e.,}\ \frac{mg}{\lambda} = \left(v_0 \sin\alpha + \frac{mg}{\lambda}\right) e^{-\lambda t^*/m}$$

$$\text{i.e.,}\ e^{\lambda t^*/m} = 1 + \frac{\lambda v_0 \sin\alpha}{mg}$$

$$\therefore\ t^* = \frac{m}{\lambda} \ln\left[1 + \frac{\lambda v_0 \sin\alpha}{mg}\right] \tag{11}$$

The maximum height would be obtained by putting $t = t^*$ from (11) into the vertical component of displacement in (10).

Thus, if H be the maximum height, then

$$\begin{aligned} H &= \frac{mv_0 \sin\alpha}{\lambda}\left(1 - e^{-\lambda t^*/m}\right) - \frac{mg}{\lambda}\left(t^* - \frac{m}{\lambda}\right) - \frac{m^2 g}{\lambda^2} e^{-\lambda t^*/m} \\ &= \frac{mv_0 \sin\alpha}{\lambda}\left(1 - \frac{mg}{mg + \lambda v_0 \sin\alpha}\right) + \frac{m^2 g}{\lambda^2}\left[1 - \frac{mg}{mg + \lambda v_0 \sin\alpha}\right] \\ &\quad - \frac{mg}{\lambda} \cdot \frac{m}{\lambda} \ln\left[1 + \frac{\lambda v_0 \sin\alpha}{mg}\right] \\ &= \frac{mv_0 \sin\alpha}{\lambda} + \frac{m^2 g}{\lambda^2}\left[1 - \frac{mg}{mg + \lambda v_0 \sin\alpha} - \frac{\lambda v_0 \sin\alpha}{mg + \lambda v_0 \sin\alpha}\right] \\ &\quad - \frac{m^2 g}{\lambda^2} \ln\left[1 + \frac{\lambda v_0 \sin\alpha}{mg}\right] \\ &= \frac{mv_0 \sin\alpha}{\lambda} - \frac{m^2 g}{\lambda^2} \ln\left[1 + \frac{\lambda v_0 \sin\alpha}{mg}\right] \end{aligned} \tag{12}$$

EXAMPLE 1.10 *A small bead of mass m is projected along a smooth circular wire under the action of a force varying inversely as the fifth power of distance from a centre of force situated on the circumference. Prove that the pressure on the wire is constant.*

Solution. The centre of force O is taken as the origin of polar coordinates (FIG. 1.15). Then equation of the circle is

$$r = 2a\cos\theta. \tag{1}$$

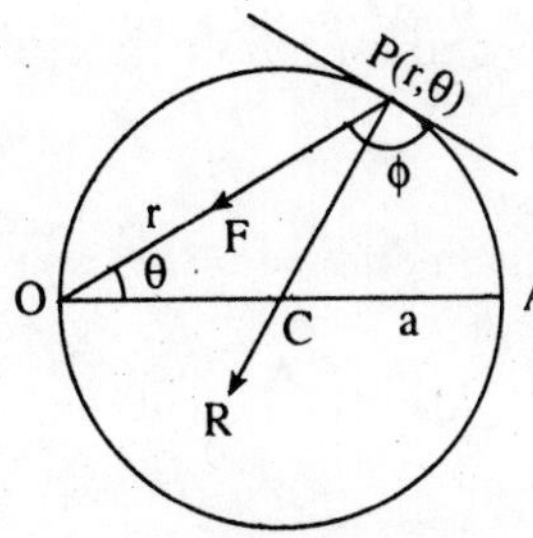

FIG. 1.15

The force on the mass is

$$F = -\frac{m\mu}{r^5}, \tag{2}$$

where $\mu =$ constant.

Equation of motion is

$$mv\frac{dv}{ds} = -\frac{m\mu}{r^5}\cos\phi = -\frac{m\mu}{r^5}\frac{dr}{ds}$$

$$\text{i.e.,}\ vdv = -\mu dr \tag{3}$$

Integrating,

$$\frac{v^2}{2} = \frac{\mu}{4r^4} + C \tag{4}$$

Initially, at A, $r = 2a$. Hence, if the initial velocity be u, then

$$\frac{u^2}{2} = \frac{\mu}{64a^4} + C, \text{ i.e., } C = \frac{u^2}{2} - \frac{\mu}{64a^4} \tag{5}$$

From (4) and (5),

$$v^2 = u^2 + \frac{\mu}{2r^4} - \frac{\mu}{32a^4} \tag{6}$$

Equation of motion along the normal is

$$\frac{v^2}{\rho} = R + \frac{\mu}{r^5}\cos(\phi - 90^\circ) = R + \frac{\mu}{r^5}\sin\phi, \tag{7}$$

where R = pressure on the wire.

Since $\rho = a$ for the circle and $\sin\phi = r\frac{d\theta}{ds}$, we get

$$R = \frac{v^2}{a} - \frac{\mu}{r^5}\cdot r\frac{d\theta}{ds} = \frac{v^2}{a} - \frac{\mu}{r^4}\frac{d\theta}{ds} \tag{8}$$

Now,

$$\frac{ds}{d\theta} = \left[r^2 + \left(\frac{dr}{d\theta}\right)^2\right]^{\frac{1}{2}} = [4a^2\cos^2\theta + 4a^2\sin^2\theta] \quad \text{[using (1)]}$$

$$\text{i.e.,} \quad \frac{ds}{d\theta} = 2a,$$

$$\therefore \frac{d\theta}{ds} = \frac{1}{2a}. \tag{9}$$

From (8) and (9),

$$R = \frac{1}{a}\left[v^2 - \frac{\mu}{2r^4}\right] = \frac{1}{a}\left[u^2 + \frac{\mu}{2r^4} - \frac{\mu}{32a^4} - \frac{\mu}{2r^4}\right] = \frac{1}{a}\left(u^2 - \frac{\mu}{32a^4}\right) \tag{10}$$

This shows that the pressure R on the wire at any point is constant.

1.22 Motion of a Charged Particle

(A) General Equations of Motion

Let a particle of mass m and charge q be under the influence of an electric field of intensity $\mathbf{E}$ as well as a magnetic field of intensity $\mathbf{B}$. Initially the particle is at rest, i.e., at $t = 0$, velocity $\mathbf{v} = 0$.

Let

$$\left.\begin{aligned} \mathbf{E} &= \mathbf{i}E_x + \mathbf{j}E_y + \mathbf{k}E_z \\ \mathbf{B} &= \mathbf{i}B_x + \mathbf{j}B_y + \mathbf{k}B_z \\ \text{and } \mathbf{v} &= \mathbf{i}\dot{x} + \mathbf{j}\dot{y} + \mathbf{k}\dot{z} \end{aligned}\right\} \tag{1.128}$$

Now, the Lorentz force acting on the particle due to electric and magnetic fields is

$$\mathbf{F} = \mathbf{F}_e + \mathbf{F}_m = q\mathbf{E} + q(\mathbf{v}\times\mathbf{B}) \tag{1.129}$$

where force due to electric field only is

$$\mathbf{F}_e = q\mathbf{E} \tag{1.130}$$

and the force due to magnetic field only is

$$\mathbf{F}_m = q(\mathbf{v}\times\mathbf{B}) \tag{1.131}$$

From Eqn (1.131), we have

$$\frac{\mathbf{F}_m}{q} = \mathbf{v} \times \mathbf{B} = \begin{vmatrix} \mathbf{i} & \mathbf{j} & \mathbf{k} \\ \dot{x} & \dot{y} & \dot{z} \\ B_x & B_y & B_z \end{vmatrix}$$

$$\text{i.e., } \mathbf{F}_m = q[\mathbf{i}(B_z\dot{y} - B_y\dot{z}) + \mathbf{j}(B_x\dot{z} - B_z\dot{x}) + \mathbf{k}(B_y\dot{x} - B_x\dot{y})]$$

Hence, the net force on the particle is

$$\mathbf{F} = q\left[\mathbf{i}\left(E_x + B_z\dot{y} - B_y\dot{z}\right) + \mathbf{j}\left(E_y + B_x\dot{z} - B_z\dot{x}\right) + \mathbf{k}\left(E_z + B_y\dot{x} - B_x\dot{y}\right)\right] \tag{1.132}$$

Again, if acceleration of the particle at any time is $\mathbf{a}$, then

$$\mathbf{F} = m\mathbf{a} = m(\mathbf{i}\ddot{x} + \mathbf{j}\ddot{y} + \mathbf{k}\ddot{z}) \tag{1.133}$$

From Eqns (1.132) and (1.133), we have

$$m\ddot{x} = q(E_x + B_z\dot{y} - B_y\dot{z}) \tag{1.134}$$

$$m\ddot{y} = q(E_y + B_x\dot{z} - B_z\dot{x}) \tag{1.135}$$

$$m\ddot{z} = q(E_z + B_y\dot{x} - B_x\dot{y}) \tag{1.136}$$

These are the equations of motion of the charged particle under the simultaneous influence of electric and magnetic fields.

Multiplying Eqn (1.134) by B_x and Eqn (1.135) by B_y, and adding the result, we get

$$m(B_x\ddot{x} + B_y\ddot{y}) = q(E_xB_x + E_yB_y) - qB_z(B_y\dot{x} - B_x\dot{y}) \tag{1.137}$$

Again, from Eqn (1.136),

$$m\ddot{z} - qE_z = q(B_y\dot{x} - B_x\dot{y}) \tag{1.138}$$

Substituting this relation into Eqn (1.137), we get

$$m(B_x\ddot{x} + B_y\ddot{y}) = q(E_xB_x + E_yB_y) - B_z(m\ddot{z} - qE_z)$$

$$\text{i.e., } m(B_x\ddot{x} + B_y\ddot{y} + B_z\ddot{z}) = q(E_xB_x + E_yB_y + E_zB_z)$$

$$\text{i.e., } m(\mathbf{B} \cdot \mathbf{a}) = q(\mathbf{B} \cdot \mathbf{E}) \text{ or, } \mathbf{B} \cdot (m\mathbf{a}) = \mathbf{B} \cdot (q\mathbf{E}) \tag{1.139}$$

Eqn (1.139) indicates that component of instantaneous acceleration vector along the direction of $\mathbf{B}$ is always constant.

(B) Some Special Cases

Let us consider a particle of mass m carrying a charge q, moving in an electromagnetic field with velocity $\mathbf{v}$. Then Lorentz force acting on the particle is $\mathbf{F} = q\left(\mathbf{E} + \mathbf{v} \times \mathbf{B}\right)$, where $\mathbf{E}$ is the intensity of electric field and $\mathbf{B}$ is the magnetic induction.

Hence, the equation of motion of the particle is

$$m\frac{d\mathbf{v}}{dt} = q\left(\mathbf{E} + \mathbf{v} \times \mathbf{B}\right) \tag{1.140}$$

We shall consider some special cases.

• Case 1 : Motion in a Constant Electric Field

We assume that the particle initially is at rest and is located at the point (x_0, y_0, z_0). If we take the constant electric field along x-axis, i.e., $\mathbf{E} = (E, 0, 0)$, then the equations of motion in Cartesian coordinate system are $m\ddot{x} = qE$, $m\ddot{y} = 0$, $m\ddot{z} = 0$.

Solutions of these equations are

$$\left.\begin{aligned} x &= x_0 + \frac{1}{2}\frac{qE}{m}t^2 \\ y &= y_0, \quad z = z_0 \end{aligned}\right\} \tag{1.141}$$

Thus, acceleration of the particle at any instant of time is

$$a = \frac{qE}{m} \tag{1.142}$$

If the particle falls through a potential difference of V volts, its initial potential energy qV is converted into kinetic energy. Hence, if v is the velocity of the particle acquired, then $\frac{1}{2}mv^2 = qV$.

$$\therefore\ v = \sqrt{\frac{2qV}{m}} \tag{1.143}$$

• Case 2 : Motion in a Constant Magnetic Field

In this case, the equation of motion of the particle is

$$m\frac{d\mathbf{v}}{dt} = q\mathbf{v} \times \mathbf{B}$$

We take dot product with velocity vector $\mathbf{v}$ on both sides. Then we get

$$m\frac{d\mathbf{v}}{dt} \cdot \mathbf{v} = q\mathbf{v} \cdot (\mathbf{v} \times \mathbf{B}), \text{ i.e., } \frac{d}{dt}\left(\frac{1}{2}m\mathbf{v}^2\right) = q(\mathbf{v} \times \mathbf{v}) \cdot \mathbf{B} = 0$$

$$\therefore\ \frac{1}{2}mv^2 = \text{constant.} \tag{1.144}$$

Thus, the kinetic energy and hence the speed of the particle remains constant. Only its direction is getting changed due to the action of the magnetic field.

Let us assume that the magnetic induction $\mathbf{B}$ is constant and is directed along x-axis. If v_1 and v_2 be component velocities of the particle perpendicular and parallel to the direction of $\mathbf{B}$ respectively, then the corresponding components of forces acting on the particle are:

$$\mathbf{F}_1 = q\mathbf{v}_1 \times \mathbf{B} \tag{1.145}$$

$$\text{and } \mathbf{F}_2 = q\mathbf{v}_2 \times \mathbf{B} = 0$$

$$\text{i.e., } \mathbf{F}_2 = m\frac{d\mathbf{v}_2}{dt} = 0, \text{ i.e., } v_2 = \text{constant.} \tag{1.146}$$

But, $v^2 = v_1^2 + v_2^2$; hence, from Eqn (1.144), we find that v_1 is also constant.

Now, from Eqn (1.140), we get

$$m\frac{d\mathbf{v}_1}{dt} = q\mathbf{v}_1 \times \mathbf{B}$$

This shows that the magnitude of v_1 is constant but its direction continuously changes.

We shall determine the trajectory of the particle. Taking $\mathbf{B} = \mathbf{i}B$, equation of motion in Cartesian coordinates is

$$m\frac{d\mathbf{v}}{dt} = q\mathbf{v} \times \mathbf{i}B$$

$$\text{i.e., } \frac{d\mathbf{v}}{dt} = \xi\mathbf{v} \times \mathbf{i}, \text{ where } \xi = \frac{qB}{m} \tag{1.147}$$

Eqn (1.147) in component form becomes

$$\frac{dv_x}{dt} = 0, \quad \frac{dv_y}{dt} = \xi v_z, \quad \frac{dv_z}{dt} = -\xi v_y$$

From the first equation, we have

$$v_x = \text{constant} = v_2 \text{ and so, } x = x_0 + v_2 t \tag{1.148}$$

Eliminating v_y from second and third equations, we get

$$\frac{d}{dt}\left(-\frac{1}{\xi}\frac{dv_z}{dt}\right) = \xi v_z, \text{ i.e., } \frac{d^2 v_z}{dt^2} + \xi^2 v_z = 0 \tag{1.149}$$

Its solution is

$$v_z = A\cos\xi t + B\sin\xi t. \tag{1.150}$$

$$\therefore \; v_y = -\frac{1}{\xi}\frac{dv_z}{dt} = -A\sin\xi t + B\cos\xi t \tag{1.151}$$

Integrating Eqns (1.150) and (1.151), we get

$$\left.\begin{aligned} \xi(y - y_0) &= A\cos\xi t + B\sin\xi t \\ \text{and } \xi(z - z_0) &= A\sin\xi t - B\cos\xi t \end{aligned}\right\} \tag{1.152}$$

where y_0 and z_0 are integration constants. It may be noted that y_0, z_0 do not correspond to initial positions at $t = 0$.

Now, squaring and adding both sides of Eqn (1.152), we get

$$(y - y_0)^2 + (z - z_0)^2 = \frac{1}{\xi^2}(A^2 + B^2) = \text{constant.} \tag{1.153}$$

This represents a circle with centre at (y_0, z_0). The trajectory of the particle is given by Eqns (1.148) and (1.152). Hence, it will be moving in circular orbit about x-axis and at the same time, it will have a translational motion with constant velocity along x-axis. Thus, the trajectory will be a helix (FIG. 1.16).

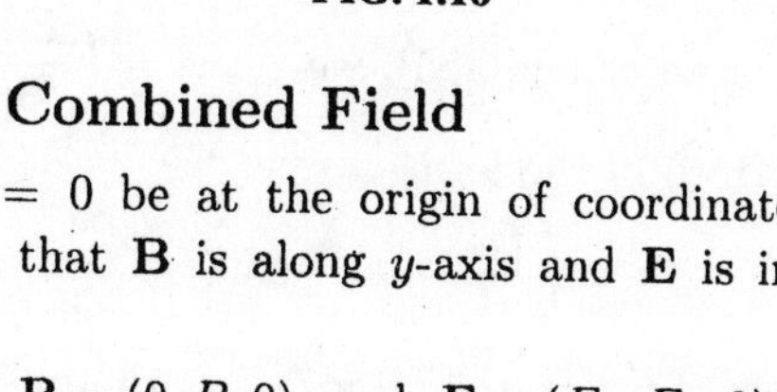

FIG. 1.16

• Case 3 : Motion in Combined Field

Let the particle at time $t = 0$ be at the origin of coordinates with velocity components (u_0, v_0, w_0). Let us assume that **B** is along y-axis and **E** is in the xy-plane making some angle with x-axis. That is

$$\mathbf{B} = (0, B, 0) \text{ and } \mathbf{E} = (E_x, E_y, 0)$$

Then the equations of motions (1.140) in the component form becomes

$$\frac{du}{dt} = a_x - \xi w, \quad \frac{dv}{dt} = a_y, \quad \frac{dw}{dt} = \xi u \tag{1.154}$$

where

$$a_x = \frac{qE_x}{m}, \quad a_y = \frac{qE_y}{m} \text{ and } \xi = \frac{qB}{m} \tag{1.155}$$

From first and third equations of the system (1.154), we have

$$\frac{d^2u}{dt^2} = -\xi\frac{dw}{dt} = -\xi^2 u, \text{ i.e., } \frac{d^2u}{dt^2} + \xi^2 u = 0$$

Its solution is $u = C\cos\xi t + A\sin\xi t$. Let us assume that at $t = 0$, $u_0 = 0$. Then, $C = 0$. Hence

$$u = A\sin\xi t \tag{1.156}$$

From third equation of (1.154), $\dfrac{dw}{dt} = \xi A\sin\xi t$.

Integrating,

$$w - w_0 = \xi A\int_0^t \sin\xi t\,dt, \text{ i.e., } w - w_0 = A(1 - \cos\xi t) \tag{1.157}$$

Integrating Eqns (1.156) and (1.157), we get

$$x = \int_0^t A\sin\xi t = \frac{A}{\xi}(1 - \cos\xi t) \tag{1.158}$$

$$\text{and } z = w_0 t + \frac{A}{\xi}(\xi t - \sin\xi t) \tag{1.159}$$

Eqns (1.158) and (1.159) give the parametric representation of a cycloid.

From second equation of the system (1.154),

$$v = v_0 + a_y t \text{ so that } y = v_0 t + \frac{1}{2}a_y t^2 \tag{1.160}$$

This shows that the particle moves in the direction of y-axis due to initial velocity component v_0 and the component of electric field E_y along the y-axis.

Let us now assume that $E_y = 0$, i.e., the fields are crossed. Then, $a_y = 0$, and hence the particle moves along y-axis with constant velocity v_0 so that its distance from origin at any time t is $v_0 t$. These are obvious from Eqn (1.160).

Now, from Eqns (1.154) and (1.156), we write

$$\frac{du}{dt} = a_x - \xi w, \text{ i.e., } A\xi\cos\xi t = a_x - \xi w$$

Thus, for $t = 0$, $A\xi = a_x - \xi w_0$,

$$\therefore\ A = \frac{1}{\xi}(a_x - \xi w_0)$$

Putting back the value of ξ from Eqn (1.155), we get $A = \left(\dfrac{E_x}{B} - w_0\right)$.

Hence, we can write Eqn (1.156) in the form

$$u = \left(\frac{E_x}{B} - w_0\right)\sin\xi t \tag{1.161}$$

Eqn (1.161) shows that, if $w_0 = E_x/B$, then $u = 0$, i.e., there is no deviation of the particle along x-axis. ***Such particles are called to be filtered out.*** Thus, the crossed combination of electric and magnetic fields has got a practical importance in that, by suitably choosing **E** and **B**, particles of desired velocity may be obtained and particles of undesired velocity may be filtered out. Such a combination of electromagnetic field works as a velocity filter for charged particles.

EXERCISE

1. Trajectory of a particle of unit mass is given by the radius vector

$$\mathbf{r} = \mathbf{i}a\cos\omega t + \mathbf{j}b\sin\omega t + \mathbf{k}c$$

where a, b, c are constants. Calculate angular momentum of the particle about the origin. Show that it is constant along k. [C.U. (HONS) 1997]

2. Prove that the force-field

$$\mathbf{F} = \mathbf{i}(yz - y) + \mathbf{j}(xz - x - 1) + \mathbf{k}(xy - 2z)$$

is conservative and find the corresponding potential function. Is the angular momentum about the origin conserved for this force? [C.U. (HONS) 1998]

3. A particle is moving under the action of a force $\mathbf{k} \times \mathbf{v}$, where $\mathbf{k}$ is unit vector along z-axis and $\mathbf{v}$ is the velocity of the particle. Show that the kinetic energy of the particle remains constant. [C.U. (HONS) 2000]

4. Estimate whether the force

$$\mathbf{F} = \mathbf{i}(3z + 5y) + \mathbf{j}(5x + 2z) + \mathbf{k}(2y + 3x + 4z)$$

is conservative.

5. A particle of unit mass moves according to the equation

$$\mathbf{r} = \mathbf{i}(2 + 3t^2) + \mathbf{j}5t^2 + \mathbf{k}t$$

Find the force $\mathbf{F}$ acting on it, torque $\mathbf{N}$ and angular momentum $\mathbf{J}$ about the origin. Hence, verify $\mathbf{N} = \dfrac{d\mathbf{J}}{dt}$. [C.U. (HONS) 2002]

6. The polar coordinates of a particle moving in a plane are given by

$$r = a\sin\omega_1 t, \ \theta = \omega_2 t$$

Obtain expressions of polar coordinates of velocity and accelerations of the particle. [C.U. (HONS) 1980]

7. A particle of mass m moves in $(x - y)$-plane and its position vector is given by

$$\mathbf{r} = \mathbf{i}a\cos\omega t + \mathbf{j}b\sin\omega t$$

where a, b, ω are constants.

(i) Show that the force acting on the particle is always directed towards the origin and the force-field is conservative.

(ii) Find the work done by the force-field in moving the particle from the point $(a, 0)$ to $(0, b)$. [C.U. (HONS) 2004]

8. Establish that the force-field given by

$$\mathbf{F} = \frac{k}{r^2}\mathbf{e_r}$$

is conservative. Calculate the corresponding potential energy at r. [C.U. (HONS) 2006]

9. Show that the total time derivative of a vector $\mathbf{P}$ written in cylindrical coordinates $\mathbf{P} = P_r\mathbf{e_r} + P_\theta\mathbf{e_\theta} + P_z\mathbf{k}$ is

$$\frac{d\mathbf{P}}{dt} = \left(\dot{P}_r - \dot{\theta}P_\theta\right)\mathbf{e_r} + \left(\dot{P}_\theta + \dot{\theta}P_r\right)\mathbf{e_\theta} + \dot{P}_z\mathbf{k}$$

10. A particle is subject to a frictional force so that the velocity decreases with time according to $v = c(t - t_1)^2$, where t_1 is the time when it stops and c is a constant. Find the force $F(v)$ and the distance travelled before it stops. **[C.U. (HONS) 2001]**

11. A particle moves under the influence of a force-field given by

$$\mathbf{F} = a(\mathbf{i}\sin\omega t + \mathbf{j}\cos\omega t)$$

If the particle be initially at rest, prove that at an instant t, work done on the particle is given by

$$\frac{a^2}{m\omega^2}(1 - \cos\omega t)$$ **[C.U. (HONS) 2002]**

12. A particle of mass m falls under gravity, the viscous drag due to air being k times the velocity. Show that the terminal velocity of the particle is mg/k. Also show that the height fallen after time t is

$$z = \frac{mg}{k}\left[1 - \frac{m}{k}\left(1 - e^{-\frac{kt}{m}}\right)\right]$$

13. A particle of mass m is projected at an angle α with the horizon with velocity u. The resistance of air is mk times the instantaneous velocity. Show that the total time of flight is given by

$$T = \left(\frac{u\sin\alpha}{g} + \frac{1}{k}\right)\left(1 - e^{-kT}\right)$$

14. A particle of mass m slides down from rest along a smooth parabolic tube placed vertex downwards in a vertical plane. Show that the reaction of the tube at any position of the particle is

$$2mg\left(\frac{h+a}{\rho}\right)$$

where $\rho =$ radius of curvature, $4a$ the latus rectum and $h =$ original vertical height of the particle above the vertex.

15. A particle of mass m rests on the top of a smooth sphere. It is displaced slightly so that it begins to slide. Show that the velocity of the particle at the point when it leaves the sphere is

$$\sqrt{\frac{2}{3}ag}, \quad a = \text{radius of the sphere.}$$

16. A particle of mass m is thrown up with velocity u along a rough inclined plane. The angle of inclination of the plane is α and the constant coefficient of friction is μ. Discuss the motion.

CHAPTER 2

Mechanics of a System of Particles

2.1 Equation of Motion

Consider a system of N particles of constant masses $m_1, m_2, \ldots, m_N$ whose position vectors with respect to some fixed coordinate axes are $\mathbf{r_1}, \mathbf{r_2}, \ldots, \mathbf{r_N}$. Let the external force acting on the i-th particle be $\mathbf{F_i^e}$ and the internal force acting on the i-th particle due to the presence of j-th particle be $\mathbf{F_{ij}}$. Then the equation of motion of the i-th particle under the action of external and internal forces is

$$m_i \ddot{\mathbf{r}}_\mathbf{i} = \mathbf{F_i^e} + \sum_{j=1}^{N} \mathbf{F_{ij}} \tag{2.1}$$

Since the internal force on i-th particle due to itself is zero, we write:

$$\mathbf{F_{ii}} = 0 \tag{2.2}$$

Taking the sum of Eqn (2.1) over all N particles, we obtain

$$\sum m_i \ddot{\mathbf{r}}_\mathbf{i} = \sum_{i=1}^{N} \mathbf{F_i^e} + \sum_{i=1}^{N} \sum_{j=1}^{N} \mathbf{F_{ij}} \; (i \neq j) \tag{2.3}$$

Now, according to Newton's third law of motion, the mutual interaction forces between i-th and j-th particles are equal and opposite. Thus,

$$\mathbf{F_{ij}} = -\mathbf{F_{ji}}$$
$$\text{i.e., } \mathbf{F_{ij}} + \mathbf{F_{ji}} = 0 \tag{2.4}$$

The second sum in Eqn (2.3) represents the internal forces in pairs and for each pair the resultant is zero by virtue of Eqn (2.4).

Thus, Eqn (2.3) is reduced to

$$\sum_{i=1}^{N} m_i \ddot{\mathbf{r}}_\mathbf{i} = \sum_{i=1}^{N} \mathbf{F_i^e} = \mathbf{F_i} \tag{2.5}$$

where $\mathbf{F_i}$ represents the sum of all external forces. Eqn (2.5) is the equation of motion of the system of particles.

2.2 Centre of Mass

The *centre of mass* of a system of particles m_i $(i = 1, \ldots, N)$ is defined as a point whose position vector $\mathbf{R}$ is given by

$$\mathbf{R} = \frac{\sum m_i \mathbf{r_i}}{\sum m_i} \tag{2.6}$$

If the total mass of the system be M, then

$$\sum m_i = M \tag{2.7}$$

Hence, Eqn (2.6) is written as

$$M\mathbf{R} = \sum m_i \mathbf{r_i} \tag{2.8}$$

Differentiating this equation twice, we get

$$M\ddot{\mathbf{R}} = \sum m_i \ddot{\mathbf{r}}_\mathbf{i} \tag{2.9}$$

In view of this equation, Eqn (2.5) becomes

$$M\ddot{\mathbf{R}} = \mathbf{F_i} \tag{2.10}$$

Thus, the acceleration of the centre of mass depends only on the external forces. Hence, under external forces applied on a system, the centre of mass of a system of particles moves as if the total mass of the system is concentrated there.

2.3 Conservation of Linear Momentum

Differentiating Eqn (2.8) once, we get

$$M\dot{\mathbf{R}} = \sum m_i \dot{\mathbf{r}}_\mathbf{i} \tag{2.11}$$

If $\mathbf{V_{cm}}$ be the velocity of the centre of mass and $\mathbf{v_i}$ the velocity of the i-th particle, then Eqn (2.11) can be written as

$$M\mathbf{V_{cm}} = \sum m_i \mathbf{v_i} \tag{2.12}$$

If $\mathbf{P}$ be the total linear momentum of all the particles of the system, then

$$\mathbf{P} = \sum m_i \mathbf{v_i} \tag{2.13}$$

Thus, from Eqns (2.12) and (2.13), we obtain

$$\mathbf{P} = M\mathbf{V_{cm}} \tag{2.14}$$

That is, ***the total linear momentum of a system of particles is equal to the product of the total mass of the system and the velocity of the centre of mass.***

Differentiating Eqn (2.14) once,

$$\dot{\mathbf{P}} = M\dot{\mathbf{V}}_\mathbf{cm} = M\ddot{\mathbf{R}} \tag{2.15}$$

In view of Eqns (2.10) and (2.15), the total external force on the system is

$$\mathbf{F_i} = \dot{\mathbf{P}} \tag{2.16}$$

Thus, if $\mathbf{F_i} = 0$, $\dot{\mathbf{P}} = 0$ or, $\mathbf{P} =$ constant. Thus, ***if the total external force on the system is zero, the total linear momentum is constant.***

This is the ***law of conservation of linear momentum*** for a system of particles.

2.4 Conservation of Angular Momentum

Let $\mathbf{J_i}$ be the angular momentum of the i-th particle about some point O. Then

$$\mathbf{J_i} = \mathbf{r_i} \times \mathbf{p_i} \tag{2.17}$$

where $\mathbf{r_i}$ is the position vector of the i-th particle from O.

The total angular momentum of the system of particles about any point will be equal to the sum of the angular momenta of all individual particles. Hence, the total angular momentum $\mathbf{J}$ of the system is

$$\mathbf{J} = \sum \mathbf{J_i} = \sum (\mathbf{r_i} \times \mathbf{p_i}) \tag{2.18}$$

If $\mathbf{N}$ be the total torque acting on the system, then

$$\mathbf{N} = \frac{d\mathbf{J}}{dt} = \frac{d}{dt}\sum (\mathbf{r_i} \times \mathbf{p_i})$$

$$\text{or, } \mathbf{N} = \sum \dot{\mathbf{r}}_\mathbf{i} \times \mathbf{p_i} + \sum \mathbf{r_i} \times \dot{\mathbf{p}}_\mathbf{i} \tag{2.19}$$

Now

$$\dot{\mathbf{r}}_\mathbf{i} \times \mathbf{p_i} = \dot{\mathbf{r}}_\mathbf{i} \times m_i \dot{\mathbf{r}}_\mathbf{i} = m_i \dot{\mathbf{r}}_\mathbf{i} \times \dot{\mathbf{r}}_\mathbf{i} = 0 \tag{2.20}$$

Hence, Eqn (2.19) assumes the form:

$$\mathbf{N} = \sum \mathbf{r_i} \times \dot{\mathbf{p}}_\mathbf{i} \tag{2.21}$$

From Eqn (2.1), we have

$$\dot{\mathbf{p}}_\mathbf{i} = \mathbf{F}_\mathbf{i}^\mathbf{e} + \sum_{j=1}^{N} \mathbf{F_{ij}} \tag{2.22}$$

Hence,

$$\mathbf{r_i} \times \dot{\mathbf{p}}_\mathbf{i} = \mathbf{r_i} \times \mathbf{F}_\mathbf{i}^\mathbf{e} + \sum_{j=1}^{N} \mathbf{r_i} \times \mathbf{F_{ij}} \tag{2.23}$$

Taking the sum over all the particles of the system, we get

$$\sum_{i=1}^{N} \mathbf{r_i} \times \dot{\mathbf{p}}_\mathbf{i} = \sum_{i=1}^{N} \mathbf{r_i} \times \mathbf{F}_\mathbf{i}^\mathbf{e} + \sum_{i=1}^{N}\sum_{j=1}^{N} \mathbf{r_i} \times \mathbf{F_{ij}} \tag{2.24}$$

The term in the right-hand side under the double sum contains pairs of terms of the form:

$$\mathbf{r_i} \times \mathbf{F_{ij}} + \mathbf{r_j} \times \mathbf{F_{ji}} = (\mathbf{r_i} - \mathbf{r_j}) \times \mathbf{F_{ij}}, \text{ since } \mathbf{F_{ij}} = -\mathbf{F_{ji}}$$

$$= \mathbf{r_{ij}} \times \mathbf{F_{ij}} \tag{2.25}$$

where $\mathbf{r_{ij}}$ denotes the vector joining the i-th and j-th particles (FIG. 2.1).

Since we are considering the forces of action and reaction only, the internal force $\mathbf{F_{ij}}$ should be along the vector $\mathbf{r_{ij}}$.

Hence, we can write,

$$\mathbf{F_{ij}} = \lambda \mathbf{r_{ij}}, \ \lambda = \text{constant} \tag{2.26}$$

Then

$$\mathbf{r_{ij}} \times \mathbf{F_{ij}} = \mathbf{r_{ij}} \times \lambda \mathbf{r_{ij}} = \lambda (\mathbf{r_{ij}} \times \mathbf{r_{ij}}) = 0 \tag{2.27}$$

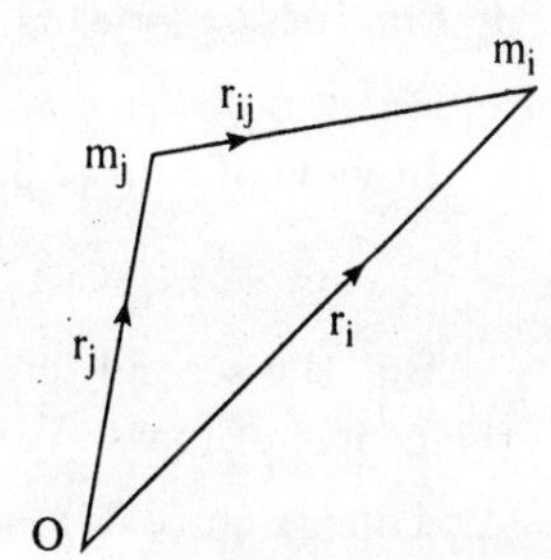

FIG. 2.1

Thus, Eqn (2.24) is reduced to

$$\sum \mathbf{r_i} \times \dot{\mathbf{p}}_\mathbf{i} = \sum \mathbf{r_i} \times \mathbf{F_i^e} \tag{2.28}$$

From Eqns (2.21) and (2.28), the total torque becomes

$$\mathbf{N} = \sum \mathbf{r_i} \times \mathbf{F_i^e} = \sum \mathbf{N_i} \tag{2.29}$$

Thus, ***the total torque on the system is equal to the vector sum of the torques acting on the individual particles.***

If $\mathbf{N} = 0$, then $\frac{d\mathbf{J}}{dt} = 0$, i.e., $\mathbf{J}$ = constant.

Thus, ***the total angular momentum of a system of particles is conserved when there is no external torque.***

This is the ***conservation theorem for total angular momentum*** of a system of particles.

2.5 Angular Momentum about Centre of Mass

Consider the i-th particle of mass m_i having position vector $\mathbf{r_i}$ with respect to a point O (FIG. 2.2). Let $\mathbf{r_i'}$ be the position vector of the particle with respect to the centre of mass.

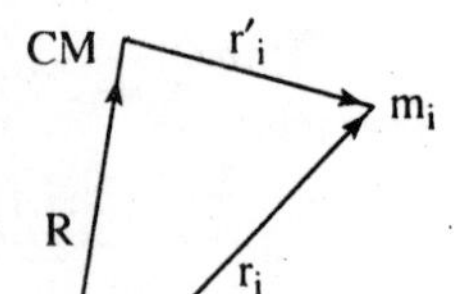

FIG. 2.2

Then

$$\mathbf{r_i} = \mathbf{R} + \mathbf{r_i'} \tag{2.30}$$

Differentiating once, we get

$$\dot{\mathbf{r}}_\mathbf{i} = \dot{\mathbf{R}} + \dot{\mathbf{r}}_\mathbf{i}'$$

$$\text{or, } \mathbf{v_i} = \mathbf{V_{cm}} + \mathbf{v_i'} \tag{2.31}$$

where $\mathbf{V_{cm}}$ = velocity of the centre of mass and $\mathbf{v_i'}$ = velocity of the particle relative to the centre of mass.

Then the total angular momentum of the system, by Eqn (2.18), is

$$\begin{aligned}
\mathbf{J} &= \sum \mathbf{r_i} \times \mathbf{p_i} \\
&= \sum m_i\,(\mathbf{R} + \mathbf{r_i'}) \times (\mathbf{V_{cm}} + \mathbf{v_i'}) \\
&= \sum m_i(\mathbf{R} \times \mathbf{V_{cm}}) + \sum m_i(\mathbf{R} \times \mathbf{v}_i') + \sum m_i(\mathbf{r_i'} \times \mathbf{V_{cm}}) + \sum m_i(\mathbf{r_i'} \times \mathbf{v_i'}) \\
&= \mathbf{R} \times \left(\sum m_i\right) \mathbf{V_{cm}} + \sum \mathbf{R} \times m\mathbf{v_i'} + \sum m_i\mathbf{r_i'} \times \mathbf{V_{cm}} + \sum \mathbf{r_i'} \times m_i\mathbf{v_i'}
\end{aligned} \tag{2.32}$$

Since the distances $\mathbf{r_i'}$ are measured with respect to the centre of mass

$$\sum m_i\mathbf{r_i'} = 0; \text{ hence, } \sum m_i\dot{\mathbf{r}}_\mathbf{i}' = 0 \tag{2.33}$$

In view of Eqn (2.33), Eqn (2.32) assumes the form:

$$\mathbf{J} = \mathbf{R} \times M\mathbf{V_{cm}} + \sum \mathbf{r_i'} \times m_i\mathbf{v_i'} \tag{2.34}$$

But the second term in Eqn (2.34) represents the angular momentum of the system about the centre of mass. We shall denote it by $\mathbf{J_{cm}}$.

Thus, Eqn (2.34) is reduced to

$$\mathbf{J} = \mathbf{J_{cm}} + \mathbf{R} \times \mathbf{P} \tag{2.35}$$

where **P** is the total linear momentum. Thus, ***the total angular momentum of a system of particles about a fixed point is equal to the sum of the angular momentum of the system about the centre of mass and the angular momentum of the total mass concentrated at the centre of mass about that point.***

2.6 Kinetic Energy of the System

The total kinetic energy of the system is

$$T = \sum \frac{1}{2} m_i v_i^2 = \sum \frac{1}{2} m_i \mathbf{v_i} \cdot \mathbf{v_i} \tag{2.36}$$

Using Eqn (2.31), we write,

$$\mathbf{v_i} \cdot \mathbf{v_i} = (\mathbf{V_{cm}} + \mathbf{v_i'}) \cdot (\mathbf{V_{cm}} + \mathbf{v_i'}) = \mathbf{V_{cm}^2} + \mathbf{v_i'^2} + 2\mathbf{V_{cm}} \cdot \mathbf{v_i'} \tag{2.37}$$

Thus, the total kinetic energy becomes

$$\begin{aligned} T &= \sum \frac{1}{2} m_i \left(V_{\text{cm}}^2 + v_i'^2 + 2\mathbf{V_{cm}} \cdot \mathbf{v_i'}\right) \\ &= \sum \frac{1}{2} m_i V_{\text{cm}}^2 + \sum \frac{1}{2} m_i v_i'^2 + \sum m_i \mathbf{V_{cm}} \cdot \mathbf{v_i'} \end{aligned} \tag{2.38}$$

Now, the third term in Eqn (2.38) is equal to

$$\begin{aligned} \sum m_i \mathbf{V_{cm}} \cdot \mathbf{v_i'} &= \mathbf{V_{cm}} \cdot \sum m_i \mathbf{v_i'} \\ &= \mathbf{V_{cm}} \cdot \sum m_i \dot{\mathbf{r}}_i' = 0 \end{aligned}$$

by Eqn (2.33).

The second term $\sum \frac{1}{2} m_i v_i'^2$ represents the kinetic energy of the system referred to the centre of mass as the origin. We denote it by T_{cm}. The first term is

$$\sum \frac{1}{2} m_i V_{\text{cm}}^2 = \frac{1}{2} V_{\text{cm}}^2 \sum m_i = \frac{1}{2} M V_{\text{cm}}^2 \tag{2.39}$$

Hence, the total kinetic energy of the system is

$$T = \frac{1}{2} M V_{\text{cm}}^2 + T_{\text{cm}} \tag{2.40}$$

Thus, ***the total kinetic energy of the system is the sum of the kinetic energy of the total mass concentrated at the centre of mass and moving with velocity*** $\mathbf{V}_{cm} = \dot{\mathbf{R}}$ ***with respect to the origin O and the kinetic energy of the system with respect to the centre of mass as the origin.***

2.7 Work

The total amount of work done by the forces acting on various particles of the system from an initial position 1 to the final position 2 is given by

$$\begin{aligned} W_{12} &= \sum \int_1^2 \mathbf{F_i} \cdot d\mathbf{r_i} = \sum \int_1^2 m_i \dot{\mathbf{v}}_i \cdot \mathbf{v_i} dt \\ &= \sum \int_1^2 d\left(\frac{1}{2} m_i v_i^2\right) = \left[\sum \frac{1}{2} m_i v_i^2\right]_1^2 = T_2 - T_1 \end{aligned} \tag{2.41}$$

Thus, the work done is equal to the change in kinetic energy of the system.

2.8 Conservation of Energy

Let the total force

$$\mathbf{F_i} = \mathbf{F_i^e} + \sum_{j=1}^{N} \mathbf{F_{ij}} \tag{2.42}$$

acting on the i-th particle be a conservative one. Then a potential function V exists such that

$$\mathbf{F_i} = -\boldsymbol{\nabla}_\mathbf{i} V_i \tag{2.43}$$

The equation of motion of the i-th particle subject to such forces is

$$\dot{\mathbf{p}}_\mathbf{i} = -\boldsymbol{\nabla}_\mathbf{i} V_i \tag{2.44}$$

Taking dot product with $\dot{\mathbf{r}}_\mathbf{i}$ on both sides of this equation, we get

$$\dot{\mathbf{p}}_\mathbf{i} \cdot \dot{\mathbf{r}}_\mathbf{i} = -\boldsymbol{\nabla}_\mathbf{i} V_i \cdot \dot{\mathbf{r}}_\mathbf{i}$$

$$\text{or,} \quad \dot{\mathbf{p}}_\mathbf{i} \cdot \frac{\mathbf{p_i}}{m_i} = -\boldsymbol{\nabla}_\mathbf{i} V_i \cdot \dot{\mathbf{r}}_\mathbf{i}$$

$$\text{or,} \quad \frac{d}{dt}\left(\frac{p_i^2}{2m_i}\right) + \boldsymbol{\nabla}_\mathbf{i} V_i \cdot \dot{\mathbf{r}}_\mathbf{i} = 0 \tag{2.45}$$

Summing this equation over all N particles, we obtain

$$\frac{d}{dt}\sum_{i=1}^{N}\left(\frac{p_i^2}{2m_i}\right) + \sum_{i=1}^{N} \boldsymbol{\nabla}_\mathbf{i} V_i \cdot \dot{\mathbf{r}}_\mathbf{i} = 0 \tag{2.46}$$

Now

$$\sum\left(\frac{p_i^2}{2m}\right) = \sum\left(\frac{m_i^2 v_i^2}{2m_i}\right) = \sum\left(\frac{1}{2} m_i v_i^2\right) = T \tag{2.47}$$

where T is the total kinetic energy of the system.

Also

$$\begin{aligned}
\sum \boldsymbol{\nabla}_\mathbf{i} V_i \cdot \dot{\mathbf{r}}_\mathbf{i} &= \sum\left(\mathbf{i}\frac{\partial V_i}{\partial x_i} + \mathbf{j}\frac{\partial V_i}{\partial y_i} + \mathbf{k}\frac{\partial V_i}{\partial z_i}\right)\cdot\left(\mathbf{i}\frac{dx_i}{dt} + \mathbf{j}\frac{dy_i}{dt} + \mathbf{k}\frac{dz_i}{dt}\right) \\
&= \sum\left(\frac{\partial V_i}{\partial x_i}\frac{dx_i}{dt} + \frac{\partial V_i}{\partial y_i}\frac{dy_i}{dt} + \frac{\partial V_i}{\partial z_i}\frac{dz_i}{dt}\right) \\
&= \sum \frac{d}{dt} V_i = \frac{d}{dt}\sum V_i = \frac{dV}{dt}
\end{aligned} \tag{2.48}$$

where V is the sum of all the potential functions. Hence, Eqn (2.46) assumes the form:

$$\frac{d}{dt}(T + V) = 0 \text{ or, } \frac{dE}{dt} = 0, \text{ i.e., } E = \text{constant}, \tag{2.49}$$

where E is the total energy of the system.

Thus, ***in a conservative force-field, the total energy of the system is conserved.***

EXAMPLE 2.1 *Three particles of masses 2, 3 and 5 units move under the influence of a force-field such that their position vectors relative to a fixed coordinate system are given by*

$$\mathbf{r_1} = 2t\mathbf{i} + 3\mathbf{j} - t^2\mathbf{k}$$

$$\mathbf{r_2} = (t+1)\mathbf{i} - 3t\mathbf{j} - 4\mathbf{k}$$

$$\mathbf{r_3} = t^2\mathbf{i} + t\mathbf{j} - (2t-1)\mathbf{k}$$

where t is the time. Determine

(a) *total linear momentum of the system;*

(b) *total kinetic energy of the particles; and hence find the total work done by the force-field in moving particles from their positions at time $t = 1$ to positions at time $t = 2$;*

(c) *total angular momentum of the system; and*

(d) *total external torque applied to the system.*

Verify that $\mathbf{N} = \dfrac{d\mathbf{J}}{dt}$.

Solution. We have

$$\mathbf{r_1} = 2t\mathbf{i} + 3\mathbf{j} - t^2\mathbf{k}$$

$$\therefore\ \dot{\mathbf{r}}_1 = 2\mathbf{i} + 0 \cdot \mathbf{j} - 2t\mathbf{k} \tag{1}$$

$$\mathbf{r_2} = (t+1)\mathbf{i} - 3t\mathbf{j} - 4\mathbf{k}$$

$$\therefore\ \dot{\mathbf{r}}_2 = \mathbf{i} - 3\mathbf{j} \tag{2}$$

$$\mathbf{r_3} = t^2\mathbf{i} + t\mathbf{j} - (2t-1)\mathbf{k}$$

$$\therefore\ \dot{\mathbf{r}}_3 = 2t\mathbf{i} + \mathbf{j} - 2\mathbf{k} \tag{3}$$

(a) Total linear momentum of the system:

$$\mathbf{P} = \sum_{i=1}^{3} m_i\dot{\mathbf{r}}_i = 2(2\mathbf{i} - 2t\mathbf{k}) + 3(\mathbf{i} - 3\mathbf{j}) + 5(2t\mathbf{i} + \mathbf{j} - 2\mathbf{k})$$

$$= (7 + 10t)\mathbf{i} - 4\mathbf{j} - (4t + 10)\mathbf{k} \tag{4}$$

(b) Total kinetic energy of the system:

$$T = \sum_{i=1}^{3} \frac{1}{2} m_i \dot{\mathbf{r}}_i \cdot \dot{\mathbf{r}}_i = \frac{1}{2} \times 2 \times (4 + 4t^2) + \frac{1}{2} \times 3 \times (10) + \frac{1}{2} \times 5 \times (4t^2 + 5)$$

$$= 14t^2 + 31.5 \tag{5}$$

The total work done by force-field in moving the particles from their position at time $t = 1$ to position at time $t = 2$ is

$$(T)_{t=2} - (T)_{t=1} = (14t^2)_{t=2} - (14t^2)_{t=1} \text{ [using Eqn (5)]}$$

$$= 42 \text{ units.}$$

(c) Total angular momentum:

$$\mathbf{J} = \sum_{i=1}^{3} \mathbf{r}_i \times m_i\dot{\mathbf{r}}_i$$

$$= 2(2t\mathbf{i} + 3\mathbf{j} - t^2\mathbf{k}) \times (2\mathbf{i} - 2t\mathbf{k}) + 3\left[(t+1)\mathbf{i} - 3t\mathbf{j} - 4\mathbf{k}\right] \times (\mathbf{i} - 3\mathbf{j})$$

$$+5\left[t^2\mathbf{i} + t\mathbf{j} - (2t-1)\mathbf{k}\right] \times (2t\mathbf{i} + \mathbf{j} - 2\mathbf{k})$$

$$= 2(4t^2\mathbf{j} - 6\mathbf{k} - 6t\mathbf{i} - 2t^2\mathbf{j}) + 3\left[-3(t+1)\mathbf{k} + 3t\mathbf{k} - 4\mathbf{j} - 12\mathbf{i}\right]$$

$$+5\left[t^2\mathbf{k} + 2t^2\mathbf{j} - 2t^2\mathbf{k} - 2t\mathbf{i} - 2t(2t-1)\mathbf{j} + (2t-1)\mathbf{i}\right]$$

$$= (-12t - 41)\mathbf{i} + (-6t^2 + 10t - 12)\mathbf{j} + (-5t^2 - 21)\mathbf{k} \tag{6}$$

(d) Total external torque applied to the system:

$$\mathbf{N} = \sum \mathbf{r}_i \times \dot{\mathbf{p}}_i = \sum \mathbf{r}_i \times m_i\ddot{\mathbf{r}}_i.$$

Hence,

$$\begin{aligned}\mathbf{N} &= 2(2t\mathbf{i} + 3\mathbf{j} - t^2\mathbf{k}) \times (-2\mathbf{k}) + 3\left[(t+1)\mathbf{i} - 3t\mathbf{j} - 4\mathbf{k}\right] \times 0 \\ &\quad +5\left[t^2\mathbf{i} + t\mathbf{j} - (2t-1)\mathbf{k}\right] \times (2\mathbf{i}) \\ &= 2(4t\mathbf{j} - 6\mathbf{i}) + 5\left[-2t\mathbf{k} - 2(2t-1)\mathbf{j}\right] \\ &= -12\mathbf{i} + (-12t + 10)\mathbf{j} - 10t\mathbf{k}. \end{aligned} \tag{7}$$

Using Eqn (6), we find that

$$\frac{d\mathbf{J}}{dt} = -12\mathbf{i} + (-12t + 10)\mathbf{j} - 10t\mathbf{k} \tag{8}$$

Thus, it is verified that $\mathbf{N} = \dfrac{d\mathbf{J}}{dt}$.

EXAMPLE 2.2 *A block of wood of mass M is suspended by a string of length L. Initially, the string is vertical and the mass is at rest. A bullet of mass m is fired in the horizontal direction into the block and becomes embedded in it. The block then swings upward and rises until the string makes an angle θ with the vertical. Find the initial speed of the bullet.* **[C.U. (HONS) 2005]**

Solution. Let v and V be the initial speed of the bullet and the block with bullet embedded in it. Then from the conservation of linear momentum in horizontal direction,

$$mv = (M+m)V \tag{1}$$

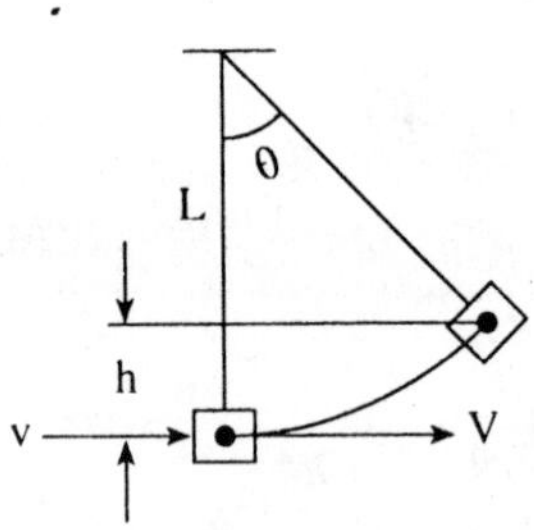

FIG. 2.3

From FIG. 2.3, we have

$$h = L(1-\cos\theta) \tag{2}$$

From conservation of the total energy of the block with bullet embedded in it, we get

$$\frac{1}{2}(M+m)V^2 = (M+m)gh = (M+m)gL(1-\cos\theta)$$

This yields

$$V = \sqrt{2gL(1-\cos\theta)} \tag{3}$$

Substituting the value of V into Eqn (1), we find the initial speed of the bullet as

$$v = \left(\frac{M+m}{m}\right)\sqrt{2gL(1-\cos\theta)} = 2\left(\frac{M+m}{m}\right)\sqrt{gL}\sin\frac{\theta}{2}.$$

2.9 Rocket Motion

This is a case of motion involving ***variable mass***. Let M be the mass of the rocket and v its speed at time t. After time $t + \Delta t$, the mass of the rocket becomes $M - \Delta M$ due to the burning of the fuel and the velocity of the rocket becomes $v + \Delta v$. Let the burnt gas products are ejected through a ***nozzle*** at the rear with a velocity u relative to the rocket (FIG. 2.4). Then, for the change in momentum of the system, we have

$$(M - \Delta M)(v + \Delta v) - \Delta M(u - v) - Mv = F\Delta t, \tag{2.50}$$

where F represents all the external forces acting on the rocket.

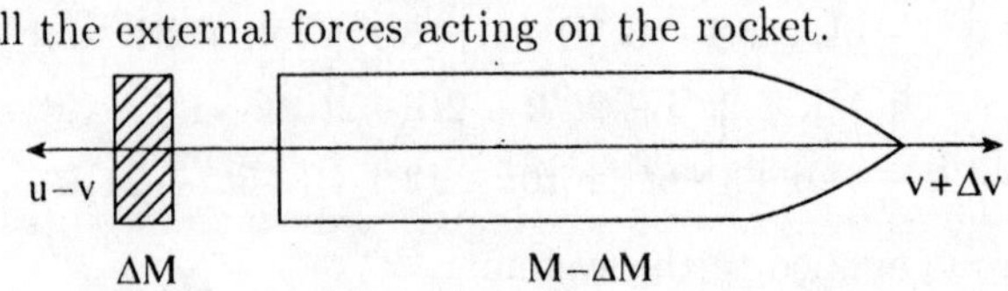

FIG. 2.4

Simplifying Eqn (2.50) to first order in infinitesimals, we obtain

$$M\Delta v - u\Delta M = F\Delta t \tag{2.51}$$

Dividing by Δt and proceeding to the limit

$$M\frac{dv}{dt} - u\frac{dM}{dt} = F \tag{2.52}$$

which is the equation of motion of the rocket.

• Case 1 : Motion in Horizontal Direction

Neglecting viscous drag due to air resistance, the equation of motion of the rocket becomes

$$M\frac{dv}{dt} = -u\frac{dM}{dt} \tag{2.53}$$

where the negative sign has been added to show that the mass decreases with increase in velocity. Eqn (2.53) is recast in the form:

$$dv = -u\frac{dM}{M} \tag{2.54}$$

Let the initial conditions be

$$\left.\begin{aligned} M &= M_0, \text{ at } t = 0 \\ v &= v_0, \text{ at } t = 0 \\ x &= x_0, \text{ at } t = 0 \end{aligned}\right\} \tag{2.55}$$

Integrating Eqn (2.54) with the use of the initial conditions above, we get

$$v - v_0 = u \ln \frac{M_0}{M} \tag{2.56}$$

Let the fuel be burnt at a constant rate. Then we can write,

$$M = M_0(1 - \alpha t), \ (\alpha = \text{constant}) \tag{2.57}$$

Eqn (2.56) assumes the form:

$$v = \frac{dx}{dt} = v_0 - u \ln(1 - \alpha t) \tag{2.58}$$

Integrating with respect to time,

$$\begin{aligned} x &= x_0 + v_0 t - u\int_0^t \ln(1-\alpha t)dt \\ &= x_0 + v_0 t - u\left[\left(t - \frac{1}{\alpha}\right)\ln(1-\alpha t) - t\right] \end{aligned} \tag{2.59}$$

Let M_f = fuel mass at time $t = 0$. Then the mass of the rocket without fuel is

$$M^* = (M_0 - M_f)$$

Hence, if at time T the fuel gets exhausted, then

$$M_0 - M_f = M_0(1 - \alpha T)$$

which yields

$$\alpha = \frac{M_f}{M_0 T} \tag{2.60}$$

Eqn (2.59) is valid up to time T. Distance travelled during this time (taking $x_0 = 0, v_0 = 0$)

$$x = \frac{u}{\alpha}\left[\frac{M_f}{M_0} - \left(\frac{M_f}{M_0} - 1\right)\ln\left(1 - \frac{M_f}{M_0}\right)\right] \tag{2.61}$$

The rocket attains maximum velocity at $t = T$. Hence, from Eqn (2.58), we find

$$v_{\max} = -u \ln\left(1 - \frac{M_f}{M_0}\right) \tag{2.62}$$

$$= u \ln \frac{M_0}{M^*} \tag{2.63}$$

• Case 2 : Motion in Vertical Direction

In this case, the gravitational force will be acting on the rocket. Neglecting the variation in the value of g, we write the equation of motion following Eqn (2.52) as

$$M\frac{dv}{dt} = -u\frac{dM}{dt} - Mg \text{ or, } \frac{dv}{dt} = -\frac{u}{M}\frac{dM}{dt} - g \tag{2.64}$$

Assuming $M = M_0(1 - \alpha t)$, we have

$$\frac{dM}{dt} = -M_0\alpha$$

Eqn (2.64) is written as

$$\frac{dv}{dt} = \frac{u\alpha}{1 - \alpha t} - g \tag{2.65}$$

Integrating w.r.t. time (taking $v = 0$ at $t = 0$)

$$v = -u\ln(1 - \alpha t) - gt = u \ln \frac{M_0}{M} - gt \tag{2.66}$$

Then the height attained by the rocket at time t

$$x = ut - \frac{1}{2}gt^2 + \frac{u}{\alpha}(1 - \alpha t)\ln(1 - \alpha t) \tag{2.67}$$

$$\text{or, } x = ut - \frac{1}{2}gt^2 - \frac{u}{\alpha}\frac{M}{M_0}\ln\frac{M_0}{M} \tag{2.68}$$

Eqn (2.68) can be given a different look if we use the relation given by Eqn (2.60). Thus,

$$x = ut - \frac{1}{2}gt^2 - u\left(t - \frac{M_0 T}{M_f}\right)\ln\left(1 - \frac{M_f}{M_0}\frac{t}{T}\right) \tag{2.69}$$

The maximum height attained by the rocket will be at $t = T$. Thus,

$$x_{\max} = uT - \frac{1}{2}gT^2 - uT\left(1 - \frac{M_0}{M_f}\right)\ln\left(1 - \frac{M_f}{M_0}\right) \tag{2.70}$$

The maximum velocity of the rocket at that time is, from Eqn (2.66)

$$v_{\max} = -u \ln\left(1 - \frac{M_f}{M_0}\right) - gT \tag{2.71}$$

EXAMPLE 2.3 *A spherical raindrop falling under constant gravity grows by absorption of moisture from the surroundings. The rate of increase of its mass is proportional to the instantaneous surface area. Assuming that the drop starts with an infinitely small size, determine its acceleration at time t.* **[C.U. (HONS) 2007]**

Solution. Let at any instant of time, the radius of the raindrop $= r$ and the distance fallen down $= z$. Then its mass

$$M = \frac{4}{3}\pi\rho r^3, \ \rho = \text{density.}$$

Hence,

$$\frac{dM}{dt} = 4\pi\rho r^2 \frac{dr}{dt} \tag{1}$$

But, by question,

$$\frac{dM}{dt} = \lambda(4\pi\rho r^2), \ (\lambda = \text{constant}) \tag{2}$$

From Eqns (1) and (2),

$$\frac{dr}{dt} = \lambda, \text{ i.e., } r = a + \lambda t \tag{3}$$

where a is the initial radius of the raindrop. Equation of motion of the raindrop is

$$\frac{d}{dt}\left(M\frac{dz}{dt}\right) = Mg \tag{4}$$

where

$$M = \frac{4}{3}\pi\rho(a + \lambda t)^3 \tag{5}$$

Substituting this value into Eqn (4), we get

$$\frac{d}{dt}\left[(a + \lambda t)^3 \frac{dz}{dt}\right] = (a + \lambda t)^3 g$$

Integrating and remembering that initial velocity is zero, we find

$$(a + \lambda t)^3 \frac{dz}{dt} = \frac{(a + \lambda t)^4}{4\lambda} g - \frac{a^4}{4\lambda} g$$

$$\text{or, } \frac{dz}{dt} = \frac{g}{4\lambda}\left[a + \lambda t - \frac{a^4}{(a + \lambda t)^3}\right] \tag{6}$$

Acceleration is obtained by differentiating Eqn (6). Thus,

$$\frac{d^2 z}{dt^2} = \frac{g}{4\lambda}\left[\lambda + 4\lambda \frac{a^4}{(a + \lambda t)^4}\right] \tag{7}$$

Since the radius of the raindrop is infinitely small, we take $a \to 0$, when the second term in Eqn (7) vanishes. Hence, the acceleration of the raindrop at any time is constant and equal to $g/4$.

2.10 Collisions

Collision is a process in which the forces of interaction between two or more particles lasts for a finite time. The nature of forces of interaction may be different in different cases, but if they satisfy Newton's third law, we can apply the laws of conservation of linear momentum and energy.

The collision process may be ***elastic*** and ***inelastic***. ***In elastic process, the total kinetic energy of all the particles before and after the collision remains conserved.*** In the case of inelastic collision, some amount of heat and sound is produced due to which a portion of total initial kinetic energy is lost. In nature, inelastic collisions mostly occur.

We intend to discuss the collision between two particles in which one particle is at rest and the other approaches and collides. Experiments are generally carried out in the laboratory with such a set up and hence we call it a laboratory system.

• Elastic Collision

Consider elastic collision of two particles of masses m_1 and m_2 in which the mass m_2 is at rest (FIG. 2.5).

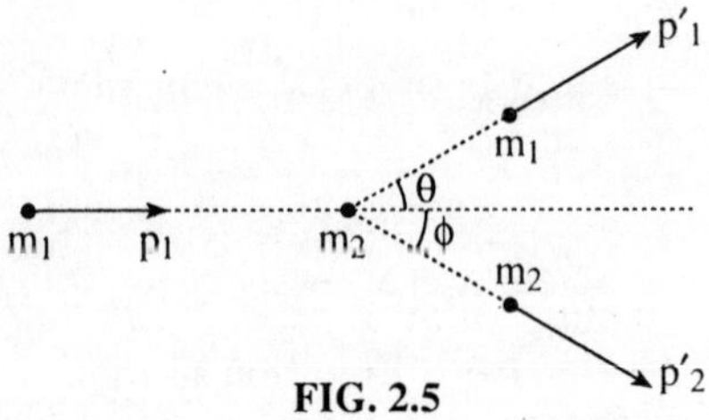

FIG. 2.5

Let p_1, p_1'—initial and final momenta respectively of the particle of mass m_1,
p_2, p_2'—initial and final momenta respectively of the particle of mass m_2,
θ, ϕ—angles through which the masses m_1 and m_2 get scattered after the collision.

We call these angles as ***scattering angles***. The conservation relation of momentum describing the process are:

$$p_1 = p_1' \cos\theta + p_2' \cos\phi \text{ or, } p_2' \cos\phi = p_1 - p_1' \cos\theta \tag{2.72}$$

$$\text{and } 0 = p_1' \sin\theta - p_2' \sin\phi \text{ or, } p_2' \sin\phi = p_1' \sin\theta \tag{2.73}$$

The conservation of kinetic energy gives

$$\frac{p_1^2}{2m_1} = \frac{p_1'^2}{2m_1} + \frac{p_2'^2}{2m_2} \tag{2.74}$$

We assume that the masses, the initial momentum of the incident particle p_1 and the scattering angle θ are known.

The final momenta and the scattering angle ϕ of the target particle can be determined in the following manner:

Squaring and adding Eqns (2.72) and (2.73), we get

$$p_2'^2 = p_1^2 + p_1'^2 - 2p_1p_1' \cos\theta \tag{2.75}$$

Substituting this value of $p_2'^2$ in Eqn (2.74), we find

$$\frac{1}{2m_1}(p_1^2 - p_1'^2) = \frac{1}{2m_2}(p_1^2 + p_1'^2 - 2p_1p_1' \cos\theta)$$

$$\text{or, } \left(1 - \frac{m_1}{m_2}\right)p_1^2 - \left(1 + \frac{m_1}{m_2}\right)p_1'^2 + \frac{2m_1}{m_2}p_1p_1' \cos\theta = 0$$

Dividing by p_1^2 and multiplying by m_2, we get

$$(m_2 - m_1) - (m_2 + m_1)\left(\frac{p_1'}{p_1}\right)^2 + 2m_1\left(\frac{p_1'}{p_1}\right)\cos\theta = 0 \tag{2.76}$$

Denoting the ratio

$$\frac{p_1'}{p_1} = \xi \tag{2.77}$$

we write Eqn (2.76) in the form:

$$(m_1 + m_2)\xi^2 - 2m_1\xi \cos\theta + (m_1 - m_2) = 0 \tag{2.78}$$

This is a quadratic equation in ξ whose solutions are

$$\xi_{1,2} = \frac{m_1}{m_1+m_2}\cos\theta \pm \left[\left(\frac{m_1}{m_1+m_2}\right)^2 \cos^2\theta - \left(\frac{m_1-m_2}{m_1+m_2}\right)\right]^{1/2} \tag{2.79}$$

The values of $\xi_{1,2}$ cannot be negative or imaginary. If $m_2 > m_1$, the value of the quantity under radical sign in Eqn (2.79) is always positive but in this case, we get only one value of ξ, since the other value being negative is discarded. Then all the values of θ from zero (no collision) to π (head-on collision) are possible. If $m_1 > m_2$, the value of the quantity under radical sign may be negative. The maximum value of θ for which the quantity under radical sign becomes zero is given by

$$\cos^2\theta_{\max} = 1 - \left(\frac{m_2}{m_1}\right)^2 \quad \text{or,} \ \sin\theta_{\max} = \frac{m_2}{m_1}, \ (0 \le \theta_{\max} \le \pi/2) \tag{2.80}$$

In such a case

$$\xi = \frac{p_1'}{p_1} = \frac{m_1}{m_1+m_2}\cos\theta_{\max} \ \text{or,} \ \xi = \frac{1}{1+(m_2/m_1)}\cos\theta_{\max} = \frac{\cos\theta_{\max}}{1+\sin\theta_{\max}} \tag{2.81}$$

Knowing the value of the ratio $\xi = p_1'/p_1$, we can find from Eqn (2.75) the value of the ratio

$$\eta = \frac{p_2'}{p_1} \tag{2.82}$$

in the form:

$$\eta = \sqrt{\xi^2 - 2\xi\cos\theta + 1} \tag{2.83}$$

From Eqns (2.72) and (2.73), we find the scattering angle ϕ as

$$\tan\phi = \frac{p_1'\sin\theta}{p_1 - p_1'\cos\theta} = \frac{\xi\sin\theta}{1-\xi\cos\theta}. \tag{2.84}$$

For scattering angles θ less than $\theta_{\max}$, we may get two values of ξ from Eqn (2.79). In such cases, as seen in Eqns (2.83) and (2.84), we shall have two values for η and ϕ also.

If $m_1 = m_2$, simple expressions are obtained from Eqns (2.79) and (2.83) for ξ and η. Thus,

$$\xi = \cos\theta, \ \eta = \sin\theta. \tag{2.85}$$

• Inelastic Scattering

Microscopic constituents of matter like nuclei, atoms and molecules possess internal potential and kinetic energy. In undergoing collisions, some of this internal energy may be released or some collision energy may be absorbed.

Inelastic collisions in which the kinetic energy is absorbed is called ***endoergic***. If the internal energy is released to add to the kinetic energy of the particles the inelastic scattering is called ***exoergic***.

Suppose, an incident particle of mass m_1 collides with a stationary target particle of mass m_2. As a result of collision, two other particles of masses m_3 and m_4 fly off with scatter angles θ and ϕ (FIG. 2.6). Then the momentum and energy conservation relations are:

$$p_1 = p_3\cos\theta + p_4\cos\phi \tag{2.86}$$

$$0 = p_3\sin\theta - p_4\sin\phi \tag{2.87}$$

$$\frac{p_1^2}{2m_1} + Q = \frac{p_3^2}{2m_3} + \frac{p_4^2}{2m_4} \tag{2.88}$$

where Q is the kinetic energy produced in the collision. If $Q < 0$, the collision is endoergic and if $Q > 0$, the collision is exoergic.

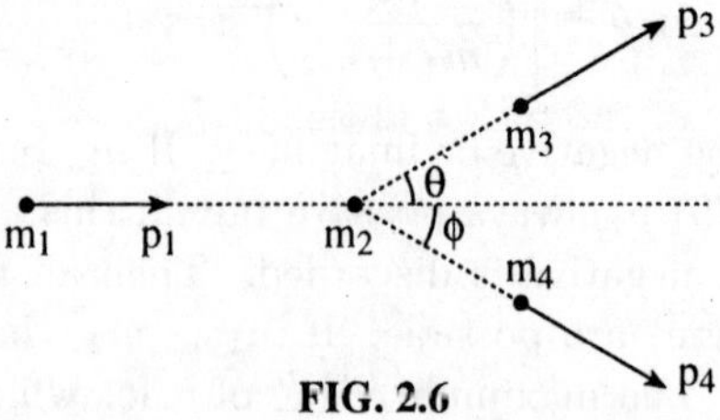

FIG. 2.6

In many experiments p_1 is known, and p_3 and θ are measured. Thus, we are to determine Q. Eliminating ϕ from Eqns (2.86) and (2.87), we get

$$p_4^2 = p_1^2 + p_3^2 - 2p_1p_3\cos\theta \tag{2.89}$$

From Eqn (2.88),

$$Q = \frac{p_3^2}{2m_3} + \frac{p_4^2}{2m_4} - \frac{p_1^2}{2m_1} \tag{2.90}$$

If T_1, T_3 be the kinetic energies of the masses m_1, m_3, then we have the following relations:

$$T_1 = \frac{p_1^2}{2m_1},\ T_3 = \frac{p_3^2}{2m_3} \tag{2.91}$$

and, from Eqn (2.89),

$$\begin{aligned}\frac{p_4^2}{2m_4} &= \frac{1}{2m_4}(p_1^2 + p_3^2 - 2p_1p_3\cos\theta)\\ &= \frac{p_1^2}{2m_1}\cdot\frac{m_1}{m_4} + \frac{p_3^2}{2m_3}\cdot\frac{m_3}{m_4} - 2\cos\theta\sqrt{\frac{p_1^2}{2m_1}\frac{p_3^2}{2m_3}\cdot\frac{4m_1m_3}{4m_4^2}}\\ &= T_1\frac{m_1}{m_4} + T_3\frac{m_3}{m_4} - 2\sqrt{\frac{m_1m_3T_1T_3}{m_4^2}}\cos\theta\end{aligned} \tag{2.92}$$

Eqn (2.90), with the use of Eqns (2.91) and (2.92), yields

$$Q = T_3\left(1 + \frac{m_3}{m_4}\right) - T_1\left(1 - \frac{m_1}{m_4}\right) - \frac{2}{m_4}\sqrt{m_1m_3T_1T_3}\cos\theta \tag{2.93}$$

EXAMPLE 2.4 *A rocket starts vertically upwards with velocity v_0. Show that its speed v at height h is given by*

$$v_0^2 - v^2 = \frac{2gh}{1 + (h/R)}$$

where R is the radius of the earth. Hence, determine the maximum height.

Solution. Let m be the mass of the rocket. Then the kinetic energy of the rocket on the earth's surface

$$= \frac{1}{2}mv_0^2 \tag{1}$$

Gravitational potential energy on the earth's surface

$$= -\frac{GMm}{R} \tag{2}$$

where M = mass of the earth and G = universal gravitational constant.

Hence, the total energy of the rocket on the earth's surface

$$E_1 = \frac{1}{2}mv_0^2 - \frac{GMm}{R} \tag{3}$$

Similarly, the kinetic energy of the rocket at height h

$$= \frac{1}{2}mv^2 \tag{4}$$

and potential energy at height h

$$= -\frac{GMm}{R+h} \tag{5}$$

Total energy at height h

$$E_2 = \frac{1}{2}mv^2 - \frac{GMm}{R+h} \tag{6}$$

From principle of energy conservation,

$$E_1 = E_2$$

$$\text{i.e., } \frac{1}{2}mv_0^2 - \frac{GMm}{R} = \frac{1}{2}mv^2 - \frac{GMm}{R+h}$$

$$\text{i.e., } v_0^2 - v^2 = 2GM\left(\frac{1}{R} - \frac{1}{R+h}\right) = \frac{2gR^2}{R} \cdot \frac{h}{R+h} \quad (\because GM = gR^2)$$

$$= \frac{2gh}{1+h/R} \tag{7}$$

Maximum height H is attained when $v = 0$. Hence, from (7), we find

$$v_0^2 = \frac{2gH}{1+\frac{H}{R}}$$

$$\text{i.e., } v_0^2\left(1 + \frac{H}{R}\right) = 2gH$$

$$\text{i.e., } H\left(\frac{v_0^2}{R} - 2g\right) = -v_0^2$$

$$\therefore \quad H = \frac{v_0^2}{2g - v_0^2/R}$$

EXAMPLE 2.5 *Integrate the rocket equation for vertical motion to obtain v as a function of M, assuming a constant time rate of loss of mass.*

Solution. Equation of motion of the rocket is

$$\frac{dv}{dt} = -g - \frac{u}{M} \cdot \frac{dM}{dt} \tag{1}$$

Now,

$$\frac{dv}{dt} = \frac{dv}{dM} \cdot \frac{dM}{dt} = -\alpha M_0 \frac{dv}{dM} \tag{2}$$

where α = constant time rate of loss of mass so that

$$M \doteq M_0(1 - \alpha t) \tag{3}$$

From (1), (2) and (3)

$$-\alpha M_0 \frac{dv}{dM} = -g - \frac{u}{M}(-M_0\alpha)$$

$$\text{i.e., } M_0\alpha \frac{dv}{dM} = g - \frac{uM_0\alpha}{M}$$

$$\text{i.e., } \frac{dv}{dM} + \frac{u}{M} = \frac{g}{\alpha M_0} \tag{4}$$

Initially at $t = 0$,

$$v = 0, M = M_0 \tag{5}$$

Let V = final velocity of the rocket. Then integrating (4) within proper limits, we get

$$\int_0^V dv = \frac{g}{\alpha M_0}\int_{M_0}^{M_e} dM - u\int_{M_0}^{M_e}\frac{dM}{M}$$

$$\text{i.e., } V = \frac{g}{\alpha M_0}(M_e - M_0) + u \ln \frac{M_0}{M_e} \tag{6}$$

where M_e = mass of empty rocket.

If initial mass of the fuel be M_f, then

$$M_f = M_0 - M_e, \text{ i.e., } M_0 = M_f + M_e \tag{7}$$

Then Eqn (6) can be written as

$$V = -\frac{gM_f}{\alpha M_0} + u \ln\left(1 + \frac{M_f}{M_e}\right)$$

$$= -\frac{g}{\alpha}\left(\frac{1}{1 + \frac{M_e}{M_f}}\right) + u \ln\left(1 + \frac{M_f}{M_e}\right)$$

$$= -\frac{g\beta}{\alpha(\beta + 1)} + u \ln(\beta + 1) \tag{8}$$

where we have denoted

$$\beta = \frac{M_f}{M_e} \tag{9}$$

Eqns (8) and (9) give a relation between the final velocity and the mass of the rocket.

For velocity v at any instant, we get from (4)

$$\int_0^v dv = \frac{g}{\alpha M_0}\int_{M_0}^{M} dM - u\int_{M_0}^{M}\frac{dM}{M}$$

$$\text{i.e., } v = \frac{g}{\alpha M_0}(M - M_0) - u \ln \frac{M}{M_0}$$

$$\text{i.e., } v = -\frac{g}{\alpha}\left(1 - \frac{M}{M_0}\right) + u \ln \frac{M_0}{M} \tag{10}$$

This gives the relation between velocity and mass at any instant.

EXAMPLE 2.6 *If a rocket, originally of mass M_0, throws off every unit of time a mass αM_0 with relative velocity u, and if M' be the mass of the case, etc., show that it cannot rise at once unless $\alpha u > g$, not at all unless $\alpha M_0 u > M'g$. If it just rises vertically at once, show that its greatest velocity is*

$$u \ln \frac{M_0}{M'} - \frac{g}{\alpha}\left(1 - \frac{M'}{M_0}\right), \quad [M_0 \textit{ includes } M']$$

Solution. Mass of the rocket at any instant is

$$M = M_0(1 - \alpha t) \tag{1}$$

Hence,

$$\frac{dM}{dt} = -M_0\alpha \tag{2}$$

Eqn (2.64) of motion of the rocket is

$$\frac{dv}{dt} = -g - \frac{u}{M} \cdot \frac{dM}{dt} \tag{3}$$

From (2) and (3),

$$\frac{dv}{dt} = -g - \frac{u}{M_0(1-\alpha t)}(-M_0\alpha), \text{ i.e., } \frac{dv}{dt} = -g + \frac{\alpha u}{1-\alpha t} \tag{4}$$

The rocket will not rise instantly, unless

$$\frac{dv}{dt} > 0, \text{ at } t = 0, \text{ i.e., } \alpha u > g \tag{5}$$

Integrating (4) with initial condition, $v = 0$ at $t = 0$, we get

$$v = -gt - u\ln(1 - \alpha t) \tag{6}$$

Let velocity is maximum at time $t = T$. At this time, only the case is left. That is

$$M' = M_0(1 - \alpha T), \text{ i.e., } \alpha T = 1 - \frac{M'}{M_0} \tag{7}$$

At $t = T$, the rocket will not rise at all,

$$\begin{aligned} &\text{if } \frac{dv}{dt} < 0 \\ &\text{i.e., if } -g + \frac{\alpha u}{1-\alpha T} < 0 \\ &\text{i.e., if } \alpha T < 1 - \frac{\alpha u}{g} \\ &\text{i.e., if } 1 - \frac{M'}{M_0} < 1 - \frac{\alpha u}{g} \quad \text{[using (7)]} \\ &\text{i.e., if } \alpha M_0 u < gM' \end{aligned} \tag{8}$$

The maximum velocity at $t = T$ is

$$v_{\max} = -gT - u\ln(1 - \alpha T) = -\frac{g}{\alpha}\left(1 - \frac{M'}{M_0}\right) + u\ln\frac{M}{M'} \tag{9}$$

EXAMPLE 2.7 *A rocket starts from rest and falls in a constant gravitational field. At the instant it starts to fall, it ejects gas at the constant rate α in the direction of the gravitational field and at speed u with respect to the rocket. Find its speed after time t.*

Solution. Let M = mass of the rocket at any instant. Then

$$M = M_0(1 - \alpha t) \tag{1}$$

where M_0 = mass at $t = 0$.

Hence,

$$\frac{dM}{dt} = -M_0\alpha \tag{2}$$

The equation of motion is

$$\frac{dv}{dt} = g + \frac{u}{M} \cdot \frac{dM}{dt}$$

$$\text{i.e.,} \quad \frac{dv}{dt} = g - \frac{M_0 \alpha u}{M_0(1-\alpha t)}, \quad \text{i.e.,} \quad \frac{dv}{dt} = g - \frac{\alpha u}{1-\alpha t} \tag{3}$$

Integrating, $v = gt + \alpha u \ln(1-\alpha t) = gt + \alpha u \ln\left(\frac{M}{M_0}\right)$.

EXAMPLE 2.8 *In the above case, find the distance travelled by the rocket in time t.*

Solution. We have from above,

$$v = \frac{dx}{dt} = gt + \alpha u \ln(1-\alpha t) \tag{1}$$

Integrating,

$$\begin{aligned} x &= \frac{1}{2}gt^2 + \alpha u \int_0^t -\frac{1}{\alpha} \ln z dz, \ z = 1-\alpha t \\ &= \frac{1}{2}gt^2 - u(1-\alpha t)\{\ln(1-\alpha t) - 1\}\Big|_0^t \\ &= \frac{1}{2}gt^2 - u[(1-\alpha t)\{\ln(1-\alpha t) - 1\} - (1)(0-1)] \\ &= \frac{1}{2}gt^2 - u[\alpha t + (1-\alpha t)\ln(1-\alpha t)] \\ &= \frac{1}{2}gt^2 - u\left[\alpha t + \frac{M}{M_0}\ln\frac{M}{M_0}\right] = \frac{1}{2}gt^2 - u\left[\alpha t - \frac{M}{M_0}\ln\frac{M_0}{M}\right] \end{aligned}$$

EXAMPLE 2.9 *A smooth sphere of mass m impinges on another of mass M at rest, the direction of motion being at an angle of 45° with the line of centres at the moment of impact.*

If the coefficient of restitution be 0.5, show that the direction of motion of m is turned through an angle $\tan^{-1}\left(\frac{3M}{M+4m}\right)$.

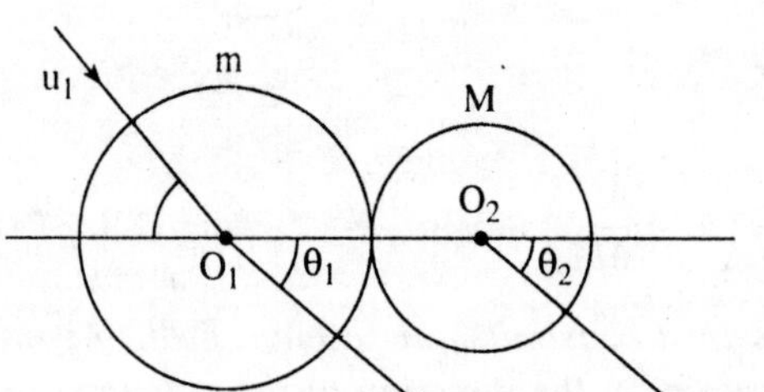

FIG. 2.7

Solution. Let u = velocity of m before impact.

The initial velocity of M is zero.

Let u_2, v_2 be the velocities of masses m, M respectively after the impact making angles θ_1, θ_2 with the line of centres O_1, O_2 (FIG. 2.7).

Since the spheres are smooth, components of velocity of m perpendicular to O_1, O_2 will remain same. Hence,

$$u_2 \sin\theta_1 = u_1 \sin 45° = \frac{u_1}{\sqrt{2}} \tag{1}$$

By Newton's law of impact,

$$v_2 \cos\theta_2 - u_2 \cos\theta_1 = \frac{1}{2}(u_1 \cos 45° - 0) = \frac{1}{2} \cdot \frac{u_1}{\sqrt{2}} \tag{2}$$

By the principle of conservation of linear momentum,

$$mu_2 \cos\theta_1 + Mv_2 \cos\theta_2 = mu_1 \cos 45° = \frac{mu_1}{\sqrt{2}} \tag{3}$$

Multiplying (2) by M and then subtracting the result from (3),

$$(m+M)u_2\cos\theta_1 = \frac{mu_1}{\sqrt{2}} - \frac{Mu_1}{2\sqrt{2}}$$

$$\therefore u_2\cos\theta_1 = \frac{\left(m-\frac{M}{2}\right)}{(m+M)\sqrt{2}}u_1 \tag{4}$$

Dividing (1) by (4),

$$\tan\theta_1 = \frac{2(m+M)}{2m-M} \tag{5}$$

Let the direction of final velocity of m be turned through an angle α. Then $\alpha = \theta_1 - 45°$.

Hence, $\tan\alpha = \tan(\theta_1 - 45°)$

$$= \frac{\tan\theta_1 - \tan 45°}{1+\tan\theta_1\tan 45°} = \frac{\frac{2(m+M)}{2m-M} - 1}{1+\frac{2(m+M)}{2m-M}} = \frac{2(M+m)-2m+M}{2m-M+2(m+M)} = \frac{3M}{4m+M}. \tag{6}$$

Hence, $\alpha = \tan^{-1}\left(\frac{3M}{4m+M}\right)$.

EXAMPLE 2.10 *A smooth sphere of mass m is tied to a fixed point by an inextensible string and another sphere of mass M impinges directly on it with velocity V in a direction making an acute angle α with the string. Show that m begins to move with a velocity* $\frac{m(1+e)\sin\alpha}{m+M\sin^2\alpha}\cdot V$.

Solution. Let u, v be the velocities of masses m, M respectively after impact. Since the string is inextensible, u will be perpendicular to the string.

Again, since impact is direct, v will be in its original direction (FIG. 2.8).

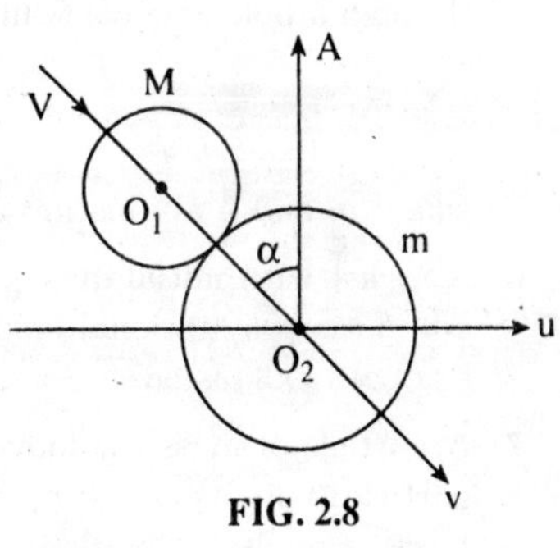

FIG. 2.8

By Newton's law of impact,

$$u\cos(90° - \alpha) - v = e(V-0)$$

$$\text{i.e., } v = u\sin\alpha - eV \tag{1}$$

Here e = coefficient of restitution.

From conservation of momentum perpendicular to the string, we have

$$MV\sin\alpha = mu + Mv\sin\alpha = mu + M\sin\alpha(u\sin\alpha - eV) \quad \text{[using (1)]}$$

$$= mu + Mu\sin^2\alpha - eMV\sin\alpha$$

$$\text{i.e., } MV\sin\alpha(1+e) = u(m+M\sin^2\alpha)$$

$$\therefore u = \frac{M\sin\alpha(1+e)}{m+M\sin^2\alpha}V \quad \textbf{Proved.}$$

NOTE. Newton's law of collision states: If two bodies impinge on one another, the relative velocity of separation of the two bodies after impact along the common normal at the point of contact bears a constant ratio to their relative velocity of approach along the same direction just before impact.

This constant ratio is called the ***coefficient of restitution***. Its value remains between zero and unity.

EXERCISE

1. Three particles of masses 2, 1 and 3 units have the following position vectors:

$$\mathbf{r_1} = 5t\mathbf{i} + 2t^2\mathbf{j} + (3t - 2)\mathbf{k}$$
$$\mathbf{r_2} = (2t - 3)\mathbf{i} + (12 - 5t^2)\mathbf{j} - (4 + 6t^2)\mathbf{k}$$
$$\mathbf{r_3} = (2t - 5)\mathbf{i} - (t^2 + 3)\mathbf{j} + 2t^3\mathbf{k}$$

where t is the time. Find

(a) velocity of the centre of mass of the system at $t = 1$,

(b) the total linear momentum of the system at $t = 1$,

(c) the total work done by the force-field in moving the particles from their position at $t = 1$ to the position at $t = 2$.

2. The distance between the centres of the earth and the moon is $60R$, where R is the radius of the earth. The centre of mass of the earth-moon system is located within the earth at a point $0.23R$ below the surface. What is the ratio of their masses?

3. Prove that, if the total linear momentum of a system is constant, then the centre of mass is either at rest or in motion with constant velocity.

4. Explain the term: centre of mass (CM) frame. Show that the total linear momentum of a system of particle in the CM frame is zero. **[C.U. (HONS) 2001]**

5. A box filled with sand is sliding down a smooth plane inclined at an angle ϕ. Sand is leaking through a hole at a constant rate. Show that its equation of motion is

$$m\frac{du}{dt} = mg\sin\phi$$

where m and u are instantaneous mass and velocity respectively of the sand box.

6. A rocket with initial mass $(M_0 + m_0)$, where m_0 is the mass of the fuel, takes off vertically up in a uniform gravitational field. What is the velocity of the rocket at an instant t_0 when all the fuel has been exhausted? **[C.U. (HONS) 2003]**

7. A particle of mass m_1 moving with velocity u undergoes a head-on elastic collision with another particle of mass $m_2 = km_1$, initially at rest. Show that after collision, the kinetic energy transferred to mass m_2 is

$$\Delta E = \frac{1}{2}m_1u^2\frac{4k}{(1+k)^2}$$ **[C.U. (HONS) 2002]**

8. A particle of mass m_1 undergoes a head-on collision with another particle of mass m_2, initially at rest. If T_1 and T_2' be the initial kinetic energy of mass m_1 and final kinetic energy of mass m_2 respectively, show that

$$\frac{T_2'}{T_1} = \frac{4m_1m_2}{(m_1+m_2)^2}$$

and that

$$\frac{m_1}{m_2} = 2\frac{T_1}{T_2'} - 1 \pm \left[\left(2\frac{T_1}{T_2'} - 1\right)^2 - 1\right]^{1/2}.$$

CHAPTER 3

Central Forces

3.1 Two-Body Problem

Let there be a system of two particles of masses m_1, m_2 whose instantaneous position vectors in an inertial frame with origin O are $\mathbf{r_1}$ and $\mathbf{r_2}$ (FIG. 3.1). Then, position vector of the centre of mass of the system is

$$\mathbf{R} = \frac{m_1\mathbf{r_1} + m_2\mathbf{r_2}}{m_1 + m_2} \tag{3.1}$$

Let the position vector of the particle m_1 relative to m_2 be $\mathbf{r}$. Then

$$\mathbf{r} = \mathbf{r_1} - \mathbf{r_2} \tag{3.2}$$

Solving Eqns (3.1) and (3.2), we find

$$\mathbf{r_1} = \mathbf{R} + \frac{m_2}{m_1 + m_2}\mathbf{r} \tag{3.3}$$

$$\text{and } \mathbf{r_2} = \mathbf{R} - \frac{m_1}{m_1 + m_2}\mathbf{r} \tag{3.4}$$

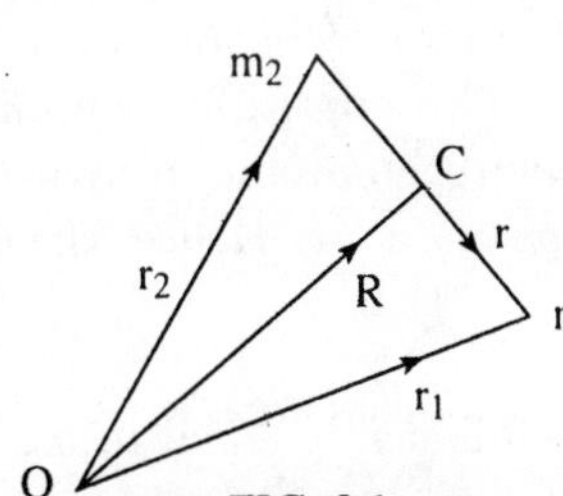

FIG. 3.1

Now, let the external forces acting on the particles be $\mathbf{F_{1e}}$ and $\mathbf{F_{2e}}$ separately. Then the total external force acting on the system is

$$\mathbf{F_e} = \mathbf{F_{1e}} + \mathbf{F_{2e}} \tag{3.5}$$

If $\mathbf{F_{12}}$ be the force exerted by the particle of mass m_2 on the particle of mass m_1, and $\mathbf{F_{21}}$ be the force exerted by the particle of mass m_1 on the particle of mass m_2, then by Newton's third law of motion,

$$\mathbf{F_{12}} = -\mathbf{F_{21}} \tag{3.6}$$

The equations of motion of the two particles separately and as a two-body system are:

$$m_1\ddot{\mathbf{r}}_1 = \mathbf{F_{1e}} + \mathbf{F_{12}}, \tag{3.7}$$

$$m_2\ddot{\mathbf{r}}_2 = \mathbf{F_{2e}} + \mathbf{F_{21}} \tag{3.8}$$

$$\text{and, } M\ddot{\mathbf{R}} = \mathbf{F_e},\ M = m_1 + m_2 \tag{3.9}$$

From Eqns (3.7) and (3.8), we get

$$m_1 m_2 (\ddot{\mathbf{r}}_1 - \ddot{\mathbf{r}}_2) = (m_2 \mathbf{F}_{12} - m_1 \mathbf{F}_{21}) + (m_2 \mathbf{F}_{1e} - m_1 \mathbf{F}_{2e})$$

$$\text{or, } m_1 m_2 \ddot{\mathbf{r}} = (m_1 + m_2)\mathbf{F}_{12} + (m_2 \mathbf{F}_{1e} - m_1 \mathbf{F}_{2e})$$

[using (3.2) and (3.6)]

$$\text{or, } \frac{m_1 m_2}{m_1 + m_2}\ddot{\mathbf{r}} = \mathbf{F}_{12} + \frac{m_1 m_2}{m_1 + m_2}\left(\frac{\mathbf{F}_{1e}}{m_1} - \frac{\mathbf{F}_{2e}}{m_2}\right)$$

$$\text{or, } \mu \ddot{\mathbf{r}} = \mathbf{F}_{12} + \mu \left(\frac{\mathbf{F}_{1e}}{m_1} - \frac{\mathbf{F}_{2e}}{m_2}\right) \tag{3.10}$$

where

$$\mu = \frac{m_1 m_2}{m_1 + m_2} \tag{3.11}$$

which is ***equivalent reduced mass*** of the system.

In absence of external forces Eqn (3.10) assumes the form:

$$\mu \ddot{\mathbf{r}} = \mathbf{F}_{12} \tag{3.12}$$

This means that the motion of particle 2 relative to particle 1 is equivalent to the situation when particle 1 is fixed and particle 2 is replaced by a particle of mass μ.

Eqn (3.10) along with (3.9) represent the motion of a two-body system under the action of internal and external forces.

If the mass of one of the particles is very large compared to the other, say $m_1 \gg m_2$, then

$$\mu = \frac{m_1 m_2}{m_1 + m_2} = \frac{m_2}{1 + \dfrac{m_2}{m_1}} \to m_2 \tag{3.13}$$

since $m_2/m_1 \to 0$.

Thus, the reduced mass is simply the mass of the other particle. In this case, the centre of the mass coincides with the centre of mass of the heavier particle which is taken to be at rest. This approximation has been made in the case of planetary motions and in Bohr's theory of hydrogen atom. In hydrogen atom, an electron of mass m revolves round a proton of much larger mass M. Hence, the equivalent reduced mass of hydrogen atom is

$$\mu = \frac{mM}{m + M} = m\left(1 + \frac{m}{M}\right)^{-1} = m\left(1 - \frac{m}{M}\right) \tag{3.14}$$

which is nearly equal to m, since m/M is equal to 1/1836.

EXAMPLE 3.1 *Consider a pair of stars of equal mass M rotating about their common centre of mass. The attraction between the stars is gravitational and they keep a separation l between them. Show that the time period of rotation of this double-star system is*

$$\pi l \sqrt{\frac{2l}{GM}}$$

where G is the universal gravitational constant. [C.U. (HONS) 2003]

Solution. The gravitational force between the stars is

$$F_1 = \frac{GM^2}{l^2} \tag{1}$$

If T = time period of rotation, then the angular velocity is

$$\omega = \frac{2\pi}{T} \tag{2}$$

By Eqn (3.12), the force acting due to rotation of the stars is

$$F_{12} = \frac{M^2}{2M}\omega^2 l = \frac{Ml}{2}\left(\frac{2\pi}{T}\right)^2 = \frac{2\pi^2 Ml}{T^2} \tag{3}$$

Equating the forces in Eqns (1) and (3), we get $\dfrac{GM^2}{l^2} = \dfrac{2\pi^2 Ml}{T^2}$

Hence, $T = \pi l\sqrt{\dfrac{2l}{GM}}$.

3.2 Motion of a Particle under Central Force

A force is said to be a ***central force*** if it is always directed towards or away from a fixed centre and its magnitude depends only upon the distance from the centre. Hence, a central force can be expressed as

$$\mathbf{F}(\mathbf{r}) = f(r)\mathbf{i}_\mathbf{r} \tag{3.15}$$

where $\mathbf{i}_\mathbf{r}$ is the unit vector along the position vector $\mathbf{r}$. The force is attractive or repulsive according as

$$f(r) < 0 \text{ or, } f(r) > 0 \tag{3.16}$$

A central force is always a conservative force. Hence, if $V(r)$ is the potential energy, then

$$f(r) = -\frac{\partial V}{\partial r} \text{ and so } \mathbf{F}(\mathbf{r}) = -\frac{\partial V}{\partial r}\mathbf{i}_\mathbf{r} \tag{3.17}$$

Again, $V(r)$ being a function of distance only, the system has a spherical symmetry. Due to this, the angular momentum of the particle remains constant, i.e.,

$$\mathbf{J} = \mathbf{r} \times \mathbf{p} = \text{constant.} \tag{3.18}$$

Here, $\mathbf{p}$ = linear momentum.

Taking dot product with $\mathbf{r}$ on both sides of Eqn (3.18), we get

$$\mathbf{r} \cdot \mathbf{J} = \mathbf{r} \cdot (\mathbf{r} \times \mathbf{p}) = (\mathbf{r} \times \mathbf{r}) \cdot \mathbf{p} = 0 \tag{3.19}$$

which shows that the vector $\mathbf{r}$ is always perpendicular to the vector $\mathbf{J}$ which is constant. Hence, ***the motion of the particle under central force always takes place in a plane.***

3.3 Equations of Motion

Using polar coordinates (r, θ), the radial and transverse accelerations can be written as (see Chapter 1)

$$\left.\begin{aligned} a_r &= \ddot{r} - r\dot{\theta}^2 \\ a_\theta &= r\ddot{\theta} + 2\dot{r}\dot{\theta} \end{aligned}\right\} \tag{3.20}$$

Hence, the equations of motion along those two directions are

$$\mu\ddot{r} - \mu r\dot{\theta}^2 = f(r) \tag{3.21}$$

$$\text{and, } \mu r\ddot{\theta} + \mu 2\dot{r}\dot{\theta} = 0 \tag{3.22}$$

respectively.

From Eqn (3.22), we get $\dfrac{d}{dt}\left(\mu r^2\dot{\theta}\right) = 0$, which on integration yields

$$\mu r^2\dot{\theta} = J = \text{constant.} \tag{3.23}$$

This is a ***first integral*** of motion.

Now, the total energy of the particle is

$$E = \text{kinetic energy} + \text{potential energy}$$
$$= \frac{1}{2}\mu\dot{r}^2 + \frac{1}{2}\mu r^2\dot{\theta}^2 + V(r) = \frac{1}{2}\mu\dot{r}^2 + \frac{J^2}{2\mu r^2} + V(r) \text{ [using (3.23)]} \tag{3.24}$$

This total energy E is constant. To prove this, we write Eqn (3.21) as

$$\mu\ddot{r} = f(r) + \mu r\dot{\theta}^2 = -\frac{dV}{dr} + \frac{J^2}{\mu r^3} \text{ [using (3.23)]}$$
$$\text{or, } \mu\ddot{r} = -\frac{d}{dr}\left(V + \frac{J^2}{2\mu r^2}\right) \tag{3.25}$$

We multiply both sides by $\dot{r}$, and get

$$\mu\dot{r}\ddot{r} = -\frac{d}{dr}\left(V + \frac{J^2}{2\mu r^2}\right)\frac{dr}{dt} \text{ or, } \frac{d}{dt}\left[\frac{1}{2}\mu\dot{r}^2 + V + \frac{J^2}{2\mu r^2}\right] = 0$$

Hence,

$$E = \frac{1}{2}\mu\dot{r}^2 + V + \frac{J^2}{2\mu r^2} = \text{constant.} \tag{3.26}$$

This is another first integral of motion. We define effective potential by

$$V_e(r) = V(r) + \frac{J^2}{2\mu r^2} \tag{3.27}$$

Then Eqn (3.26) assumes the form:

$$E = \frac{1}{2}\mu\dot{r}^2 + V_e \tag{3.28}$$

which shows that it is a one-dimensional motion in r with radial kinetic energy $\frac{1}{2}\mu\dot{r}^2$ and effective potential energy V_e.

From Eqn (3.28), we get

$$\dot{r}^2 = \frac{2}{\mu}(E - V_e) \tag{3.29}$$

Thus, for any possible radial motion, we must have

$$E \geq V_e(r) \tag{3.30}$$

For $E = V_e$, $\dot{r}$ vanishes for extreme values of $r = r_1$ and $r = r_2$. If both the bounds r_1 and r_2 exist, the motion is called ***bounded*** and the distances $r_1$and r_2 are called ***apsidal distances***.

3.4 Equation of the Orbit

Equation of motion of a particle of reduced mass μ can be written as

$$\mu\ddot{r} = f(r) + \frac{J^2}{\mu r^3} \tag{3.31}$$

Now

$$\dot{r} = \frac{dr}{dt} = \frac{dr}{d\theta}\cdot\frac{d\theta}{dt} = \frac{J}{\mu r^2}\frac{dr}{d\theta} \quad \text{[using (3.23)].}$$

Hence

$$\ddot{r} = \frac{d}{dt}\left(\frac{J}{\mu r^2}\frac{dr}{d\theta}\right) = \frac{d}{d\theta}\left(\frac{J}{\mu r^2}\frac{dr}{d\theta}\right)\frac{d\theta}{dt} = \frac{J}{\mu r^2}\frac{d}{d\theta}\left(\frac{J}{\mu r^2}\frac{dr}{d\theta}\right) \tag{3.32}$$

We introduce a new variable

$$u = \frac{1}{r} \tag{3.33}$$

Then

$$\frac{du}{d\theta} = -\frac{1}{r^2}\frac{dr}{d\theta} \tag{3.34}$$

Hence, from Eqn (3.32), we get

$$\ddot{r} = -\frac{J^2u^2}{\mu^2}\frac{d}{d\theta}\left(\frac{du}{d\theta}\right) = -\frac{J^2u^2}{\mu^2}\frac{d^2u}{d\theta^2} \tag{3.35}$$

With the use of this relation, Eqn (3.31) becomes

$$-\frac{J^2u^2}{\mu}\frac{d^2u}{d\theta^2} - \frac{J^2u^3}{\mu} = f\left(\frac{1}{u}\right) \quad \text{or,} \quad \frac{J^2u^2}{\mu}\left(\frac{d^2u}{d\theta^2} + u\right) = -f\left(\frac{1}{u}\right) \tag{3.36}$$

This is the differential equation of the orbit and can be solved if the law of force $f(r)$ is given. At apsidal distances, r is maximum or minimum. Hence, at an apse, u is minimum or maximum, so that

$$\frac{du}{d\theta} = 0 \tag{3.37}$$

3.5 Areal Velocity

The rate of change of the area traced out by the radius vector joining the particle to the centre of force is called the ***areal velocity*** of the particle.

The elementary area swept out by the radius in time dt is (FIG. 3.2)

$$dA = \frac{1}{2}r \cdot r d\theta$$

Hence

$$\frac{dA}{dt} = \frac{1}{2}r^2\frac{d\theta}{dt} = \frac{J}{2\mu} = \text{constant.} \tag{3.38}$$

Thus, areal velocity of the particle in a central force-field is constant. This is ***Kepler's second law*** of planetary motion.

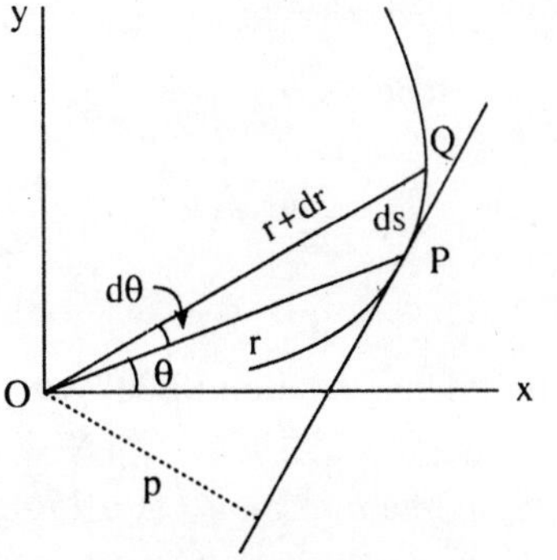

FIG. 3.2

Now, let v = linear velocity of the particle and p = length of the perpendicular from O on the tangent at P.

Then areal velocity can be written as

$$\frac{dA}{dt} = \frac{1}{2}\frac{ds}{dt} \times p = \frac{1}{2}v \times p \tag{3.39}$$

Equating the areal velocities in Eqn (3.38) and (3.39), we get

$$\frac{1}{2}v \times p = \frac{J}{2\mu}, \text{ i.e., } vp = \frac{J}{\mu} \tag{3.40}$$

Also, from differential calculus, we know that

$$\frac{1}{p^2} = \frac{1}{r^2} + \frac{1}{r^4}\left(\frac{dr}{d\theta}\right)^2 = u^2 + \left(\frac{du}{d\theta}\right)^2 \tag{3.41}$$

Eqns (3.40) and (3.41) yield

$$v^2 = h^2\left[u^2 + \left(\frac{du}{d\theta}\right)^2\right] \tag{3.42}$$

where we have denoted

$$h^2 = \left(\frac{J}{\mu}\right)^2. \tag{3.43}$$

Eqn (3.42) determines the linear velocity of the particle.

EXAMPLE 3.2 *A particle of mass m describes the curve*

$$r = a\tanh\left(\frac{\theta}{\sqrt{2}}\right).$$

Find the law of force acting on the particle.

Solution. We are given

$$r = a\tanh\left(\frac{\theta}{\sqrt{2}}\right)$$

$$\text{or, } au = \coth\left(\frac{\theta}{\sqrt{2}}\right), \text{ where } u = \frac{1}{r} \tag{1}$$

$$\therefore\ a\frac{du}{d\theta} = -\frac{1}{\sqrt{2}}\operatorname{cosech}^2\left(\frac{\theta}{\sqrt{2}}\right)$$

$$\text{and, } a\frac{d^2u}{d\theta^2} = \left(-\frac{1}{\sqrt{2}}\right)\left(\frac{-2}{\sqrt{2}}\right)\operatorname{cosech}^2\left(\frac{\theta}{\sqrt{2}}\right)\coth\left(\frac{\theta}{\sqrt{2}}\right)$$

$$= \operatorname{cosech}^2\left(\frac{\theta}{\sqrt{2}}\right)\coth\left(\frac{\theta}{\sqrt{2}}\right) \tag{2}$$

Hence

$$a\left(\frac{d^2u}{d\theta^2} + u\right) = \coth\left(\frac{\theta}{\sqrt{2}}\right)\left[1 + \operatorname{cosech}^2\left(\frac{\theta}{\sqrt{2}}\right)\right]$$

$$= \coth\left(\frac{\theta}{\sqrt{2}}\right)\coth^2\left(\frac{\theta}{\sqrt{2}}\right) = a^3u^3, \text{ [using Eqn (1)]} \tag{3}$$

Thus, by Eqn (3.36), we get

$$-f\left(\frac{1}{u}\right) = \frac{J^2}{m}u^2\left(\frac{d^2u}{d\theta^2} + u\right) = \frac{J^2}{m}u^2(a^2u^3), \text{ [using Eqn (3)]}$$

$$= \frac{J^2a^2}{m}u^5$$

Hence, $-f(r) = \dfrac{\lambda}{r^5}, \lambda = \dfrac{J^2a^2}{m} = \text{constant.}$

Thus, the force acting on the particle varies inversely as r^5.

EXAMPLE 3.3 *A particle of mass m moves under a force*

$$f(r) = m\lambda\left\{\frac{3a}{r^4} - 2(a^2 - b^2)\frac{1}{r^5}\right\}, \quad a > b$$

and is projected from an apse at a distance $(a + b)$ *with velocity* $\dfrac{\sqrt{\lambda}}{a+b}$. *Show that its orbit is* $r = a + b\cos\theta$.

Solution. We have

$$\frac{J^2u^2}{m}\left(\frac{d^2u}{d\theta^2}+u\right)=-f\left(\frac{1}{u}\right)=m\lambda\left\{3au^4-2(a^2-b^2)u^5\right\}$$

$$\text{or, } h^2\left(\frac{d^2u}{d\theta^2}+u\right)=\lambda\left\{3au^2-2(a^2-b^2)u^3\right\} \tag{1}$$

Multiplying both sides by $2\dfrac{du}{d\theta}$ and integrating

$$h^2\left[u^2+\left(\frac{du}{d\theta}\right)^2\right]=\lambda\left[2au^3-(a^2-b^2)u^4\right]+A \tag{2}$$

$$\text{or, } v^2=\lambda\left[2au^3-(a^2-b^2)u^4\right]+A \quad \text{[by Eqn (3.42)]} \tag{3}$$

We have the initial conditions that, for $r=a+b$,

$$v^2=\frac{\lambda}{(a+b)^2} \text{ and } \frac{du}{d\theta}=0 \text{ (at an apse)}. \tag{4}$$

From Eqn (3) by condition (4), we get

$$\frac{\lambda}{(a+b)^2}=\lambda\left[\frac{2a}{(a+b)^3}-\frac{a^2-b^2}{(a+b)^4}\right]+A, \quad \therefore\ A=0$$

Again, from Eqn (2) by condition (4), we get

$$h^2\left[\frac{1}{(a+b)^2}+0\right]=\lambda\left[\frac{2a}{(a+b)^3}-\frac{a^2-b^2}{(a+b)^4}\right]$$

This gives

$$h^2=\lambda \tag{5}$$

Putting the values of h^2 and A in Eqn (2), we get

$$\lambda\left[u^2+\left(\frac{du}{d\theta}\right)^2\right]=\lambda\left[2au^3-(a^2-b^2)u^4\right]$$

$$\text{or, } \left(-\frac{1}{r^2}\frac{dr}{d\theta}\right)^2=\frac{2a}{r^3}-\frac{a^2-b^2}{r^4}-\frac{1}{r^2} \quad \left[\because\ u=\frac{1}{r}\right]$$

$$\text{or, } \left(\frac{dr}{d\theta}\right)^2=2ar-(a^2-b^2)-r^2=b^2-(r-a)^2$$

$$\therefore\ \frac{dr}{d\theta}=\left[b^2-(r-a)^2\right]^{1/2}$$

$$\text{or, } \frac{dr}{[b^2-(r-a)^2]^{1/2}}=d\theta$$

Integrating, we get

$$\sin^{-1}\left(\frac{r-a}{b}\right)=\theta+\text{const.}=\theta+B \text{ (say)} \tag{6}$$

Initially, at $\theta=0$, $r=a+b$ (given).

Hence, we find

$$B=\sin^{-1}(1)=\pi/2 \tag{7}$$

Thus, from Eqns (6) and (7), we have

$$\sin^{-1}\left(\frac{r-a}{b}\right) = \theta + \frac{\pi}{2} \text{ or, } \frac{r-a}{b} = \sin\left(\theta + \frac{\pi}{2}\right) = \cos\theta$$

Hence, $r = a + b\cos\theta$ which is the required equation of the orbit.

3.6 Inverse Square Law of Force

We now consider the case when the mutual force between two particles varies inversely as the square of the distance between them.

Then

$$f(r) = \frac{k}{r^2}, \; k = \text{constant}. \tag{3.44}$$

The most important examples of such type of central forces are gravitational and Coulomb forces. In the case of gravitational force between two masses m_1, m_2

$$f(r) = -\frac{Gm_1m_2}{r^2}, \text{ so that } k = -Gm_1m_2 \tag{3.45}$$

where G is the universal constant of gravitation.

In the case of Coulomb force between two charges q_1 and q_2,

$$f(r) = \frac{1}{4\pi\epsilon_0}\frac{q_1q_2}{r^2}, \text{ so that } k = \frac{1}{4\pi\epsilon_0} \tag{3.46}$$

where ϵ_0 = permittivity of free space. In this case, k is positive when the force is of repulsion and is negative if the force is of attraction.

Since $f(r) = -\dfrac{dV}{dr}$, we get from Eqn (3.44)

$$V(r) = \frac{k}{r} \tag{3.47}$$

Hence, the effective potential energy is given by

$$V_e = \frac{k}{r} + \frac{J^2}{2\mu r^2} \tag{3.48}$$

From Eqn (3.48), it is clear that, for $k > 0$, the effective potential energy tends to zero at $r \to \infty$ only. But, for $k < 0$, V_e becomes zero at two points given by

$$\frac{1}{r}\left(-k + \frac{J^2}{2\mu r}\right) = 0, \text{ i.e., } r = \frac{J^2}{2\mu k} \text{ and } r = \infty.$$

Thus, the value of r varies in the range $\frac{J^2}{2\mu k} \le r \le \infty$.

In (FIG. 3.3) we have shown the variation of effective potential energy $V_e(r)$ with radial distance r for $k < 0$. Three cases of total energy E are considered.

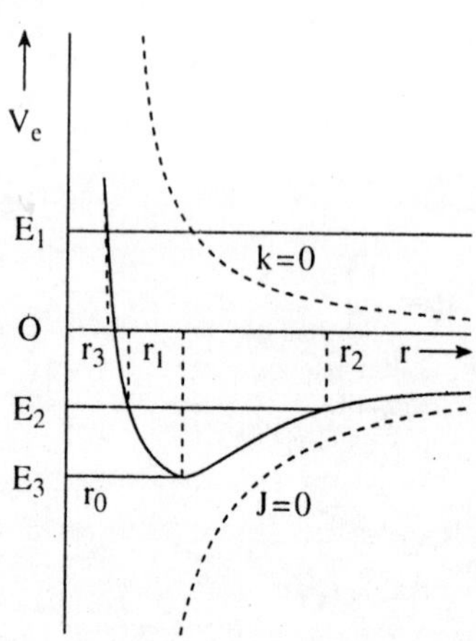

FIG. 3.3

• Case - I

Let the total energy E be positive and $E = E_1$. This is shown by thick line in the FIG. 3.3. The point of intersection of this line with V_e-curve at $r = r_3$ gives the turning point of the particle. There being no other turning point, the particle moves away so that the motion becomes unbounded. We call this phenomenon as ***scattering***. The particle will have similar type of motion when the total energy $E = 0$.

• Case - II

Let the total energy E be negative and its magnitude be E_2, which is shown in the figure by thick line. Here we find two points at $r = r_1$ and r_2 of intersections of the E_2-line with V_e-curve. Hence, the particle has two turning points and the motion is bounded. The orbits are closed and elliptic.

• Case - III

Let the total energy be negative and its magnitude E_3 be such that its graph touches the V_e-curve. In this case, the two points of intersection coincide resulting in a circular orbit. At the point of contact, $\frac{d}{dr}(V_e) = 0$, i.e., $\frac{d}{dr}\left(-\frac{k}{r} + \frac{J^2}{2\mu r^2}\right) = 0$, which gives

$$r = r_0 = \frac{J^2}{\mu k} \tag{3.49}$$

3.7 Determination of Orbit

The inverse square law of force of attraction is given by

$$f\left(\frac{1}{u}\right) = -ku^2, \; u = \frac{1}{r} \tag{3.50}$$

Hence, the differential equation of the orbit is

$$\frac{J^2u^2}{\mu}\left(\frac{d^2u}{d\theta^2} + u\right) = ku^2 \text{ or, } \frac{d^2u}{d\theta^2} + u - \frac{\mu k}{J^2} = 0 \tag{3.51}$$

We substitute

$$\xi = u - \frac{\mu k}{J^2} \tag{3.52}$$

Then, Eqn (3.51) assumes the form:

$$\frac{d^2\xi}{d\theta^2} + \xi = 0 \tag{3.53}$$

Its solution is

$$\xi = A\cos(\theta - \alpha) \tag{3.54}$$

where A and α are constants of integration. By virtue of Eqn (3.52), the solution in Eqn (3.54) can be expressed as

$$u - \frac{\mu k}{J^2} = A\cos(\theta - \alpha) \text{ or, } \frac{1}{r} = \frac{\mu k}{J^2} + A\cos(\theta - \alpha) \tag{3.55}$$

$$\text{or, } \frac{J^2/\mu k}{r} = 1 + \frac{J^2 A}{\mu k}\cos(\theta - \alpha) \text{ or, } \frac{l}{r} = 1 + e\cos(\theta - \alpha) \tag{3.56}$$

where

$$l = \frac{J^2}{\mu k} \text{ and } e = \frac{J^2 A}{\mu k} \tag{3.57}$$

Eqn (3.56) represents a general conic section with one focus at the origin and having eccentricity equal to e. Now, we differentiate Eqn (3.55) with respect to time, and get

$$-\frac{\dot{r}}{r^2} = -A\sin(\theta - \alpha)\dot{\theta} = -A\sin(\theta - \alpha)\frac{J}{\mu r^2}, \text{ or, } \dot{r} = \frac{AJ}{\mu}\sin(\theta - \alpha). \tag{3.58}$$

From energy Eqn (3.26), we have $\frac{1}{2}\mu\dot{r}^2 - \frac{k}{r} + \frac{J^2}{2\mu r^2} = E$.

Using expressions for r in Eqn (3.55) and $\dot{r}$ in Eqn (3.58),

$$\frac{1}{2}\mu\frac{A^2J^2}{\mu^2}\sin^2(\theta-\alpha)-k\left[\frac{\mu k}{J^2}+A\cos(\theta-\alpha)\right]+\frac{J^2}{2\mu}\left[\frac{\mu k}{J^2}+A\cos(\theta-\alpha)\right]^2=E$$

$$\text{or,}\quad \frac{A^2J^2}{2\mu}=E+\frac{\mu k^2}{2J^2}\ \text{ or, }\ A^2=\frac{2\mu}{J^2}\left(E+\frac{\mu k^2}{2J^2}\right)$$

$$\therefore\ A=\frac{\mu k}{J^2}\left(1+\frac{2J^2E}{\mu k^2}\right)^{1/2} \tag{3.59}$$

Hence, from Eqns (3.57) and (3.59), we get

$$e=\frac{J^2A}{\mu k}=\left(1+\frac{2EJ^2}{\mu k^2}\right)^{1/2} \tag{3.60}$$

Eqn (3.60) determines the eccentricity of the conic section which will be ellipse, parabola or hyperbola according as $e<1$ or $e=1$ or $e>1$ respectively.

From Eqn (3.60), it is obvious that

$$\left.\begin{aligned} e<1, &\quad \text{when}\quad E<0\\ e=1, &\quad \text{when}\quad E=0\\ e>1, &\quad \text{when}\quad E>0\end{aligned}\right\} \tag{3.61}$$

For a circular orbit, $e=0$, hence,

$$E=-\frac{\mu k^2}{2J^2} \tag{3.62}$$

For planetary motions, $E<0$ and so the orbit is elliptical. This conforms to ***Kepler's first law*** of planetary motion which states:

> **Every planet moves around the sun in elliptic orbits, the sun being at one of the foci.**

• Periodic Time

Let r_1, r_2 be the apsidal distances and a, b the semi-major and semi-minor axes of the ellipse respectively. From Eqn (3.56), r will be maximum when $\cos(\theta-\alpha)$ is equal to -1, hence,

$$\frac{l}{r_1}=1-e,\ \text{i.e.,}\ r_1=\frac{l}{1-e} \tag{3.63}$$

Similarly, for minimum value of r, $\cos(\theta-\alpha)=1$, and so

$$\frac{l}{r_2}=1+e,\ \text{i.e.,}\ r_2=\frac{l}{1+e} \tag{3.64}$$

But

$$a=\frac{1}{2}(r_1+r_2)=\frac{l}{2}\left(\frac{1}{1-e}+\frac{1}{1+e}\right)=\frac{l}{1-e^2} \tag{3.65}$$

From Eqn (3.60)

$$1-e^2=1-\left(1+\frac{2EJ^2}{\mu k^2}\right)=-\frac{2EJ^2}{\mu k^2} \tag{3.66}$$

Putting the value of l from Eqn (3.57) and using the relation (3.66), we get from Eqn (3.65)

$$a=-\frac{J^2}{\mu k}\cdot\frac{\mu k^2}{2EJ^2}=-\frac{k}{2E}$$

Hence,

$$E = -\frac{k}{2a} \tag{3.67}$$

Thus, we get the value of total energy for an elliptic orbit. ***Its value depends only on the major axis of the ellipse.***

We now substitute the value of E from Eqn (3.67) into Eqn (3.66), and get

$$1 - e^2 = -\frac{2EJ^2}{\mu k^2} = \left(-\frac{k}{2a}\right)\left(-\frac{2J^2}{\mu k^2}\right) = \frac{J^2}{\mu k a} \tag{3.68}$$

Hence, area of the ellipse is

$$\begin{aligned} A &= \pi ab = \pi a \cdot a\sqrt{1 - e^2} \\ &= \pi a^2 \frac{J}{\sqrt{\mu k a}} = \pi a^{3/2} \frac{J}{\sqrt{\mu k}} \text{ [using Eqn (3.68)].} \end{aligned} \tag{3.69}$$

Now, if T be the time to complete one revolution when the radius vector sweeps out area of the ellipse, then by virtue of Eqn (3.38), we have

$$A = \int_0^T \left(\frac{1}{2} r^2 \dot{\theta}\right) dt = \int_0^T \frac{J}{2\mu} dt = \frac{JT}{2\mu} \tag{3.70}$$

Equating the values in Eqns (3.69) and (3.70), we get

$$\pi a^{3/2} \frac{J}{\sqrt{\mu k}} = \frac{JT}{2\mu} \quad \text{or, } T = 2\pi a^{3/2} \sqrt{\frac{\mu}{k}} \tag{3.71}$$

Thus, we have obtained the periodic time in an elliptical orbit. Squaring both sides of Eqn (3.71), we get

$$T^2 = 4\pi^2 a^3 \frac{\mu}{k}, \text{ i.e., } T^2 \propto a^3 \tag{3.72}$$

Thus, we have found that:

The square of the periodic time of revolution of a planet around the sun varies as the cube of semi-major axis of the elliptic orbit. This is ***Kepler's third law*** of planetary motion.

EXAMPLE 3.4 *A particle of mass m describes an elliptical orbit about a centre of attractive force at one of its foci given by $-k/r^2$. Show that the speed of the particle at any point of the orbit is given by $v^2 = \frac{k}{m}\left(\frac{2}{r} - \frac{1}{a}\right)$, where 'a' is the semi-major axis.*

Solution. Since the orbit is elliptical, the total energy is

$$E = -\frac{k}{2a} \tag{1}$$

Now, the total energy = KE + PE

$$\text{i.e., } E = \frac{1}{2} m v^2 + V \tag{2}$$

Since

$$f(r) = -\frac{k}{r^2}, \text{ i.e., } -\frac{dV}{dr} = -\frac{k}{r^2} \text{ (given),}$$

we get

$$V(r) = -\frac{k}{r} \tag{3}$$

From Eqns (1), (2) and (3), we have

$$-\frac{k}{2a}=\frac{1}{2}mv^2-\frac{k}{r}$$

Hence,

$$v^2=\frac{k}{m}\left(\frac{2}{r}-\frac{1}{a}\right)$$

EXAMPLE 3.5 *According to Yukawa's theory of nuclear forces, the attractive force between two nucleons inside a nucleus is given by the potential*

$$V(r)=k\frac{e^{-ar}}{r},\quad a=\textit{const.}$$

Find the law of force. If the particle moves in a circle of radius r_0, *determine* (i) *total energy* E, (ii) *angular momentum* J *and* (iii) *period of circular motion.*

Solution. Since, $-f(r)=\frac{dV}{dr}$, we get

$$f(r)=-\frac{dV}{dr}=-\frac{d}{dr}\left(k\frac{e^{-ar}}{r}\right)\quad \text{or, } f(r)=ke^{-ar}\left(\frac{a}{r}+\frac{1}{r^2}\right) \tag{1}$$

(i) If v be the linear velocity of the particle of mass m around circle of radius r_0, then

$$\frac{mv^2}{r_0}=-f(r_0)=-ke^{-ar_0}\left(\frac{a}{r_0}+\frac{1}{r_0^2}\right)\quad \text{or, } v^2=-\frac{k}{m}e^{-ar_0}\left(a+\frac{1}{r_0}\right) \tag{2}$$

Total energy is

$$E=\frac{1}{2}mv^2+V(r_0)$$

$$\text{i.e., } E=-\frac{k}{2}e^{-ar_0}\left(a+\frac{1}{r_0}\right)+ke^{-ar_0}\cdot\frac{1}{r_0}=\frac{ke^{-ar_0}}{2r_0}(1-ar_0) \tag{3}$$

(ii) Angular momentum is

$$J=r\times p=r_0\times mv=mr_0\sqrt{-\frac{k}{m}e^{-ar_0}\left(a+\frac{1}{r_0}\right)},\quad \text{[using Eqn (2)]}$$

$$\text{i.e., } J=\left[-kmr_0e^{-ar_0}(1+ar_0)\right]^{1/2} \tag{4}$$

(iii) Time period of circular motion is

$$T=\frac{2\pi r}{v}=2\pi r_0\left[-\frac{k}{m}e^{-ar_0}\left(a+\frac{1}{r_0}\right)\right]^{-1/2},\quad \text{[from Eqn (2)]}$$

$$=2\pi\left[-\frac{k}{mr_0^3}e^{-ar_0}(1+ar_0)\right]^{-1/2}.$$

3.8 Scattering in a Central Force-field

We intend to consider the problem of scattering of micro-particles in a central force-field by classical method, not by quantum mechanical approach. We make an assumption that the incident as well as target particles remain sufficiently apart from one another. This will keep the process of scattering of one incident particle by one target particle unaffected by the presence of other particles. Hence, we can consider the interaction between a single incident particle and a single target particle, and the results obtained can be used for a beam of particles incident on a collection of target particles. Let us define some terminologies.

• Scattering

Let an α-particle with positive charge approach a target particle. The α-particle will experience repulsion from the +vely charged nucleus of the target particle. This will cause a deviation of the path of α-particle. The repulsive force on the α-particle will gradually decrease and ultimately vanish when the α-particle will move in a straight line but in a different direction. We say that the α-particle has been ***scattered***.

• Flux Density

We consider a uniform beam of particles, all the particles having same mass and energy. The ***flux density*** or ***intensity*** of the incident beam is defined as the number of particles crossing unit area per second normal to the direction of the beam. We shall denote it by N_0.

• Scattering Cross Section

Let a uniform beam of flux density N_0 approach target particles which are at rest. Then the number of particles incident perpendicularly upon a small area ds of the scattering centre per second is $N_0 ds$. All these particles will be scattered in a region bounded by a small solid angle $d\Omega$. We define the intensity of scattered particles as the number of particles scattered per second per unit solid angle, and denote it as N. Then, we can write,

$$N_0 ds = N d\Omega \text{ or, } \frac{ds}{d\Omega} = \frac{N}{N_0} \tag{3.73}$$

The quantity in the left-hand side of Eqn (3.73) is called ***differential scattering cross section*** scattered in Ω-direction. We shall denote it by $\sigma(\Omega)$. It indicates probability of scattering of some of the particles in specified direction of Ω.

• Impact Parameter

The length of the perpendicular drawn from the scattering centre to the direction of incident particle is called the ***impact parameter***. We shall denote it by p.

Let us now consider a uniform beam of particles moving along X-X' direction. The force at the scattering centre being a central force, there will be complete symmetry of the incident beam about X-X' axis. We consider particles within impact parameters p and $p + dp$ (FIG. 3.4).

The number of particles crossing through this annular ring of area $2\pi p dp$ is

$$N_0 2\pi p dp \tag{3.74}$$

All these particles will be scattered in solid angle $d\Omega$ given by

$$d\Omega = \frac{ds}{r^2} = \frac{r \sin\phi d\theta r d\phi}{r^2} = \sin\phi d\theta d\phi$$

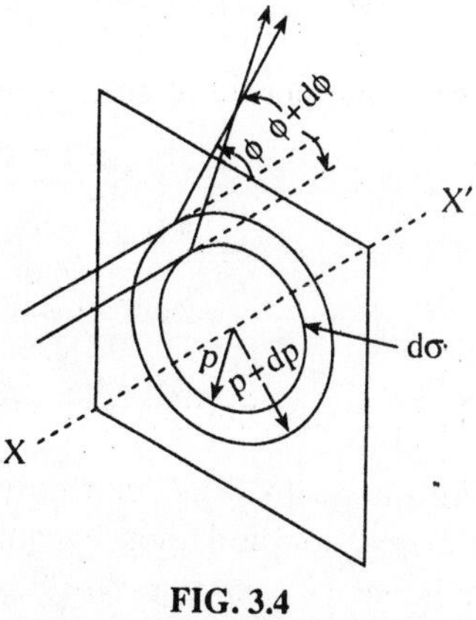

FIG. 3.4

Due to symmetry about X-X', this can be integrated over $d\theta$ from $\theta = 0$ to 2π, so that

$$d\Omega = 2\pi \sin\phi d\phi \tag{3.75}$$

where ϕ is the scattering angle.

Hence, the number of particles scattered into the solid angle $d\Omega$ is

$$N d\Omega = N_0 \sigma(\phi) 2\pi \sin\phi d\phi \tag{3.76}$$

where $\sigma(\phi)$ is the differential cross section for the direction ϕ. Equating the expressions in (3.74) and (3.76), we get

$$N_0 2\pi p dp = -N_0 \sigma(\phi) 2\pi \sin\phi d\phi, \text{ i.e., } \sigma(\phi) = -\frac{p}{\sin\phi}\frac{dp}{d\phi} \tag{3.77}$$

The negative sign in the right-hand side has been taken, since an increase in p will cause a decrease in ϕ. Eqn (3.77) is useful for us to calculate the differential cross section.

3.9 The Rutherford Formula

Let a particle with positive charge $Z_1 e$ be moving with initial velocity v_0 towards a heavy nucleus of charge $Z_2 e$. Here e represents the magnitude of electronic charge. The repulsive force between the two particles is

$$f(r) = \frac{(Z_1 e)(Z_2 e)}{4\pi\varepsilon_0 r^2} = \frac{Z_1 Z_2 e^2}{4\pi\varepsilon_0} \cdot \frac{1}{r^2} = \frac{k}{r^2}, \tag{3.78}$$

where

$$k = \frac{Z_1 Z_2 e^2}{4\pi\varepsilon_0} \tag{3.79}$$

The initial velocity of the particle being v_0, its total energy is given by

$$E = \frac{1}{2}\mu v_0^2 \tag{3.80}$$

which is positive. But E remains constant during motion under central force, which shows that the orbit of the charged particle moving in the central force-field is a hyperbola.

We take the equation of the orbit as

$$\frac{l}{r} = 1 + e\cos\theta, \text{ taking } \alpha = 0 \text{ in Eqn (3.56).} \tag{3.81}$$

The scattering angle ϕ is given by (FIG. 3.5)

$$\phi = \pi - 2\beta, \text{ so that, } \beta = \frac{1}{2}(\pi - \phi) \tag{3.82}$$

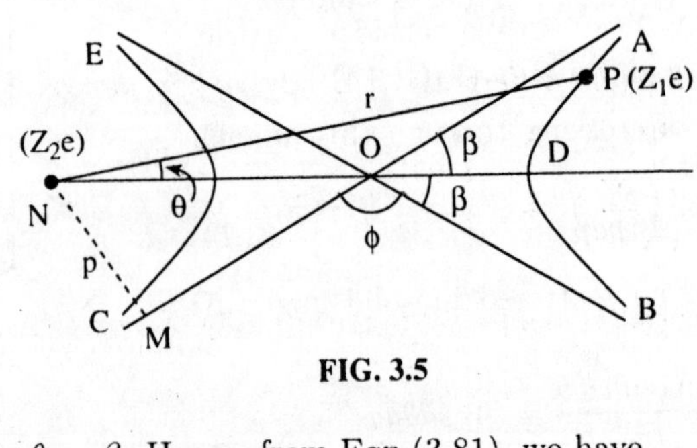

FIG. 3.5

Again, when $r \to \infty$, $\theta \to \beta$. Hence, from Eqn (3.81), we have

$$0 = 1 + e\cos\beta, \text{ i.e., } \cos\beta = -\frac{1}{e}$$

$$\therefore \cos\frac{1}{2}(\pi - \phi) = -\frac{1}{e}, \text{ i.e., } \operatorname{cosec}\frac{\phi}{2} = -e$$

Squaring this, we get

$$\operatorname{cosec}^2\frac{\phi}{2} = (e)^2 = 1 + \frac{2EJ^2}{\mu k^2} \quad \text{[using Eqn (3.60)]}$$

$$\text{or, } 1 + \cot^2\frac{\phi}{2} = 1 + \frac{2EJ^2}{\mu k^2}$$

Hence,

$$\cot\frac{\phi}{2} = \left(\frac{2EJ^2}{\mu k^2}\right)^{1/2} \tag{3.83}$$

Eqn (3.83) determines the scattering angle. Now, by virtue of conservation of angular momentum,

$$J = \mu v_0 p = p\mu\sqrt{2E/\mu} \text{ [using Eqn (3.80)]}$$

$$\text{i.e., } J = p\sqrt{2\mu E} \tag{3.84}$$

Putting this value of J into Eqn (3.83), we get

$$\cot\frac{\phi}{2} = \frac{2Ep}{k}, \text{ i.e., } p = \frac{k}{2E}\cot\frac{\phi}{2} \tag{3.85}$$

$$\therefore \quad \frac{dp}{d\phi} = -\frac{k}{4E}\operatorname{cosec}^2\frac{\phi}{2} \tag{3.86}$$

Hence, from Eqns (3.77), (3.85) and (3.86), we get

$$\sigma(\phi) = \left(-\frac{k}{2E}\cot\frac{\phi}{2}\right)\cdot\frac{1}{\sin\phi}\left(-\frac{k}{4E}\operatorname{cosec}^2\frac{\phi}{2}\right) = \frac{1}{4}\left(\frac{k}{2E}\right)^2\operatorname{cosec}^4\frac{\phi}{2} \tag{3.87}$$

$$= \frac{1}{4}\left[\frac{Z_1Z_2e^2}{(4\pi\varepsilon_0)2E}\right]^2\operatorname{cosec}^4\frac{\phi}{2} \text{ [using Eqn (3.79)]} \tag{3.88}$$

This is the well-known ***Rutherford scattering formula***. Hence, if $N(\phi)$ is the number of particles scattered along the direction ϕ, we can write:

$$N(\phi) = C\operatorname{cosec}^4\frac{\phi}{2}, \tag{3.89}$$

where

$$C = \text{constant.}$$

EXAMPLE 3.6 *Show that the differential equation of the motion of a particle in a central force-field can be written as*

$$\frac{\mu h^2}{2r^4}\left[\left(\frac{dr}{d\theta}\right)^2 + r^2\right] - \int f(r)dr = E, \textit{ where } h = r^2\dot{\theta}$$

Solution. We write Eqns (3.21) and (3.22) for the motion of a particle in a central force-field as

$$\mu\left(\ddot{r} - r\dot{\theta}^2\right) = f(r) \tag{1}$$

$$\text{and, } \mu\left(r\ddot{\theta} + 2\dot{r}\dot{\theta}\right) = 0 \tag{2}$$

Multiplying Eqn (1) by $\dot{r}$ and (2) by $r\dot{\theta}$ and adding, we get

$$\mu\left(\dot{r}\ddot{r} + r^2\dot{\theta}\ddot{\theta} + r\dot{r}\dot{\theta}^2\right) = f(r)\cdot\dot{r}$$

Integrating both sides, we have

$$\frac{1}{2}\mu\left(\dot{r}^2 + r^2\dot{\theta}^2\right) - \int f(r)dr = \text{constant.} \tag{3}$$

The LHS of this equation represents the total energy E of the particle. Hence, we write:

$$\frac{1}{2}\mu\left(\dot{r}^2 + r^2\dot{\theta}^2\right) - \int f(r)dr = E \tag{4}$$

Now

$$\dot{r} = \frac{dr}{dt} = \frac{dr}{d\theta}\cdot\frac{d\theta}{dt} = \frac{dr}{d\theta}\dot{\theta} \tag{5}$$

In view of Eqn (5), Eqn (4) becomes

$$\frac{1}{2}\mu\left[\left(\frac{dr}{d\theta}\right)^2 + r^2\right]\dot{\theta}^2 - \int f(r)dr = E$$

$$\text{or,}\quad \frac{\mu h^2}{2r^4}\left[\left(\frac{dr}{d\theta}\right)^2 + r^2\right] - \int f(r)dr = E\ [\because\ h = r^2\dot{\theta}]$$

EXAMPLE 3.7 *Show that, a formal solution for the orbit of a particle moving under a central force can be given as*

$$\theta - \theta_0 = \int_{r_0}^{r} \frac{J dr}{r^2\left[2\mu\left(E - V - \dfrac{J^2}{2\mu r^2}\right)\right]^{1/2}}$$

Hence, find an expression for the angle of scattering.

Solution. From Eqn (3.24), we have $\frac{1}{2}\mu\dot{r}^2 + \frac{J^2}{2\mu r^2} + V = E.$

Hence,

$$\dot{r} = \left[\frac{2}{\mu}\left(E - V - \frac{J^2}{2\mu r^2}\right)\right]^{1/2} \tag{1}$$

Now

$$\dot{r} = \frac{dr}{dt} = \frac{dr}{d\theta}\frac{d\theta}{dt} = \frac{dr}{d\theta}\frac{J}{\mu r^2}\quad\left[\because\ \mu r^2\dot{\theta} = J\right].$$

Thus, Eqn (1) can be written as

$$\frac{dr}{d\theta} = \frac{\mu r^2}{J}\left[\frac{2}{\mu}\left(E - V - \frac{J^2}{2\mu r^2}\right)\right]^{1/2}$$

Hence,

$$d\theta = \frac{J dr}{r^2\left[2\mu\left(E - V - \dfrac{J^2}{2\mu r^2}\right)\right]^{1/2}}$$

Integrating, we get

$$\theta - \theta_0 = \int_{r_0}^{r} \frac{J dr}{r^2\psi(r)} \tag{2}$$

where we have denoted

$$\psi(r) = \left[2\mu\left(E - V - \frac{J^2}{2\mu r^2}\right)\right]^{1/2} \tag{3}$$

Rotating the coordinates to make $\theta_0 = 0$, we write Eqn (2) as

$$\theta = \int_{r_0}^{r} \frac{J dr}{r^2 \psi(r)} \tag{4}$$

We put $J = p\sqrt{2\mu E}$ [see Eqn (3.84)]. Then

$$\theta = \int_{r_0}^{r} \frac{p dr}{r^2 \psi'(r)} \tag{5}$$

where

$$\psi'(r) = \left(1 - \frac{V}{E} - \frac{p^2}{r^2}\right)^{1/2} \tag{6}$$

Hence, the angle of scattering is (taking r_0 as turning point)

$$\phi = \pi - 2\theta = \pi - 2\int_{\infty}^{r_0} \frac{p dr}{r^2 \psi'(r)} \tag{7}$$

If we use the variable $r = \dfrac{1}{u}$, so that $dr = -\dfrac{du}{u^2}$, then Eqn (7) takes the form:

$$\phi = \pi + 2\int_{0}^{u_0} \frac{p du}{\left(1 - \dfrac{V}{E} - p^2u^2\right)^{1/2}} \tag{8}$$

which is the required expression for scattering angle.

NOTE. For motions of particles under inverse square law, the central force-field is given as

$$f(r) = \frac{|k|}{r^2}, \text{ so that } V = \frac{|k|}{r}$$

In this case, expression (8) becomes

$$\phi = \pi + 2\int_{0}^{u_0} \frac{p du}{\left[1 - u\left(\dfrac{|k|}{E} + p^2 u\right)\right]^{1/2}}$$

EXAMPLE 3.8 *Find the time of description of an arc of the elliptic orbit starting from nearer end of major axis.*

Solution. We have $\mu r^2\dot{\theta} = J$ and $\dfrac{l}{r} = 1 + e\cos\theta$. Hence,

$$\int_0^t \frac{J}{\mu} dt = \int_0^{\theta} \frac{l^2 d\theta}{(1 + e\cos\theta)^2} \tag{1}$$

We introduce a new variable ϕ, so that

$$\cos\theta = \frac{\cos\phi - e}{1 - e\cos\phi} \tag{2}$$

Hence,

$$\sin\theta d\theta = \frac{\sin\phi(1 - e^2)}{(1 - e\cos\phi)^2} d\phi \tag{3}$$

Then

$$\int_0^\theta \frac{d\theta}{(1+e\cos\theta)^2} = \int_0^\phi \frac{1-e\cos\phi}{(1-e^2)^{3/2}}d\phi = \frac{1}{(1-e^2)^{3/2}}[\phi - e\sin\phi] \tag{4}$$

Now

$$\tan^2\frac{\phi}{2} = \frac{1-\cos\phi}{1+\cos\phi} = \frac{1-e}{1+e}\cdot\frac{1-\cos\theta}{1+\cos\theta} = \frac{1-e}{1+e}\tan^2\frac{\theta}{2}$$

Hence,

$$\phi = 2\tan^{-1}\left[\sqrt{\frac{1-e}{1+e}}\tan^2\frac{\theta}{2}\right] \tag{5}$$

From Eqn (2), we can find

$$\sin\phi = \frac{\sqrt{1-e^2}\sin\theta}{1+e\cos\theta} \tag{6}$$

Thus, from Eqns (1), (4), (5) and (6), we get

$$t = \frac{\mu}{J(1-e^2)^{3/2}}\left[2\tan^{-1}\left\{\sqrt{\frac{1-e}{1+e}}\tan^2\frac{\theta}{2} - e\sqrt{1-e^2}\cdot\frac{\sin\theta}{1+e\cos\theta}\right\}\right]$$

3.10 Stability of Circular Orbits

Consider a particle of mass m moving along a circular orbit of radius R under a central attractive force $F(r)$. The centrifugal force is balanced by the attractive force for equilibrium. Hence,

$$-R\dot{\theta}^2 = F(R) \tag{3.90}$$

The radial component of equation of motion is

$$m\left(\ddot{r} - r\dot{\theta}^2\right) = mF(r) \tag{3.91}$$

We eliminate $\dot{\theta}$ from the relation $J = mr^2\dot{\theta}$ and write Eqn (3.91) in the form:

$$\ddot{r} - \frac{J^2}{m^2r^3} = F(r) \tag{3.92}$$

Let there be a small radial displacement ξ in the motion of the particle so that

$$r = R + \xi \tag{3.93}$$

where $\xi \ll R$.

Then

$$\ddot{r} = \ddot{\xi} \tag{3.94}$$

$$\text{and}\quad \frac{J^2}{m^2r^3} = \frac{J^2}{m^2(R+\xi)^3} = \frac{J^2}{m^2R^3}\left(1+\frac{\xi}{R}\right)^{-3} = \frac{J^2}{m^2R^3}\left(1-\frac{3\xi}{R}\right) \tag{3.95}$$

where we have neglected the higher terms.

We also expand the force function $F(r)$ by Taylor series around R and get

$$F(r) = F(R) + \xi F'(R) + \cdots \tag{3.96}$$

Using Eqns (3.94)–(3.96) into Eqn (3.92), we get

$$\ddot{\xi} - \frac{J^2}{m^2R^3}\left(1-\frac{3\xi}{R}\right) = F(R) + \xi F'(R) \tag{3.97}$$

From Eqn (3.90), we have

$$F(R) = -R\dot{\theta}^2 = -R\frac{J^2}{m^2R^4} = -\frac{J^2}{m^2R^3} \tag{3.98}$$

Eqn (3.97) in view of Eqn (3.98) is reduced to

$$\ddot{\xi} - \left[\frac{3}{R}F(R) + F'(R)\right]\xi = 0 \tag{3.99}$$

Eqn (3.99) will represent an oscillatory motion only if the quantity

$$-\left[\frac{3}{R}F(R) + F'(R)\right]$$

is positive. Thus, for ***stable oscillations*** we must have

$$\left[\frac{3}{R}F(R) + F'(R)\right] < 0 \tag{3.100}$$

For a central force given by

$$F(r) = -\frac{k}{r^n} \tag{3.101}$$

where k and n are constants, the condition in Eqn (3.100) yields

$$\frac{F(R)}{R}(n-3) < 0 \tag{3.102}$$

Thus, for central forces given by Eqn (3.101), the circular orbits will be stable in only those cases for which $n < 3$.

3.11 Tidal Force-fields

The phenomenon of tides in oceans occurs due to tidal force-fields arising due to the unbalanced force of attraction in an inhomogeneous gravitational fields. Consider the sun and the earth system. We take a coordinate system at the centre of the earth and call it ***absolute geocentric reference frame***. The earth is rotating relative to this reference frame and is subjected to the gravitational field of the sun.

For simplicity, we consider points lying on the plane containing the centres of the sun and the earth. If the acceleration in the gravitational field of the sun be a, then all the points in the absolute geocentric reference frame will have an acceleration a towards the centre of the sun. We take a particle of mass m in this reference frame at a distance R (FIG. 3.6).

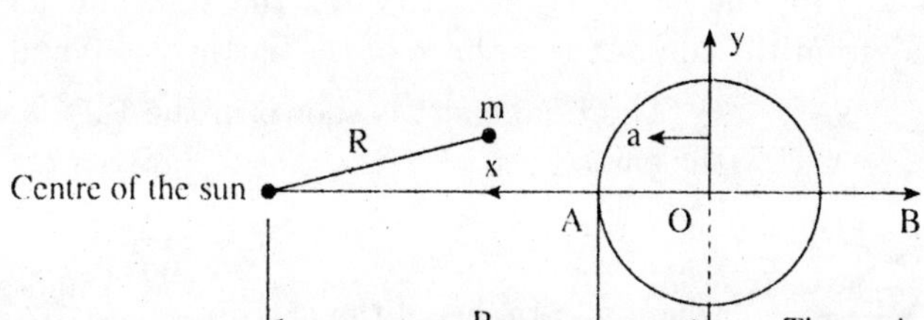

FIG. 3.6

Then the motion of this particle relative to inertial frame will be as if a uniform gravitational acceleration ($-a$) acts in addition to the natural forces. Thus, the total force acting on the particle as seen from the geocentric reference frame is

$$F = -ma + mg \tag{3.103}$$

where g = acceleration on the particle due to the gravitational field of the sun. If M be the mass of the sun and R_0 be the distance between the centres of the sun and the earth, then,

$$\left.\begin{aligned} a &= -\frac{GM}{R_0^2} \\ g &= -\frac{GM}{R^2} \end{aligned}\right\} \tag{3.104}$$

Hence,

$$F = GMm\left(\frac{1}{R_0^2} - \frac{1}{R^2}\right) \tag{3.105}$$

This is the tidal force-field from the sun. If the particle were at the centre of the earth, $R = R_0$, so that $F = 0$. For other positions of the particle, $F \neq 0$ and this force causes tides in oceans.

The acceleration f acting on the particle of mass m corresponding to the force F is

$$f = GM\left(\frac{1}{R_0^2} - \frac{1}{R^2}\right) \tag{3.106}$$

The expression for acceleration f can be written as

$$f = \frac{GM}{R_0^2}\left(1 - \frac{R_0^2}{R^2}\right) = -a\left(1 - \frac{R_0^2}{R^2}\right) \tag{3.107}$$

Let the particle lie on x-axis. Then

$$R = (R_0 - x) \tag{3.108}$$

From Eqn (3.107)

$$f = -a\left[1 - \frac{R_0^2}{(R_0 - x)^2}\right] = -a\left[1 - \frac{1}{\left(1 - \frac{x}{R_0}\right)^2}\right] \tag{3.109}$$

For points near the surface of the earth, $x \ll R_0$, and hence

$$f = -a\left[1 - \left(1 + \frac{2x}{R_0}\right)\right] = \frac{2ax}{R_0} \tag{3.110}$$

Thus, for points $x > 0$, the acceleration is towards the sun, and for points $x < 0$, the acceleration f is away from the sun. At the centre of the earth, $x = 0$ and hence $f = 0$.

The tidal effect is shown in the FIG. 3.7. On the surface of the earth,

$$f = \frac{2ar}{R_0} \tag{3.111}$$

where r = radius of the earth.

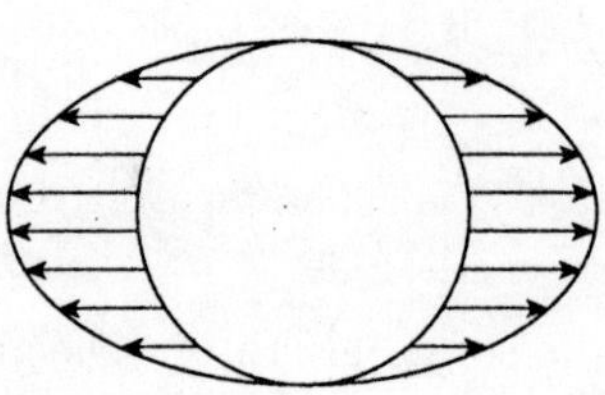

FIG. 3.7

Substituting the value of a from Eqn (3.104) into Eqn (3.111), we get

$$f = \frac{2GMr}{R_0^3} \tag{3.112}$$

This implies that the tidal acceleration is directly proportional to the mass of the sun and inversely proportional to the cube of the distance between the two centres. That is

$$\left.\begin{aligned} f &\propto M \\ \text{and,}\quad f &\propto \frac{1}{R_0^3} \end{aligned}\right\} \tag{3.113}$$

Similar will be the case for earth-moon system. We now consider this case. Let there be a rock of mass m' on the surface of the moon of mass m, the mass of the earth being M.

The tidal force on the particle is

$$F = \frac{2GMm'\rho}{D^3} \tag{3.114}$$

where ρ = radius of the moon and D is the distance between the centres. The gravitational force from the moon on the particle lying on equatorial plane is

$$F' = \frac{Gm'm}{\rho^2} \tag{3.115}$$

The rock will just remain on the moon's surface, if

$$F = F', \text{ i.e., } \frac{2GMm'\rho}{D^3} = \frac{Gmm'}{\rho^2} \tag{3.116}$$

This yields

$$D^3 = \rho^3 \frac{2M}{m} \tag{3.117}$$

If the orbital radius of the moon is less than the value of D given by Eqn (3.117), the rock would be lifted away by the tidal force.

If the densities of the earth and the moon be respectively d_e and d_m, then

$$\left.\begin{aligned} M &= \frac{4}{3}\pi R^3 d_e \\ m &= \frac{4}{3}\pi \rho^3 d_m \end{aligned}\right\} \tag{3.118}$$

where R = radius of the earth.

Then by Eqn (3.117),

$$D = R\left(\frac{2d_e}{d_m}\right)^{1/3} = 1.26R\left(\frac{d_e}{d_m}\right)^{1/3} \tag{3.119}$$

The value of the orbital radius D given by this equation is known as ***Roche limit***.

It may be noted that we have neglected the deformation of the participating bodies. If this be considered, the numerical coefficient 1.26 will change to 2.46.

3.12 Artificial Satellites

Any object launched from the earth into space so as to move around it is called an ***artificial satellite***. The satellite placed in the desired orbit moves around the earth under its attraction.

Let m = mass of the satellite, M = mass of the earth, R = radius of the earth, h = height of the satellite from earth's surface, ω = angular velocity of the satellite and g = acceleration due to gravity.

When the satellite is on the earth's surface,

$$mg = \frac{GmM}{R^2}, \text{ i.e., } GM = gR^2 \tag{3.120}$$

The centripetal force on the satellite when it is moving at height h in circular orbit is

$$F_1 = m\omega^2(R+h) \tag{3.121}$$

The gravitational attraction of the earth on the satellite is

$$F_2 = \frac{GMm}{(R+h)^2} = \frac{gR^2m}{(R+h)^2} \quad \text{[using (3.120)]} \tag{3.122}$$

The forces F_1 and F_2 must be equal for the motion of the satellite along the orbit. Hence,

$$m\omega^2(R+h) = \frac{gR^2m}{(R+h)^2}$$

$$\therefore \omega = \left[\frac{gR^2}{(R+h)^3}\right]^{1/2} \tag{3.123}$$

This determines the minimum angular velocity of the satellite. The minimum linear velocity of the satellite is

$$v = \omega(R+h) = \left(\frac{gR^2}{R+h}\right)^{1/2} \tag{3.124}$$

The time period T of the satellite is

$$T = \frac{2\pi}{\omega} = 2\pi\left[\frac{(R+h)^3}{gR^2}\right]^{1/2} \tag{3.125}$$

Squaring, we get $\frac{4\pi^2}{T^2} = \frac{gR^2}{(R+h)^3}$, which yields

$$h = \left(\frac{gR^2T^2}{4\pi^2}\right)^{1/3} - R \tag{3.126}$$

This determines the height of the satellite when the time period is given.

If the satellite moves round the earth with the same angular velocity and in the same sense as that of the earth about its own axis of rotation, it is called ***geo-stationary satellite***. Thus, the time period of a geo-stationary satellite is 24 hours.

Generally, the satellites move around the earth in an elliptic orbit. If the satellite be given a horizontal velocity v at a height h from the earth's surface, then its

$$\left.\begin{aligned} \text{kinetic energy} &= \frac{1}{2}mv^2 \\ \text{and potential energy} = V &= \frac{GMm}{R+h} = \frac{mgR^2}{R+h} \end{aligned}\right\} \tag{3.127}$$

so that its total energy

$$E = \frac{1}{2}mv^2 - \frac{mgR^2}{R+h} \tag{3.128}$$

From Eqn (3.61), we know that $E < 0$ for elliptic orbits. From Eqn (3.60), we have

$$e^2 = 1 + \frac{2EJ^2}{\mu k^2}, \quad k = mgR^2 \tag{3.129}$$

where μ = equivalent reduced mass of the system and J = angular momentum of the satellite.

Then

$$J = mv(R+h) \tag{3.130}$$

From Eqns (3.128), (3.129) and (3.130), we get

$$\begin{aligned} e^2 &= 1 + \frac{1}{\mu m^2 g^2 R^4} \cdot \left(mv^2 - \frac{2gR^2 m}{R+h} \right) m^2 v^2 (R+h)^2 \\ &= 1 + \left\{ \frac{v^2(R+h)}{gR^2} \right\}^2 - 2\frac{v^2(R+h)}{gR^2} \quad [\because \mu = m] \\ &= \left[1 - \frac{v^2(R+h)}{gR^2} \right]^2 \end{aligned}$$

$$\therefore e = 1 - \frac{v^2(R+h)}{gR^2} \tag{3.131}$$

This determines the eccentricity of the elliptic orbit. For circular orbits, $e = 0$. Putting $e = 0$ in Eqn (3.131), we get Eqn (3.124) for the velocity v of the satellite.

Velocity of escape

The velocity with which a body is projected vertically so as to escape from the gravitational attraction of the earth is called ***escape velocity***.

If m be the mass of the body and V_e the escape velocity, its kinetic energy is

$$T = \frac{1}{2} m V_e^2 \tag{3.132}$$

Work done in projecting the body from the surface of the earth to an infinite distance is

$$W = \int_R^\infty \frac{GMm}{x^2} dx = \frac{GMm}{R} \tag{3.133}$$

The body will escape from the influence of the earth's gravitational attraction if

$$T \geq W, \text{ i.e., } \frac{1}{2} m V_e^2 \geq \frac{GMm}{R} \tag{3.134}$$

Hence, the minimum value of the escape velocity is

$$V_e = \sqrt{\frac{2GM}{R}} = \sqrt{2gR} \tag{3.135}$$

Thus, using the values $R = 6.4 \times 10^6$ m, $g = 9.8\,\text{m/sec}^2$, we find

$$V_e = \sqrt{2 \times 9.8 \times 6.4 \times 10^6} = 1.12 \times 10^4 \text{ m/sec}$$

EXAMPLE 3.9 *If the gravitational force between two bodies were inversely proportional to the fourth power of distance between them, then find the escape velocity at the surface of the earth. Given:* $g = 10\,m/sec^2$, $R = 6.4 \times 10^6$.

Solution. The force of attraction on a mass m distant x from the centre of the earth is

$$F = \frac{GM}{x^4} \tag{1}$$

Hence, work done in projecting the body from the surface of the earth to infinity is

$$W = \int_R^\infty \frac{GM}{x^4} dx = \frac{1}{3} \frac{GMm}{R} \tag{2}$$

Equating this to the kinetic energy of the body

$$\frac{1}{2}mV_e^2 = \frac{1}{3}\frac{GMm}{R}$$

$$\therefore V_e = \sqrt{\frac{2}{3}\frac{GM}{R}} = \sqrt{\frac{2gR}{3}} \tag{3}$$

Putting here the given values

$$V_e = \sqrt{\frac{2}{3} \times 10 \times 6.4 \times 10^6} = 6.532 \times 10^3 \text{ m/sec.}$$

EXAMPLE 3.10 *An artificial satellite revolves round the earth at a height h from the surface. Determine orbital speed and orbital period so that a man in the satellite will be in a state of weightlessness.*

Solution. On the surface of the earth, acceleration due to gravity is given by

$$gR^2 = GM, \text{ i.e., } g = \frac{GM}{R^2} \tag{1}$$

Similarly, acceleration due to gravity at height h is

$$g' = \frac{GM}{(R+h)^2} = g\left(\frac{R}{R+h}\right)^2 \tag{2}$$

Hence, attraction of the earth on the satellite is

$$F = mg' = mg\left(\frac{R}{R+h}\right)^2 \tag{3}$$

Centripetal force on the satellite is

$$F_e = \frac{mv^2}{(R+h)} \tag{4}$$

Hence, for the state of weightlessness,

$$\frac{mv^2}{R+h} = mg\left(\frac{R}{r+h}\right)^2$$

$$\therefore v = R\sqrt{\frac{g}{R+h}} \tag{5}$$

If h be small,

$$v = R\sqrt{g}(R+h)^{-1/2} = R\sqrt{g}\left[R\left(1+\frac{h}{R}\right)\right]^{-1/2} = \sqrt{gR}\left(1-\frac{h}{2R}\right) \text{ (approx.)} \tag{6}$$

Time period is

$$T = \frac{2\pi(R+h)}{v} = \frac{2\pi(R+h)}{1} \cdot \frac{\sqrt{R+h}}{R\sqrt{g}} = \frac{2\pi}{R\sqrt{g}}(R+h)^{3/2} \tag{7}$$

For small h,

$$T = \frac{2\pi}{R\sqrt{g}} \cdot R^{3/2}\left(1+\frac{h}{R}\right)^{3/2} = 2\pi\sqrt{\frac{R}{g}}\left(1-\frac{3h}{2R}\right) \text{ (approx.)} \tag{8}$$

EXAMPLE 3.11 *An artificial satellite is to be placed in a stable elliptic orbit of eccentricity 1/2 and minimum distance R/2 from the surface of the earth (where R is the radius of the earth) by using a launching rocket which raises the satellite to an altitute R above the surface of the earth, and stops after imparting to it a horizontal velocity V. Find the required value of V.*

Solution. If the mass of the satellite be taken as unity, the force of attraction on it is

$$F = \frac{GM}{r^2} = \frac{gR^2}{r^2} \tag{1}$$

where r = distance at any instant from the centre of the earth (FIG. 3.8). Taking OA as the initial line, the equation of the path is

$$\frac{d^2u}{d\theta^2} + u = \frac{F}{h^2u^2} = \frac{gR^2}{r^2} \cdot \frac{1}{h^2u^2} = \frac{gR^2}{h^2} \tag{2}$$

where $u = 1/r$ and $h = vp$ = constant $= V \cdot 2R$.

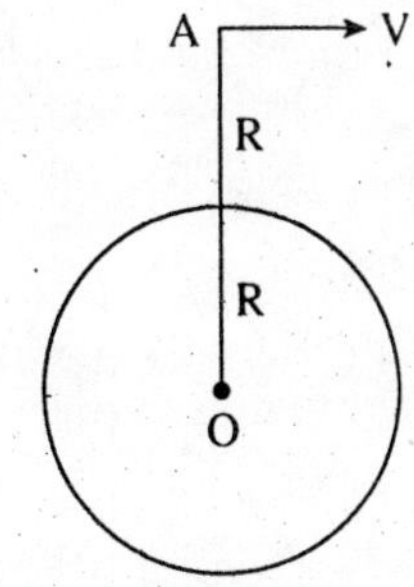

FIG. 3.8

Hence, Eqn (2) becomes

$$\frac{d^2u}{d\theta^2} + u = \frac{gR^2}{4V^2R^2} = \frac{g}{4V^2} \tag{3}$$

Defining a variable,

$$\xi = u - \frac{g}{4V^2} \tag{4}$$

Eqn (3) is written as

$$\frac{d^2\xi}{d\theta^2} + \xi = 0 \tag{5}$$

Its solution is

$$\xi = A\cos(\theta - \alpha), \text{ i.e., } \frac{1}{r} = \frac{g}{4V^2} + A\cos(\theta - \alpha) \tag{6}$$

where A and α are constants.

Eqn (6) is written as

$$\frac{4V^2/g}{r} = 1 + \frac{4AV^2}{g}\cos(\theta - \alpha) \tag{7}$$

Since eccentricity of the given orbit is 1/2, we have

$$\frac{4AV^2}{g} = \frac{1}{2}, \text{ i.e., } A = \frac{g}{8V^2} \tag{8}$$

From (7), we observe that r is minimum when $\cos(\theta - \alpha)$ is maximum = 1. Hence,

$$\frac{4V^2/g}{r_{\min}} = 1 + \frac{4AV^2}{g} = \frac{g + 4AV^2}{g}$$

$$\therefore r_{\min} = \frac{4V^2}{g + 4AV^2} = \frac{4V^2}{g + g/2} \quad \text{[using (8)]}$$

$$= \frac{8V^2}{3g} \tag{9}$$

By the condition given, $r_{\min} = R + \frac{R}{2} = \frac{3R}{2}$.

Hence, $\frac{8V^2}{3g} = \frac{3R}{2}$, $\quad \therefore V = \frac{3}{4}\sqrt{gR}$.

EXAMPLE 3.12 *A launching rocket carries an artificial satellite to an altitude R above the surface of the earth (R = radius of the earth) and stops after imparting to the satellite a velocity $\sqrt{gR/2}$ inclined at an angle of 45° above the horizontal. Find maximum distance of the satellite from the surface of the earth, and investigate whether it crashes on the earth or goes into orbit.*

Solution. The satellite is given a velocity $\sqrt{gR/2}$ from the point A at an angle 45° above horizontal (FIG. 3.9). Taking mass of the satellite to be unity, the force of attraction on it is

$$F = \frac{gR^2}{r^2} \tag{1}$$

The equation of the orbit with OA as the initial line is

$$\frac{d^2u}{d\theta^2} + u = \frac{gR^2}{r^2}\frac{1}{h^2u^2} = \frac{gR^2}{h^2} \tag{2}$$

where $u = 1/r$.

From the figure,

$$p = 2R\cos 45° = \sqrt{2}R \tag{3}$$

Also,

$$h = vp = \text{constant} = \sqrt{\frac{gR}{2}}\cdot\sqrt{2}R = \sqrt{gR^3} \tag{4}$$

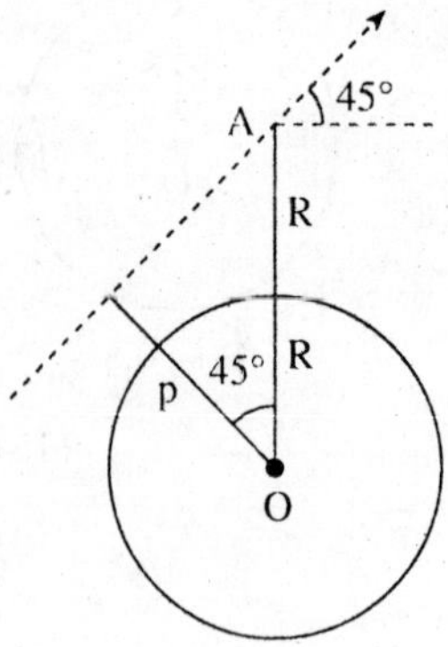

FIG. 3.9

From (2) and (4),

$$\frac{d^2u}{d\theta^2} + u = \frac{gR^2}{gR^3} = \frac{1}{R} \tag{5}$$

Its solution is

$$u = \frac{1}{R} + A\cos(\theta - \alpha), \tag{6}$$

where A and α are constants.

Then

$$\frac{du}{d\theta} = -A\sin(\theta - \alpha) \tag{7}$$

Velocity v at any point on the orbit is given by

$$v^2 = h^2\left[\left(\frac{du}{d\theta}\right)^2 + u^2\right] = gR^3\left[\left(\frac{du}{d\theta}\right)^2 + u^2\right] \tag{8}$$

Initially at the point A, we have

$$v = \sqrt{\frac{gR}{2}},\quad u = \frac{1}{2R} \tag{9}$$

From (8) and (9),

$$\frac{gR}{2} = gR^3\left[\left(\frac{du}{d\theta}\right)_A^2 + \frac{1}{4R^2}\right] \tag{10}$$

Hence,

$$\left(\frac{du}{d\theta}\right)_A = \sqrt{\frac{1}{4R^2}} = -\frac{1}{2R} \tag{11}$$

We have taken $-$ve sign since $u = 1/r$ decreases with θ. Inserting initial conditions (9) and (11) into (6) and (7), we get

$$\left.\begin{aligned} \frac{1}{2R} &= \frac{1}{R} + A\cos\alpha \\ -\frac{1}{2R} &= A\sin\alpha \end{aligned}\right\} \tag{12}$$

Solving,

$$A = \frac{1}{\sqrt{2}\,R}, \quad \alpha = -135° \tag{13}$$

Then, from (6),

$$u = \frac{1}{R} + \frac{1}{\sqrt{2}R}\cos(\theta + 135°), \text{ i.e., } \frac{R}{r} = 1 + \frac{1}{\sqrt{2}}\cos(\theta + 135°) \tag{14}$$

Eccentricity of the orbit $= \frac{1}{\sqrt{2}} < 1$. Hence, the orbit is an ellipse.

From (14),

$$r = \frac{R}{1 + \left(\frac{1}{\sqrt{2}}\right)\lambda}, \quad \lambda = \cos(\theta + 135°) \tag{15}$$

For r_{max}, $\lambda = -1$ and for r_{min}, $\lambda = 1$.

Hence,

$$\left.\begin{aligned} r_{max} &= \frac{R}{1 - 1/\sqrt{2}} = \frac{\sqrt{2}R}{\sqrt{2} - 1} = \left(2 + \sqrt{2}\right)R \\ r_{min} &= \frac{R}{1 + 1/\sqrt{2}} = \frac{\sqrt{2}R}{\sqrt{2} + 1} = \left(2 - \sqrt{2}\right)R \end{aligned}\right\} \tag{16}$$

Since $r_{min} < R$, the satellite obviously crashes on the surface of the earth.

EXERCISE

1. The masses of the sun and the earth are respectively M and m. If the earth moves around the sun in an elliptical orbit of semi-major exis a, then show that the periodic time is given by

$$T = \frac{2\pi a^{3/2}}{\sqrt{G(M+m)}}$$

2. A particle of mass m moves under a central force $F = -k/r^3$. If E be the total energy supplied to the particle, then show that its speed is given by

$$v = \sqrt{\frac{2E}{m} + \frac{k}{mr^2}}$$

3. Writing $u = \dfrac{1}{r}$, show that the conservation of energy equation becomes

$$\left(\frac{du}{d\theta}\right)^2 + u^2 = \frac{2(E - V)}{mh^2}$$

Assume an inverse square force to be $f(r) = \dfrac{-k}{r^2}$ $(k > 0)$. Prove that the speed v of a particle moving in that field is given by

$$v^2 = \frac{k}{m}\left(\frac{2}{r} \pm \frac{1}{a}\right), \quad (a = \text{constant})$$

when the upper and lower signs correspond to an elliptical and a hyperbolical path. Hence, find the escape velocity. [C.U. (HONS) 1996]

4. For a particle moving under a central force-field, show that
   (a) the torque about the force is zero,
   (b) the central orbit is a planar path. [C.U. (HONS) 2002]

5. A particle of mass m moves under the action of a central force $F \equiv f(r)$. Show that the equation for the path of the particle is

$$\frac{d^2u}{d\theta^2} + u = -\frac{f(1/u)}{mL^2u^2}$$

where L is a constant of motion and $u = 1/r$. For the special case of $F = -k/r^2$ $(k > 0)$, prove that the path is a conic section. [C.U. (HONS) 1998]

6. A particle describes a circular orbit under the influence of an attractive central force directed towards a point on the circle. Show that the force varies as the inverse of fifth power of distance.

7. A particle moves with a central acceleration $\mu(r^5 - c^4 r)$, being projected from an apse at a distance c with a velocity $c^3\sqrt{2\mu/3}$. Show that its path is

$$x^4 + y^4 = c^4$$

8. A particle is projected from the earth's surface with velocity v. Show that, if diminution of gravity is taken into account, but the resistance of air neglected, the path is an ellipse of major axis

$$\frac{2gR^3}{2gR - v^2}, \quad R = \text{radius of the earth.}$$

9. A particle moves with a central acceleration of inverse square law. It is projected with a velocity V at a distance R. Show that its path is a rectangular hyperbola if the angle of projection is

$$\sin^{-1}\left[\frac{\mu}{VR(V^2 - 2\mu/R)^{1/2}}\right]$$

10. A particle is projected at a smooth hard sphere of radius R. The impact parameter for the scattering is $p < R$. Assuming equal angles of incidence and reflection, prove that

$$\sigma = \int \sigma(\phi)d\phi = \pi R^2$$

11. Show that the differential scattering cross section for the scattering of particles by the potential

$$V(r) = \lambda\left(\frac{1}{r} - \frac{1}{R}\right), \quad \text{for } r < R$$

$$\text{and } V(r) = 0, \quad \text{for } r > R$$

is given by

$$\sigma(\phi) = \frac{R^2}{4}\frac{1+\xi}{\{1+\xi\sin^2\left(\frac{\phi}{2}\right)\}^2}, \quad \text{where } \xi = \frac{4ER}{\lambda^2}(\lambda + RE)$$

12. A particle moves in a circular orbit of radius R under the action of an attractive force

$$F(r) = -\left(\frac{\lambda}{r^2} + \frac{\mu}{r^4}\right)$$

where λ, μ are positive constants. Show that the orbit is stable to small radial perturbation if $\lambda R^2 > \mu$.

CHAPTER 4

Moments and Products of Inertia

4.1 Definitions

Let there be a particle of mass m at a point $P(x, y, z)$ with respect to Cartesian coordinate system (FIG. 4.1). Then we define the quantity

$$I_{xx} = m(y^2 + z^2) \tag{4.1}$$

as the ***moment of inertia*** of the particle about x-axis.

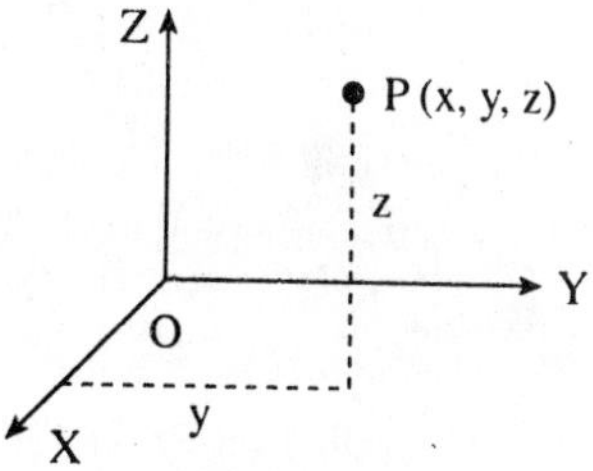

FIG. 4.1

Similarly, the quantities

$$I_{yy} = m(x^2 + z^2) \tag{4.2}$$

$$I_{zz} = m(x^2 + y^2) \tag{4.3}$$

are respectively defined as the moment of inertia of the particle about y- and z-axes.

The quantities

$$\left.\begin{aligned} I_{xy} &= -mxy \\ I_{yz} &= -myz \\ I_{zx} &= -mzx \end{aligned}\right\} \tag{4.4}$$

are called ***products of inertia***.

It is obvious from above definition that

$$\left.\begin{aligned} I_{xy} &= I_{yx} \\ I_{yz} &= I_{zy} \\ I_{zx} &= I_{xz} \end{aligned}\right\} \tag{4.5}$$

We note the following properties:

(i) $I_{xx} + I_{yy} - I_{zz} = m(y^2 + z^2) + m(z^2 + x^2) - m(x^2 + y^2) = 2mz^2$, which is always positive. Hence

$$I_{xx} + I_{yy} > I_{zz} \tag{4.6}$$

Thus, ***the sum of the moments of inertia about any two of three mutually perpendicular axes is greater than the moment of inertia about the third.***

(ii)
$$\begin{aligned} I_{xx} + I_{yy} + I_{zz} &= m(y^2 + z^2) + m(z^2 + x^2) + m(x^2 + y^2) \\ &= 2m(x^2 + y^2 + z^2) = 2mr^2 \end{aligned} \tag{4.7}$$

which is constant. Thus, ***the sum of the moments of inertia about any three mutually perpendicular axes of coordinates is always constant and is equal to twice the moment of inertia with respect to the origin.***

4.2 Moment of Inertia of a System of Particles

Consider a system of N particles of masses $m_1, m_2, \ldots, m_N$ placed at positions $(x_1, y_1, z_1), \ldots,$ (x_N, y_N, z_N) with respect to Cartesian coordinate system. Then the moments and products of inertia of the system are:

$$I_{xx} = \sum_{i=1}^{N} m_i \left(y_i^2 + z_i^2\right). \tag{4.8}$$

$$I_{yy} = \sum_{i=1}^{N} m_i \left(x_i^2 + z_i^2\right) \tag{4.9}$$

$$I_{zz} = \sum_{i=1}^{N} m_i \left(x_i^2 + y_i^2\right) \tag{4.10}$$

$$I_{xy} = -\sum_{i=1}^{N} m_i x_i y_i \tag{4.11}$$

$$I_{yz} = -\sum_{i=1}^{N} m_i y_i z_i \tag{4.12}$$

$$I_{zx} = -\sum_{i=1}^{N} m_i z_i x_i \tag{4.13}$$

For a continuous distribution of mass, we consider an elementary mass dm situated at a point (x, y, z) and replace the summation in the above equations by integration which is to be taken so as to cover the whole body.

4.3 Radius of Gyration

Suppose that the total mass M of a system of particles or of a continuous body is concentrated at a point such that the moment of inertia of this concentrated mass about an axis is same as the moment of inertia of the system about that axis. Then the distance of the concentrated mass from that axis is called ***radius of gyration***. Thus, if K be the radius of gyration of a system of particles or of a body, then

$$MK^2 = \Sigma m_i r_i^2, \text{ i.e., } K^2 = \frac{\Sigma m_i r_i^2}{M} = \frac{\Sigma m_i r_i^2}{\Sigma m_i} \tag{4.14}$$

4.4 Determination of Moment of Inertia

We consider below some standard cases:

(a) Rectangular Lamina

Let M be the total mass of a rectangular lamina of sides $2a$ and $2b$ (FIG. 4.2). Consider an elementary mass $dm = \rho dx dy$ at a point (x, y) with respect to rectangular coordinates. Here ρ is the surface density and $\rho = \dfrac{M}{4ab}$.

Hence, $dm = \dfrac{M}{4ab} dx dy$

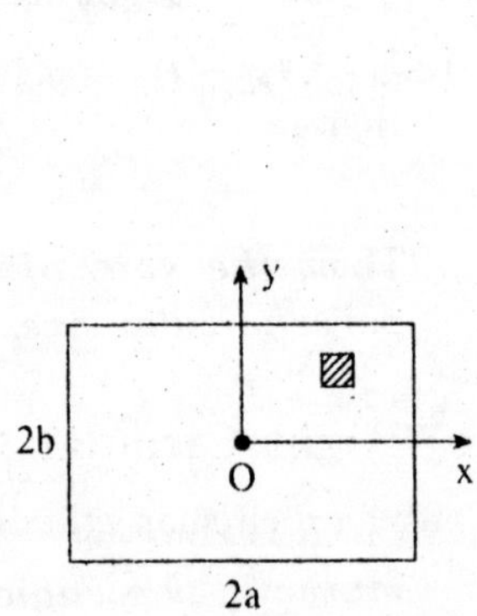

FIG. 4.2

The moment of inertia about Ox-axis is

$$I_{xx} = \int_{-a}^{a}\int_{-b}^{b} (dm)y^2 = 4\int_0^a\int_0^b \frac{M}{4ab}y^2 dx dy = \frac{M}{ab}x\frac{y^3}{3}\Bigg|_{\substack{x=0\\y=0}}^{\substack{x=a\\y=b}} = \frac{1}{3}Mb^2 \tag{4.15}$$

Similarly, the moment of inertia about Oy-axis is

$$I_{yy} = \frac{1}{3}Ma^2 \tag{4.16}$$

The moment of inertia about an axis passing through the centre and perpendicular to the plane of the lamina is

$$\begin{aligned} I_{zz} &= \iint dm(y^2 + x^2) \\ &= \text{sum of the moments of inertia about } Ox\text{- and } Oy\text{-axes} \\ &= \frac{1}{3}M(a^2 + b^2) \end{aligned} \tag{4.17}$$

(b) Circular Disc

Consider a circular disc of radius R and total mass M. We take polar coordinate system through its centre (FIG. 4.3). The surface density of the disc is

$$\rho = \frac{M}{\pi R^2}$$

Consider an elementary mass

$$dm = \rho r d\theta dr = \frac{M}{\pi R^2} r d\theta dr$$

at distance r from the centre. Its distance from x-axis is $r\sin\theta$.

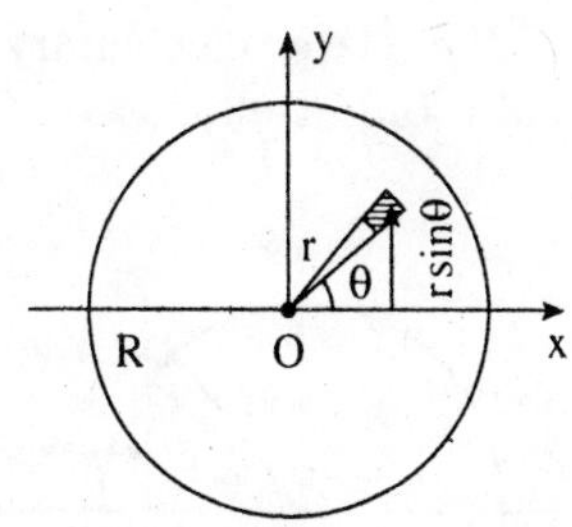

FIG. 4.3

Hence, the moment of inertia about x-axis, i.e., about a diameter is

$$\begin{aligned} \frac{M}{\pi R^2}\int_0^{2\pi}\int_0^R (r d\theta dr)(r\sin\theta)^2 &= \frac{M}{\pi R^2}\left(\int_0^R r^3 dr\right)\int_0^{2\pi} \sin^2\theta d\theta \\ &= \frac{M}{\pi R^2}\left(\frac{R^4}{4}\right)4\int_0^{\pi/2} \sin^2\theta d\theta \\ &= \frac{MR^2}{\pi}\frac{\pi}{4} = \frac{1}{4}MR^2 \end{aligned} \tag{4.18}$$

Similarly, the moment of inertia about y-axis is

$$I_{yy} = \frac{1}{4}MR^2 \tag{4.19}$$

The moment of inertia about z-axis is

$$I_{zz} = I_{xx} + I_{yy} = \frac{1}{4}MR^2 + \frac{1}{4}MR^2 = \frac{1}{2}MR^2 \tag{4.20}$$

(c) Right Circular Cylinder

Consider a circular cylinder of height h and radius of the base R (FIG. 4.4). Then mass of the cylinder is $M = \pi R^2 \rho h$, where ρ = density of the material. This yields

$$\rho = \frac{M}{\pi R^2 h}$$

We take a circular disc element of thickness dx at a distance x from the base. The mass of the elemental disc is

$$\rho\pi R^2 dx = \frac{M}{h}dx$$

and its moment of inertia about the line through its centre perpendicular to its plane is, by Eqn (4.20),

$$\frac{1}{2}\left(\frac{M dx}{h}\right) R^2$$

Hence, the moment of inertia of the cylinder about its axis is

$$\frac{MR^2}{2h}\int_0^h dx = \frac{1}{2}MR^2 \tag{4.21}$$

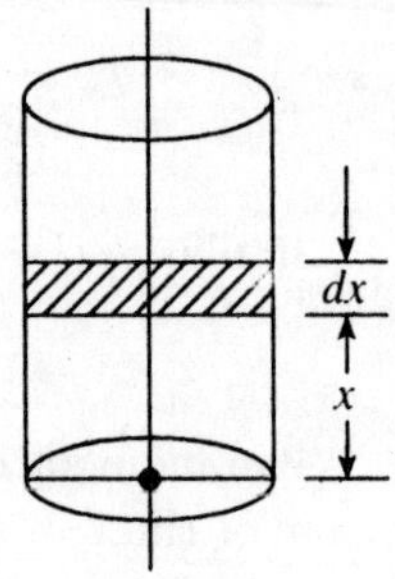

FIG. 4.4

(d) Spherical Shell

Consider a spherical thin shell of radius R. If ρ be the surface density, then its mass

$$M = 4\pi R^2\rho, \text{ i.e., } \rho = \frac{M}{4\pi R^2}$$

We take an elementary arc of length $Rd\theta$ at a point having polar coordinates (R, θ) (FIG. 4.5) and rotate it about axis Ox. Then an elementary circular ring of radius $R\sin\theta$ and thickness $Rd\theta$ will be generated. Mass of this elemental ring is

$$(2\pi R\sin\theta)\rho R d\theta = \frac{M}{4\pi R^2}(2\pi R\sin\theta)Rd\theta$$
$$= \frac{M}{2}\sin\theta d\theta$$

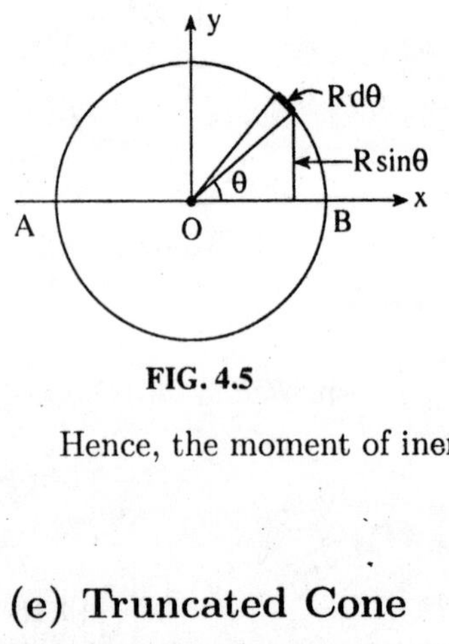

FIG. 4.5

The distance of this ring from diameter $AB = R\sin\theta$.

Hence, the moment of inertia of the elemental ring about the diameter is

$$\left(\frac{M}{2}\sin\theta d\theta\right)(R\sin\theta)^2 = \frac{MR^2}{2}\sin^3\theta d\theta$$

Hence, the moment of inertia of the spherical shell about the diameter AB is

$$\frac{1}{2}MR^2\int_0^{\pi}\sin^3\theta d\theta = \frac{2}{3}MR^2 \tag{4.22}$$

(e) Truncated Cone

Consider the frustrum of a cone with vertical angle 2ϕ, and R, r the radii at the base and top faces respectively (FIG. 4.6). The distances from the vertex of the top and bottom surfaces are $h_1 = r\cot\phi$, $h_2 = R\cot\phi$.

We take a thin circular disc of thickness dx at a distance x from its vertex O. Then total mass of the frustrum is

$$M = \pi\rho\tan^2\phi\int_{h_1}^{h_2} x^2 dx = \pi\rho\tan^2\phi\frac{x^3}{3}\bigg|_{h_1}^{h_2}$$
$$= \pi\rho\tan^2\phi\frac{1}{3}(R^3 - r^3)\cot^3\phi = \frac{1}{3}\pi\rho\cot\phi(R^3 - r^3)$$

Hence, $\pi\rho\cot\phi = \dfrac{3M}{R^3 - r^3}$.

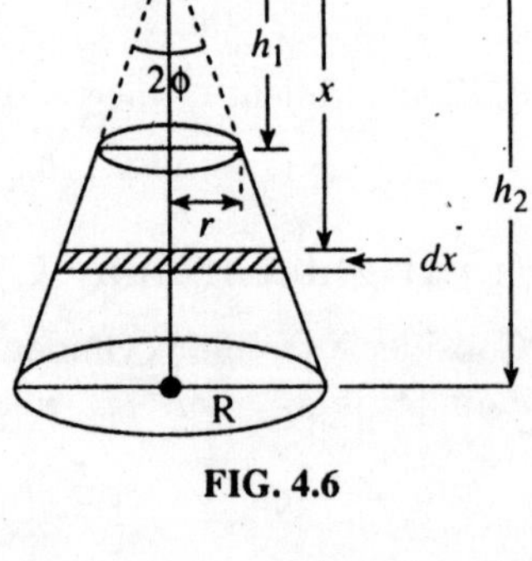

FIG. 4.6

Now the moment of inertia of the elemental disc about the axis of the cone is, by Eqn (4.20),

$$\frac{1}{2}(\text{mass}) \times (\text{radius})^2 = \frac{1}{2}\pi\rho(x\tan\phi)^2 dx \times (x\tan\phi)^2$$

$$= \frac{1}{2}\pi\rho\tan^4\phi x^4 dx$$

Hence, the moment of inertia of the whole frustrum is

$$\frac{1}{2}\pi\rho\tan^4\phi\int_{h_1}^{h_2} x^4 dx = \frac{\pi\rho}{10}\tan^4\phi x^5\Big|_{h_1}^{h_2} = \frac{\pi\rho}{10}\tan^4\phi\cot^5\phi\,(R^5 - r^5)$$

$$= \frac{\pi\rho}{10}\cot\phi(R^5 - r^5) = \frac{3M}{10}\frac{R^5 - r^5}{R^3 - r^3} \tag{4.23}$$

If $r = 0$, we get the moment of inertia of a right circular cone as

$$\frac{3M}{10}R^2 \tag{4.24}$$

4.5 Theorems on Moments of Inertia

THEOREM 4.1 : Parallel Axes Theorem

Let I_A be the moment of inertia of a system of particles (or of a continuous body) about an axis OA and let I_C be the moment of inertia of the same system about an axis parallel to OA and passing through the centre of mass C of the system. Then, by the theorem of parallel axes

$$I_A = I_C + Mh^2 \tag{4.25}$$

where M is the total mass of the system and h is the distance between the axes.

Proof. Let $P(x_i, y_i, z_i)$ be the position of the particle of mass m_i with respect to Cartesian coordinate system with z-axis taken along axis OA (FIG. 4.7). If (a, b, c) be the coordinates of the centre of mass C, then

$$h^2 = a^2 + b^2 \tag{4.26}$$

Also, let x_i', y_i', z_i' be the coordinates of P with respect to a rectangular coordinate system parallel to the previous one with origin at C. Then we have

$$\left.\begin{aligned} x_i &= x_i' + a \\ y_i &= y_i' + b \\ z_i &= z_i' + c \end{aligned}\right\} \tag{4.27}$$

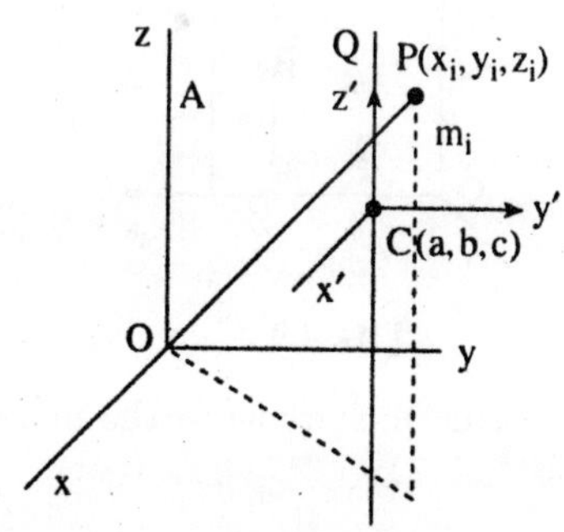

FIG. 4.7

The moment of inertia of the system about axis OA is

$$I_A = \Sigma m_i\,(y_i^2 + x_i^2)$$

$$= \Sigma m_i\left[(y_i' + b)^2 + (x_i' + a)^2\right]$$

$$= \Sigma m_i\,(y_i'^2 + x_i'^2) + \Sigma m_i(b^2 + a^2) + 2b\Sigma m_i y_i' + 2a\Sigma m_i x_i' \tag{4.28}$$

The first term in Eqn (4.28) is the moment of inertia of the system about the axis CQ, and the third and fourth terms vanish since the coordinates of the centre of mass with respect to (x', y', z') axes are zero. Hence, we get

$$I_A = I_C + \Sigma m_i(b^2 + a^2) = I_C + Mh^2 \tag{4.29}$$

THEOREM 4.2 : Perpendicular Axes Theorem
Let there be a system of particles on the xy-plane with respect to a Cartesian coordinate system. Let I_{xx}, I_{yy} and I_{zz} be the moments of inertia of the system about the x-, y- and z-axes respectively. Then by the theorem of perpendicular axes

$$I_{zz} = I_{xx} + I_{yy} \tag{4.30}$$

Proof. Let the position of the mass m_i on the xy-plane be at point $P(x_i, y_i)$ (FIG. 4.8). Then the moment of inertia of the mass about z-axis is

$$m_i(x_i^2 + y_i^2)$$

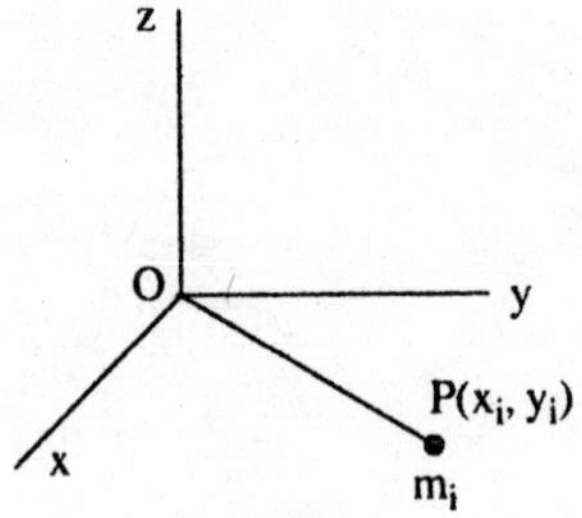

FIG. 4.8

Hence, the total moment of inertia of the system about z-axis is

$$\begin{aligned} I_{zz} &= \Sigma m_i(x_i^2 + y_i^2) \\ &= \Sigma m_i x_i^2 + \Sigma m_i y_i^2 \\ &= I_{xx} + I_{yy} \text{ (by definition).} \end{aligned}$$

EXAMPLE 4.1 *Find the moments and products of inertia of a square plate of side a, the x- and y-axes being taken to be two adjacent sides of the plate and z-axis perpendicular to the plate.* [C.U. (HONS) 1999]

Solution. Let M be the total mass of the square plate of side a (FIG. 4.9). Consider an elementary mass $\rho dxdy$ at a point $P(x, y)$, where ρ is the surface density and equal to M/a^2.

Hence,

$$dm = \rho dxdy = \frac{M}{a^2}dxdy \tag{1}$$

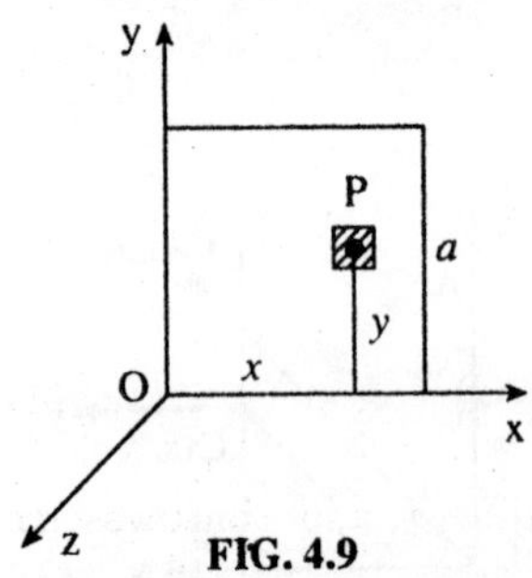

FIG. 4.9

Then the moment of inertia of the plate about x-axis is

$$\begin{aligned} I_{xx} &= \int_0^a \int_0^a (dm)y^2 \\ &= \int_0^a \int_0^a \frac{M}{a^2} y^2 dxdy \\ &= \frac{M}{a^2}(a)\frac{a^3}{3} = \frac{1}{3}Ma^2 \end{aligned} \tag{2}$$

Due to symmetry, the moment of inertia of the plate about y-axis should also be equal to

$$I_{yy} = \frac{1}{3}Ma^2 \tag{3}$$

The moment of inertia of the plate about z-axis is

$$\begin{aligned} I_{zz} &= I_{xx} + I_{yy} \text{ (by perpendicular axes theorem)} \\ &= \frac{1}{3}Ma^2 + \frac{1}{3}Ma^2 = \frac{2}{3}Ma^2 \end{aligned} \tag{4}$$

The product of inertia of the plate about x- and y-axes is

$$\begin{aligned} I_{xy} &= \int_0^a \int_0^a (dm)xy = \int_0^a \int_0^a \frac{M}{a^2}xydxdy \\ &= \frac{M}{a^2}\left(\frac{a^2}{2}\right)\left(\frac{a^2}{2}\right) = \frac{1}{4}Ma^2 \end{aligned} \tag{5}$$

The products of inertia I_{yz} and I_{zx} are zero.

EXAMPLE 4.2 *Find out the ratio of the moments of inertia of a thick spherical shell and a solid sphere of same mass and same outer radius when the inner radius of the spherical shell is half of the radius of the outer. Deduce the formula you use.* [C.U. (HONS) 2001]

Solution. Let M = mass of both the solid sphere and spherical shell,
R = outer radius of both the solid sphere and spherical shell,
r = inner radius of the spherical shell.

We deduce the formulae for the moments of inertia of the two bodies.

(i) **Solid sphere.** If ρ be the density of the material, then

$$M = \frac{4}{3}\pi R^3 \rho, \text{ i.e., } \rho = \frac{3M}{4\pi R^3} \tag{1}$$

Consider an elementary mass

$$dm = \rho r d\theta dr = \frac{3M}{4\pi R^3} r d\theta dr \tag{2}$$

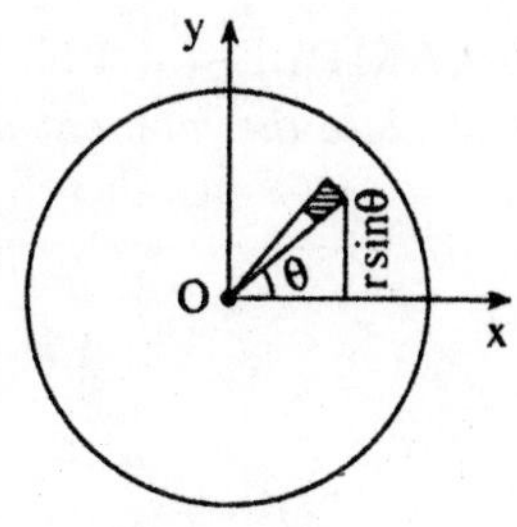

FIG. 4.10

at a point (r, θ) and rotate it about x-axis (FIG. 4.10).

Then an elementary ring of radius $r\sin\theta$ will be generated. So the mass of the ring is

$$dm = \rho(2\pi r\sin\theta) r d\theta dr = \frac{3M}{2R^3} r^2 \sin\theta d\theta dr \tag{3}$$

Every point of this ring will be at a distance of $r\sin\theta$ from x-axis. Hence, the moment of inertia of the ring about x-axis is

$$\frac{3M}{2R^3}(r^2 \sin\theta d\theta dr)(r\sin\theta)^2$$

Therefore, the moment of inertia of the sphere is

$$I_1 = \frac{3M}{2R^3}\int_0^{\pi}\int_0^{R} r^4 \sin^3\theta d\theta dr = \frac{3M}{2R^3}\cdot\frac{R^5}{5}\int_0^{\pi}\sin^3\theta d\theta$$
$$= \frac{3MR^2}{10}\cdot 2\int_0^{\pi/2}\sin^3\theta d\theta = \frac{3MR^2}{5}\left(\frac{2}{3}\right) = \frac{2}{5}MR^2 \tag{4}$$

(ii) **Thick spherical shell.** Consider a spherical shell of radius x and thickness dx (FIG. 4.11). Mass of this elementary shell is $(\rho 4\pi x^2 dx)$, where ρ is the density. Then the total mass of the shell is

$$M = 4\pi\rho\int_r^R x^2 dx = \frac{4}{3}\pi\rho(R^3 - r^3) \tag{5}$$

The moment of inertia of a shell of radius x is

$$\frac{2}{3}(\text{mass}) \times (\text{radius})^2 \text{ [by Eqn (4.22)]}$$
$$= \frac{2}{3}(4\pi\rho x^2 dx)x^2$$

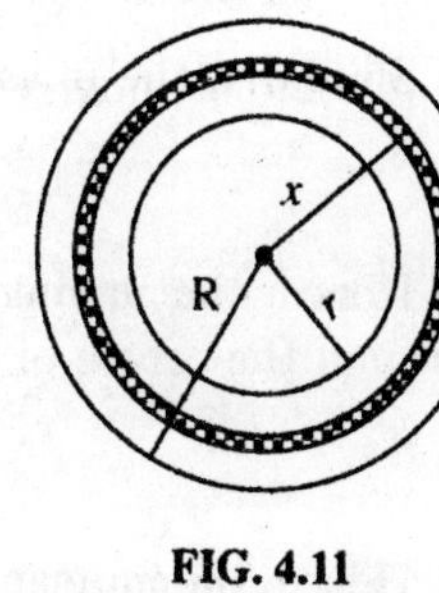

FIG. 4.11

Hence, the moment of inertia of the spherical shell is

$$I_2 = \frac{8\pi\rho}{3}\int_r^R x^4 dx = \frac{8\pi}{3}\frac{3M}{4\pi(R^3 - r^3)}\frac{R^5 - r^5}{5} \text{ [using Eqn (5)]}$$
$$= \frac{2M}{5}\frac{R^5 - r^5}{R^3 - r^3} \tag{6}$$

Since $r = R/2$ (given), Eqn (6) assumes the form:

$$I_2 = \frac{2M}{5}\left(\frac{31}{32}R^5\right)\left(\frac{8}{7R^3}\right) = \frac{31}{70}MR^2 \tag{7}$$

Hence, from Eqns (4) and (7),

$$\frac{I_2}{I_1} = \left(\frac{31}{70}R^2\right)\left(\frac{5}{2R^2}\right) = \frac{31}{28}$$

EXAMPLE 4.3 FIG. (4.12) *shows a uniform wire of mass M and radius of each ring R. Calculate the moment of inertia of the system about a normal axis through O.*

[C.U. (HONS) 2004]

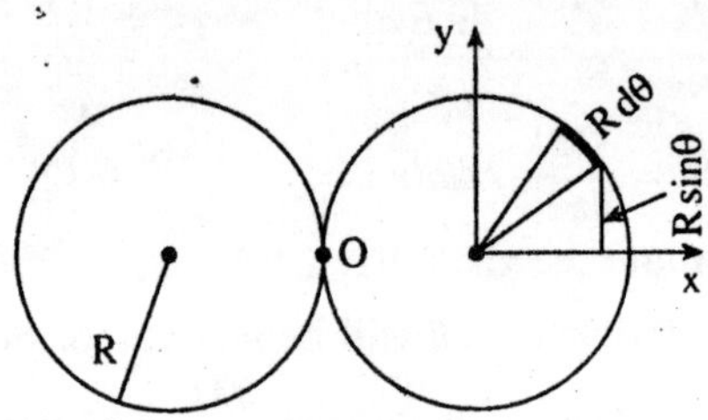

FIG. 4.12

Solution. Consider a single ring of radius R and mass $M/2$. We take an elementary arc $Rd\theta$ at a point (r, θ). Its mass is

$$dm = \frac{M}{4\pi R}Rd\theta = \frac{M}{4\pi}d\theta \tag{1}$$

Its distance from x-axis $= R\sin\theta$. Hence, its moment of inertia about x-axis is

$$\left(\frac{M}{4\pi}d\theta\right)(R\sin\theta)^2 = \frac{MR^2}{4\pi}\sin^2\theta d\theta \tag{2}$$

Hence, the moment of inertia of the ring about x-axis is

$$\frac{MR^2}{4\pi}\int_0^{2\pi}\sin^2\theta d\theta = \frac{MR^2}{4\pi}\frac{1}{2}\int_0^{2\pi}(1-\cos 2\theta)d\theta = \frac{MR^2}{8\pi}\cdot 2\pi = \frac{1}{4}MR^2 \tag{3}$$

Similarly, the moment of inertia of the ring about y-axis is

$$\frac{1}{4}MR^2 \tag{4}$$

Hence, the moment of inertia of the ring about an axis normal to xy-plane and passing through the centre of the ring (by perpendicular axes theorem) is

$$I_{zz} = I_{xx} + I_{yy} = \frac{1}{4}MR^2 + \frac{1}{4}MR^2 = \frac{1}{2}MR^2 \tag{5}$$

Hence, the moment of inertia of the ring about a normal axis through O is (by parallel axes theorem)

$$I_O = I_{zz} + \frac{M}{2}R^2 = \frac{1}{2}MR^2 + \frac{1}{2}MR^2 = MR^2 \tag{6}$$

Hence, the moment of inertia of the system of two rings about a normal axis through O is

$$2I_O = 2MR^2$$

EXAMPLE 4.4 *Determine the moment of inertia of a solid cylinder of mass M, length L and radius R about the axis perpendicular to its length and passing through the centre of mass.*

[C.U. (HONS) 2007]

Solution. We take an elementary disc of thickness dx at a distance x from the axis AA (FIG. 4.13). If ρ be the density of material, then

$$M = \pi R^2 L\rho \text{ or, } \rho = \frac{M}{\pi R^2 L} \tag{1}$$

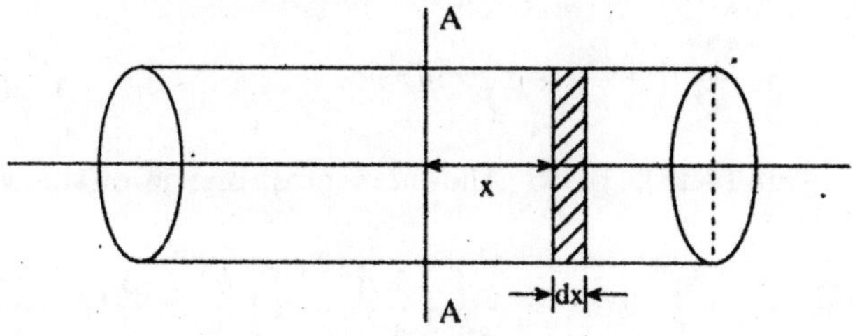

FIG. 4.13

Then mass of elementary disc is

$$dm = (\pi R^2 dx)\rho = \pi R^2 dx \frac{M}{\pi R^2 L} = \frac{M}{L} dx \tag{2}$$

Hence, the moment of inertia of the elementary disc about a diameter parallel to AA-axis is

$$(dm)\frac{R^2}{4} = \frac{MR^2}{4L} dx. \tag{3}$$

By theorem of parallel axes, the moment of inertia of the elementary disc about AA-axis is

$$\frac{MR^2}{4L} dx + (dm)x^2 = \frac{MR^2}{4L} dx + \frac{M}{L} x^2 dx \tag{4}$$

Hence, the moment of inertia of the whole cylinder about AA-axis is

$$\frac{MR^2}{4L}\int_{-L/2}^{L/2} dx + \frac{M}{L}\int_{-L/2}^{L/2} x^2 dx = \frac{MR^2}{4} + \frac{ML^2}{12} = M\left(\frac{R^2}{4} + \frac{L^2}{12}\right)$$

EXAMPLE 4.5 *Find the moment of inertia of a solid uniform right circular cone about an axis which is perpendicular to the axis of the cone and passes through its vertex.*

[C.U. (HONS) 2000]

Solution. Let M = mass of the cone, h = height of the cone, R = radius of the circular base and ρ = density of the material. Then

$$M = \frac{1}{3}\pi R^2 h\rho$$

$$\text{i.e., } \rho = \frac{3M}{\pi R^2 h} \tag{1}$$

We take an elementary disc of thickness dx at a distance x from the vertex (FIG. 4.14). The radius of this elementary disc is

$$r = \frac{R}{h} x \tag{2}$$

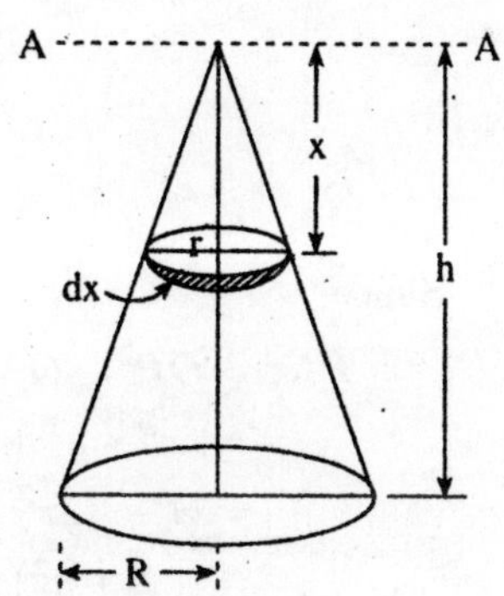

FIG. 4.14

Mass of the elementary disc is

$$\pi r^2 \rho dx = \pi \left(\frac{R}{h}\right)^2 x^2 \frac{3M}{\pi R^2 h} dx = \frac{3M}{h^3} x^2 dx \tag{3}$$

By the theorem of parallel axes, the moment of inertia of this elementary disc about the axis AA = its moment of inertia about the diameter parallel to AA + its mass $\times\ x^2$

$$= \left(\frac{3M}{h^3} x^2 dx\right) \frac{1}{4} \left(\frac{R}{h} x\right)^2 + \left(\frac{3M}{h^3} x^2 dx\right) \times x^2$$

$$= \frac{3M}{h^3} \left(\frac{1}{4}\frac{R^2}{h^2} + 1\right) x^4 dx \tag{4}$$

Here we have used Eqn (4.18). Hence, the moment of inertia of the whole cone about the axis AA is

$$\frac{3M}{h^3}\left(\frac{1}{4}\frac{R^2}{h^2} + 1\right)\int_0^h x^4 dx = \frac{3M}{h^3}\left(\frac{1}{4}\frac{R^2}{h^2} + 1\right)\frac{h^5}{5} = \frac{3}{5}M\left(\frac{R^2}{4} + h^2\right)$$

4.6 Moment of Inertia about a Line

Consider a particle of mass m' located at a point $P(x, y, z)$ with respect to a rectangular frame of axes. We are to determine the moment of inertia of the body about a line OQ having direction cosines (l, m, n).

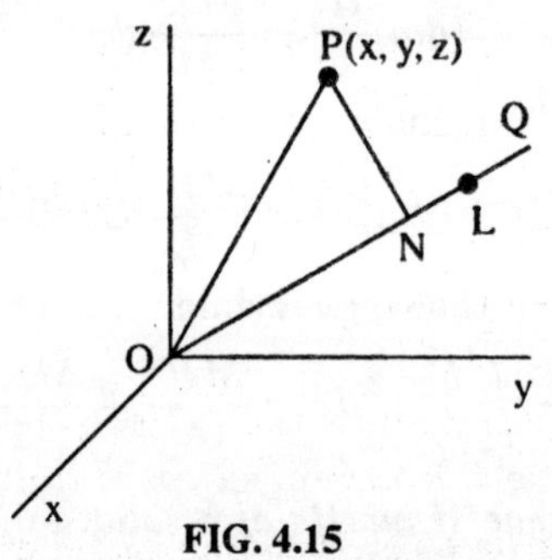

FIG. 4.15

We draw a perpendicular PN from P on the line OQ (FIG. 4.15). Then we have

$$l^2 + m^2 + n^2 = 1 \tag{4.31}$$

$$\text{and } ON = \text{projection of } OP \text{ on } OQ$$

$$= lx + my + nz \tag{4.32}$$

Hence,

$$\begin{aligned} PN^2 &= OP^2 - ON^2 \\ &= (x^2 + y^2 + z^2) - (lx + my + nz)^2 \\ &= (1 - l^2)x^2 + (1 - m^2)y^2 + (1 - n^2)z^2 - 2lmxy - 2mnyz - 2nlzx \\ &= (m^2 + n^2)x^2 + (l^2 + n^2)y^2 + (l^2 + m^2)z^2 - 2lmxy - 2mnyz - 2nlzx \\ &= l^2(y^2 + z^2) + m^2(z^2 + x^2) + n^2(x^2 + y^2) - 2lmxy - 2mnyz - 2nlzx \end{aligned} \tag{4.33}$$

Hence, the moment of inertia of the body about OQ is

$$\begin{aligned}I &= \Sigma m' PN^2 \\ &= \Sigma m' \left[l^2(y^2+z^2)+m^2(z^2+x^2)+n^2(x^2+y^2)-2lmxy-2mnyz-2nlzx\right] \\ &= l^2\Sigma m'(y^2+z^2)+m^2\Sigma m'(z^2+x^2)+n^2\Sigma m'(x^2+y^2) \\ &\quad -2lm\Sigma m'xy-2mn\Sigma m'yz-2nl\Sigma m'zx \\ &= l^2 I_{xx}+m^2 I_{yy}+n^2 I_{zz}+2lmI_{xy}+2mnI_{yz}+2nlI_{zx} \qquad (4.34)\end{aligned}$$

where I_{xx}, I_{yy}, I_{zz} are the moments of inertia of the system about x-, y- and z-axes respectively, and I_{xy}, I_{yz}, I_{zx} are the products of inertia of the system referred to $x-y$, $y-z$ and $z-x$ axes respectively.

4.7 Momental Ellipsoid

We have seen that the moment of inertia of a system about a line OQ having direction cosines l, m, n is given by Eqn (4.34).

Now we choose a point $L(x, y, z)$ on OQ such that:

(i) $OL = r$, so that

$$x = lr, \; y = mr \text{ and } z = nr \qquad (4.35)$$

and (ii) $I \propto \dfrac{1}{r^2}$, so that

$$I = \frac{MK^4}{r^2} \qquad (4.36)$$

where M is the total mass and K is the radius of gyration.

Then, from Eqns (4.34)–(4.36), we find

$$l^2 I_{xx}+m^2 I_{yy}+n^2 I_{zz}+2lmI_{xy}+2mnI_{yz}+2nlI_{zx} = \frac{MK^4}{r^2}$$

$$\text{or, } I_{xx}x^2+I_{yy}y^2+I_{zz}z^2+2I_{xy}xy+2I_{yz}yz+2I_{zx}zx = MK^4 = \text{constant}, \qquad (4.37)$$

which is the locus of the point L.

Eqn (4.37) represents an ellipsoid, since the coefficients of x^2, y^2 and z^2 are essentially positive. This ellipsoid is called the ***momental ellipsoid*** of the system about the point O.

From properties of an ellipsoid we know that Eqn (4.37) can always be reduced to the form:

$$Ax^2+By^2+Cz^2 = \text{constant} \qquad (4.38)$$

with respect to three mutually perpendicular diameters as axes of coordinates. These new axes of coordinates are called ***principal axes of coordinates***. Thus:

> **A set of mutually perpendicular axes at any point are said to be the principal axes if the products of inertia of a given system about them taken in pairs vanish.**

The coefficients A, B, C in Eqn (4.38) represent the moments of inertia of the system about the three principal axes of coordinates and are known as ***principal moments of inertia***.

4.8 Setting the Principal Axes

If we take a line passing through the centre of mass of the system as one axis of coordinates, say x-axis, and any two other mutually perpendicular lines which are at right angles with the former as y- and z-axes, then the coordinates of the centre of mass will be $(x, 0, 0)$. Hence, the products of inertia about these axes taken in pairs will be zero and as such these axes will be the principal axes.

In this way we can set the principal axes in a system. The task becomes simpler in the case of symmetrical bodies. We consider below some simple cases:

(i) Sphere

A sphere is symmetrical about any three mutually perpendicular lines with centre of mass as the origin. These lines will be the principal axes.

(ii) Rectangular Parallelepiped

If we take origin at the centre of mass, then the axes parallel to its sides will be the principal axes. If we take the origin at a corner, then the line joining the opposite corner may be taken as x-axis and any two other mutually perpendicular lines which are also at right angles with the diagonal may be taken as y- and z-axes. Then these axes will be the principal axes at the corner of the body.

(iii) Circular Cylinder

Any rectangular coordinate system with the axis of the cylinder taken as one coordinate axis will constitute principal axes.

(iv) Right Circular Cone

Here also the principal axes are obtained by taking the axis of the cone as one coordinate axis, other two coordinate axes being mutually perpendicular and also at right angles with the axis of the cone.

EXAMPLE 4.6 *Show that the equation of the momental ellipsoid at the corner of a cube of mass M and side 2a referred to its principal axes is given by*

$$2x^2 + 11(y^2 + z^2) = \text{constant}.$$

[DELHI (HONS) 1963]

Solution. Let G be the centre of mass of the cube. We take x-axis along the line OG (FIG. 4.16). The distance OG is equal to $\sqrt{3}a$. The z- and y-axes are taken as shown in the figure. Then the coordinates of G in this system is $(\sqrt{3}a, 0, 0)$. Hence, the products of inertia of the cube about these axes taken in pairs are zero. So the axes chosen are principal axes.

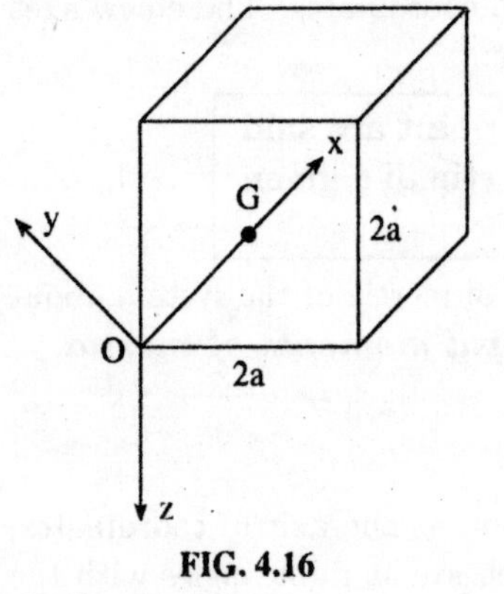

FIG. 4.16

The moments of inertia of the cube about axes through G parallel to its edges are

$$I_{xx} = I_{yy} = I_{zz} = \frac{2}{3}Ma^2 \tag{1}$$

$$\text{and } I_{xy} = I_{yz} = I_{zx} = 0 \tag{2}$$

Hence, the moment of inertia about any line through G having direction cosines l, m, n will be

$$I = \frac{2}{3}Ma^2(l^2 + m^2 + n^2) = \frac{2}{3}Ma^2 \tag{3}$$

which is constant.

Thus, the moment of inertia of the cube about OG is

$$A = \frac{2}{3}Ma^2 \tag{4}$$

The moment of inertia of the cube about y-axis is

$$B = \frac{2}{3}Ma^2 + M(OG)^2 = \frac{11}{3}Ma^2 \tag{5}$$

The moment of inertia of the cube about z-axis is

$$C = \frac{11}{3}Ma^2 \tag{6}$$

Hence, the equation of the momental ellipsoid at O is

$$Ax^2 + By^2 + Cz^2 = \text{constant}$$

$$\text{or, } \frac{2}{3}Ma^2x^2 + \frac{11}{3}Ma^2y^2 + \frac{11}{3}Ma^2z^2 = \text{constant}$$

$$\text{or, } 2x^2 + 11(y^2 + z^2) = \text{constant.}$$

NOTE. We have noted that the moment of inertia of a cube about any axis passing through the centre of mass is constant.

EXAMPLE 4.7 *Calculate the moment of inertia of a homogeneous right circular solid cone about its generating line.* [C.U. (HONS) 1975]

Solution. Let M = mass, R = radius and h = height of the cone. We take a rectangular coordinate system with origin at the vertex and the axis of the cone as x-axis (FIG. 4.17). Since the centre of mass of the cone lies on x-axis, the products of inertia about these axes taken in pairs will be zero. Hence, the coordinate axes will constitute the principal axes.

The moment of inertia of the cone about x-axis is

$$A = \frac{3}{10}MR^2 \tag{1}$$

The moments of inertia of the cone about y- and z-axes are

$$B = C = \frac{3M}{5}\left(\frac{R^2}{4} + h^2\right) \tag{2}$$

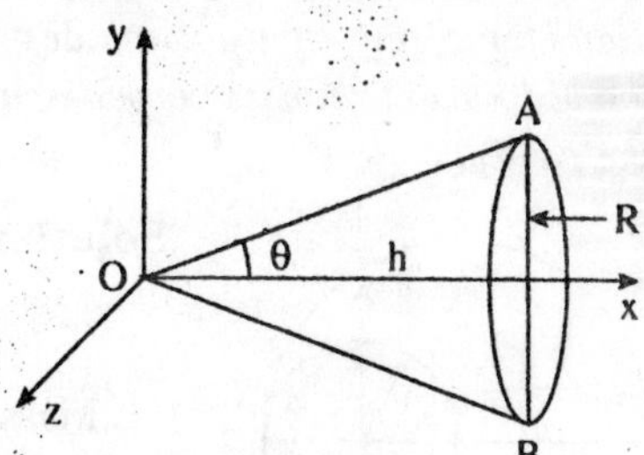

FIG. 4.17

Now we are to find the moment of inertia about the generator OA lying on xy-plane. The direction cosines of the line OA are

$$l = \cos\theta = \frac{h}{\sqrt{R^2 + h^2}}, \quad m = \sin\theta = \frac{R}{\sqrt{R^2 + h^2}}, \quad n = 0 \tag{3}$$

Hence, the moment of inertia about OA axis is

$$\begin{aligned} I &= Al^2 + Bm^2 + Cn^2 \\ &= \frac{3}{10}MR^2\frac{h^2}{R^2 + h^2} + \frac{3}{5}M\left(\frac{R^2}{4} + h^2\right)\frac{R^2}{R^2 + h^2} \\ &= \frac{3}{10}\frac{MR^2h^2}{R^2 + h^2}(1 + 2) + \frac{3}{20}M\frac{R^4}{R^2 + h^2} \\ &= \frac{3}{20}\frac{MR^2}{R^2 + h^2}(R^2 + 6h^2) \\ &= \frac{3}{20}MR^2\left(1 + \frac{5h^2}{R^2 + h^2}\right) \end{aligned}$$

EXAMPLE 4.8 *Find the equation of momental ellipsoid of a beam of mass M and length $2l$ at one of its ends.*

Solution. We take x-axis along the beam (FIG. 4.18). OA is the beam of length $2l$ and mass M. Taking an element dx at a distance x from origin, the moment of inertia about z-axis is

$$C = \int_0^{2l} \frac{M}{2l} x^2 dx = \frac{M}{2l}\left(\frac{x^3}{3}\right)_0^{2l} = \frac{4}{3}Ml^2 \tag{1}$$

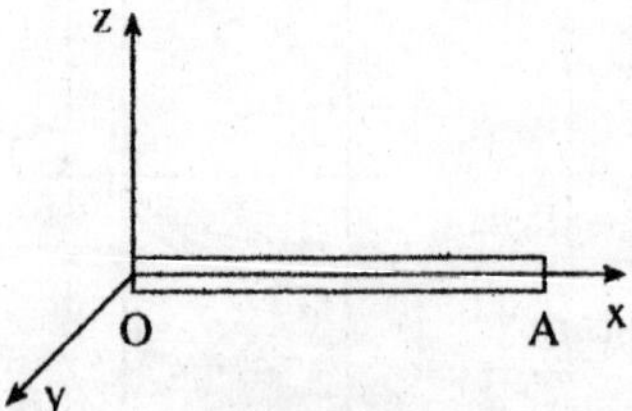

FIG. 4.18

The moment of inertia of the beam about y-axis is also

$$B = \frac{4}{3}Ml^2 \tag{2}$$

The moment of inertia about x-axis is

$$A = 0 \tag{3}$$

The products of inertia will be zero since the coordinates of the centre of mass are $(l, 0, 0)$. Hence, the equation of momental ellipsoid of the beam at O is

$$By^2 + Cz^2 = \text{constant}$$

$$\text{or, } \frac{4}{3}Ml^2(y^2 + z^2) = \text{constant}$$

$$\text{or, } y^2 + z^2 = \text{constant,}$$

which represents a right circular cylinder with x-axis as its own axis.

EXAMPLE 4.9 *A sphere of radius $2r$ and density ρ has a spherical cavity of radius r. The diameter of the cavity coincides with the radius of the sphere. Find the moment of inertia of the body about an axis perpendicular to the common diameter and passing through the centre of the cavity.* **[C.U. (HONS.)]**

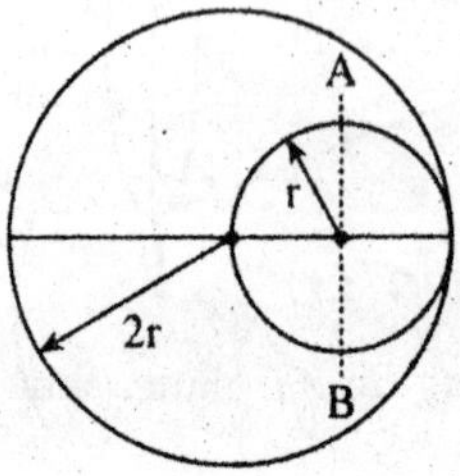

FIG. 4.19

Solution. Mass of the sphere

$$M = \frac{4}{3}\pi(2r)^3\rho = \frac{32}{3}\pi r^3 \rho. \tag{1}$$

Mass of the cavity

$$M_c = \frac{4}{3}\pi r^3 \rho. \tag{2}$$

Moment of inertia of the sphere about AB axis (FIG. 4.19)

$$I_1 = \frac{2}{5}M(2r)^2 + Mr^2 = \frac{13}{5}Mr^2 = \frac{13}{5}\cdot\frac{32}{3}\pi\rho r^5. \tag{3}$$

Moment of inertia of the cavity about AB axis

$$I_2 = \frac{2}{5}M_c r^2 = \frac{2}{5}\cdot\frac{4}{3}\pi\rho r^5 \tag{4}$$

Hence, required moment of inertia

$$I = I_1 - I_2 = \frac{13}{5}\cdot\frac{32}{3}\pi\rho r^5 - \frac{8}{15}\pi\rho r^5 = \frac{8}{15}(52-1)\pi\rho r^5 = \frac{136}{5}\pi\rho r^5.$$

EXAMPLE 4.10 *The density of a cylinder of length L from the centre of its base is given by*

$$\rho = \rho_0\left(1 - \frac{z}{2L}\right).$$

Find the moment of inertia of the cylinder about its axis. Also prove that it is equal to $\frac{1}{2}MR^2$, M = mass of the cylinder.

Solution. We take an elementary disc of thickness dz at a height z from the base (FIG. 4.20). Mass of this disc is

$$m = \pi R^2 \rho dz = \pi R^2 \rho_0 \left(1 - \frac{z}{2l}\right) dz, \quad (1)$$

where R = radius of the cylinder.

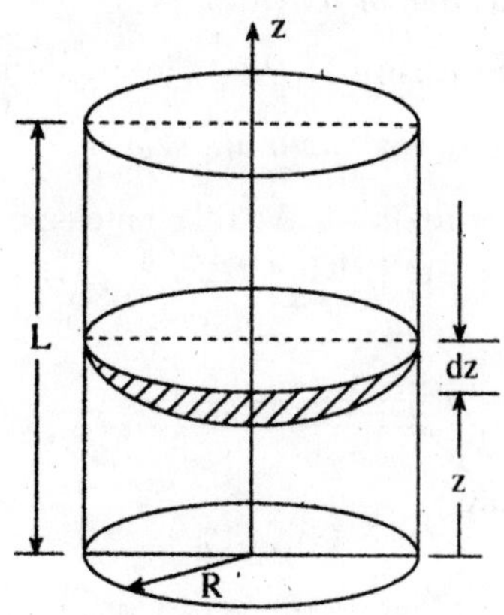

FIG. 4.20

Moment of inertia of this disc about z-axis

$$dI = \frac{1}{2} m R^2 = \frac{1}{2}\left\{\pi R^2 \rho_0 \left(1 - \frac{z}{2L}\right) dz\right\} R^2$$

$$= \frac{1}{2}\pi R^4 \rho_0 \left(1 - \frac{z}{2L}\right) dz \quad (2)$$

Hence, the moment of inertia of the whole cylinder is

$$I = \frac{1}{2}\pi R^4 \rho_0 \int_0^L \left(1 - \frac{z}{2L}\right) dz$$

$$= \frac{1}{2}\pi R^4 \rho_0 \left(L - \frac{L}{4}\right) = \frac{3}{8}\pi R^4 \rho_0 L \quad (3)$$

Now, mass of the whole cylinder is

$$M = \pi R^2 \rho_0 \int_0^L \left(1 - \frac{z}{2L}\right) dz = \pi R^2 \rho_0 \cdot \left(\frac{3}{4}\right) L \quad (4)$$

From (3) and (4), we get

$$I = \frac{1}{2} M R^2. \quad (5)$$

Hence proved.

EXAMPLE 4.11 *The density ρ at a point of a sphere of radius R is given by*

$$\rho = a + br^\lambda,$$

where a, b and λ are constants. Find the moment of inertia of the sphere about a diameter.

Solution. We take an elementary disc of thickness dr at a distance r from the centre. The plane of the disc is perpendicular to the diameter AB about which the moment of inertia is required (FIG. 4.21).

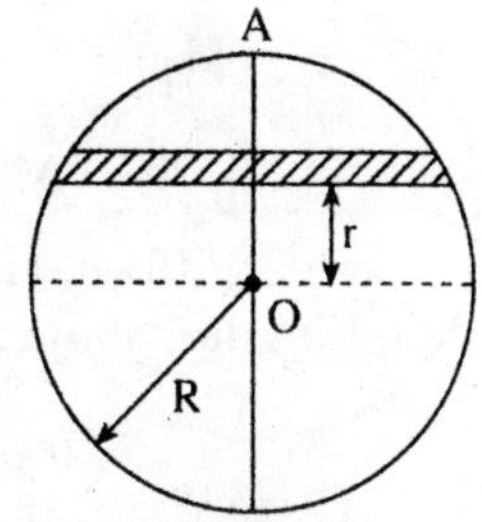

FIG. 4.21

Mass of the disc

$$m = \pi(R^2 - r^2)dr \cdot \rho = \pi(R^2 - r^2)(a + br^\lambda)dr \quad (1)$$

Moment of inertia of this disc about AB

$$dI = \frac{1}{2} m (R^2 - r^2) = \frac{1}{2}\pi \left(R^2 - r^2\right)^2 (a + br^\lambda)dr \quad (2)$$

Hence, moment of inertia of the entire sphere is

$$I = \frac{1}{2}\pi \int_{-R}^{R} (R^4 - 2r^2R^2 + r^4)(a + br^\lambda)dr$$

$$= \pi \int_0^R a(R^4 - 2r^2R^2 + r^4)dr + b\pi \int_0^R \left(R^4 r^\lambda - 2R^2 r^{\lambda+2} + r^{\lambda+4}\right) dr$$

$$= \pi a \left[R^5 - \frac{2}{3}R^5 + \frac{R^5}{5}\right] + \pi b \left[R^4 \cdot \frac{R}{\lambda+1} - 2R^2 \cdot \frac{R^{\lambda+3}}{\lambda+3} + \frac{R^{\lambda+5}}{\lambda+5}\right]$$

$$= \frac{8}{15}\pi a R^5 + \pi b R^{\lambda+5}\left[\frac{1}{\lambda+1} - \frac{2}{\lambda+3} + \frac{1}{\lambda+5}\right]$$

EXAMPLE 4.12 *Show that the equation of the momental ellipsoid at the centre of an elliptic lamina is*

$$\frac{x^2}{a^2} + \frac{y^2}{b^2} + z^2\left(\frac{1}{a^2} + \frac{1}{b^2}\right) = constant,$$

where a and b are semi-major and semi-minor axes.

Solution. We take an elementary strip of length $2y$ and thickness dx at a distance x from the centre (FIG. 4.22).

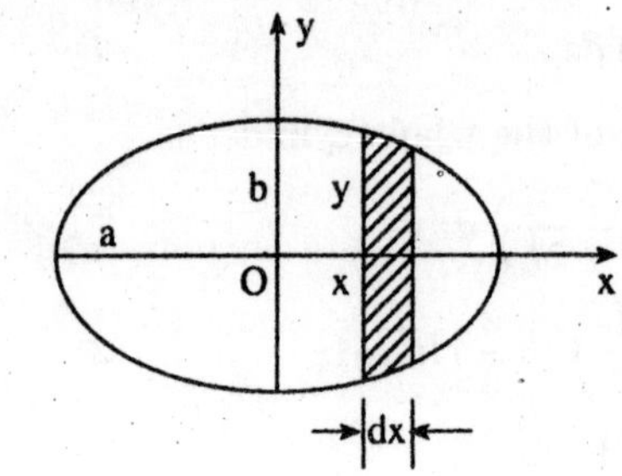

FIG. 4.22

Area of the strip $= 2ydx$ and its mass m is $\rho(2ydx) = \dfrac{M}{\pi ab}(2ydx)$, where M = mass of the lamina.

Moment of inertia of this strip about x-axis is

$$dA = \frac{1}{3}my^2 = \frac{1}{3}\frac{M}{\pi ab}(2ydx)y^2 = \frac{2M}{3\pi ab}y^3dx \tag{1}$$

Equation of the ellipse is

$$\frac{x^2}{a^2} + \frac{y^2}{b^2} = 1 \tag{2}$$

This yields

$$y = \frac{b}{a}\sqrt{a^2 - x^2} \tag{3}$$

Inserting (3) in (1) and integrating within proper limits, we find the moment of inertia of the entire disc about x-axis

$$A = \frac{2M}{3\pi ab}\int_{-a}^{a}\frac{b^3}{a^3}(a^2 - x^2)^{3/2}dx = \frac{2M}{3\pi ab}\cdot 2\int_{0}^{a}\frac{b^3}{a^3}(a^2 - x^2)^{3/2}dx \tag{4}$$

To integrate this, we put

$$x = \sin\theta, \text{ i.e., } dx = \cos\theta d\theta \tag{5}$$

Then

$$A = \frac{4M}{3\pi ab}\int_{0}^{\pi/2}\frac{b^3}{a^3}\cdot a^4\cos^4\theta d\theta = \frac{4M}{3\pi}b^2\left(\frac{3\cdot 1}{4\cdot 2}\cdot\frac{\pi}{2}\right) = \frac{1}{4}Mb^2 \tag{6}$$

Similarly, the moment of inertia of the lamina about y-axis is

$$B = \frac{1}{4}Ma^2 \tag{7}$$

and about z-axis is

$$C = A + B = \frac{1}{4}M(a^2 + b^2) \tag{8}$$

Hence, the equation of the momental ellipsoid at the centre is

$$Ax^2 + By^2 + Cz^2 = \text{constant}$$

$$\text{i.e., } \frac{1}{4}M\left[b^2x^2 + a^2y^2 + (a^2+b^2)z^2\right] = \text{constant}$$

$$\therefore \frac{x^2}{a^2} + \frac{y^2}{b^2} + \left(\frac{1}{a^2} + \frac{1}{b^2}\right)z^2 = \text{constant. } \textbf{(Proved)}$$

EXERCISE

1. Find the moment of inertia of a solid rectangular parallelepiped of mass M and sides $2a, 2b, 2c$ about an axis through the centre and parallel to one of its edges.
2. Find the moments of inertia of an elliptical disc about its axes.
3. Show that the moment of inertia of a circular ring of mass M and radius R about a tangent in its plane is $\frac{3}{2}MR^2$.
4. Find the moment of inertia of a hollow circular cylinder of mass M with internal and external radii a, b respectively about an axis passing through its centre of mass and perpendicular to its length.
5. Show that the moment of inertia of a triangular lamina of mass M and height h about its base is
$$\frac{1}{6}Mh^2$$
6. Find the moment of inertia of the area bounded by the curve $r^2 = a^2 \cos 2\theta$ about its axis. [DELHI (HONS) 1960]
7. Find the moments of inertia of a hollow circular cone of mass M and radius R about (i) its axis of symmetry and (ii) an axis passing through the vertex and perpendicular to axis of symmetry.
8. Show that the moment of inertia of a thin hemispherical shell of mass M and radius R about the diameter of circular base is $\frac{2}{3}MR^2$.
9. Show that the moment of inertia of a rectangular plate of mass M and sides a, b about a diagonal is
$$\frac{1}{6}Ma^2b^2/(a^2+b^2)$$
10. Find the moment of inertia of a solid hemisphere of mass M and radius R about (i) the diameter at circular base and (ii) the vertical axis passing through its centre and centre of mass.
11. Find the equation of the momental ellipsoid at a point on the rim of a solid hemisphere.
12. Find the equation of the momental ellipsoid at a point on the circular edge of a solid cone.
13. Set up principal axes and principal moments of inertia of a square plate of mass m and sides of length a. [C.U. (HONS) 2002]
14. Show that, for a homogeneous sphere (i) any three mutually perpendicular diameters form a set of principal axis and (ii) the ellipsoid of inertia about the centre is a sphere [C.U. (HONS) 1997]

15. Show that the moment of inertia of three equal masses 120° apart as shown in the FIG. 4.23 about any axis in its plane is $\frac{3}{2}ml^2$.

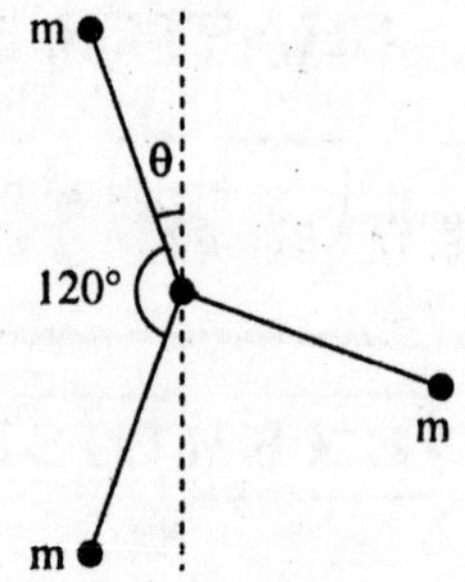

FIG. 4.23

16. Show that the moment of inertia of a triangular lamina of mass M and height h about its base is $\frac{1}{6}Mh^2$ and about an axis through its centre of mass is $\frac{1}{18}Mh^2$.

CHAPTER 5

Plane Motion of Rigid Bodies

5.1 Introduction

A rigid body is a system of particles whose mutual distances do not change regardless of the forces acting on it. When a rigid body moves from one position to another, it undergoes two types of motion: (1) rotation about a fixed axis and (2) translation of the centre of mass. In the first case, the rigid body has only one degree of freedom and in the second case, it has two degrees of freedom. In the translational motion, we consider the total mass M of the rigid body to be concentrated at the centre of mass and all the external forces are transferred parallel to themselves at the centre of mass. For rotational motion, the sum of the moments of effective forces about the centre of mass should be equal to the sum of moments of external forces.

We intend to discuss both the types of motion in a plane.

5.2 Motion about a Fixed Axis

Consider a rigid body of mass M rotating about a fixed axis AB with uniform angular velocity ω (FIG. 5.1). A particle of mass m_i at distance r_i from the axis will describe a circle with linear velocity $v_i = \omega r_i$. Then the ***kinetic energy*** of the particle due to rotation about the axis AB is

$$\frac{1}{2}m_i v_i^2 = \frac{1}{2}m_i \omega^2 r_i^2 \tag{5.1}$$

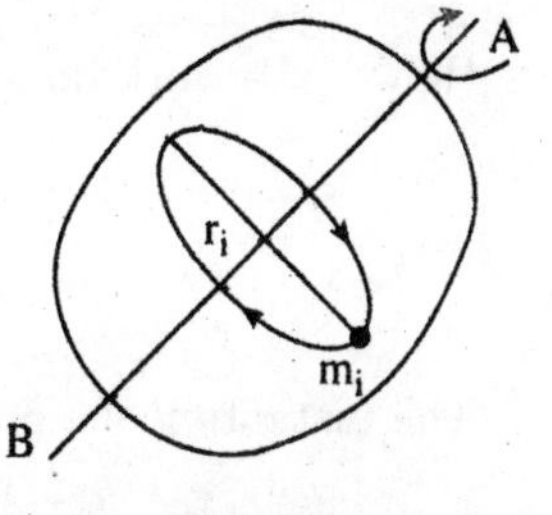

FIG. 5.1

Hence, the total kinetic energy of the rigid body is

$$T = \Sigma\frac{1}{2}m_i\omega^2 r_i^2 = \frac{1}{2}\omega^2\Sigma m_i r_i^2$$

$$\text{or, } T = \frac{1}{2}I\omega^2 \tag{5.2}$$

where I is the moment of inertia of the body about AB.

Again, the linear momentum of the particle m_i being $p_i = m_i\omega r_i$, its ***angular momentum*** about axis $AB = m_i\omega r_i^2$. Hence, the total angular momentum of the rigid body is

$$J = \Sigma m_i\omega r_i^2 = \omega\Sigma m_i r_i^2 = I\omega \tag{5.3}$$

Let N be the ***total torque*** or moment of all the external forces about axis AB. Since a rigid body is a particular case of a system of particles, by the principle of angular momentum,

$$N = \frac{dJ}{dt} = \frac{d}{dt}(I\omega) = I\dot{\omega} \tag{5.4}$$

If the net torque $N = 0$, then from Eqn (5.4),

$$\frac{d}{dt}(I\omega) = 0, \text{ i.e., } I\omega = \text{constant}.$$

Thus, if the net torque is zero, the angular momentum $I\omega$ of the body is conserved.

If the forces acting on the system be conservative, then the principle of conservation of energy holds good. Thus, if T and V be the total kinetic and potential energies of the system, then

$$T + V = \text{constant} = E, \text{ i.e., } \frac{1}{2}I\omega^2 + V = E \tag{5.5}$$

• Work and Power

Consider a rigid body which is rotating in a plane about an axis through a point O perpendicular to the plane (FIG. 5.2).

Let P be any point with position vector $\mathbf{r}$ at time t. After time $t + dt$, the point rotates by an angle $d\theta$ under the action of external force $\mathbf{F}$. If ω be the angular velocity of the body, then

$$\omega = \frac{d\theta}{dt} = \dot{\theta} \tag{5.6}$$

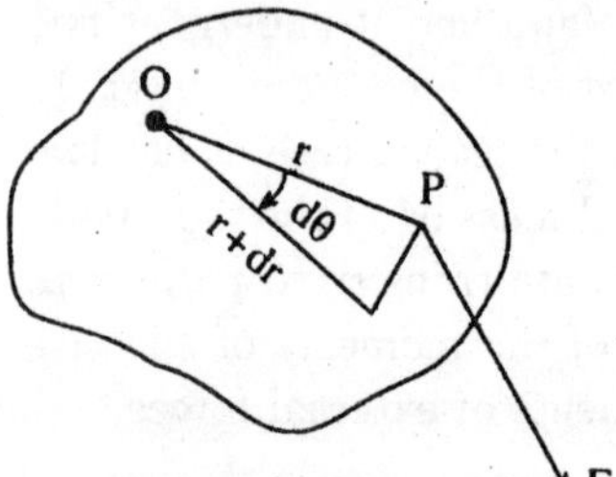

FIG. 5.2

Taking a unit vector $\mathbf{k}$ in the direction of the axis of rotation, we write:

$$\boldsymbol{\omega} = \omega\mathbf{k}, \mathbf{N} = N\mathbf{k} \tag{5.7}$$

where $\mathbf{N}$ is the torque acting on the body under the influence of the external force $\mathbf{F}$.

Then the work done by the force is

$$dW = \mathbf{F}\cdot d\mathbf{r} = \mathbf{F}\cdot\frac{d\mathbf{r}}{dt}dt = \mathbf{F}\cdot\mathbf{v}dt \tag{5.8}$$

where $\mathbf{v}$ is the linear velocity and is given by

$$\mathbf{v} = \boldsymbol{\omega}\times\mathbf{r} \tag{5.9}$$

Hence, the work done is

$$\begin{aligned} dW &= \mathbf{F}\cdot(\boldsymbol{\omega}\times\mathbf{r})dt = (\mathbf{r}\times\mathbf{F})\cdot\boldsymbol{\omega}dt \\ &= \mathbf{N}\cdot\boldsymbol{\omega}dt = N\mathbf{k}\cdot\omega\mathbf{k}dt = N\omega dt \end{aligned}$$

$$\text{i.e., } dW = Nd\theta \text{ [using Eqn (5.6)]} \tag{5.10}$$

The instantaneous power developed is

$$P = \frac{dW}{dt} = N\omega \tag{5.11}$$

The total work done in rotating a rigid body from an angle θ_1 to an angle θ_2, from Eqn (5.10), is

$$W = \int_{\theta_1}^{\theta_2} Nd\theta \tag{5.12}$$

Since $N = I\dot{\omega}$, we write Eqn (5.12) as

$$W = \int_{\theta_1}^{\theta_2} I\frac{d\omega}{dt}d\theta = \int_{\theta_1}^{\theta_2} I\omega d\omega \tag{5.13}$$

Let ω_1 and ω_2 be the angular velocities at the angles of rotations θ_1 and θ_2 respectively. Then Eqn (5.13) can be written as

$$W = \int_{\omega_1}^{\omega_2} I\omega d\omega = \frac{1}{2} I \left(\omega_2^2 - \omega_1^2\right) \tag{5.14}$$

Eqn (5.14) represents the difference in kinetic energies of the rigid body at ω_1 and ω_2. Thus, ***the total work done in rotating a rigid body about an axis from an angle θ_1 with angular speed ω_1 to the angle θ_2 with angular speed ω_2 is the difference in the kinetic energies of rotation at the two positions.***

5.3 Compound Pendulum

The compound pendulum is a rigid body which is free to oscillate in a vertical plane about a fixed horizontal axis under the influence of gravity.

Let C be the centre of mass of the rigid body of mass M which oscillates about a horizontal axis through the point O (FIG. 5.3). OA is the vertical line through O. Let $OC = h$. The only force acting on the body is its weight Mg acting vertically downwards. Then the total external torque about the axis of rotation through O is

$$N = -Mgh \sin\theta \tag{5.15}$$

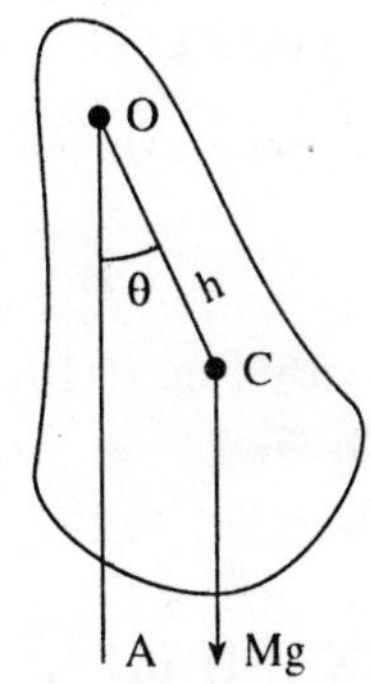

FIG. 5.3

Substituting this into Eqn (5.4), we get

$$I\dot{\omega} = -Mgh \sin\theta \tag{5.16}$$

where I = moment of inertia of the body about the axis of rotation and ω = angular speed. But $\omega = \dot{\theta}$.

Hence, Eqn (5.16) assumes the form:

$$I\ddot{\theta} + Mgh \sin\theta = 0 \tag{5.17}$$

For small oscillations, $\sin\theta \simeq \theta$. Hence, Eqn (5.17) is reduced to

$$\ddot{\theta} + \frac{Mgh}{I}\theta = 0 \tag{5.18}$$

which represents a simple harmonic motion. The period of oscillation is given by

$$P = 2\pi\sqrt{\frac{I}{Mgh}} \tag{5.19}$$

The period of oscillation of a simple pendulum of length l is

$$P = 2\pi\sqrt{\frac{l}{g}} \tag{5.20}$$

Hence, the length of the equivalent simple pendulum is

$$l = \frac{I}{Mh} \tag{5.21}$$

EXAMPLE 5.1 *In a simple pendulum, the string supporting the bob is replaced by a uniform rod of mass M and length L. Find the new time period for small oscillations. Mass of the bob = m.* **[C.U. (HONS) 2005]**

Solution. The centre of mass C of the rod is at $L/2$ from the point of suspension O (FIG. 5.4). Let x be the distance of the centre of mass of the system from C.

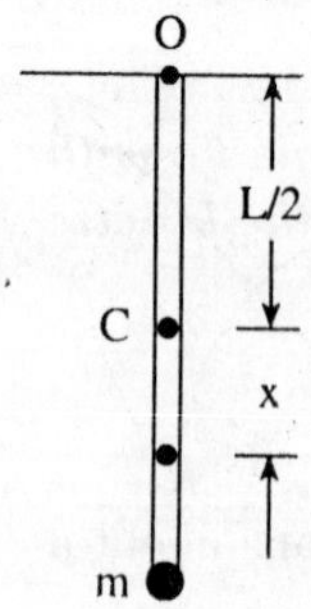

FIG. 5.4

Then

$$Mx = m\left(\frac{L}{2} - x\right), \text{ i.e., } x = \frac{m}{M+m}\frac{L}{2} \tag{1}$$

Hence, the distance of the centre of mass from O is

$$h = x + \frac{L}{2} = \left(1 + \frac{m}{M+m}\right)\frac{L}{2} = \frac{M+2m}{M+m}\frac{L}{2} \tag{2}$$

The moment of inertia of the rod about the axis of rotation through O is

$$I_1 = \frac{1}{3}ML^2 \tag{3}$$

The moment of inertia of the bob about the same axis is

$$I_2 = mL^2 \tag{4}$$

Hence, the moment of inertia of the system is

$$I = I_1 + I_2 = \frac{1}{3}ML^2 + mL^2 = \frac{L^2}{3}(M+3m) \tag{5}$$

By Eqn (5.19), the time period for small oscillation of the system is

$$P = 2\pi\sqrt{\frac{I}{(M+m)gh}} = 2\pi\sqrt{\frac{L^2(M+3m)}{3(M+2m)}\frac{2}{Lg}} = 2\pi\sqrt{\frac{2(M+3m)L}{3(M+2m)g}}.$$

EXAMPLE 5.2 *A solid homogeneous cone of mass M, height H and radius at base R oscillates about a horizontal axis through its vertex. Find the time period and also the length of the equivalent simple pendulum.*

Solution. The moment of inertia of the cone about the horizontal axis through its vertex is (see Ex. 4.5 in the previous chapter)

$$I = \frac{3M}{20}(R^2 + 4H^2) \tag{1}$$

Also, the distance of centre of mass from the vertex is $h = \frac{3}{4}H$. Hence, the period of oscillation by Eqn (5.19) is

$$P = 2\pi\sqrt{\frac{I}{Mgh}} = 2\pi\sqrt{\frac{3M(R^2+4H^2)4}{20(Mg)3H}} = 2\pi\sqrt{\frac{R^2+4H^2}{5gH}}$$

The length of the equivalent simple pendulum, by Eqn (5.20), is

$$l = \frac{I}{Mh} = \frac{3M}{20}\frac{(R^2+4H^2)}{M}\frac{4}{3H} = \frac{1}{5H}(R^2+4H^2).$$

EXAMPLE 5.3 *A rigid body rotates about an axis with direction cosines l, m, n with angular velocity ω. Show that the kinetic energy of rotation of the rigid body is given by*

$$T_{\text{rot}} = \frac{1}{2}I_{l,m,n}\omega^2$$

where $I_{l,m,n}$ is the moment of inertia of the rigid body about the axis of rotation.

[C.U. (HONS) 2002]

Solution. Let the axis of rotation AB have direction cosines l, m, n (FIG. 5.5). We consider a particle of mass m_i at a distance r_i. Due to rotation of the rigid body about the axis AB, the mass m_i will describe a circle of radius r_i with linear velocity $v_i = \omega r_i$. Hence, its kinetic energy of rotation is

$$\frac{1}{2} \cdot m_i \cdot v_i^2 = \frac{1}{2} m_i \omega^2 r_i^2 \qquad (1)$$

The total kinetic energy of the rigid body is

$$T = \Sigma \frac{1}{2} m_i \omega^2 r_i^2 = \frac{1}{2} \omega^2 \Sigma m_i r_i^2 \text{ (since } \omega = \text{constant).} \qquad (2)$$

The quantity under summation represents the moment of inertia of the rigid body about the axis with direction cosines l, m, n. Denoting it as $I_{l,m,n}$ we write Eqn (2) as

$$T = \frac{1}{2} I_{l,m,n} \omega^2. \text{ Hence proved.}$$

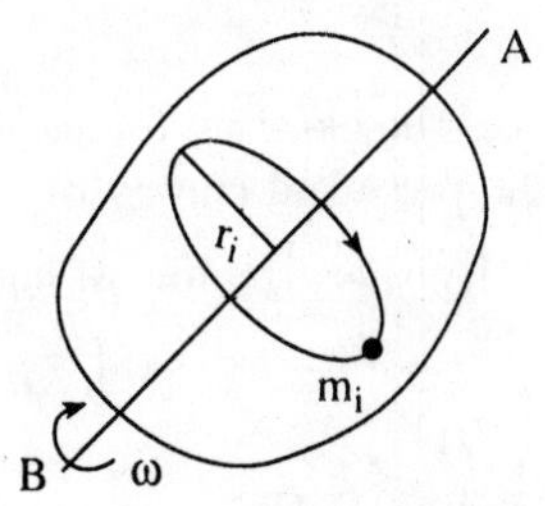

FIG. 5.5

5.4 General Plane Motion

In this case, the rigid body undergoes a translational motion parallel to a plane and a rotational motion about a suitable axis perpendicular to the plane. Hence, the total number of degrees of freedom is three.

Consider a rigid body of mass M moving under the net external force $\mathbf{F}$. Let $\mathbf{r}$ be the position vector of the centre of mass at any instant. Since the rigid body is a particular case of a system of particles, all the theorems concerning the motion of a system of particles are equally applicable to the motion of rigid bodies. Thus, we have the following important principles:

A. Principle of Linear Momentum

The momentum of the centre of mass at any instant is

$$p = M\dot{r} \qquad (5.22)$$

By the principle of linear momentum

$$\frac{d}{dt}(p) = F, \text{ i.e., } M\ddot{r} = F \qquad (5.23)$$

B. Principle of Angular Momentum

Let I_C = moment of inertia of the rigid body about an axis passing through the centre of mass and perpendicular to the plane, ω = angular velocity of rotation and N_C = total moment of external forces about the centre of mass (i.e., torque). Then by the principle of angular momentum,

$$N_C = \frac{d}{dt}(I_C \omega) = I_C \dot{\omega} \qquad (5.24)$$

C. Principle of Conservation of Energy

The total kinetic energy of the body has two parts—one due to translation and other due to rotation. The kinetic energy due to translation is

$$T_1 = \frac{1}{2} M v^2 = \frac{1}{2} M \dot{r}^2 \qquad (5.25)$$

and the kinetic energy due to rotation is

$$T_2 = \frac{1}{2} I_C \omega^2 \tag{5.26}$$

Thus, the total kinetic energy of the rigid body is

$$T = T_1 + T_2 = \frac{1}{2} M\dot{r}^2 + \frac{1}{2} I_C \omega^2 \tag{5.27}$$

If the external forces acting on the body are conservative, then a potential function V exists. The potential energy of the rigid body is V.

By the principle of conservation of energy, the total energy is

$$T + V = \frac{1}{2} M\dot{r}^2 + \frac{1}{2} I_C \omega^2 + V = E = \text{constant.} \tag{5.28}$$

EXAMPLE 5.4 *A cylinder has a mass distribution that varies with distance from its axis. It rolls without slipping on a smooth horizontal plane. It is found that its translational kinetic energy = its rotational kinetic energy. Find the moment of inertia of the cylinder. Show that the mass distribution is that of a thin cylindrical shell.* **[C.U. (HONS) 2002]**

Solution. Let the cylinder move a distance x when it has rotated by an angle θ (FIG. 5.6). Since the cylinder has rolled without slipping

$$x = R\theta, \text{ i.e., } \dot{x} = R\dot{\theta} \tag{1}$$

where R is its radius.

If M = mass of the cylinder and I = moment of inertia about its axis, then its translational kinetic energy $= \frac{1}{2} M\dot{x}^2$ and rotational kinetic energy $= \frac{1}{2} I\dot{\theta}^2$. Hence, by condition,

$$\frac{1}{2} M\dot{x}^2 = \frac{1}{2} I\dot{\theta}^2 = \frac{1}{2} I \left(\frac{\dot{x}}{R}\right)^2 \quad \text{[by Eqn (1)].}$$

Hence,

$$I = MR^2 \tag{2}$$

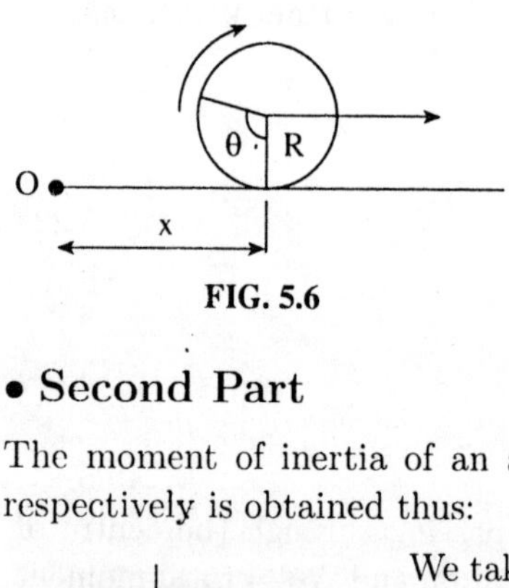

FIG. 5.6

• Second Part

The moment of inertia of an annular disc of mass m, and internal and external radii a, R respectively is obtained thus:

We take a ring of radius x, of thickness dx. Then its mass is (FIG. 5.7)

$$2\pi x \rho = 2\pi x \cdot \frac{m}{\pi(R^2 - a^2)} \tag{3}$$

The moment of inertia of this elementary ring about the axis passing through O and perpendicular to its plane is

$$\frac{2mx}{R^2 - a^2} \cdot x^2 = \frac{2m}{R^2 - a^2} x^3 \tag{4}$$

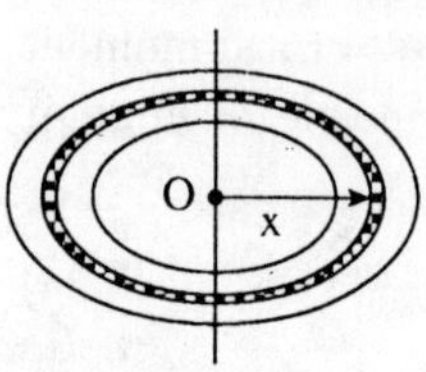

FIG. 5.7

Hence, the moment of inertia of the whole disc is

$$\frac{2m}{R^2 - a^2} \int_a^R x^3 dx = \frac{m(R^4 - a^4)}{2(R^2 - a^2)} = \frac{m}{2}(R^2 + a^2) \tag{5}$$

Now a hollow cylinder can be assumed to consist of a large number of annular discs placed one above the other. Hence, the moment of inertia of a hollow cylinder about its axis is

$$I = \Sigma m \left(\frac{R^2 + a^2}{2}\right) = \frac{R^2 + a^2}{2} \Sigma m = \frac{1}{2} M \left(R^2 + a^2\right) \quad (6)$$

If the cylinder is very thin, then $a = R$. Hence,

$$I = \frac{1}{2} M (R^2 + R^2) = MR^2. \quad (7)$$

From Eqns (2) and (7), we find that the moment of inertia of the given cylinder is the same as that of a hollow thin cylindrical shell of same radius. Thus, their mass distributions are same.

EXAMPLE 5.5 *A yo-yo consists of a cylinder of mass* 90 gm *around which a massless string of length* 30 cm *is wound. If one end of the string is kept fixed and the yo-yo is allowed to fall vertically starting from rest, find its speed when it reaches the end of the string. Show that the speed is independent of the mass and radius of the yo-yo* [$g = 980$ cm/sec^2]. **[C.U. (HONS) 1997]**

Solution. Let M = mass and R = radius of the yo-yo. Let x be the distance fallen after time t (FIG. 5.8) and θ be the angle of rotation. Then

$$x = R\theta, \text{ i.e., } \ddot{x} = R\ddot{\theta} \quad (1)$$

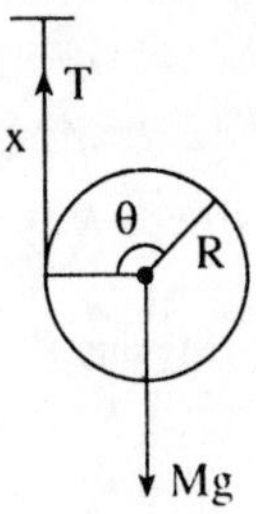

FIG. 5.8

Equation for translational motion is

$$M\ddot{x} = Mg - T \quad (2)$$

where T is the tension in the string.

Equation for rotational motion is

$$I\ddot{\theta} = T \cdot R \text{ or, } \frac{1}{2} MR^2 \left(\frac{\ddot{x}}{R}\right) = TR$$

$$\therefore \; M\ddot{x} = 2T \quad (3)$$

From Eqns (2) and (3), by eliminating T, we get

$$\frac{3}{2} M\ddot{x} = Mg \text{ or, } \ddot{x} = \frac{2}{3} g \quad (4)$$

Integrating once with condition $\dot{x} = 0$ at $t = 0$, we get

$$\dot{x} = \frac{2}{3} gt \quad (5)$$

Eqn (5) shows that the speed is independent of mass and radius of the yo-yo.

Integrating Eqn (5) once more with condition $x = 0$ at $t = 0$,

$$x = \frac{1}{3} gt^2 \quad (6)$$

If l = total length of string, the time taken to reach its end is given by

$$l = \frac{1}{3} gt^2, \text{ i.e., } t = \sqrt{\frac{3l}{g}} \quad (7)$$

The speed at this time is, from Eqn (5),

$$v = \frac{2}{3} gt = \frac{2}{3} g \sqrt{\frac{3l}{g}} = 2\sqrt{\frac{gl}{3}} \quad (8)$$

We are given here $l = 30$ cm, $g = 980$ cm/sec^2. Hence,

$$v = 2 \times \sqrt{\frac{980 \times 30}{3}} = 19.8 \text{ cm/sec.}$$

EXAMPLE 5.6 *A uniform sphere of mass M and radius R rolls without slipping down an inclined plane. Show that its acceleration is constant and is given by*

$$\frac{R^2}{R^2 + K^2} g \sin\alpha$$

where K = radius of gyration about its axis and α = angle of inclination of the plane. Also find the speed of the sphere.

Solution. Let x be distance along the plane traversed by the centre of mass (FIG. 5.9). Then

$$x = R\theta, \text{ i.e., } \ddot{x} = R\ddot{\theta} \tag{1}$$

Let N = normal reaction on the plane and f = frictional force.

Then equation for translational motion is

$$M\ddot{x} = Mg\sin\alpha - f \tag{2}$$

and equation for rotational motion is

$$MK^2\ddot{\theta} = f \cdot R$$

$$\text{i.e., } MK^2\frac{\ddot{x}}{R} = f \cdot R$$

$$\text{i.e., } MK^2\ddot{x} = fR^2 \tag{3}$$

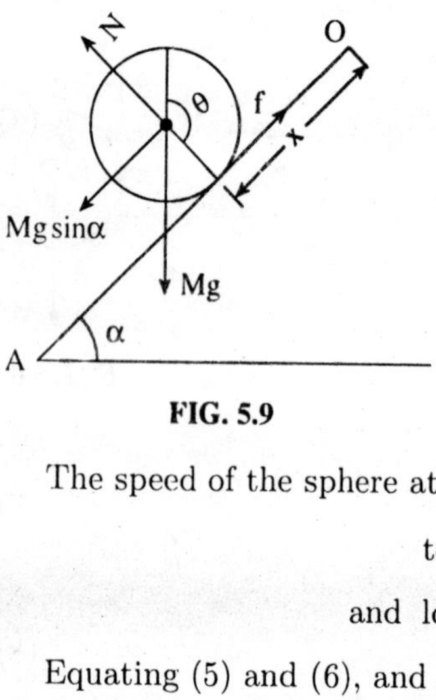

FIG. 5.9

Eliminating the frictional force f from Eqns (2) and (3), we get

$$\ddot{x} = \frac{R^2 g\sin\alpha}{R^2 + K^2} \tag{4}$$

which is the desired result.

The speed of the sphere at any distance x is obtained from energy principle. Thus,

$$\text{total kinetic energy} = \frac{1}{2}Mv^2 + \frac{1}{2}MK^2\dot{\theta}^2 \tag{5}$$

$$\text{and loss in potential energy} = Mgx\sin\alpha \tag{6}$$

Equating (5) and (6), and remembering that $v = R\dot{\theta}$, we find

$$v^2 = \frac{2gxR^2\sin\alpha}{R^2 + K^2}$$

NOTE. Eqn (4) provides the value of the acceleration for different solids. Thus,

(1) for a solid sphere, $K^2 = \frac{2}{5}R^2$, so that $\ddot{x} = \frac{5}{7}g\sin\alpha$

(2) for a hollow sphere, $K^2 = \frac{2}{3}R^2$, so that $\ddot{x} = \frac{3}{5}g\sin\alpha$

(3) for a solid cylinder or a circular disc, $K^2 = \frac{1}{2}R^2$, so that $\ddot{x} = \frac{2}{3}g\sin\alpha$

(4) for a thin cylindrical shell or a circular ring, $K^2 = R^2$, and so $\ddot{x} = \frac{1}{2}g\sin\alpha$.

EXAMPLE 5.7 *A solid circular disc of mass M and radius R moving with constant speed V comes to an incline of angle α. Prove that if there is no slipping, it will rise a distance*

$$\frac{3V^2}{4g\sin\alpha}$$

up the incline.

Solution. Let x = distance up the plane covered by the disc after time t (FIG. 5.10). Since there is no slipping, we have

$$x = R\theta, \text{ i.e., } \ddot{x} = R\ddot{\theta} \quad (1)$$

The forces acting on the disc are its weight along vertically downward direction, the frictional force along upward direction parallel to the plane and the normal reaction N.

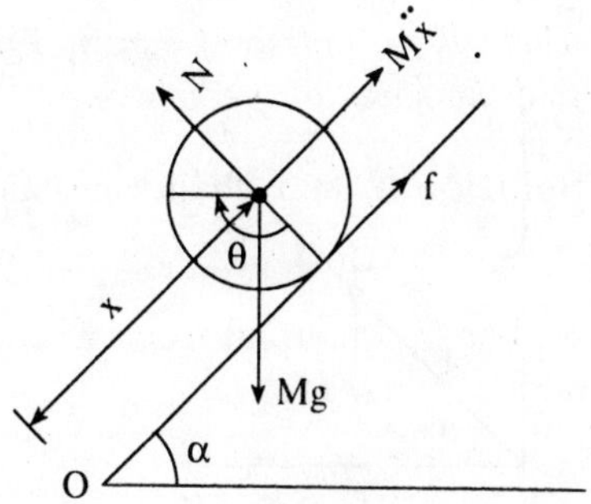

FIG. 5.10

Equation for translational motion is

$$M\ddot{x} = -Mg\sin\alpha + f \quad (2)$$

Equation for rotational motion is

$$MK^2\ddot{\theta} = -f \cdot R \text{ or, } MK^2\ddot{x} = -fR^2 \quad (3)$$

Eliminating the frictional force f from Eqns (2) and (3), we get

$$M\left(1 + \frac{K^2}{R^2}\right)\ddot{x} = -Mg\sin\alpha \text{ or, } \ddot{x} = -g\lambda\sin\alpha \quad (4)$$

where we have denoted

$$\lambda = \frac{R^2}{R^2 + K^2} \quad (5)$$

Integrating once Eqn (4) using the condition that $\dot{x} = V$ at $t = 0$, we find

$$\dot{x} = V - g\lambda t\sin\alpha \quad (6)$$

The body will stop after a time t_0 given by

$$0 = V - g\lambda t_0\sin\alpha, \text{ i.e., } t_0 = \frac{V}{g\lambda\sin\alpha} \quad (7)$$

Integrating once more Eqn (6) using the condition that $x = 0$ at $t = 0$, we get

$$x = Vt - \frac{g\lambda}{2}t^2\sin\alpha \quad (8)$$

The body will rise a distance x_0 during time t_0 given by

$$x_0 = V \cdot \frac{V}{g\lambda\sin\alpha} - \frac{g\lambda\sin\alpha}{2}\left(\frac{V}{g\lambda\sin\alpha}\right)^2 = \frac{V^2}{2g\lambda\sin\alpha} \quad (9)$$

For a cylindrical disc, $K^2 = \frac{1}{2}R^2$; hence,

$$\lambda = \frac{R^2}{R^2 + \frac{1}{2}R^2} = \frac{2}{3}$$

so that Eqn (9) assumes the form:

$$x_0 = \frac{3V^2}{4g\sin\alpha} \quad (10)$$

which is the required result.

EXAMPLE 5.8 *A uniform rod is held at an angle α to the horizon with one end in contact with a horizontal table whose coefficient of friction is μ. If it be then released, show that it will commence to slide if*

$$\mu < \frac{3\sin\alpha\cos\alpha}{1+3\sin^2\alpha}$$

Solution. Let at any instant of time, the rod of length $2l$ and mass M makes an angle θ with the horizon (FIG. 5.11). The coordinates of the centre of mass in rectangular coordinate system are:

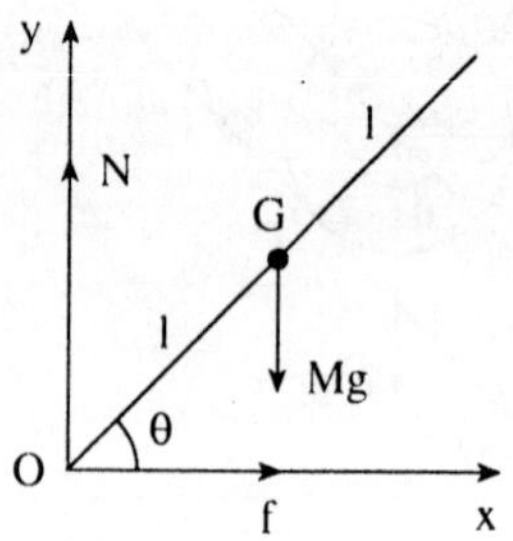

FIG. 5.11

$$x = l\cos\theta,\ y = l\sin\theta \tag{1}$$

Hence,

$$\dot{x} = -l\sin\theta\dot{\theta},\ \dot{y} = l\cos\theta\dot{\theta} \tag{2}$$

$$\left.\begin{aligned}\ddot{x} &= -l\cos\theta\dot{\theta}^2 - l\sin\theta\ddot{\theta}\\ \ddot{y} &= -l\sin\theta\dot{\theta}^2 + l\cos\theta\ddot{\theta}\end{aligned}\right\} \tag{3}$$

The forces acting on the system are: (1) weight of the rod vertically downwards, (2) the frictional force f acting horizontally and (3) the normal reaction N.

Since the rod is rotating about O, equation of rotational motion is

$$M\left(l^2 + \frac{l^2}{3}\right)\ddot{\theta} = -Mgl\cos\theta \text{ or, } \ddot{\theta} = -\frac{3g}{4l}\cos\theta \tag{4}$$

The negative sign appears since θ decreases with time.

Multiplying by $2\dot{\theta}$ on both sides of Eqn (4) and integrating with the condition that $\dot{\theta} = 0$, when $\theta = \alpha$, we get

$$\dot{\theta}^2 = \frac{3g}{2l}(\sin\alpha - \sin\theta) \tag{5}$$

Now, the equations of motion of the centre of mass are

$$\left.\begin{aligned}M\ddot{x} &= f\\ M\ddot{y} &= N - Mg\end{aligned}\right\} \tag{6}$$

With the use of Eqns (3), we find

$$M\left(-l\cos\theta\dot{\theta}^2 - l\sin\theta\ddot{\theta}\right) = f \tag{7}$$

$$\text{and } M\left(-l\sin\theta\dot{\theta}^2 + l\cos\theta\ddot{\theta}\right) = N - Mg \tag{8}$$

Substituting the values of $\ddot{\theta}$ and $\dot{\theta}^2$ from Eqns (4) and (5) into Eqns (7) and (8), we find

$$f = M\left[-l\cos\theta\cdot\frac{3g}{2l}(\sin\alpha - \sin\theta) - l\sin\theta\left(-\frac{3g}{4l}\cos\theta\right)\right]$$

$$= \frac{3}{4}Mg\cos\theta(3\sin\theta - 2\sin\alpha) \tag{9}$$

$$N = Mg + M\left[-l\sin\theta\cdot\frac{3g}{2l}(\sin\alpha - \sin\theta) + l\cos\theta\left(-\frac{3g}{4l}\cos\theta\right)\right]$$

$$= \frac{1}{4}Mg\left(4 - 6\sin\theta\sin\alpha + 6\sin^2\theta - 3\cos^2\theta\right) \tag{10}$$

Initially, $\theta = \alpha$. For this value of θ, Eqns (9) and (10) yield

$$f = \frac{3}{4}Mg\cos\alpha\sin\alpha \tag{11}$$

$$\text{and } N = \frac{1}{4}Mg(4 - 3\cos^2\alpha) = \frac{1}{4}Mg(1 + 3\sin^2\alpha) \tag{12}$$

The end O of the rod will commence to slide if $\mu < \dfrac{f}{N}$, which yields the result.

EXAMPLE 5.9 *A uniform rod of length 2l and mass M slides down in a vertical plane with its ends on a smooth vertical wall and a smooth horizontal floor. It slides starting from rest when its end on the floor was at a distance d from the wall. Prove that the other end will leave the wall when its distance from the wall is given by*

$$\frac{2}{3}\sqrt{5l^2 + d^2}.$$

Solution. The forces acting on the system are shown in FIG. 5.12. $AB = 2l$ is the rod.

Initially, $OA = d$ so that if α were the angle made by the rod with the floor, then

$$\sin\alpha = \frac{1}{2l}\sqrt{4l^2 - d^2} \tag{1}$$

At any instant of time, let the rod make an angle θ with the floor. Then the coordinates of the centre of mass of the rod are

$$x = l\cos\theta,\ y = l\sin\theta \tag{2}$$

Hence,

$$\dot{x} = -l\sin\theta\dot{\theta},\ \dot{y} = l\cos\theta\dot{\theta} \tag{3}$$

$$\text{and } \ddot{x} = -l\cos\theta\dot{\theta}^2 - l\sin\theta\ddot{\theta}$$

$$\ddot{y} = -l\sin\theta\dot{\theta}^2 + l\cos\theta\ddot{\theta} \tag{4}$$

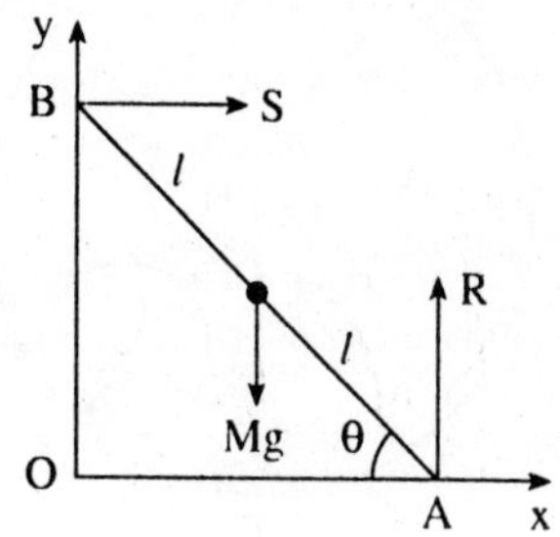

FIG. 5.12

The velocity of the centre of mass at any instant of time is

$$v^2 = \dot{x}^2 + \dot{y}^2 = l^2(\sin^2\theta + \cos^2\theta)\dot{\theta}^2 = l^2\dot{\theta}^2 \tag{5}$$

Change in the potential energy of the centre of mass is

$$Mgl(\sin\alpha - \sin\theta) \tag{6}$$

Hence, by conservation of energy,

$$\frac{1}{2}Mv^2 + \frac{1}{2}I\dot{\theta}^2 = Mgl(\sin\alpha - \sin\theta)$$

$$\text{or, } \frac{1}{2}M\left(l^2\dot{\theta}^2 + \frac{l^2}{3}\dot{\theta}^2\right) = Mgl(\sin\alpha - \sin\theta)$$

$$\text{or, } \dot{\theta}^2 = \frac{3g}{2l}(\sin\alpha - \sin\theta) \tag{7}$$

Differentiating w.r.t. time,

$$2\dot{\theta}\ddot{\theta} = -\frac{3g}{2l}\cos\theta\dot{\theta} \text{ or, } \ddot{\theta} = -\frac{3g}{4l}\cos\theta \tag{8}$$

Equation of motion in the horizontal direction is

$$M\ddot{x} = S \text{ or, } M(-l\cos\theta\dot{\theta}^2 - l\sin\theta\ddot{\theta}) = S \tag{9}$$

Substituting the values of $\dot{\theta}^2$ and $\ddot{\theta}$ from Eqns (7) and (8), we get after simplification

$$S = \frac{3}{4}Mg\cos\theta(3\sin\theta - 2\sin\alpha) \tag{10}$$

When the rod leaves the wall at B, the normal reaction $S = 0$. This yields

$$\sin\theta = \frac{2}{3}\sin\alpha = \frac{1}{3l}\sqrt{4l^2 - d^2} \tag{11}$$

If $OB = a$ at that time, then $a = \frac{2}{3}\sqrt{4l^2 - d^2}$.

If b = distance of end A from the wall, then

$$b = \sqrt{4l^2 - a^2} = \sqrt{4l^2 - \frac{4}{9}(4l^2 - d^2)} = \frac{2}{3}\sqrt{5l^2 + d^2}$$

EXAMPLE 5.10 *A circular cylinder of radius r rolls without slipping inside a fixed hollow cylinder of radius R. Assuming the oscillations to be small, find the length of the equivalent simple pendulum.*

Solution. Let M = mass of the rolling cylinder and K^2 its radius of gyration. At any instant of time, let θ be angle of the line joining the centres of cylinders with the vertical (FIG. 5.13).

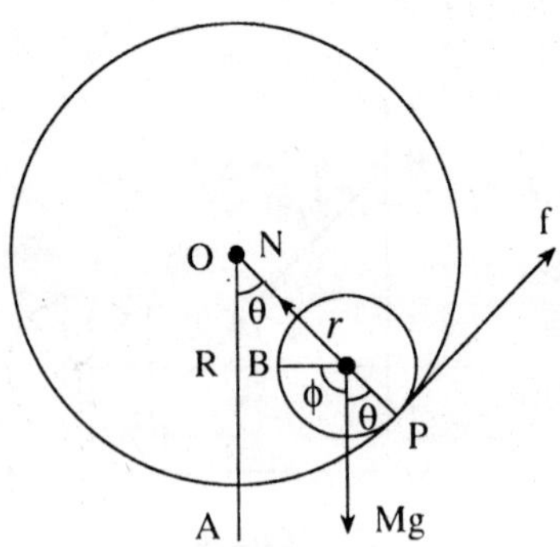

FIG. 5.13

At the lowest point A, the point B on the rolling cylinder coincided. Then from the figure,

$$R\theta = r(\theta + \phi) \text{ or, } r\ddot{\phi} = (R - r)\ddot{\theta} \tag{1}$$

The forces acting on the system are: (i) the weight Mg acting vertically downwards, (ii) the frictional force f acting tangentially at the point of contact and (iii) the normal reaction N.

The equations of motion of the rolling cylinder are

$$M(R - r)\ddot{\theta} = f - Mg\sin\theta \tag{2}$$

$$\text{and } MK^2\ddot{\phi} = -f \cdot r \tag{3}$$

From Eqns (2) and (3),

$$MK^2\ddot{\phi} = -r[Mg\sin\theta + M(R - r)\ddot{\theta}]$$

$$\text{or, } MK^2\left(\frac{R - r}{r}\right)\ddot{\theta} = -r[Mg\sin\theta + M(R - r)\ddot{\theta}], \text{ [using Eqn (1)]}$$

$$\text{or, } (R - r)\left(1 + \frac{K^2}{r^2}\right)\ddot{\theta} = -g\sin\theta \tag{4}$$

For small oscillations, $\sin\theta \simeq \theta$. Hence,

$$\ddot{\theta} = -\frac{g}{\lambda}\theta \tag{5}$$

where we have denoted

$$\lambda = (R - r)\left(1 + \frac{K^2}{r^2}\right) \tag{6}$$

Eqn (5) represents a simple harmonic motion with time period

$$P = 2\pi\sqrt{\frac{\lambda}{g}} \tag{7}$$

Hence, the length of the equivalent simple pendulum is

$$\lambda = (R-r)\left(1+\frac{K^2}{r^2}\right) \tag{8}$$

which is the required result.

EXAMPLE 5.11 *A sphere of radius a is suspended by a fine wire from a fixed point at a distance l from the centre. Show that the time of a small oscillation is given by*

$$2\pi\sqrt{\frac{2a^2+5l}{5gl}}\left(1+\frac{1}{4}\sin^2\frac{\alpha}{2}\right)$$

where α represents the amplitude of vibration.

Solution. Let the mass of the sphere = M, and its radius of gyration = K about axis of rotation through O (FIG. 5.14).

Then

$$K^2 = \left(\frac{2}{5}a^2 + l^2\right) \tag{1}$$

Equation of motion gives

$$MK^2\ddot{\theta} = -Mgl\sin\theta \; [l = OC] \tag{2}$$

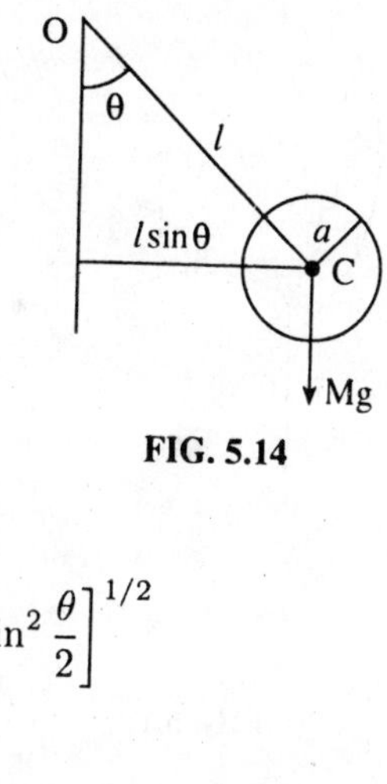

FIG. 5.14

Multiplying this equation by $2\dot{\theta}$ both sides and integrating,

$$K^2\left(\dot{\theta}\right)^2 = 2gl\cos\theta + A \tag{3}$$

When $\theta = \alpha$, $\dot{\theta} = 0$; hence, the integration constant

$$A = -2gl\cos\alpha. \tag{4}$$

Thus, from Eqns (3) and (4), we have

$$K^2\left(\dot{\theta}\right)^2 = 2gl(\cos\theta - \cos\alpha) \text{ or, } \frac{d\theta}{dt} = \sqrt{\frac{2gl}{K^2}}\left[2\sin^2\frac{\alpha}{2} - 2\sin^2\frac{\theta}{2}\right]^{1/2}$$

Thus,

$$\frac{d\theta}{\sqrt{\sin^2\frac{\alpha}{2} - \sin^2\frac{\theta}{2}}} = \frac{2}{K}\sqrt{gl}\,dt \tag{5}$$

$$\text{or, } \int_0^{\alpha}\frac{d\theta}{\sqrt{\sin^2\frac{\alpha}{2} - \sin^2\frac{\theta}{2}}} = \frac{2}{K}\sqrt{gl}\,t \tag{6}$$

We substitute

$$\sin\frac{\theta}{2} = \sin\frac{\alpha}{2}\sin\phi \tag{7}$$

so that

$$d\theta = \frac{2\sin\frac{\alpha}{2}\cos\phi d\phi}{\cos\frac{\theta}{2}} \tag{8}$$

Hence, the integral in the LHS of Eqn (6) is

$$I = \int_0^{\pi/2} \frac{2d\phi}{\sqrt{1-\sin^2\frac{\alpha}{2}\sin^2\phi}} = \int_0^{\pi/2} 2(1-\sin^2\frac{\alpha}{2}\sin^2\phi)^{-1/2}d\phi$$

$$= 2\int_0^{\pi/2}\left[1+\frac{1}{2}\sin^2\frac{\alpha}{2}\sin^2\phi - \cdots\right]d\phi = 2\left[\frac{\pi}{2}+\frac{1}{2}\sin^2\frac{\alpha}{2}\left(\frac{1}{2}\frac{\pi}{2}\right)\right]$$

$$= \pi\left(1+\frac{1}{4}\sin^2\frac{\alpha}{2}\right) \tag{9}$$

Hence, from Eqns (6) and (9), we get

$$\frac{2}{K}\sqrt{gl}\cdot t = \pi\left(1+\frac{1}{4}\sin^2\frac{\alpha}{2}\right) \text{ or, } t = \frac{\pi}{2}\frac{K}{\sqrt{gl}}\left(1+\frac{1}{4}\sin^2\frac{\alpha}{2}\right) \tag{10}$$

Hence, time of complete oscillation is

$$4t = 2\pi\frac{K}{\sqrt{gl}}\left(1+\frac{1}{4}\sin^2\frac{\alpha}{2}\right) = 2\pi\sqrt{\frac{2a^2+5l^2}{5gl}}\left(1+\frac{1}{4}\sin^2\frac{\alpha}{2}\right) \text{ (Proved).}$$

EXAMPLE 5.12 *A uniform circular disc of radius R and mass M is making a complete revolution in a vertical plane about a horizontal axis through a point on its circumference. If the pressure on the axis vanishes when the centre is vertically above the axis, show that when the centre is vertically below the axis, the angular velocity is $\sqrt{11g/3R}$ and pressure on the axis is $\frac{14}{3}Mg$.*

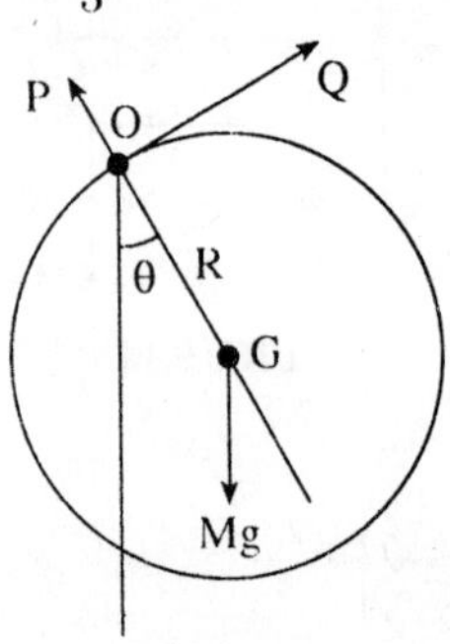

FIG. 5.15

Solution. In the FIG. 5.15, we have shown all the forces acting on the system. The centre of inertia G is describing a circle of radius R. Hence, the equations of motion are

$$MR\left(\dot{\theta}\right)^2 = P - Mg\cos\theta \tag{1}$$

$$\text{and } MR\ddot{\theta} = Q - Mg\sin\theta \tag{2}$$

Taking moments about O, we have

$$MK^2\ddot{\theta} = -MgR\sin\theta$$

$$\text{i.e., } \ddot{\theta} = -\frac{gR}{K^2}\sin\theta = -\frac{2g}{3R}\sin\theta \quad \left[\because K^2 = \frac{3}{2}R^2\right] \tag{3}$$

Integrating, we get

$$\left(\dot{\theta}\right)^2 = \frac{4g}{3R}\cos\theta + C \tag{4}$$

Now, when $\theta = 0$,

$$\frac{d\theta}{dt} = \omega \text{ (say)}$$

Hence,

$$C = \omega^2 - \frac{4g}{3R} \tag{5}$$

From Eqns (4) and (5),

$$\left(\dot{\theta}\right)^2 = \omega^2 - \frac{4g}{3R}(1-\cos\theta) \tag{6}$$

From Eqns (1) and (6),

$$P = MR\left(\dot{\theta}\right)^2 + Mg\cos\theta = MR\left[\omega^2 - \frac{4g}{3R}(1-\cos\theta)\right] + Mg\cos\theta$$

$$= M\left(R\omega^2 - \frac{4}{3}g + \frac{7}{3}g\cos\theta\right) \tag{7}$$

From Eqns (2) and (3),

$$Q = MR\ddot{\theta} + Mg\sin\theta = MR\left(-\frac{2g}{3R}\sin\theta\right) + Mg\sin\theta = \frac{1}{3}Mg\sin\theta \tag{8}$$

When the centre is vertical above the axis, i.e., $\theta = \pi$, the pressure is zero,

i.e., $P = 0$, $Q = 0$.

Putting $\theta = \pi$ in Eqn (7),

$$0 = M\left(R\omega^2 - \frac{4}{3}g - \frac{7}{3}g\right), \text{ i.e., } \omega^2 = \frac{11}{3}\frac{g}{R} \tag{9}$$

$\therefore$ angular velocity, $\omega = \sqrt{11g/3R}$.

Hence, from (7),

$$P = \frac{7}{3}Mg(1+\cos\theta) \tag{10}$$

Pressure on the axis when the centre is vertically below it is given by $\theta = 0$.

Hence, putting $\theta = 0$ in (8) and (10), we get $P = \frac{14}{3}Mg$, $Q = 0$.

Thus, total pressure on the axis for $\theta = 0$ is $P = \frac{14}{3}Mg$.

EXAMPLE 5.13 *A circular cylinder whose CG is at a distance 'c' from axis rolls on a horizontal plane. If it be just started from a position of unstable equilibrium, show that the normal reaction of the plane when the centre of mass is in its lowest position is*

$$1 + \frac{4c^2}{(R-c)^2 + K^2}$$

times its weight, where K = radius of gyration about an axis through the CG, and R = radius of the cylinder.

Solution. In the FIG. 5.16, O is the starting point when the centre of mass G was on the vertical line through centre C. After time t, the vertical line moved by angle θ, so that the arc $AP = R\theta = OA$. We take origin at O. Then, the coordinates of G are:

$$x = OA + AB = PA + AB = R\theta + c\sin\theta \tag{1}$$

$$y = GB = AC + GN = R + c\cos\theta \tag{2}$$

The kinetic energy is

$$T = \frac{1}{2}m\left[(\dot{x}^2 + \dot{y}^2) + K^2\dot{\theta}^2\right]$$

$$= \frac{1}{2}m\left[\left(R\dot{\theta} + c\cos\theta\dot{\theta}\right)^2 + \left(-c\sin\theta\dot{\theta}\right)^2 + K^2\dot{\theta}^2\right] \tag{3}$$

The work function is

$$W = mg(c - c\cos\theta) \tag{4}$$

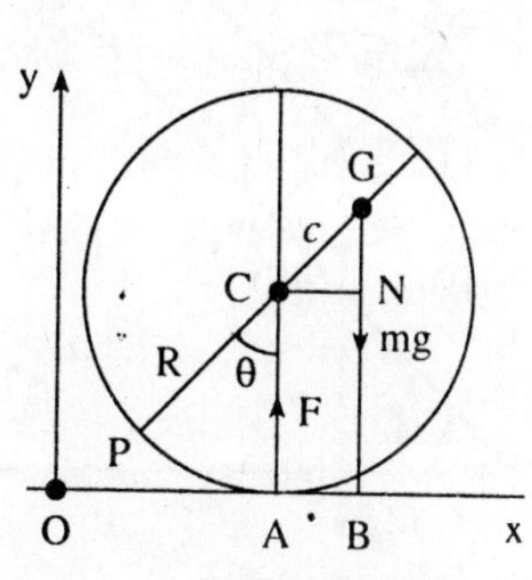

FIG. 5.16

Equating (3) and (4), we get

$$\frac{1}{2}m\left[\left(R\dot{\theta}+c\cos\theta\dot{\theta}\right)^2+c^2\sin^2\theta\dot{\theta}^2+K^2\dot{\theta}^2\right]=mg(c-c\cos\theta) \tag{5}$$

When $\theta=\pi$, i.e., the point G is at the lowest position, let $\dot{\theta}=\omega_0$. Then, from Eqn (5),

$$\frac{1}{2}m\left[(R\omega_0-c\omega_0)^2+K^2\omega_0^2\right]=mgc[1-(-1)]=2mgc$$

Hence,

$$\omega_0^2=\frac{4gc}{(R-c)^2+K^2} \tag{6}$$

Now, the equation of motion along vertical direction is

$$m\frac{d^2y}{dt^2}=F-mg \tag{7}$$

where F = normal reaction.

Eqn (7) gives

$$F=mg+m\ddot{y}=mg-mc\left(\cos\theta\dot{\theta}^2+\sin\theta\ddot{\theta}\right) \tag{8}$$

In the lowest position, $\theta=\pi$ and $\dot{\theta}=\omega_0$. Hence,

$$F=mg+mc\omega_0^2=mg+mc\frac{4gc}{(R-c)^2+K^2} \quad \text{[by Eqn (6)]}$$

$$\text{or, } F=mg\left[1+\frac{4c^2}{(R-c)^2+K^2}\right].$$

EXAMPLE 5.14 *A sphere of radius R is projected up an inclined plane with velocity V and angular velocity ω in the sense which would cause it to roll up. If $V>R\omega$ and coefficient of friction $\mu>\frac{2}{7}\tan\alpha$, show that the sphere will cease to ascend at the end of a time*

$$\frac{5V+2R\omega}{5g\sin\alpha}$$

where α is the inclination of the plane.

Solution. In the FIG. 5.17, we have shown all the forces acting in the system. The line CQ was initially perpendicular to the plane; CP is its position after time t so that rotation of the sphere is $\angle PCQ=\theta$.

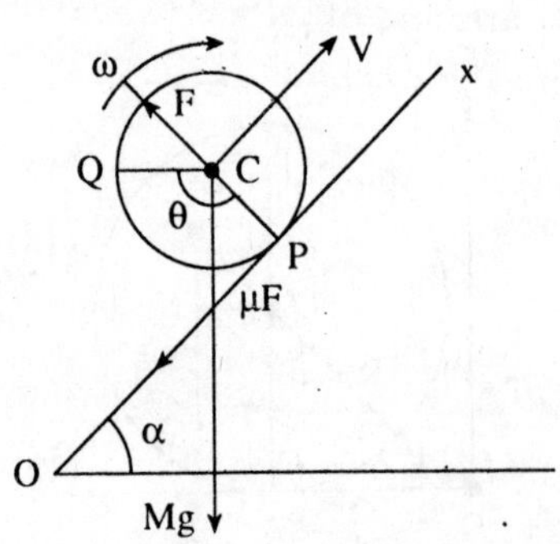

FIG. 5.17

Now, the initial velocity of the point P up the plane

= velocity of the centre C + velocity relative to C

= $V-R\omega$, which is positive by given condition.

Thus, the sphere slides as well as rolls up, so that the force of friction acts down the plane. Equations of motion along and perpendicular to the plane are

$$M\ddot{x}=-Mg\sin\alpha-\mu F \tag{1}$$

$$\text{and } M\ddot{y}=0=F-Mg\cos\alpha \tag{2}$$

From (2),

$$F=Mg\cos\alpha. \tag{3}$$

Using (3) in (1),

$$M\ddot{x} = -Mg\sin\alpha - \mu Mg\cos\alpha \quad \text{or, } \ddot{x} = -g(\sin\alpha + \mu\cos\alpha) \tag{4}$$

Integrating (4) with initial condition, $\dot{x} = V$ for $t = 0$,

$$\dot{x} = -g(\sin\alpha + \mu\cos\alpha)t + V \tag{5}$$

For rotational motion

$$MK^2\ddot{\theta} = \mu FR, \text{ i.e., } M\frac{2}{5}R^2\ddot{\theta} = \mu(Mg\cos\alpha)R \text{ [using (3)]}$$

$$\text{or, } R\ddot{\theta} = \frac{5}{2}\mu g\cos\alpha \tag{6}$$

Integrating (6) with initial condition $\dot{\theta} = \omega$ for $t = 0$,

$$R\dot{\theta} = \frac{5}{2}\mu gt\cos\alpha + R\omega \tag{7}$$

Now, when rolling commences, the point of contact remains at rest. Hence, its velocity is zero,

$$\text{i.e., } \dot{x} - R\dot{\theta} = 0 \tag{8}$$

The corresponding time t_1 is given by using (5) and (7) in (8), so that

$$-g(\sin\alpha + \mu\cos\alpha)t_1 + V - \frac{5}{2}\mu g t_1\cos\alpha - R\omega = 0 \text{ or, } t_1 = \frac{2(V - R\omega)}{g(2\sin\alpha + 7\mu\cos\alpha)} \tag{9}$$

Translational velocity along the plane at time $t = t_1$ is obtained from (5) as

$$\begin{aligned} V_1 &= V - g(\sin\alpha + \mu\cos\alpha)\frac{2(V - R\omega)}{g(2\sin\alpha + 7\mu\cos\alpha)} \\ &= \frac{5\mu V\cos\alpha + 2R\omega(\sin\alpha + \mu\cos\alpha)}{g(2\sin\alpha + 7\mu\cos\alpha)} \end{aligned} \tag{10}$$

After rolling commences, let F_1 be the force of friction just sufficient to keep the point of contact at rest. At this stage, the friction will act upwards, so that equations of motion become:

$$M\ddot{x} = F_1 - Mg\sin\alpha \tag{11}$$

$$F = Mg\cos\alpha \tag{12}$$

$$\text{and, } MK^2\ddot{\phi} = -F_1R \tag{13}$$

The point of contact being always at rest,

$$\dot{x} - R\dot{\phi} = 0, \text{ i.e., } \ddot{x} = R\ddot{\phi} \text{ or, } \frac{F_1 - Mg\sin\alpha}{M} = -\frac{5}{2M}F_1 \quad \text{[from (11) and (13)].}$$

Hence,

$$F_1 = \frac{2}{7}Mg\sin\alpha \tag{14}$$

Now, we are given that $\mu > \frac{2}{7}\tan\alpha$. Hence,

$$\mu F = \mu Mg\cos\alpha > \frac{2}{7}\tan\alpha \cdot Mg\cos\alpha, \text{ i.e., } \mu F > \frac{2}{7}Mg\sin\alpha = F_1 \tag{15}$$

Thus, the condition for pure rolling (that is the maximum force of friction will not act) is satisfied.

From (11) and (14), we get

$$M\ddot{x} = F_1 - Mg\sin\alpha = \left(\frac{2}{7} - 1\right) Mg\sin\alpha$$

$$\text{or, } \ddot{x} = -\frac{5}{7}g\sin\alpha$$

Integrating with condition, $\dot{x} = V_1$ for $t = 0$, we get

$$\dot{x} = -\frac{5}{7}gt\sin\alpha + V_1 \tag{16}$$

If the sphere ceases to ascend after time t_2, then from (16),

$$0 = -\frac{5}{7}gt_2\sin\alpha + V_1 \quad \text{or, } t_2 = \frac{7V_1}{5g\sin\alpha} \tag{17}$$

Hence, the total time of ascent for the sphere is

$$t = t_1 + t_2 \tag{18}$$

Using Eqns (9), (10) and (17), we get after some elementary algebraic manipulations

$$t = \frac{5V + 2R\omega}{5g\sin\alpha}.$$

which is the desired result.

EXAMPLE 5.15 *A solid homogeneous sphere, resting on the top of another fixed sphere is slightly displaced and begins to roll down it. Prove that the upper sphere will leave when*

$$\theta = \cos^{-1}\frac{10}{17}$$

where θ is the angle made by the common normal with the vertical.

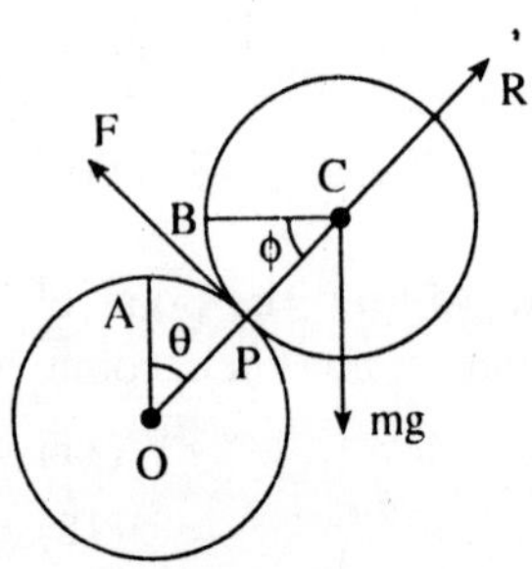

FIG. 5.18

Solution. In the FIG. 5.18, we have shown the fixed sphere with centre at O and radius $= a$ (say). The moving sphere with centre at C and radius $= b$ (say) was initially at the top, so that CB was vertical, B coinciding with A. After time t, the point of contact of two spheres is P. Then

$$a\theta = b\phi$$

$$\text{i.e., } a\dot{\theta} = b\dot{\phi} \tag{1}$$

Three forces are acting on the system; normal reaction R, force of friction F and weight of the top sphere mg.

Since the centre C of the moving sphere describes a circle of radius $(a+b)$, its equations of motion are

$$m(a+b)\dot{\theta}^2 = mg\cos\theta - R \text{ (along } CO) \tag{2}$$

$$\text{and } m(a+b)\ddot{\theta} = mg\sin\theta - F \text{ (perpendicular to } CO). \tag{3}$$

Coordinates of centre C with origin at O are

$$x = (a+b)\sin\theta \text{ and } y = (a+b)\cos\theta \tag{4}$$

Hence, kinetic energy of the system is

$$T = \frac{1}{2}m\left(\dot{x}^2 + \dot{y}^2\right) + \frac{1}{2}mK^2\left(\dot{\phi} + \dot{\theta}\right)^2$$

$$= \frac{1}{2}m(a+b)^2\dot{\theta}^2 + \frac{1}{2}mK^2\left(\frac{a+b}{b}\right)^2\dot{\theta}^2 \text{ [using (1) and (4)].} \tag{5}$$

Work function is

$$W = mg(a+b)(1-\cos\theta) \tag{6}$$

since initially C was at height $(a+b)$ and now it is at height of $(a+b)\cos\theta$.

Equating (5) and (6), and putting $K^2 = \dfrac{2b^2}{5}$, we get

$$(a+b)^2\dot{\theta}^2 + \frac{2b^2}{5}\left(\frac{a+b}{b}\right)^2\dot{\theta}^2 = 2g(a+b)(1-\cos\theta)$$

$$\text{or, } \dot{\theta}^2 = \frac{10g}{7(a+b)}(1-\cos\theta) \tag{7}$$

Hence, differentiating w.r.t. time and cancelling $2\dot{\theta}$ from both sides,

$$\ddot{\theta} = \frac{5g}{7(a+b)}\sin\theta \tag{8}$$

From Eqns (2) and (7),

$$R = mg\cos\theta - m\frac{10g}{7}(1-\cos\theta), \text{ i.e., } R = \frac{mg}{7}(17\cos\theta - 10) \tag{9}$$

The upper sphere will leave when $R = 0$; that is, when

$$17\cos\theta - 10 = 0, \text{ i.e., } \theta = \cos^{-1}\frac{10}{17} \textbf{ (Proved)}. \tag{10}$$

EXERCISE

1. A cube of mass M and sides a is suspended vertically from one edge. Show that the period of small vibrations is $P = 2\sqrt{2}\pi\sqrt{2a/3g}$.

2. Find the time period of oscillations of a compound pendulum consisting of a rod of mass m and length l, carrying at one end a sphere of mass M and radius R, the other end of the rod being fixed.

3. A solid homogeneous cone of height H and base radius R oscillates about a diameter of the base. Show that the length of the equivalent simple pendulum is

$$\frac{1}{5H}(2H^2 + 3R^2)$$

4. Prove that the period of small vibrations of a compound pendulum is minimum when the distance from the point of suspension to the centre of mass is equal to the radius of gyration of the body about a horizontal axis through the centre of mass.

5. A flywheel of radius R and mass M rotates about a fixed axis perpendicular to it and is passing through its centre. It starts rotating under a constant force F acting tangentially. Show that it will acquire an angular speed ω after a time

$$\frac{MK^2\omega}{FR}$$

where K is its radius of gyration.

6. A body of mass m hangs from one end of a light rope which is wound on the surface of a horizontal cylinder of radius a, which is free to rotate about its axis. Prove that the body descends with an acceleration

$$f = \frac{g}{1 + \dfrac{I}{ma^2}}$$

where I is the moment of inertia of the cylinder about its axis. **[C.U. (HONS) 1980]**

7. A body of mass m is attached to a rope which is wound round a fixed pulley of mass M and radius of gyration K which can rotate freely about its axis. Find the acceleration of the mass m if it is released.

8. A uniform sphere of radius R and radius of gyration K rolls down without slipping an inclined plane of length L. Show that the speed of the sphere at the bottom of the plane when starting from the top is

$$v = \left[\frac{2gL\sin\theta}{\left(1 + \frac{K^2}{R^2}\right)}\right]^{1/2}$$

where θ = angle of inclination of the plane.

9. Two masses m_1 and m_2 are connected by an inextensible string of negligible mass. They pass over a frictionless pulley of mass M, radius R and radius of gyration K which can freely rotate about its horizontal axis. Find the common acceleration of the masses.

10. A cylinder of mass M rolls down without slipping an inclined, unrapping as it goes, a fine massless string fixed to the highest point of the plane. Show that the tension of the string is $\frac{1}{3}Mg\sin\alpha$, where α is the inclination of the plane.

11. A uniform rod of length $2l$ is held in vertical position with one end resting upon a rough table. When it is released, it rotates about the end in contact with the table. Show that when it makes an angle θ with the vertical, its angular velocity is

$$\sqrt{\frac{3g}{2L}(1 - \cos\theta)}$$

12. A solid sphere of radius r rolls on the inside of a smooth hollow sphere of radius R. Prove that the period of small oscillations is

$$P = 2\pi\sqrt{\frac{7(R-r)}{5g}}$$

CHAPTER 6

Moving Axes of Coordinates

We have so far discussed the motion of particles with reference to inertial frame of axes. We now intend to find how Newton's second law of motion gets modified when the frame of axes is under continuous translational or rotational motion.

6.1 Axes Moving with Constant Velocity

Consider two frames of Cartesian axes S_1 and S_2 initially coincident with one another. Let now the coordinate system S_2 move with constant velocity $\mathbf{v}$ in the direction of positive x-axis (FIG. 6.1). Let $\mathbf{r_1}$ and $\mathbf{r_2}$ be the position vectors of any point P after time t.

Then

$$\mathbf{r_1} = \mathbf{r_2} + \mathbf{v}t \tag{6.1}$$

Hence,

$$\dot{\mathbf{r}}_1 = \dot{\mathbf{r}}_2 + \mathbf{v} \tag{6.2}$$

$$\text{and } \ddot{\mathbf{r}}_1 = \ddot{\mathbf{r}}_2 \tag{6.3}$$

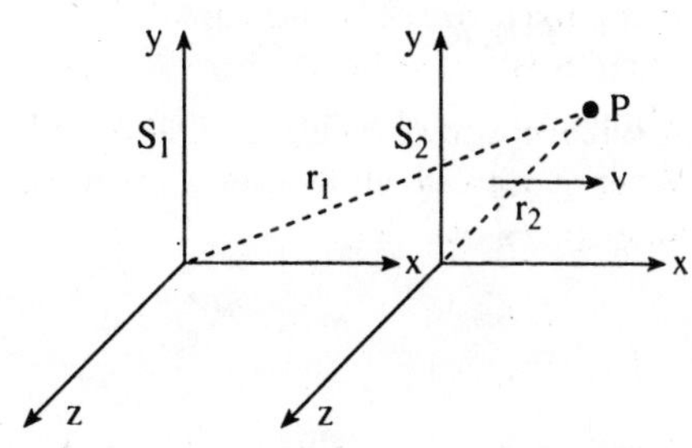

FIG. 6.1

Eqn (6.2) shows that the absolute velocity with respect to the fixed coordinate axes S_1 is equal to the vector sum of the relative velocity and the velocity of the S_2 axes. But Eqn (6.3) shows that the accelerations in reference to the two coordinate systems are same.

6.2 Axes Moving with Constant Accelerations

Consider, as before, two frames of Cartesian axes S_1 and S_2 initially coincident with one another. Let now the frame of axes S_2 move with constant acceleration $\mathbf{a}$ in the positive x-direction. If $\mathbf{r_1}$ and $\mathbf{r_2}$ be the position vectors of any point P at time t with reference to the frames of axes S_1 and S_2 respectively, then

$$\mathbf{r_1} = \mathbf{r_2} + \frac{1}{2}\mathbf{a}t^2 \tag{6.4}$$

Hence,

$$\dot{\mathbf{r}}_1 = \dot{\mathbf{r}}_2 + \mathbf{a}t \tag{6.5}$$

$$\text{and } \ddot{\mathbf{r}}_1 = \ddot{\mathbf{r}}_2 + \mathbf{a} \tag{6.6}$$

Thus, the absolute velocity $\dot{\mathbf{r}}_1$ goes on increasing along with time. Considering a particle of mass m, we find from Eqn (6.6) that the impressed force in the inertial frame of axes S_1 is

$$\mathbf{F} = m\ddot{\mathbf{r}}_1 = m(\ddot{\mathbf{r}}_2 + \mathbf{a}) \tag{6.7}$$

However, the impressed force in the moving frame of axes is

$$\mathbf{F} = m\ddot{\mathbf{r}}_2 = m\ddot{\mathbf{r}}_1 - m\mathbf{a} \tag{6.8}$$

6.3 Uniformly Rotating Frame of Axes

Consider a fixed Cartesian coordinate axes (xyz) having $\mathbf{i}, \mathbf{j}, \mathbf{k}$ as unit vectors along the axes. Let there be another Cartesian coordinate system $(x_1y_1z_1)$ having the same origin and moving with uniform angular velocity $\boldsymbol{\omega}$ about an axis passing through the origin (FIG. 6.2). If $\mathbf{r}$ be the position vector of any point P at any instant of time, then

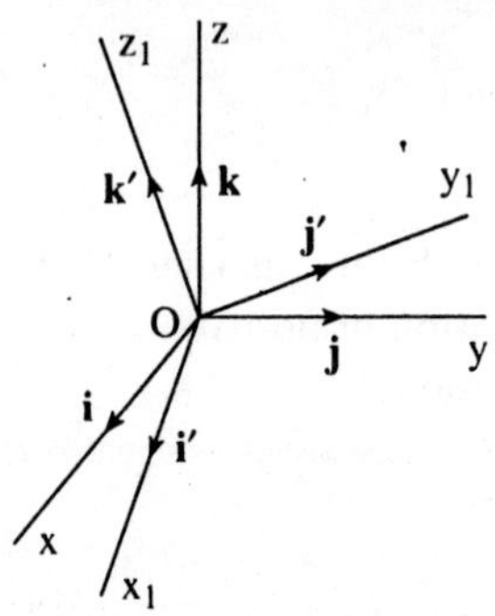

FIG. 6.2

$$\left.\begin{aligned} \mathbf{r} &= \mathbf{i}x + \mathbf{j}y + \mathbf{k}z \\ \text{and also,}\quad \mathbf{r} &= \mathbf{i}'x_1 + \mathbf{j}'y_1 + \mathbf{k}'z_1 \end{aligned}\right\} \tag{6.9}$$

where $\mathbf{i}', \mathbf{j}'$ and $\mathbf{k}'$ are unit vectors along x_1-, y_1- and z_1-directions. These unit vectors are continuously changing the direction and hence are not constants. The rate of change of the vector $\mathbf{r}$ (i.e., the velocity of the particle) in the moving system $(x_1y_1z_1)$ is

$$\begin{aligned} \frac{d\mathbf{r}}{dt} &= \left(\mathbf{i}'\frac{dx_1}{dt} + \mathbf{j}'\frac{dy_1}{dt} + \mathbf{k}'\frac{dz_1}{dt}\right) \\ &\quad + \left(x_1\frac{d\mathbf{i}'}{dt} + y_1\frac{d\mathbf{j}'}{dt} + z_1\frac{d\mathbf{k}'}{dt}\right) \end{aligned} \tag{6.10}$$

Since $\mathbf{i}', \mathbf{j}'$ and $\mathbf{k}'$ are rotating unit vectors, their derivatives are given by (ref. Section 1.3)

$$\left.\begin{aligned} \frac{d\mathbf{i}'}{dt} &= \boldsymbol{\omega} \times \mathbf{i}' \\ \frac{d\mathbf{j}'}{dt} &= \boldsymbol{\omega} \times \mathbf{j}' \\ \frac{d\mathbf{k}'}{dt} &= \boldsymbol{\omega} \times \mathbf{k}' \end{aligned}\right\} \tag{6.11}$$

Using these expressions, we find

$$\begin{aligned} x_1\frac{d\mathbf{i}'}{dt} + y_1\frac{d\mathbf{j}'}{dt} + z_1\frac{d\mathbf{k}'}{dt} &= x_1(\boldsymbol{\omega} \times \mathbf{i}') + y_1(\boldsymbol{\omega} \times \mathbf{j}') + z_1(\boldsymbol{\omega} \times \mathbf{k}') \\ &= \boldsymbol{\omega} \times (x_1\mathbf{i}' + y_1\mathbf{j}' + z_1\mathbf{k}') = \boldsymbol{\omega} \times \mathbf{r} \end{aligned} \tag{6.12}$$

Again, if the unit vectors $\mathbf{i}', \mathbf{j}', \mathbf{k}'$ were constants, then

$$\begin{aligned} \mathbf{i}'\frac{dx_1}{dt} + \mathbf{j}'\frac{dy_1}{dt} + \mathbf{k}'\frac{dz_1}{dt} &= \frac{d}{dt}(\mathbf{i}'x_1 + \mathbf{j}'y_1 + \mathbf{k}'z_1) \\ &= \frac{d'\mathbf{r}}{dt} \end{aligned} \tag{6.13}$$

where we have denoted the operator d/dt in rotating frame by d'/dt.

Thus, Eqn (6.10) with the use of Eqns (6.12) and (6.13) assumes the form:

$$\frac{d\mathbf{r}}{dt} = \frac{d'\mathbf{r}}{dt} + \boldsymbol{\omega} \times \mathbf{r} \tag{6.14}$$

Eqn (6.14) determines the velocity of the particle, called ***true velocity***. The velocity given by Eqn (6.13) is called the ***apparent velocity***.

The differential operator

$$\frac{d}{dt} = \frac{d'}{dt} + \boldsymbol{\omega} \times \tag{6.15}$$

gives the relation between time derivatives of any vector in reference to a fixed and a uniformly rotating frame of axes.

If we operate it on $\boldsymbol{\omega}$, we get

$$\frac{d\boldsymbol{\omega}}{dt} = \frac{d'\boldsymbol{\omega}}{dt} + \boldsymbol{\omega} \times \boldsymbol{\omega} = \frac{d'\boldsymbol{\omega}}{dt} \tag{6.16}$$

Eqn (6.16) shows that the angular acceleration is the same on both the fixed and rotating coordinate systems.

• Acceleration

Acceleration of the particle at any instant of time is given by

$$\frac{d}{dt}\left(\frac{d\mathbf{r}}{dt}\right) = \left(\frac{d'}{dt} + \boldsymbol{\omega}\times\right)\left(\frac{d'\mathbf{r}}{dt} + \boldsymbol{\omega}\times\mathbf{r}\right) = \frac{d'^2\mathbf{r}}{dt^2} + 2\left(\boldsymbol{\omega}\times\frac{d'\mathbf{r}}{dt}\right) + \boldsymbol{\omega}\times(\boldsymbol{\omega}\times\mathbf{r}) \tag{6.17}$$

The first term on the RHS of Eqn (6.17) represents the acceleration in the rotating frame of axes and is known as ***apparent acceleration***. Third term on the RHS of Eqn (6.17) represents the acceleration in reference to the fixed frame of a point which is fixed in the rotating system. The second term, i.e.,

$$2\left(\boldsymbol{\omega}\times\frac{d'\mathbf{r}}{dt}\right) \tag{6.18}$$

is called the ***Coriolis acceleration***.

If m be the mass of the particle, then from Eqn (6.17) we get the impressed force as

$$\mathbf{F} = m\frac{d^2\mathbf{r}}{dt^2} = m\frac{d'^2\mathbf{r}}{dt^2} + 2m\left(\boldsymbol{\omega}\times\frac{d'\mathbf{r}}{dt}\right) + m\left[\boldsymbol{\omega}\times(\boldsymbol{\omega}\times\mathbf{r})\right] \tag{6.19}$$

Hence, the equation of motion of the particle in rotating coordinate system is

$$m\frac{d'^2\mathbf{r}}{dt^2} = \mathbf{F} - 2m\left(\boldsymbol{\omega}\times\frac{d'\mathbf{r}}{dt}\right) - m\left[\boldsymbol{\omega}\times(\boldsymbol{\omega}\times\mathbf{r})\right] \tag{6.20}$$

where $\mathbf{F}$ is the resultant of all forces acting on the particle in reference to the fixed or inertial coordinate system.

The inertial force $-m[\boldsymbol{\omega}\times(\boldsymbol{\omega}\times\mathbf{r})]$ is the ***centrifugal force*** arising due to the rotation of the coordinate axes.

6.4 Non-uniform Rotation of Axes

In this case, the angular velocity $\boldsymbol{\omega}$ of the rotating frame of axes is not constant so that

$$\frac{d'\boldsymbol{\omega}}{dt} \neq 0 \tag{6.21}$$

Hence,

$$\begin{aligned}\frac{d}{dt}\left(\frac{d\mathbf{r}}{dt}\right) &= \left(\frac{d'}{dt} + \boldsymbol{\omega}\times\right)\left(\frac{d'\mathbf{r}}{dt} + \boldsymbol{\omega}\times\mathbf{r}\right)\\ &= \frac{d'^2\mathbf{r}}{dt^2} + 2\left(\boldsymbol{\omega}\times\frac{d'\mathbf{r}}{dt}\right) + \frac{d'\boldsymbol{\omega}}{dt}\times\mathbf{r} + \boldsymbol{\omega}\times(\boldsymbol{\omega}\times\mathbf{r})\end{aligned} \tag{6.22}$$

In view of this, Eqn (6.20) in this case assumes the form:

$$m\frac{d'^2\mathbf{r}}{dt^2} = \mathbf{F} - 2m\left(\boldsymbol{\omega}\times\frac{d'\mathbf{r}}{dt}\right) - m\left(\frac{d'\boldsymbol{\omega}}{dt}\times\mathbf{r}\right) - m[\boldsymbol{\omega}\times(\boldsymbol{\omega}\times\mathbf{r})] \tag{6.23}$$

NOTE. The centrifugal and Coriolis forces are not basic physical forces like gravitational or electrostatic forces. These forces have been introduced in an artificial manner in order to extend the Newton's second law of motion in a non-inertial frame of axes. In this sense, the concept of the centrifugal and Coriolis forces is very important.

6.5 Effect of the Earth's Motion on Acceleration due to Gravity

The earth is continuously rotating about its own axis with an angular velocity ω. A coordinate system on the surface of the earth will also be rotating with same angular speed.

Let there be a particle of mass m situated on the surface of the earth under the action of external forces $\mathbf{F}$ in addition to the gravitational force $m\mathbf{g}$. Then the equation of motion of the particle in the rotating frame of axes will be

$$m\frac{d'^2\mathbf{r}}{dt^2} = \mathbf{F} + m\,[\mathbf{g} - \boldsymbol{\omega}\times(\boldsymbol{\omega}\times\mathbf{r})] - 2m\left(\boldsymbol{\omega}\times\frac{d'\mathbf{r}}{dt}\right) \tag{6.24}$$

We denote

$$\mathbf{g_e} = \mathbf{g} - \boldsymbol{\omega}\times(\boldsymbol{\omega}\times\mathbf{r}) \tag{6.25}$$

Since the centrifugal force $-\boldsymbol{\omega}\times(\boldsymbol{\omega}\times\mathbf{r})$ always acts away from the centre, the term $\mathbf{g_e}$ given by Eqn (6.25) is the effective value of acceleration due to gravity.

At the poles, the vectors $\boldsymbol{\omega}$ and $\mathbf{r}$ are parallel so that $\boldsymbol{\omega}\times\mathbf{r} = 0$. Hence, $\mathbf{g_e} = \mathbf{g}$. At the equator, the vectors $\boldsymbol{\omega}$ and $\mathbf{r}$ are at right angles so that $\boldsymbol{\omega}\times\mathbf{r}$ is maximum. Hence, the value of $\mathbf{g_e}$ is minimum at the equator.

6.6 Effect of Coriolis Force

Consider a particle of mass m moving on the surface of the earth with a velocity v. At any instant of time, let its position be at a point having latitude λ. We construct rotating frame of axes at P such that: (i) the x-axis is directed towards the east, (ii) the y-axis is directed towards the north and (iii) the z-axis is in the direction of the plumb line at the geographical latitude (FIG. 6.3). We consider the motion in northern hemisphere.

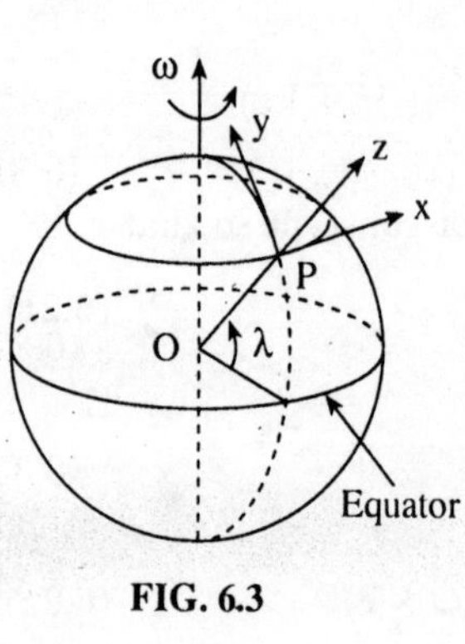

FIG. 6.3

Then the components of the angular velocity ω of the earth in the northern hemisphere along the coordinate axes are:

$$\left.\begin{aligned} \omega_x &= 0 \\ \omega_y &= \omega\cos\lambda \\ \omega_z &= \omega\sin\lambda \end{aligned}\right\} \tag{6.26}$$

The Coriolis force acting on the particle due to the rotation of the frame of axes along with the earth is

$$-2m\left(\boldsymbol{\omega}\times\frac{d'\mathbf{r}}{dt}\right) = -2m\boldsymbol{\omega}\times\mathbf{v} \tag{6.27}$$

This force will be directed at right angles to the plane containing the vectors $\boldsymbol{\omega}$ and $\mathbf{v}$. Thus, if the particle be moving in a plane perpendicular to the axis of the earth's rotation, then the particle will be deflected due to the Coriolis force eastward in northern hemisphere and westward in southern hemisphere. The maximum value of this force will be at poles and equal to $2m\omega v$. Its value is zero at the equator.

The numerical value of the earth's angular velocity ω is 7.272×10^{-5} radians/sec. Thus, the maximum value of Coriolis acceleration is

$$2\omega v = 2 \times 7.272 \times 10^{-5} v = 1.454 \times 10^{-4} v, \tag{6.28}$$

where v is the velocity in the horizontal plane.

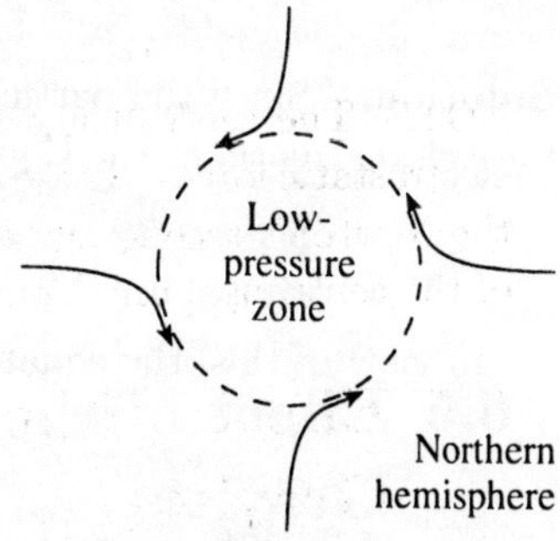

FIG. 6.4

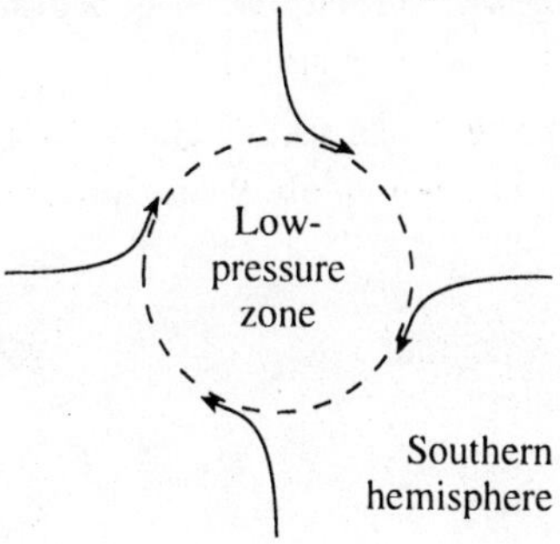

FIG. 6.5

The effect of Coriolis force plays an important role in guided missiles due to their large velocity and long time of flight.

The Coriolis force also plays an important role in determining the motion of the earth's atmosphere. When a mass of air rushes to any low-pressure zone, it gets deflected due to Coriolis force. This deflection is towards the right in the northern hemisphere and towards the left in the southern hemisphere. As a result, a vortex motion in anticlockwise direction in northern hemisphere and in clockwise direction in southern hemisphere is created. This is shown in FIG. 6.4 and FIG. 6.5.

6.7 Motion of a Particle on the Earth

Let a particle of mass m move on or near the surface of the earth such that there is no resistance due to air. Let $P(x, y, z)$ be the position of the particle at any instant of time with respect to a rotating frame of axes. If $\mathbf{i}, \mathbf{j}, \mathbf{k}$ be the unit vectors along the coordinate axes, then the equation of motion of the particle under action of gravitational force only can be written as

$$m\ddot{\mathbf{r}} = -mg_e\mathbf{k} - 2m(\boldsymbol{\omega} \times \dot{\mathbf{r}}) \tag{6.29}$$

where

$$\mathbf{r} = \mathbf{i}x + \mathbf{j}y + \mathbf{k}z \tag{6.30}$$

Eqn (6.29) after using the value of $\mathbf{r}$ assumes the form:

$$\begin{aligned} m(\mathbf{i}\ddot{x} + \mathbf{j}\ddot{y} + \mathbf{k}\ddot{z}) &= -mg_e\mathbf{k} - 2m(\mathbf{i}\omega_x + \mathbf{j}\omega_y + \mathbf{k}\omega_z) \times (\mathbf{i}\dot{x} + \mathbf{j}\dot{y} + \mathbf{k}\dot{z}) \\ &= -mg_e\mathbf{k} - 2m(\mathbf{j}\omega\cos\lambda + \mathbf{k}\omega\sin\lambda) \times (\mathbf{i}\dot{x} + \mathbf{j}\dot{y} + \mathbf{k}\dot{z}) \end{aligned} \tag{6.31}$$

Thus, in component forms, the equations of motion are:

$$m\frac{d^2x}{dt^2} = -2m\omega\left(\cos\lambda\frac{dz}{dt} - \sin\lambda\frac{dy}{dt}\right) \tag{6.32}$$

$$m\frac{d^2y}{dt^2} = -2m\omega\sin\lambda\frac{dx}{dt} \tag{6.33}$$

$$\text{and } m\frac{d^2z}{dt^2} = -mg_e + 2m\omega\cos\lambda\frac{dx}{dt} \tag{6.34}$$

EXAMPLE 6.1 *A particle of mass m is thrown vertically downwards with a velocity v. Discuss the effect of Coriolis force on its motion when variation of gravity is neglected.*

Solution. Since the particle moves along z-axis, the Coriolis force along other axes can be neglected. Hence, in the Eqns (6.32)–(6.34), we take

$$\frac{dx}{dt} = 0, \; \frac{dy}{dt} = 0 \tag{1}$$

In view of this, the equations of motion of the particle are:

$$\frac{d^2x}{dt^2} = -2\omega \cos\lambda \frac{dz}{dt} \tag{2}$$

$$\frac{d^2y}{dt^2} = 0 \tag{3}$$

$$\frac{d^2z}{dt^2} = -g \tag{4}$$

The initial conditions are:

$$\left.\begin{aligned} \text{At } t = 0, \quad x = 0 \text{ and } \frac{dx}{dt} = 0 \\ \text{and } z = 0 \text{ and } \frac{dz}{dt} = -v \end{aligned}\right\} \tag{5}$$

Integrating Eqn (4), we get

$$\frac{dz}{dt} = -gt + C \tag{6}$$

Using initial condition, we find $C = -v$, so that

$$\frac{dz}{dt} = -gt - v \tag{7}$$

From Eqns (2) and (7), we get

$$\frac{d^2x}{dt^2} = 2\omega(gt + v)\cos\lambda \tag{8}$$

Integrating we find

$$\frac{dx}{dt} = 2\omega\left(g\frac{t^2}{2} + vt\right)\cos\lambda \tag{9}$$

The constant of integration is zero by virtue of the initial conditions.

Integrating Eqn (9), we get, after using initial conditions,

$$x = \frac{1}{3}\omega g t^3 \cos\lambda + \omega v t^2 \cos\lambda \tag{10}$$

This gives the amount of shift at any time due to the Coriolis force.

EXAMPLE 6.2 *A particle of mass m is thrown vertically upwards with a velocity v at a northern latitude λ. Determine the shift of the particle when it returns. Neglect the variation of gravity.*

Solution. The equations of motion of the particle are:

$$\frac{d^2x}{dt^2} = -2\omega \cos\lambda \frac{dz}{dt} \tag{1}$$

$$\frac{d^2y}{dt^2} = 0 \tag{2}$$

$$\frac{d^2z}{dt^2} = -g \tag{3}$$

Initial conditions are:

$$\left.\begin{aligned} \text{At } t = 0,\ x = 0,\ \frac{dx}{dt} = 0 \\ \text{and } z = 0,\ \frac{dz}{dt} = v \end{aligned}\right\} \tag{4}$$

Integrating Eqn (3) with the use of the initial condition, we get.

$$\frac{dz}{dt} = -gt + v \tag{5}$$

Substituting this expression for dz/dt into Eqn (1), we get

$$\frac{d^2x}{dt^2} = 2\omega(gt - v)\cos\lambda \tag{6}$$

Integrating this twice with the use of the initial conditions, we find

$$x = 2\omega\left(\frac{1}{6}gt^3 - \frac{1}{2}vt^2\right)\cos\lambda \tag{7}$$

From Eqn (5), the time to reach the greatest height where the velocity is zero is given by

$$t' = \frac{v}{g} \tag{8}$$

Same time will be taken by the particle to come down to the ground. Hence, the total time of flight is

$$T = 2t' = \frac{2v}{g} \tag{9}$$

Putting this value for t in Eqn (7), we find the shift of the particle as

$$x = 2\omega\left[\frac{1}{6}g\left(\frac{2v}{g}\right)^3 - \frac{1}{2}v\left(\frac{2v}{g}\right)^2\right]\cos\lambda$$

$$\text{or, } x = -\frac{4v^3}{3g^2}\omega\cos\lambda \tag{10}$$

If h be the greatest height reached, then from energy consideration, we have

$$\frac{1}{2}mv^2 = mgh, \text{ i.e., } v = \sqrt{2gh} \tag{11}$$

Feeding this value of v into Eqn (10), we get

$$x = -\frac{4\omega}{3g^2}\cos\lambda(2gh)^{3/2}$$

$$= -\frac{8}{3}\omega\sqrt{\frac{2h^3}{g}}\cos\lambda \tag{12}$$

Eqn (12) gives the amount of shift towards the west.

6.8 Foucault's Pendulum

The French physicist Foucault in 1851 performed an experiment to demonstrate that the earth rotates about its axis. He took a pendulum with a very heavy bob suspended by means of a long strong wire the upper end of which was attached to a rigid support. This pendulum is called ***Foucault's pendulum***.

The plane of oscillation of the pendulum will rotate from east to west in the northern hemisphere, and from west to east in the southern hemisphere of the earth. This rotation is actually due to Coriolis force.

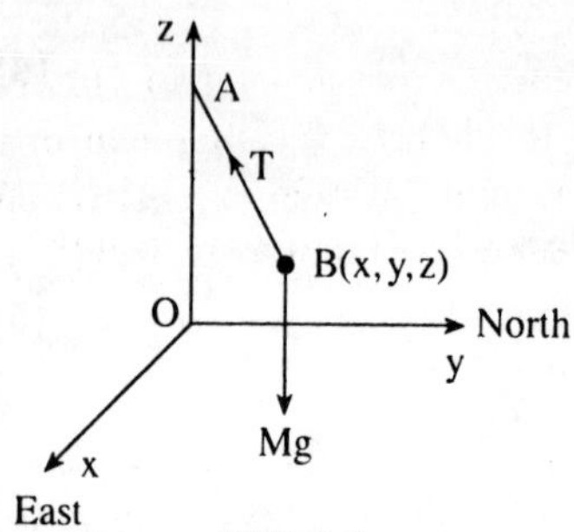

FIG. 6.6

Let a bob of mass M is suspended from a point A with a strong wire of length L (FIG. 6.6). At any instant of time, let (x, y, z) be the coordinates of the bob B. Then components of the tension $\mathbf{T}$ in the string along the axes are:

$$\left.\begin{aligned} T_x &= -T\frac{x}{L} \\ T_y &= -T\frac{y}{L} \\ T_z &= T\frac{L-z}{L} \end{aligned}\right\} \tag{6.35}$$

Equation of motion of the bob is

$$M\ddot{\mathbf{r}} = \mathbf{T} + M\mathbf{g} - 2M\,(\boldsymbol{\omega} \times \dot{\mathbf{r}}) - M\boldsymbol{\omega} \times (\boldsymbol{\omega} \times \mathbf{r}) \tag{6.36}$$

We neglect the higher order term $M\boldsymbol{\omega} \times (\boldsymbol{\omega} \times \mathbf{r})$ in ω in Eqn (6.36) and put $\mathbf{g} = -g\mathbf{k}$. Then the equation of motion along axes of coordinates are:

$$M\ddot{x} = -T\frac{x}{L} + 2M\omega\dot{y}\sin\lambda - 2M\omega\dot{z}\cos\lambda \tag{6.37}$$

$$M\ddot{y} = -T\frac{y}{L} - 2M\omega\dot{x}\sin\lambda \tag{6.38}$$

$$M\ddot{z} = T\frac{L-z}{L} - Mg + 2M\omega\dot{x}\cos\lambda \tag{6.39}$$

We assume that the oscillations are small so that the vertical motion of the bob is neglected. That is

$$\left.\begin{aligned} \dot{z} = 0;\quad \ddot{z} &= 0 \\ \text{Also,}\quad (L-z)/L &\simeq 1 \end{aligned}\right\} \tag{6.40}$$

In view of this, Eqn (6.39) yields

$$T = Mg - 2M\omega\dot{x}\cos\lambda \tag{6.41}$$

Substituting this value of T into Eqns (6.37) and (6.38), we get

$$\ddot{x} = -\frac{g}{L}x + \frac{2\omega}{L}x\dot{x}\cos\lambda + 2\omega\dot{y}\sin\lambda \tag{6.42}$$

$$\ddot{y} = -\frac{g}{L}y + \frac{2\omega}{L}y\dot{x}\cos\lambda - 2\omega\dot{x}\sin\lambda \tag{6.43}$$

These equations contain non-linear terms $x\dot{x}$ and $y\dot{x}$ which can be neglected due to smallness of ω, x and y.

Then equations of motion of the bob are simplified to

$$\ddot{x} = -\frac{g}{L}x + 2\omega\dot{y}\sin\lambda \tag{6.44}$$

$$\ddot{y} = -\frac{g}{L}y - 2\omega\dot{x}\sin\lambda \tag{6.45}$$

Writing $\omega_z = \omega\sin\lambda$, the equations of motion are recast in the form:

$$\ddot{x} - 2\omega_z\dot{y} + \frac{g}{L}x = 0 \tag{6.46}$$

$$\text{and } \ddot{y} + 2\omega_z\dot{x} + \frac{g}{L}y = 0 \tag{6.47}$$

It is obvious that ω_z is the local vertical component of the earth's rotation vector. In order to decouple the oscillatory motion in a vertical plane from the local vertical component of the earth's rotation vector ω_z, we introduce another Cartesian coordinate system (x_1, y_1, z_1) such that its origin and z_1-axis coincide with the origin and z-axis of the laboratory frame, but rotates with frequency ω_z. Then the transformation relation is given by

$$\left.\begin{aligned} x &= x_1\cos\omega_z t + y_1\sin\omega_z t \\ y &= -x_1\sin\omega_z t + y_1\cos\omega_z t \end{aligned}\right\} \tag{6.48}$$

Using these relations in either of Eqns (6.46) or (6.47) and neglecting terms in ω_z^2, we finally get

$$\left(\ddot{x}_1 + \frac{g}{L}x_1\right)\cos\omega_z t + \left(\ddot{y}_1 + \frac{g}{L}y_1\right)\sin\omega_z t = 0 \tag{6.49}$$

Since this equation should be true for all values of $\omega_z t$, we must have

$$\ddot{x}_1 + \frac{g}{L}x_1 = 0 \tag{6.50}$$

$$\text{and } \ddot{y} + \frac{g}{L}y_1 = 0 \tag{6.51}$$

Thus, the effect of decoupling the oscillatory motion is a precession of the plane of oscillation at a frequency ω_z. The precession is clockwise in northern hemisphere and anticlockwise in southern hemisphere.

Solutions of Eqns (6.50) and (6.51) can be written as

$$x_1 = A\cos\left(\sqrt{\frac{g}{L}}t\right) \text{ and } y_1 = B\cos\left(\sqrt{\frac{g}{L}}t\right) \tag{6.52}$$

Substituting these values in Eqn (6.48), we get the solution in old variables (x, y) as

$$\left.\begin{aligned} x &= A\cos\left(\sqrt{\frac{g}{L}}t\right)\cos\omega_z t + B\cos\left(\sqrt{\frac{g}{L}}t\right)\sin\omega_z t \\ \text{and } y &= -A\cos\left(\sqrt{\frac{g}{L}}t\right)\sin\omega_z t + B\cos\left(\sqrt{\frac{g}{L}}t\right)\cos\omega_z t \end{aligned}\right\} \tag{6.53}$$

If initially the bob is released at a distance a along y-axis from origin, then the conditions are

$$\left.\begin{aligned} x &= 0,\ \dot{x} = 0 \\ \text{and } y &= a,\ \dot{y} = 0 \end{aligned}\right\} \text{ at } t = 0 \tag{6.54}$$

Using these conditions in Eqn (6.53), we find

$$A = 0,\ B = a \tag{6.55}$$

Hence, the solutions are:

$$\left.\begin{aligned} x &= a\cos\left(\sqrt{\frac{g}{L}}\,t\right)\sin\omega_z t \\ \text{and } y &= a\cos\left(\sqrt{\frac{g}{L}}\,t\right)\cos\omega_z t \end{aligned}\right\} \tag{6.56}$$

Replacing the value of $\omega_z = \omega\sin\lambda$, we write:

$$\left.\begin{aligned} x &= a\cos\left(\sqrt{\frac{g}{L}}\,t\right)\sin[(\omega\sin\lambda)t] \\ \text{and } y &= a\cos\left(\sqrt{\frac{g}{L}}\,t\right)\cos[(\omega\sin\lambda)t] \end{aligned}\right\} \tag{6.57}$$

6.9 Euler's Angles

The rotation of rigid bodies are conveniently studied by three angular coordinates which are obtained by giving three successive rotations in a given sequence to a fixed Cartesian coordinate system in the body. These three successive angles of rotation are called ***Euler's angles***.

Let XYZ be a coordinate system fixed in a rigid body. The first rotation by an angle ϕ in anticlockwise direction is given to the XYZ system about Z-axis (FIG. 6.7). Let OX_1, OY_1, OZ_1 be the new coordinate axes. The axis OX_1 is called the ***line of nodes***.

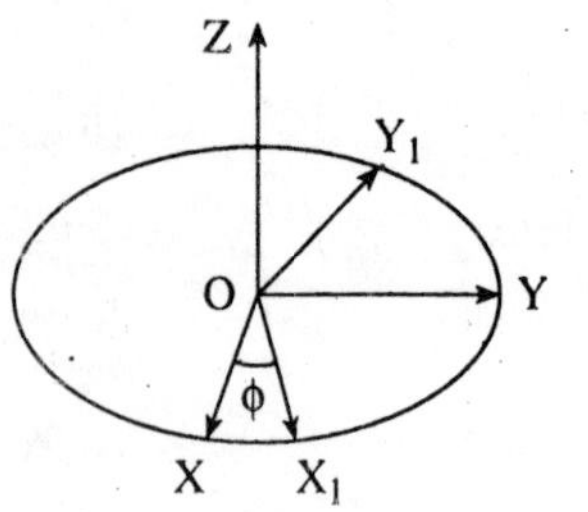

FIG. 6.7

Any point (x_1, y_1, z_1) in this system is related to the old system by the equations:

$$\left.\begin{aligned} x_1 &= x\cos\phi + y\sin\phi \\ y_1 &= -x\sin\phi + y\cos\phi \\ z_1 &= z \end{aligned}\right\} \tag{6.58}$$

In matrix notation, we write:

$$\begin{bmatrix} x_1 \\ y_1 \\ z_1 \end{bmatrix} = \begin{bmatrix} \cos\phi & \sin\phi & 0 \\ -\sin\phi & \cos\phi & 0 \\ 0 & 0 & 1 \end{bmatrix} \begin{bmatrix} x \\ y \\ z \end{bmatrix} \quad \text{or,} \quad \begin{bmatrix} x_1 \\ y_1 \\ z_1 \end{bmatrix} = [R_1] \begin{bmatrix} x \\ y \\ z \end{bmatrix} \tag{6.59}$$

where the matrix

$$[R_1] = \begin{bmatrix} \cos\phi & \sin\phi & 0 \\ -\sin\phi & \cos\phi & 0 \\ 0 & 0 & 1 \end{bmatrix} \tag{6.60}$$

is called the ***rotation matrix***.

The second rotation is given to the XY-plane by an angle θ about the line of node OX_1 in anticlockwise direction (FIG. 6.8). This causes the OZ-axis to turn through an angle θ. Any point (x_2, y_2, z_2) in this new coordinate system is related to the previous system by the equation

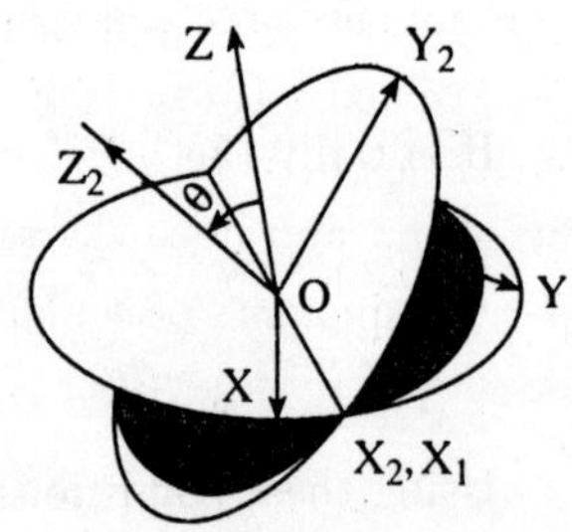

FIG. 6.8

$$\begin{bmatrix} x_2 \\ y_2 \\ z_2 \end{bmatrix} = [R_2] \begin{bmatrix} x_1 \\ y_1 \\ z_1 \end{bmatrix} \tag{6.61}$$

where the rotation matrix $[R_2]$ is given by

$$[R_2] = \begin{bmatrix} 1 & 0 & 0 \\ 0 & \cos\theta & \sin\theta \\ 0 & -\sin\theta & \cos\theta \end{bmatrix} \tag{6.62}$$

Finally, we give a spin about OZ_2 axis by an angle ψ in anticlockwise direction. This causes the OX_2 axis to rotate through an angle ψ. The rotation matrix is similar to the first rotation (FIG. 6.9). Any point in this new system can be expressed as

$$\begin{bmatrix} x_3 \\ y_3 \\ z_3 \end{bmatrix} = [R_3] \begin{bmatrix} x_2 \\ y_2 \\ z_2 \end{bmatrix} \tag{6.63}$$

where

$$[R_3] = \begin{bmatrix} \cos\psi & \sin\psi & 0 \\ -\sin\psi & \cos\psi & 0 \\ 0 & 0 & 1 \end{bmatrix} \tag{6.64}$$

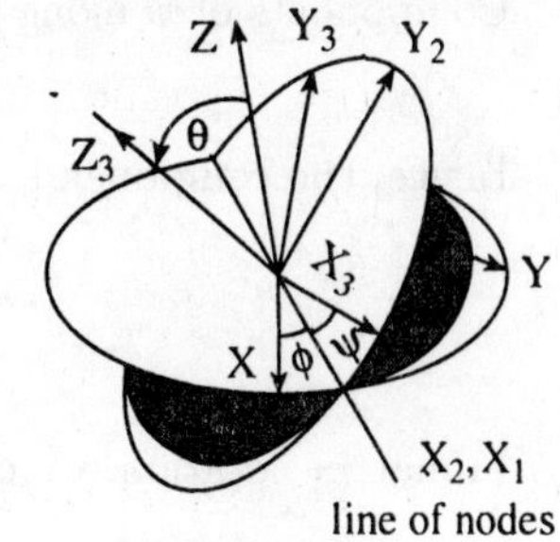

FIG. 6.9

Thus, the sequence of transformation is represented as

$$\begin{bmatrix} x_3 \\ y_3 \\ z_3 \end{bmatrix} = [R_3] \begin{bmatrix} x_2 \\ y_2 \\ z_2 \end{bmatrix} = [R_3][R_2] \begin{bmatrix} x_1 \\ y_1 \\ z_1 \end{bmatrix} = [R_3][R_2][R_1] \begin{bmatrix} x \\ y \\ z \end{bmatrix} = [R] \begin{bmatrix} x \\ y \\ z \end{bmatrix} \tag{6.65}$$

where

$$[R] = [R_3][R_2][R_1] \tag{6.66}$$

The non-zero elements of the matrix $[R]$ are obtained by multiplication of matrices in the RHS of Eqn (6.66) in the following form:

$$R_{11} = \cos\psi\cos\phi - \sin\psi\sin\phi\cos\theta \tag{6.67}$$

$$R_{12} = \cos\psi\sin\phi + \sin\psi\cos\phi\cos\theta \tag{6.68}$$

$$R_{13} = \sin\psi\sin\theta \tag{6.69}$$

$$R_{21} = -\sin\psi\cos\phi - \cos\psi\sin\phi\cos\theta \tag{6.70}$$

$$R_{22} = -\sin\psi\sin\phi + \cos\psi\cos\theta\cos\phi \tag{6.71}$$

$$R_{23} = \cos\psi\sin\theta \tag{6.72}$$

$$R_{33} = \cos\theta \tag{6.73}$$

• Angular Velocities

Let the angular velocity of the body be ω. We shall find the components of ω along OX_3-, OY_3- and OZ_3-axes.

We have shown the angles θ, ϕ and ψ and also the angular velocities $\dot{\theta}, \dot{\phi}, \dot{\psi}$ in the FIG. 6.10. The angle θ takes values from zero to π and both the angles ϕ and ψ vary from zero to 2π.

Components of $\dot{\phi}$ along OX_3, OY_3 and OZ_3 are:

$$\left.\begin{aligned} \dot{\phi}_x &= \dot{\phi}\sin\theta\sin\psi \\ \dot{\phi}_y &= \dot{\phi}\sin\theta\cos\psi \\ \dot{\phi}_z &= \dot{\phi}\cos\theta \end{aligned}\right\} \tag{6.74}$$

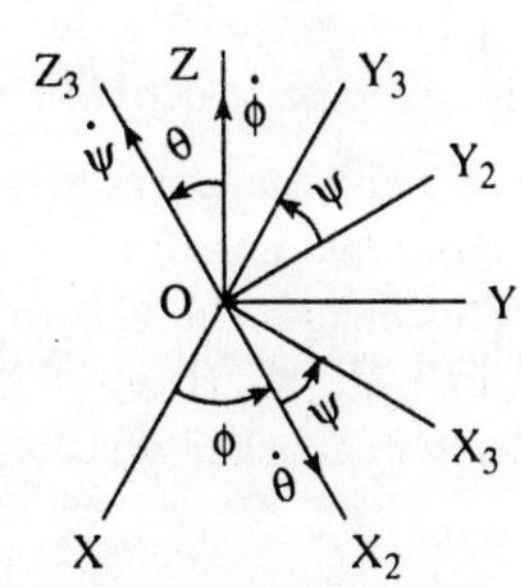

FIG. 6.10

Components of $\dot{\theta}$ along OX_3, OY_3 and OZ_3 are:

$$\left.\begin{aligned}\dot{\theta}_x &= \dot{\theta}\cos\psi \\ \dot{\theta}_y &= -\dot{\theta}\sin\psi \\ \dot{\theta}_z &= 0\end{aligned}\right\} \qquad (6.75)$$

Components of $\dot{\psi}$ along OX_3, OY_3 and OZ_3 are:

$$\dot{\psi}_x = 0, \quad \dot{\psi}_y = 0, \quad \dot{\psi}_z = \dot{\psi} \qquad (6.76)$$

Hence, the components of ω along OX_3-, OY_3- and OZ_3-axes are:

$$\left.\begin{aligned}\omega_x &= \dot{\phi}\sin\theta\sin\psi + \dot{\theta}\cos\psi \\ \omega_y &= \dot{\phi}\sin\theta\cos\psi - \dot{\theta}\sin\psi \\ \omega_z &= \dot{\phi}\cos\theta + \dot{\psi}\end{aligned}\right\} \qquad (6.77)$$

In matrix notation, we can write Eqn (6.77) in the form:

$$\begin{bmatrix}\omega_x \\ \omega_y \\ \omega_z\end{bmatrix} = \begin{bmatrix}\sin\theta\sin\psi & 0 & \cos\psi \\ \sin\theta\cos\psi & 0 & -\sin\psi \\ \cos\theta & 1 & 0\end{bmatrix}\begin{bmatrix}\dot{\phi} \\ \dot{\psi} \\ \dot{\theta}\end{bmatrix} \qquad (6.78)$$

EXAMPLE 6.3 *Determine the magnitude of the effective value of the acceleration due to gravity at a place of latitude λ. Also determine its direction.*

Solution. Let R be the radius of the earth and P be a point on its surface at latitude λ (FIG. 6.11). The effective acceleration due to gravity is

$$\mathbf{g_e} = \mathbf{g} - \boldsymbol{\omega} \times (\boldsymbol{\omega} \times \mathbf{r}) \qquad (1)$$

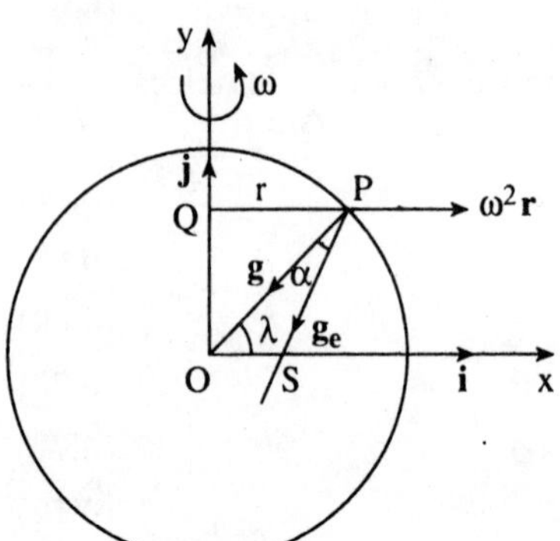

FIG. 6.11

The vector acts along PO. Then

$$\mathbf{g} = -g(\mathbf{i}\cos\lambda + \mathbf{j}\sin\lambda) \qquad (2)$$

$$\text{and } \boldsymbol{\omega} = \mathbf{j}\omega \qquad (3)$$

The radius of the circle on which the point P rotates is

$$\mathbf{r} = QP = \mathbf{i}R\cos\lambda \qquad (4)$$

Hence,

$$\begin{aligned}\mathbf{g_e} &= -g(\mathbf{i}\cos\lambda + \mathbf{j}\sin\lambda) - \mathbf{j}\omega \times (\mathbf{j}\omega \times \mathbf{i}R\cos\lambda) \\ &= -g(\mathbf{i}\cos\lambda + \mathbf{j}\sin\lambda) + \mathbf{i}\omega^2 R\cos\lambda \\ &= -\mathbf{i}(g\cos\lambda - \omega^2 R\cos\lambda) - \mathbf{j}g\sin\lambda \end{aligned} \qquad (5)$$

The magnitude of this vector is given by

$$\begin{aligned} g_e^2 &= (g\cos\lambda - \omega^2 R\cos\lambda)^2 + g^2\sin^2\lambda \\ &= g^2\cos^2\lambda - 2g\omega^2 R\cos^2\lambda + \omega^4 R^2\cos^2\lambda + g^2\sin^2\lambda \\ &= g^2\left(1 - \frac{2\omega^2 R}{g}\cos^2\lambda\right) \end{aligned} \qquad (6)$$

where we have neglected the term $(R\omega^2/g)^2$ due to smallness. Thus,

$$g_e = g\left(1 - \frac{2\omega^2 R}{g}\cos^2\lambda\right)^{1/2}, \text{ i.e., } g_e \approx g - \omega^2 R\cos^2\lambda$$

The direction of g_e will not be along PO, it will be shifted from the vertical by a small angle, say α.

From triangle POS, we have

$$\frac{g_e}{\sin\lambda} = \frac{\omega^2 R\cos\lambda}{\sin\alpha}$$

$$\therefore \sin\alpha = \frac{\omega^2 R\sin\lambda\cos\lambda}{g_e} = \frac{\omega^2 R}{2g_e}\sin 2\lambda$$

$$\therefore \alpha = \frac{\omega^2 R}{2g_e}\sin 2\lambda$$

since for small $\alpha, \sin\alpha \simeq \alpha$.

EXAMPLE 6.4 *A particle is fired due east from a point on the earth's surface at northern latitute λ with velocity v_0 at an angle α to the horizon. Show that the lateral deflection of the projectile when it hits the ground is given by*

$$d = \frac{4v_0^3}{g^2}\omega\sin\lambda\sin^2\alpha\cos\alpha,$$

where ω is the rotational frequency of the earth.

Solution. In the equations of motion (6.32 – 6.34) we take $\frac{dy}{dt} = 0$, since the motion is in xz-plane. We have at $t = 0$ the following conditions:

$$\left.\begin{aligned} x = 0;\ \dot{x} &= v_0\cos\alpha \\ y = 0;\ \dot{y} &= 0 \\ z = 0;\ \dot{z} &= v_0\sin\alpha \end{aligned}\right\} \tag{1}$$

The equations of motion become

$$\left.\begin{aligned} \frac{d^2x}{dt^2} &= -2\omega\cos\lambda\frac{dz}{dt} \\ \frac{d^2y}{dt^2} &= -2\omega\sin\lambda\frac{dx}{dt} \\ \frac{d^2z}{dt^2} &= -g + 2\omega\cos\lambda\frac{dx}{dt} \end{aligned}\right\} \tag{2}$$

Integrating the first equation,

$$\frac{dx}{dt} = -2\omega\cos\lambda(z) + \text{constant}. \tag{3}$$

Using initial conditions, this becomes

$$\frac{dx}{dt} = (v_0\cos\alpha - 2\omega z\cos\lambda) \tag{4}$$

Inserting this in the third of (2), we get

$$\frac{d^2z}{dt^2} = -g + 2\omega\cos\lambda(v_0\cos\alpha - 2\omega z\cos\lambda) \tag{5}$$

We neglect the second term in RHS of (5) since ω is very small compared to g. Thus, we write:

$$\frac{d^2z}{dt} = -g \tag{6}$$

Integrating with due care of initial conditions, we get

$$\frac{dz}{dt} = -gt + v_0\sin\alpha \tag{7}$$

The maximum height is reached at $t = t^*$ given by

$$t^* = \frac{v_0 \sin\alpha}{g} \tag{8}$$

From the second equation in (2) and using (4),

$$\frac{d^2y}{dt^2} = -2\omega \sin\lambda(v_0 \cos\alpha - 2\omega z \cos\lambda) \tag{9}$$

Integrating and using initial condition,

$$\frac{dy}{dt} = -2\omega v_0 t \sin\lambda \cos\alpha \tag{10}$$

where we have neglected the term containing ω^2.

Integrating (10) once more, we get

$$y = -\omega v_0 t^2 \sin\lambda \cos\alpha \tag{11}$$

The projectile reaches the ground after time

$$t = 2t^* = \frac{2v_0 \sin\alpha}{g} \tag{12}$$

During this time the lateral displacement

$$y = -\omega v_0 \sin\lambda \cos\alpha \left(\frac{2v_0 \sin\alpha}{g}\right)^2 = -\frac{4\omega v_0^3}{g^2} \sin\lambda \cos\alpha \sin^2\alpha$$

towards the south. **(Proved)**

EXAMPLE 6.5 *A river runs southward along a channel of width D in northern latitude λ. It is found that the water level at the east bank is different from the level of the west bank. Which one will be higher? Calculate the difference in height.* **[C.U. (HONS.) 2004]**

Solution. Since the river runs southward, Eqn (6.33) of motion becomes

$$m\ddot{y} = 2m\omega \sin\lambda \dot{x} \tag{1}$$

If v be the velocity of the river, the Coriolis force on water element of mass m is

$$F = 2m\omega v \sin\lambda \tag{2}$$

This force acts towards west and so the water level at western bank will be higher (FIG. 6.12). Let H be the difference in water level. The forces acting on the water element m at western bank are F acting horizontally, mg acting vertically downward and the resultant R of forces F and mg. The force R is balanced by the reaction force N acting normally to the water level. Then by Lami's theorem,

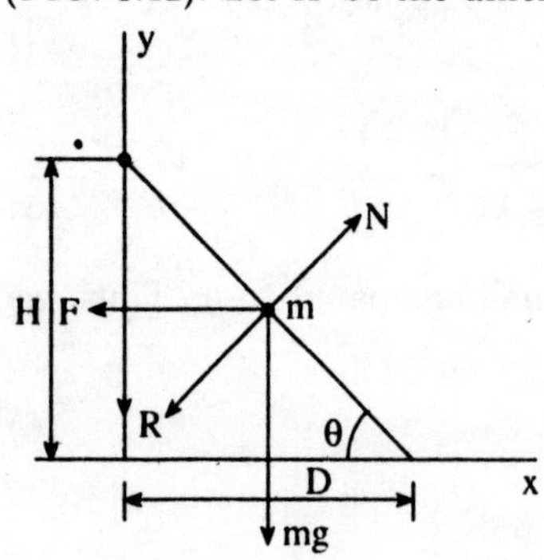

FIG. 6.12

$$\frac{F}{\sin(180° - \theta)} = \frac{mg}{\sin(90° + \theta)}$$

$$\text{i.e., } \tan\theta = \frac{F}{mg} \tag{3}$$

Again, from the FIG. 6.12, we have

$$\tan\theta = \frac{H}{D} \tag{4}$$

From (3) and (4),

$$\frac{H}{D} = \frac{F}{mg}, \text{ i.e., } H = \frac{FD}{mg} \tag{5}$$

Using (2), we get

$$H = \frac{2m\omega v}{mg}\sin\lambda = \frac{2\omega v D}{g}\sin\lambda$$

EXAMPLE 6.6 *The mass m of a conical pendulum of length l moves in a horizontal circle of radius R. Prove that the speed is*

$$\frac{R\sqrt{g}}{\sqrt[4]{l^2 - R^2}}$$

and the tension in the string is $\dfrac{mgl}{\sqrt{l^2 - R^2}}$.

Solution. We take unit vector **i** and **k** on the plane of the paper and the unit vector **j** perpendicular to it (FIG. 6.13). The position vector of the mass from O is

$$\mathbf{r} = \mathbf{i}R \tag{1}$$

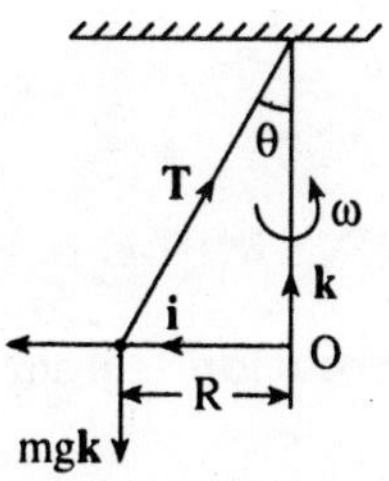

FIG. 6.13

Centrifugal force on the mass

$$= -m[\boldsymbol{\omega} \times (\boldsymbol{\omega} \times \mathbf{r})]$$
$$= -m[\mathbf{k}\omega \times (\mathbf{k}\omega \times \mathbf{i}R)] = \mathbf{i}m\omega^2 R \tag{2}$$

Tension **T** acting on the mass is

$$\mathbf{T} = -\mathbf{i}T\sin\theta + \mathbf{k}T\cos\theta \tag{3}$$

Then, in equilibrium position, we must have

$$\left.\begin{aligned} -\mathbf{i}T\sin\theta + \mathbf{k}T\cos\theta + \mathbf{i}m\omega^2 R - \mathbf{k}mg = 0 \\ \text{i.e., } \mathbf{i}(m\omega^2 R - T\sin\theta) + \mathbf{k}(T\cos\theta - mg) = 0 \end{aligned}\right\} \tag{4}$$

Hence,

$$\left.\begin{aligned} m\omega^2 R - T\sin\theta = 0 \\ T\cos\theta - mg = 0 \end{aligned}\right\} \tag{5}$$

Solving,

$$T = \frac{mg}{\cos\theta} = \frac{mgl}{\sqrt{l^2 - R^2}} \tag{6}$$

and

$$m\omega^2 R = T\sin\theta = mg\frac{\sin\theta}{\cos\theta} = mg\tan\theta$$

$$\text{i.e., } \omega^2 R = g\frac{R}{\sqrt{l^2 - R^2}}$$

$$\text{i.e., } \omega = \sqrt{g}\frac{1}{\sqrt[4]{l^2 - R^2}} \tag{7}$$

Hence, the angular speed of the mass

$$= \omega R = \frac{\sqrt{g}R}{\sqrt[4]{l^2 - R^2}} \quad \textbf{(Proved)} \tag{8}$$

EXAMPLE 6.7 *A projectile of mass m is thrown vertically down with velocity v from a height H. Determine the shift of the projectile when it hits the ground.*

Solution. Proceeding as in Example 6.1, we find shift of the projectile at any instant

$$x = \omega \cos \lambda \left(vt^2 + \frac{1}{3} gt^3 \right) \tag{1}$$

If t^* be the time to reach the ground, then

$$H = vt^* + \frac{1}{2} g(t^*)^2 \tag{2}$$

This yields

$$t^* = \frac{1}{g}\left(\sqrt{v^2 + 2gH} - v\right) \tag{3}$$

Hence, the shift of the projectile at $t = t^*$ is

$$\begin{aligned} x &= \frac{\omega \cos \lambda}{g^2}\left[v\left(\sqrt{v^2+2gH} - v\right)^2 + \frac{1}{3}\left(\sqrt{v^2+2gH} - v\right)^3 \right] \\ &= \frac{\omega \cos \lambda}{3g^2}\left(\sqrt{v^2+2gH} - v\right)^2 \left(\sqrt{v^2+2gH} + 2v\right) \end{aligned}$$

EXAMPLE 6.8 *Prove that the trajectory of the Foucault's pendulum is, in general, an ellipse.*

Solution. From Eqns (6.46) and (6.47), we have the equations of motion of the pendulum

$$\ddot{x} - 2\omega_z \dot{y} + \xi^2 x = 0 \tag{1}$$

$$\ddot{y} + 2\omega_z \dot{x} + \xi^2 y = 0 \tag{2}$$

where

$$\left.\begin{aligned} \omega_z &= \omega \sin \lambda \\ \xi^2 &= g/L \end{aligned}\right\} \tag{3}$$

Multiplying (2) by complex number $i = \sqrt{-1}$ and adding with (1), we get

$$(\ddot{x} + i\ddot{y}) + 2\omega_z(i\dot{x} - \dot{y}) + \xi^2(iy + x) = 0$$

$$\text{i.e., } (\ddot{x} + i\ddot{y}) + 2\omega_z i(\dot{x} + i\dot{y}) + \xi^2(x + iy) = 0 \tag{4}$$

Introducing the complex function

$$u = x + iy \tag{5}$$

we write (4) in the form:

$$\ddot{u} + 2\omega_z i\dot{u} + \xi^2 u = 0 \tag{6}$$

Let its solution be

$$u = ce^{-\alpha t} \tag{7}$$

Then the characteristic equation is

$$\alpha^2 + 2i\omega_z\alpha + \xi^2 = 0 \tag{8}$$

This yields

$$\alpha = \frac{1}{2}\left[-2i\omega_z \pm \sqrt{-4\omega_z^2 - 4\xi^2}\right] = -i\omega_z \pm i\sqrt{\omega_z^2 + \xi^2} \tag{9}$$

Since $\omega_z^2 = \omega^2 \sin^2 \lambda$ is small compared to $\xi^2 = g/L$, we write:

$$\alpha = -i\omega_z \pm i\xi \tag{10}$$

Hence, the general solution of (6) is

$$u = \left(C_1 e^{i\xi t} + C_2 e^{-i\xi t}\right) e^{-i\omega_z t} \approx C_1 e^{i\xi t} + C_2 e^{-i\xi t}$$

$$\text{i.e., } x + iy = C_1(\cos \xi t + i \sin \xi t) + C_2(\cos \xi t - i \sin \xi t)$$

$$= (C_1 + C_2) \cos \xi t + (C_1 - C_2) i \sin \xi t$$

Hence,

$$\left.\begin{array}{l} x = A \cos \xi t,\ A = C_1 + C_2 \\ y = B \sin \xi t,\ B = C_1 - C_2 \end{array}\right\} \tag{11}$$

Eliminating t,

$$\frac{x^2}{A^2} + \frac{y^2}{B^2} = 1 \tag{12}$$

which is the equation of an ellipse with its centre at the origin. Hence proved.

EXERCISE

1. An xyz coordinate system is fixed in space while an $x'y'z'$ coordinate system with same origin is rotating with angular velocity $\omega = 2t\mathbf{i} + t^2\mathbf{j} + (2 + 4t)\mathbf{k}$ relative to the fixed coordinate system. The position vector of a particle as observed in $x'y'z'$ system is given by

   $$r = \left(2 + 3t^2\right) \mathbf{i} - 3t\mathbf{j} + 5t^3\mathbf{k}$$

   Find (a) apparent velocity and acceleration and (b) true velocity and acceleration at time $t = 1$.
2. Prove that the centrifugal force acting on a particle of mass m on the earth's surface is a vector directed away from the earth and perpendicular to angular velocity vector ω. Determine its magnitude.
3. Prove that the observed acceleration due to gravity at the latitude λ is related to its value g by the equation

   $$g_\lambda^2 = \left(g \cos \lambda - \omega^2 R \cos \lambda\right)^2 + (g \sin \lambda)^2$$

4. A river of width D flows northward with a speed V at latitude λ. Prove that the left bank of the river will be higher than the right by an amount

   $$(2D\omega V \sin \lambda) \left(g^2 + 4\omega^2 V^2 \sin^2 \lambda\right)^{-1/2}$$

   where ω is the angular speed of the earth.
5. In a rotating frame of reference, deduce the operator

   $$\frac{d}{dt} = \frac{d'}{dt} + \omega \times$$

6. A bead slides along a smooth circular wire in a vertical plane. The wire is rotating about its vertical diameter with uniform angular velocity ω. If the bead were released on the top of the wire, show that it will not reach the lowest point, if

   $$\omega > \sqrt{\frac{2g}{R}}$$

   where R is the radius of the wire.

7. A particle of mass m falls freely from rest under gravity. Neglecting the variation of g, show that the amount of shift towards east when it reaches the ground is given by

$$\frac{2}{3}\omega\cos\lambda\left(\frac{2h^3}{g}\right)^{1/2}.$$

8. A projectile is fired with velocity v at an angle α with the horizon from a place having latitude λ. Show that its time of flight is

$$\frac{2v\sin\alpha}{g}\left(1+\frac{2\omega v}{g}\cos\lambda\cos\alpha\right)$$

CHAPTER 7

Lagrangian Formulation

In Newtonian mechanics, we have to depend on the coordinate systems for specific problem. Moreover, the forces of constraints appearing in the equations of motion complicate the problem and finding the solution may be difficult and even impossible.

In Lagrangian mechanics, the above difficulties are removed by using generalised coordinates. Here the Newtonian theory is reformulated in terms of scalars like potential and kinetic energies.

7.1 Basic Concepts

• Generalised Coordinates

A minimum number of independent coordinates required to describe completely the state of configuration of a dynamical system at any given time is called ***generalised coordinates***. These coordinates may not necessarily have the dimension of length; these may be angles or other convenient quantities. We shall denote these coordinates as $q_1, q_2, q_3, \ldots, q_n$, where n is the total number of coordinates. For example:

(i) For a particle moving on the surface of a sphere, the generalised coordinates are $q_1 = \theta$ and $q_2 = \phi$, θ and ϕ being polar coordinates on the surface.

(ii) For a simple pendulum, the only generalised coordinate is $q_1 = \theta$, where θ is the angle of inclination with the vertical.

The time derivatives of generalised coordinates are called ***generalised velocities***. The generalised velocities need not have the dimension of velocity.

• Degrees of Freedom

The minimum number of independent coordinates required to specify a dynamical system consistent with the constraints is called the ***degrees of freedom*** of that system. For example:

(i) For a particle constrained to move freely in a plane, two coordinates $q_1 = x$, $q_2 = y$ (or $q_1 = r$, $q_2 = \theta$) are sufficient to describe the motion. Hence, the particle has two degrees of freedom.

(ii) For a particle moving freely in space, three coordinates are sufficient to describe the motion. Hence, the particle has three degrees of freedom. If a dynamical system consisting of N particles move freely in space, $3N$ independent coordinates are required to describe the motion. Thus, the system has $3N$ degrees of freedom.

• Constraints

Any restriction imposed on the motion of a system in some way is called ***constraint*** and the motion is said to be ***constrained***. When constraints are imposed on a system, its number of degrees of freedom is reduced. For example, if a bead slides down a circular wire, then only one coordinate $q_1 = \theta$ is sufficient to describe the motion. Similarly, if a sphere rolls down an inclined plane, then only one coordinate $q_1 = \theta$ (angle of rotation of the sphere) is needed to describe the motion.

• Holonomic Constraints

A constraint which can be expressed in the form of an equation relating the coordinates of the system and time is called ***holonomic***; otherwise, it is ***non-holonomic***. For a system of N particles, the general form of such equation is

$$f_m(\mathbf{r}_1, \mathbf{r}_2, \mathbf{r}_3, \ldots, \mathbf{r}_N, t) = 0 \tag{7.1}$$

where m is the number of constraints. Using these equations, it is possible to eliminate m-numbers of coordinates so that the number of independent coordinates become

$$n = 3N - m \tag{7.2}$$

which is now the degree of freedom of the system.

Constraints are also classified as (i) ***scleronomous*** and (ii) ***rheonomous***. In the former types, the constraints are independent of time, while in the latter type the constraints are explicitly dependent on time.

For a pendulum with fixed support, the constraint is scleronomous; but if the point of support moves, or if the length changes with time, the constraint is rheonomous.

• Virtual Work

Any arbitrary instantaneous and infinitesimal displacement which is independent of time and consistent with the constraints of a dynamical system is called ***virtual displacement***. The work done by the system due to virtual displacement is called ***virtual work***.

7.2 Principle of Virtual Work

We consider a system consisting of n particles. If the system is in equilibrium, then the resultant force on each particle is zero, that is

$$\mathbf{F}_i = 0, \; i = 1, 2, \ldots, n \tag{7.3}$$

We now give virtual displacement $\delta\mathbf{r}_i$ to each particle. Then the total virtual work for all the particles is

$$\delta W = \sum_{i=1}^{n} \mathbf{F}_i \cdot \delta\mathbf{r}_i = 0 \tag{7.4}$$

Thus, work done for an arbitrary virtual displacement of a system from the position of equilibrium is zero. This is the ***principle of virtual work***.

Now, if the particle is constrained, the total force $\mathbf{F}_i$ on the i-th particle can be expressed as

$$\mathbf{F}_i = \mathbf{F}_i^a + \mathbf{f}_i \tag{7.5}$$

where $\mathbf{F}_i^a$ is the applied force and $\mathbf{f}_i$ the force of constraints.

Then, Eqn (7.4) can be written as

$$\sum_{i=1}^{n} \mathbf{F}_i^a \cdot \delta \mathbf{r}_i + \sum_{i=1}^{n} \mathbf{f}_i \cdot \delta \mathbf{r}_i = 0. \tag{7.6}$$

We consider only such systems where the virtual work of forces of constraints is zero. This is possible when frictional forces are not present; because these forces act along a direction opposite to that of displacement. Then, Eqn (7.6) assumes the form:

$$\sum_{i=1}^{n} \mathbf{F}_i^a \cdot \delta \mathbf{r}_i = 0 \tag{7.7}$$

which shows that, for equilibrium of a system, the virtual work of applied forces is zero.

The above principle of virtual work relates to a static system. It was extended to dynamical system by D'Alembert. It is to be noted that the virtual displacements $\delta \mathbf{r}_i$ are not independent of each other so that we cannot treat them as completely arbitrary and equate their coefficients to zero.

7.3 D'Alembert's Principle

By Newton's second law of motion, the equation of motion of i-th particle is

$$\mathbf{F}_i = \dot{\mathbf{p}}_i \tag{7.8}$$

where $\mathbf{p}_i$ is the linear momentum of i-th particle. We can write Eqn (7.8) in the form:

$$\mathbf{F}_i - \dot{\mathbf{p}}_i = 0 \tag{7.9}$$

which states that the particle in the system will be in equilibrium under a force equal to actual force $\mathbf{F}_i$ plus a reversed effective force $\dot{\mathbf{p}}_i$. Thus, for virtual displacements $\delta \mathbf{r}_i$,

$$\sum_{i=1}^{n} (\mathbf{F}_i - \dot{\mathbf{p}}_i)\, \delta \mathbf{r}_i = 0 \tag{7.10}$$

For a constrained particle, $\mathbf{F}_i = \mathbf{F}_i^a + \mathbf{f}_i$. Then the Eqn (7.10) can be written as

$$\sum_{i=1}^{n} (\mathbf{F}_i^a - \dot{\mathbf{p}}_i) \cdot \delta \mathbf{r}_i + \sum_{i=1}^{n} \mathbf{f}_i \cdot \delta \mathbf{r}_i = 0 \tag{7.11}$$

As before, we consider the system for which the constraints do no virtual work, so that $\Sigma \mathbf{f}_i \cdot \delta \mathbf{r}_i = 0$. Hence,

$$\sum_{i=1}^{n} (\mathbf{F}_i^a - \dot{\mathbf{p}}_i) \cdot \delta \mathbf{r}_i = 0 \tag{7.12}$$

This is known as ***D'Alembert's principle*** of virtual work.

EXAMPLE 7.1 *A uniform rod OA of length $2l$, free to turn about its end O, revolves with uniform angular velocity ω about vertical axis through O. If it is inclined at constant angle α with the vertical, show that $\cos\alpha = \dfrac{3g}{4\omega^2 l}$.*

Solution. We take a point P on the rod at distance $OP = x$ (FIG. 7.1). For a small element dx, acceleration of P is $\omega^2 PQ$ along PQ.

Hence, the reversed effective force is $dF = \left(\dfrac{dx}{2l} \cdot m\right) \omega^2 x \sin\alpha$ along QP.

Taking moments about O of all the reversed effective forces acting on the rod and the weight mg of the rod, we get

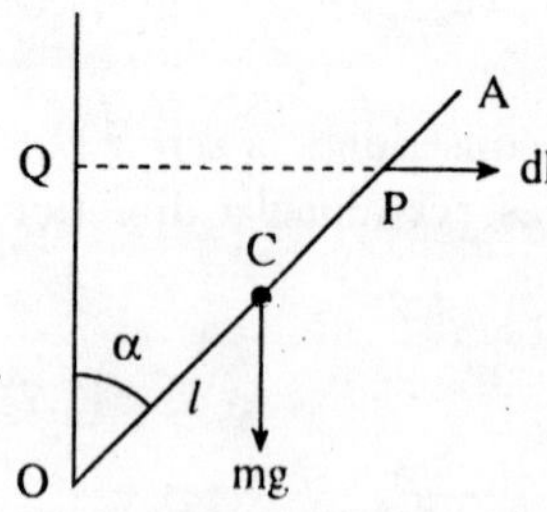

FIG. 7.1

$$mg \cdot l \sin\alpha = \int \left(\frac{dx}{2l} m\right) \omega^2 x \sin\alpha (x \cos\alpha)$$

$$= \frac{m\omega^2}{2l} \sin\alpha \cos\alpha \int_0^{2l} x^2 dx$$

$$\text{or, } gl \sin\alpha = \frac{\omega^2}{2l} \cos\alpha \sin\alpha \frac{8l^3}{3}$$

which gives

$$\cos\alpha = \frac{3g}{4\omega^2 l}$$

We have rejected the case for $\sin\alpha = 0$, i.e., $\alpha = 0$, because in that case the rod hangs vertically. For a real value of $\cos\alpha$, $3g < 4\omega^2 l$.

EXAMPLE 7.2 *Two uniform spheres, each of mass M and radius a are firmly fixed to the ends of two uniform thin rods, each of mass m and length l, and the other ends of the rods are freely hinged to a point. The whole system revolves, as in the Governor of a steam engine about a vertical line through O with angular velocity ω. Show that, when the motion is steady, the rods are inclined to the vertical at an angle θ given by*

$$\cos\theta = \frac{g}{\omega^2} \frac{M(l+a) + \frac{1}{2}lm}{M(l+a)^2 + \frac{1}{3}l^2 m}$$

Solution. The arrangement is shown in the FIG. 7.2.

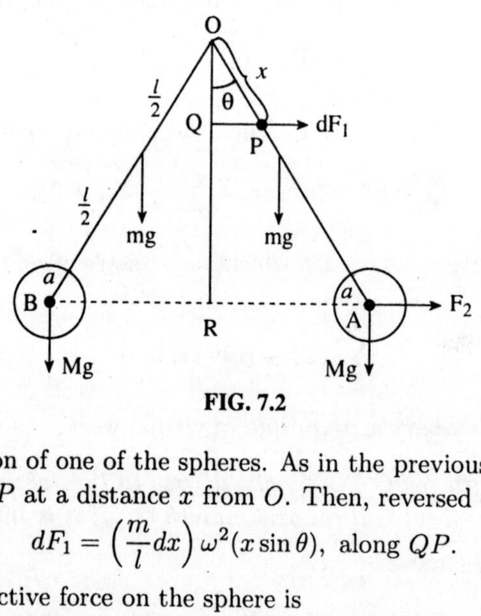

FIG. 7.2

We consider the motion of one of the spheres. As in the previous problem, we take a small element dx of the rod at P at a distance x from O. Then, reversed effective force at P is

$$dF_1 = \left(\frac{m}{l} dx\right) \omega^2 (x \sin\theta), \text{ along } QP. \tag{1}$$

Also, the reversed effective force on the sphere is

$$F_2 = M\omega^2 (l+a) \sin\theta, \text{ along } RA. \tag{2}$$

The external forces are Mg and mg.

All the forces are shown in the figure.

Taking moments of all forces about O for the system remaining on one side of the vertical, we have

$$\int dF_1 x\cos\theta + F_2(l+a)\cos\theta = Mg(l+a)\sin\theta + mg\frac{l}{2}\sin\theta$$

or, $$\int_0^l \frac{m}{l}\omega^2\sin\theta\cos\theta x^2 dx + M\omega^2(l+a)^2\sin\theta\cos\theta = Mg(l+a)\sin\theta + mg\frac{l}{2}\sin\theta$$

or, $$\omega^2\sin\theta\cos\theta\left[\frac{1}{3}ml^2 + M(l+a)^2\right] = g\sin\theta\left[M(l+a)+\frac{ml}{2}\right] \tag{3}$$

Assuming that $\sin\theta \neq 0$, i.e., $\theta \neq 0$, we get

$$\cos\theta = \frac{g}{\omega^2}\frac{M(l+a)+ml/2}{M(l+a)^2+ml^2/3}$$

7.4 Generalised Force in Holonomic System

We consider a system of N particles with m-number of constraints, so that the system can be described by the generalised coordinates

$$q_1, q_2, q_3, \dots, q_n, \quad n = 3N - m \tag{7.13}$$

Let the position vector of the i-th particle in terms of generalised coordinates be given by

$$\mathbf{r}_i = \mathbf{r}_i(q_1, q_2, \dots, q_n, t) \tag{7.14}$$

Then the velocity of the i-th particle is

$$\dot{\mathbf{r}}_i = \sum_{k=1}^{n}\frac{\partial \mathbf{r}_i}{\partial q_k}\dot{q}_k + \frac{\partial \mathbf{r}_i}{\partial t} \tag{7.15}$$

where $\dot{q}_k$ are the ***generalised velocities***.

The virtual displacement $\delta\mathbf{r}_i$ at a particular instant of time (hence, variation w.r.t. time is zero) is

$$\delta\mathbf{r}_i = \sum_{k=1}^{n}\frac{\partial \mathbf{r}_i}{\partial q_k}\delta q_k \tag{7.16}$$

Virtual work done by the applied forces $\mathbf{F}_i$ is

$$\delta W = \sum_{i=1}^{N}\mathbf{F}_i\cdot\delta\mathbf{r}_i = \sum_{i=1}^{N}\mathbf{F}_i\cdot\sum_{k=1}^{n}\frac{\partial \mathbf{r}_i}{\partial q_k}\delta q_k \text{ [using (7.16)]}$$

$$= \sum_{k=1}^{n}\sum_{i=1}^{N}\left[\mathbf{F}_i\cdot\frac{\partial \mathbf{r}_i}{\partial q_k}\right]\delta q_k = \sum_{k=1}^{n}Q_k\delta q_k \tag{7.17}$$

where

$$Q_k = \sum_{i=1}^{N}\mathbf{F}_i\cdot\frac{\partial \mathbf{r}_i}{\partial q_k} \tag{7.18}$$

The coefficients Q_k are called ***generalised forces*** associated with coordinates q_k. The generalised forces need not have the dimension of force but the quantity $Q_k\delta q_k$ should have the dimension of work.

For a conservative system with potential energy

$$V = V(q_1, q_2, \dots, q_n) \tag{7.19}$$

we have

$$\delta W = -\delta V, \text{ or, } \delta W = -\sum_{k=1}^{n} \frac{\partial V}{\partial q_k} \delta q_k \tag{7.20}$$

Thus, from Eqns (7.17) and (7.20), we have

$$\sum_{k=1}^{n} Q_k \delta q_k = -\sum_{k=1}^{n} \frac{\partial V}{\partial q_k} \delta q_k \tag{7.21}$$

The quantities δq_k are independent and arbitrary. Hence, from Eqn (7.21), we have

$$Q_k = -\frac{\partial V}{\partial q_k} \tag{7.22}$$

7.5 Lagrange's Equation

We consider a holonomic system of N particles with $q_1, q_2, \ldots, q_n$ as generalised coordinates. The mass of i-th particle is m_i, its position vector is $\mathbf{r_i}$ and applied force acting on it is $\mathbf{F_i}$. We restrict ourselves to a system where constraints do no virtual work. Then, by D'Alembert's principle,

$$\sum_{i=1}^{N} (\mathbf{F_i} - m_i \ddot{\mathbf{r}}_\mathbf{i})\, \delta \mathbf{r_i} = 0 \tag{7.23}$$

Now, we have

$$\mathbf{r_i} = \mathbf{r_i}(q_1, q_2, \ldots, q_n, t) \tag{7.24}$$

Hence,

$$\begin{aligned} \dot{\mathbf{r}}_\mathbf{i} &= \frac{d\mathbf{r_i}}{dt} \\ &= \frac{\partial \mathbf{r_i}}{\partial q_1}\frac{dq_1}{dt} + \frac{\partial \mathbf{r_i}}{\partial q_2}\frac{dq_2}{dt} + \cdots + \frac{\partial \mathbf{r_i}}{\partial q_n}\frac{dq_n}{dt} + \frac{\partial \mathbf{r_i}}{\partial t} \end{aligned}$$

$$\text{i.e., } \dot{\mathbf{r}}_\mathbf{i} = \frac{\partial \mathbf{r_i}}{\partial q_1}\dot{q}_1 + \frac{\partial \mathbf{r_i}}{\partial q_2}\dot{q}_2 + \cdots + \frac{\partial \mathbf{r_i}}{\partial q_n}\dot{q}_n + \frac{\partial \mathbf{r_i}}{\partial t} \tag{7.25}$$

Differentiating both sides of Eqn (7.25) partially w.r.t. $\dot{q}_k$, we get

$$\frac{\partial \dot{\mathbf{r}}_\mathbf{i}}{\partial \dot{q}_k} = \frac{\partial \mathbf{r_i}}{\partial q_k}, \; (k = 1, 2, \ldots, n) \tag{7.26}$$

Now

$$\frac{d}{dt}\left(\frac{\partial \mathbf{r_i}}{\partial q_j}\right) = \sum_{k=1}^{n} \frac{\partial^2 \mathbf{r_i}}{\partial q_k \partial q_j}\dot{q}_k + \frac{\partial^2 \mathbf{r_i}}{\partial q_j \partial t} \tag{7.27}$$

which has been obtained by considering $\frac{\partial \mathbf{r_i}}{\partial q_j}$ as a function of generalised coordinates q_k and time t.

Again, from Eqn (7.25), we have

$$\frac{\partial}{\partial q_j}\left(\frac{d\mathbf{r_i}}{dt}\right) = \sum_{k=1}^{n} \frac{\partial^2 \mathbf{r_i}}{\partial q_k \partial q_j}\dot{q}_k + \frac{\partial^2 \mathbf{r_i}}{\partial q_j \partial t} \tag{7.28}$$

From Eqns (7.27) and (7.28), we get

$$\frac{d}{dt}\left(\frac{\partial \mathbf{r_i}}{\partial q_j}\right) = \frac{\partial \dot{\mathbf{r}}_\mathbf{i}}{\partial q_j}, \; (j = 1, 2, \ldots, n) \tag{7.29}$$

Let us now consider Eqn (7.23). We already have from Eqn (7.17)

$$\sum_{i=1}^{N} \mathbf{F_i} \cdot \delta \mathbf{r_i} = \sum_{k=1}^{n} Q_k \delta q_k \tag{7.30}$$

Now

$$\sum_{i=1}^{N} \dot{\mathbf{p}}_\mathbf{i} \cdot \delta \mathbf{r_i} = \sum_{i=1}^{N} m_i \ddot{\mathbf{r}}_\mathbf{i} \cdot \sum_{k=1}^{n} \frac{\partial \mathbf{r_i}}{\partial q_k} \delta q_k$$

$$\text{or,} \quad \sum_{i=1}^{N} \dot{\mathbf{p}}_\mathbf{i} \cdot \delta \mathbf{r_i} = \sum_{k=1}^{n} \left[\sum_{i=1}^{N} m_i \ddot{\mathbf{r}}_\mathbf{i} \cdot \frac{\partial \mathbf{r_i}}{\partial q_k} \right] \delta q_k \tag{7.31}$$

The kinetic energy T of the system is

$$T = \sum_{i=1}^{N} \frac{1}{2} m_i \dot{\mathbf{r}}_\mathbf{i}^2 \tag{7.32}$$

Hence,

$$\frac{\partial T}{\partial \dot{q}_k} = \sum_{i=1}^{N} m_i \dot{\mathbf{r}}_\mathbf{i} \frac{\partial \dot{\mathbf{r}}_\mathbf{i}}{\partial \dot{q}_k} = \sum_{i=1}^{N} m_i \dot{\mathbf{r}}_\mathbf{i} \cdot \frac{\partial \mathbf{r_i}}{\partial q_k} \quad \text{[using Eqn (7.26)].} \tag{7.33}$$

Also

$$\frac{\partial T}{\partial q_k} = \sum_{i=1}^{N} m_i \dot{\mathbf{r}}_\mathbf{i} \cdot \frac{\partial \dot{\mathbf{r}}_\mathbf{i}}{\partial q_k} \tag{7.34}$$

From Eqn (7.33), we get

$$\begin{aligned}
\frac{d}{dt}\left(\frac{\partial T}{\partial \dot{q}_k}\right) &= \sum_{i=1}^{N} m_i \ddot{\mathbf{r}}_\mathbf{i} \frac{\partial \mathbf{r_i}}{\partial q_k} + \sum_{i=1}^{N} m_i \dot{\mathbf{r}}_\mathbf{i} \frac{d}{dt}\left(\frac{\partial \mathbf{r_i}}{\partial q_k}\right) \\
&= \sum_{i=1}^{N} m_i \ddot{\mathbf{r}}_\mathbf{i} \frac{\partial \mathbf{r_i}}{\partial q_k} + \sum_{i=1}^{N} m_i \dot{\mathbf{r}}_\mathbf{i} \frac{\partial \dot{\mathbf{r}}_\mathbf{i}}{\partial q_k} \quad \text{[using Eqn (7.29)]} \\
&= \sum_{i=1}^{N} m_i \ddot{\mathbf{r}}_\mathbf{i} \frac{\partial \mathbf{r_i}}{\partial q_k} + \frac{\partial T}{\partial q_k} \quad \text{[using Eqn (7.34)]}
\end{aligned}$$

Hence,

$$\sum_{i=1}^{N} m_i \ddot{\mathbf{r}}_\mathbf{i} \frac{\partial \mathbf{r_i}}{\partial q_k} = \frac{d}{dt}\left(\frac{\partial T}{\partial \dot{q}_k}\right) - \frac{\partial T}{\partial q_k} \tag{7.35}$$

Substituting the expression for LHS of Eqn (7.35) into Eqn (7.31), we get

$$\sum_{i=1}^{N} \dot{\mathbf{p}}_\mathbf{i} \cdot \delta \mathbf{r_i} = \sum_{k=1}^{n} \left[\frac{d}{dt}\left(\frac{\partial T}{\partial \dot{q}_k}\right) - \frac{\partial T}{\partial q_k} \right] \delta q_k \tag{7.36}$$

From Eqns (7.23), (7.30) and (7.36), we have

$$\sum_{k=1}^{n} \left[\frac{d}{dt}\left(\frac{\partial T}{\partial \dot{q}_k}\right) - \frac{\partial T}{\partial q_k} - Q_k \right] \delta q_k = 0 \tag{7.37}$$

Since the constraints are holonomic, virtual displacements $\delta q_1, \delta q_2, \ldots$, etc. are independent and arbitrary. Hence, the quantity within the brackets in Eqn (7.37) should vanish for every δq_k.

Thus,

$$\frac{d}{dt}\left(\frac{\partial T}{\partial \dot{q}_k}\right)-\frac{\partial T}{\partial q_k}-Q_k=0$$

$$\text{or,}\quad \frac{d}{dt}\left(\frac{\partial T}{\partial \dot{q}_k}\right)-\frac{\partial T}{\partial q_k}=Q_k,\ (k=1,2,\dots,n) \tag{7.38}$$

This is ***Lagrange's equation for holonomic system.***

For a conservative system,

$$Q_k=-\frac{\partial V}{\partial q_k},\ (k=1,2,\dots,n) \tag{7.39}$$

Hence, in this case, equation assumes the form:

$$\frac{d}{dt}\left(\frac{\partial T}{\partial \dot{q}_k}\right)-\frac{\partial T}{\partial q_k}=-\frac{\partial V}{\partial q_k}\ \text{or,}\ \frac{d}{dt}\left(\frac{\partial T}{\partial \dot{q}_k}\right)-\frac{\partial}{\partial q_k}(T-V)=0 \tag{7.40}$$

The scalar potential V is a function of generalised coordinates only and does not depend on time. Hence, we can write Eqn (7.40) as

$$\frac{d}{dt}\left[\frac{\partial (T-V)}{\partial \dot{q}_k}\right]-\frac{\partial (T-V)}{\partial q_k}=0 \tag{7.41}$$

We define a new function

$$L=T-V \tag{7.42}$$

which is called the ***Lagrangian*** of the system. Then the equation assumes the form:

$$\frac{d}{dt}\left(\frac{\partial L}{\partial \dot{q}_k}\right)-\frac{\partial L}{\partial q_k}=0,\ (k=1,2,\dots,n) \tag{7.43}$$

These equations are known as ***Lagrange's equations*** for a conservative and holonomic dynamical system.

Lagrange's equation can also be put in the form of Eqn (7.43) when the system is not conservative, provided the generalised forces are obtained from a function $U(q_k,\dot{q}_k)$ given by

$$Q_k=-\frac{\partial U}{\partial q_k}+\frac{d}{dt}\left(\frac{\partial U}{\partial \dot{q}_k}\right) \tag{7.44}$$

In such a case, the Lagrangian is given by

$$L=T-U \tag{7.45}$$

where $U(q_k,\dot{q}_k)$ is called velocity-dependent potential or ***generalised potential.***

We come across such a potential when we consider a charge moving in an electromagnetic force-field.

EXAMPLE 7.3 *Obtain the equations of motion of a particle in space by Lagrangian method in Cartesian coordinates.*

Solution. If $P(x,y,z)$ be the position of a particle of mass m at any instant of time t, then kinetic energy is

$$T=\frac{1}{2}m\left(\dot{x}^2+\dot{y}^2+\dot{z}^2\right) \tag{1}$$

Hence,

$$\frac{\partial T}{\partial x} = \frac{\partial T}{\partial y} = \frac{\partial T}{\partial z} = 0 \tag{2}$$

$$\text{and } \frac{\partial T}{\partial \dot{x}} = m\dot{x}, \frac{\partial T}{\partial \dot{y}} = m\dot{y}, \frac{\partial T}{\partial \dot{z}} = m\dot{z} \tag{3}$$

Here generalised coordinates are $q_1 = x, q_2 = y$ and $q_3 = z$.

The generalised force components are

$$Q_1 = F_x,\ Q_2 = F_y,\ Q_3 = F_z \tag{4}$$

Lagrange's equations of motion are

$$\frac{d}{dt}\left(\frac{\partial T}{\partial \dot{q}_k}\right) - \frac{\partial T}{\partial q_k} = Q_k \tag{5}$$

For $k = 1$,

$$\frac{d}{dt}\left(\frac{\partial T}{\partial \dot{x}}\right) - \frac{\partial T}{\partial x} = F_x \text{ or, } \frac{d}{dt}(m\dot{x}) = F_x \text{ [from (2) and (3)]}$$

$$\text{i.e., } m\ddot{x} = F_x \tag{6}$$

Similarly, we can get the other two equations of motion as $m\ddot{y} = F_y$ and $m\ddot{z} = F_z$.

EXAMPLE 7.4 *Obtain the equation of motion of a simple pendulum by Lagrangian method.*

Solution. Here we have only one generalised coordinate $q_1 = \theta$, where θ is the angular displacement of the simple pendulum from the vertical.

If l is the length of the simple pendulum and m is the mass of the bob, then kinetic energy

$$T = \frac{1}{2}mv^2 = \frac{1}{2}ml^2\dot{\theta}^2 \tag{1}$$

Potential energy at any position of the bob (taking zero at its vertical position) is

$$V = mgl(1 - \cos\theta) \tag{2}$$

Hence, the Lagrangian is

$$L = T - V = \frac{1}{2}ml^2\dot{\theta}^2 - mgl(1 - \cos\theta) \tag{3}$$

Thus,

$$\frac{\partial L}{\partial \dot{\theta}} = ml^2\dot{\theta}, \frac{\partial L}{\partial \theta} = -mgl\sin\theta \tag{4}$$

We substitute these values in Lagrange's equation of motion

$$\frac{d}{dt}\left(\frac{\partial L}{\partial \dot{\theta}}\right) - \frac{\partial L}{\partial \theta} = 0 \tag{5}$$

and get

$$\frac{d}{dt}(ml^2\dot{\theta}) + mgl\sin\theta = 0, \text{ i.e., } \ddot{\theta} + \frac{g}{l}\sin\theta = 0 \tag{6}$$

This is the equation of motion of a simple pendulum. If θ is taken to be small, then $\sin\theta = \theta$; the equation of motion assumes the form:

$$\ddot{\theta} + \frac{g}{l}\theta = 0. \tag{7}$$

EXAMPLE 7.5 *Derive equations of motion of a particle of mass m in plane polar coordinates.*

Solution. In plane polar coordinates,

$$x = r\cos\theta \text{ and } y = r\sin\theta \tag{1}$$

Hence,

$$\left.\begin{aligned}\dot{x} &= \dot{r}\cos\theta - r\sin\theta\dot{\theta}\\ \dot{y} &= \dot{r}\sin\theta + r\cos\theta\dot{\theta}\end{aligned}\right\} \tag{2}$$

The generalised coordinates are $q_1 = r, q_2 = \theta$. Kinetic energy is

$$T = \frac{1}{2}m\left(\dot{x}^2 + \dot{y}^2\right) = \frac{1}{2}m\left(\dot{r}^2 + r^2\dot{\theta}^2\right) \tag{3}$$

Generalised forces are $Q_k = \sum_{i=1}^{N} F_i \cdot \frac{\partial r_i}{\partial q_k}$ [see Eqn (7.18)].

Thus,

$$Q_r = \mathbf{F}\cdot\frac{\partial \mathbf{r}}{\partial r} = F_r \tag{4}$$

$$Q_\theta = \mathbf{F}\cdot\frac{\partial \mathbf{r}}{\partial \theta} = \mathbf{F}\cdot r\mathbf{n} = rF_\theta \tag{5}$$

Here **n** is unit vector perpendicular to the radial direction.

Hence, equation of motion in radial direction is

$$\frac{d}{dt}\left(\frac{\partial T}{\partial \dot{r}}\right) - \frac{\partial T}{\partial r} = Q_r, \text{ i.e., } \frac{d}{dt}(m\dot{r}) - mr\dot{\theta}^2 = F_r \tag{6}$$

Similarly, equation of motion in tangential direction is

$$\frac{d}{dt}\left(\frac{\partial T}{\partial \dot{\theta}}\right) - \frac{\partial T}{\partial \theta} = Q_\theta, \text{ i.e., } \frac{d}{dt}\left(mr^2\dot{\theta}\right) = rF_\theta \tag{7}$$

EXAMPLE 7.6 *A bead of mass m slides on a smooth uniform circular wire of radius R which is rotating with a constant angular velocity ω about a fixed vertical diameter. Find equation of motion of the bead.*

If the bead is released with no vertical velocity at the highest point, show that it remains steady at a point where the radius makes with vertical an angle α given by

$$\alpha = 2\cos^{-1}\left(\frac{g}{\omega^2 R}\right)^{1/2}$$

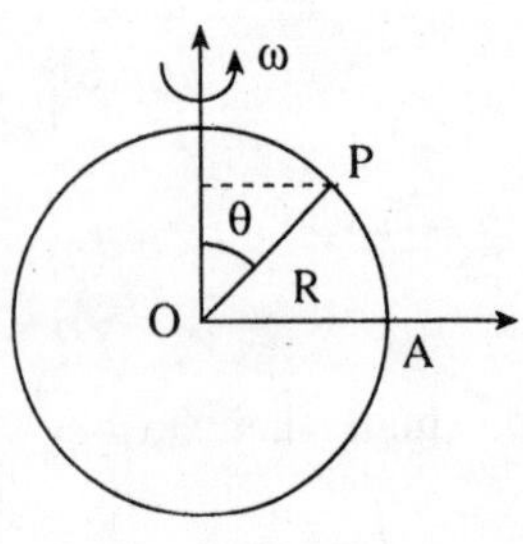

FIG. 7.3

Solution. The position of the bead at any instant is shown in FIG. 7.3. The velocity along the tangent is $R\dot{\theta}$; but the point P revolves around vertical axis, so that its velocity is $\omega R\sin\theta$. Hence, kinetic energy

$$T = \frac{1}{2}m\left(R^2\dot{\theta}^2 + \omega^2R^2\sin^2\theta\right)$$

Potential energy $V = mgR\cos\theta$ (OA is zero level)

Hence, Lagrangian is

$$L = T - V = \frac{1}{2}m\left(R^2\dot{\theta}^2 + \omega^2R^2\sin^2\theta\right) - mgR\cos\theta$$

Thus, we have only one generalised coordinate θ. Hence, equation of motion is given by

$$\frac{d}{dt}\left(\frac{\partial L}{\partial \dot{\theta}}\right) - \frac{\partial L}{\partial \theta} = 0$$

$$\text{i.e., } \frac{d}{dt}\left(mR^2\dot{\theta}\right) - \frac{m}{2}\omega^2 R^2 \sin 2\theta - mgR\sin\theta = 0$$

$$\text{or, } \ddot{\theta} - \frac{1}{2}\omega^2 \sin 2\theta - \frac{g}{R}\sin\theta = 0 \qquad (1)$$

• Second Part

Initial total energy of the particle is mgR.

Total energy at any position is

$$T + V = \frac{1}{2}m\left(R^2\dot{\theta}^2 + \omega^2 R^2 \sin^2\theta\right) + mgR\cos\theta$$

Hence, by conservation of energy,

$$\frac{1}{2}m\left(R^2\dot{\theta}^2 + \omega^2 R^2 \sin^2\theta\right) + mgR\cos\theta = mgR \qquad (2)$$

At the steady state of motion, the tangential velocity $R\dot{\theta}$ is zero. Hence, for steady state, we have from (2) for $\theta = \alpha$

$$\frac{1}{2}m\left(\omega^2 R^2 \sin^2\alpha\right) = mgR(1 - \cos\alpha)$$

$$\text{or, } 2\omega^2 R^2 \sin^2\frac{\alpha}{2}\cos^2\frac{\alpha}{2} = 2gR\sin^2\frac{\alpha}{2}$$

$$\text{or, } \omega^2 R^2 \cos^2\frac{\alpha}{2} = gR \quad \left[\text{assuming } \frac{\alpha}{2} \neq 0\right]$$

Hence,

$$\cos\frac{\alpha}{2} = \left(\frac{g}{\omega^2 R}\right)^{1/2}, \text{ i.e., } \alpha = 2\cos^{-1}\left(\frac{g}{\omega^2 R}\right)^{1/2}.$$

EXAMPLE 7.7 *Derive the equations of motion of a particle of mass m in a conservative force-field in cylindrical polar coordinates.*

Solution. In cylindrical polar coordinates,

$$x = r\cos\theta,\ y = r\sin\theta,\ z = z \qquad (1)$$

Hence,

$$\dot{x} = \dot{r}\cos\theta - r\sin\theta\dot{\theta},\ \dot{y} = \dot{r}\sin\theta + r\cos\theta\dot{\theta},\ \dot{z} = \dot{z} \qquad (2)$$

Thus, the kinetic energy is

$$T = \frac{1}{2}m\left(\dot{x}^2 + \dot{y}^2 + \dot{z}^2\right) = \frac{1}{2}m\left(\dot{r}^2 + r^2\dot{\theta}^2 + \dot{z}^2\right) \qquad (3)$$

For a conservative system,

$$V = V(r, \theta, z) \qquad (4)$$

Hence, the Lagrangian is

$$L = T - V = \frac{1}{2}m\left(\dot{r}^2 + r^2\dot{\theta}^2 + \dot{z}^2\right) - V(r, \theta, z) \qquad (5)$$

Here, we have three generalised coordinates

$$q_1 = r,\ q_2 = \theta,\ q_3 = z$$

Lagrange's equations of motion are

$$\frac{d}{dt}\left(\frac{\partial L}{\partial \dot{q}_k}\right) - \frac{\partial L}{\partial q_k} = 0, \ k = 1, 2, 3 \tag{6}$$

Hence, from (5) and (6), we get

$$m\left(\ddot{r} - r\dot{\theta}^2\right) = -\frac{\partial V}{\partial r}, \ \frac{d}{dt}(r^2\dot{\theta}) = -\frac{\partial V}{\partial \theta} \text{ and } \ddot{z} = -\frac{\partial V}{\partial z}.$$

EXAMPLE 7.8 *Derive the equations of motion for a spherical pendulum where the bob of mass m is constrained to move on a smooth spherical surface of radius R.*

Solution. We take spherical coordinates with origin at the centre of the sphere. The length of the pendulum is equal to the radius of the sphere.

The Cartesian coordinates of the bob are

$$x = R\sin\theta\cos\phi, \ y = R\sin\theta\sin\phi, \ z = R\cos\theta \tag{1}$$

The generalised coordinates are

$$q_1 = \theta, \ q_2 = \phi \tag{2}$$

Kinetic energy is

$$T = \frac{1}{2}m\left(\dot{x}^2 + \dot{y}^2 + \dot{z}^2\right) = \frac{1}{2}mR^2\left(\dot{\theta}^2 + \dot{\phi}^2\sin^2\theta\right) \tag{3}$$

Potential energy relative to horizontal plane through origin is

$$V = -mgR\cos\theta \tag{4}$$

Hence, the Lagrangian is

$$L = T - V = \frac{1}{2}mR^2\left(\dot{\theta}^2 + \dot{\phi}^2\sin^2\theta\right) + mgR\cos\theta \tag{5}$$

Equation of motion in θ direction is

$$\frac{d}{dt}\left(\frac{\partial L}{\partial \dot{\theta}}\right) - \frac{\partial L}{\partial \theta} = 0$$

$$\text{i.e., } mR^2\ddot{\theta} - mR^2\dot{\phi}^2\sin\theta\cos\theta + mgR\sin\theta = 0$$

$$\text{or, } \ddot{\theta} - \frac{1}{2}\dot{\phi}^2\sin 2\theta + \frac{g}{R}\sin\theta = 0 \tag{6}$$

Equation of motion in ϕ direction is

$$\frac{d}{dt}\left(\frac{\partial L}{\partial \dot{\phi}}\right) - \frac{\partial L}{\partial \phi} = 0$$

$$\text{i.e., } \frac{d}{dt}\left(mR^2\dot{\phi}\sin^2\theta\right) = 0, \text{ i.e., } mR^2\dot{\phi}\sin^2\theta = \text{const.}$$

Hence,

$$\dot{\phi}\sin^2\theta = \text{const.} \tag{7}$$

Eqns (6) and (7) are the equations of motion of spherical pendulum.

EXAMPLE 7.9 *Derive the equation of motion of a particle of mass m under a central force*

$$F = -k/r^2$$

Solution. Under central force, the particle moves in a plane, and the force is conservative, so that

$$F = -\frac{\partial V}{\partial r} = -\frac{k}{r^2} \tag{1}$$

Taking polar coordinates, the kinetic energy is

$$T = \frac{1}{2}m\left(\dot{r}^2 + r^2\dot{\theta}^2\right) \tag{2}$$

Hence, the Lagrangian is

$$L = T - V = \frac{1}{2}m\left(\dot{r}^2 + r^2\dot{\theta}^2\right) - V(r) \tag{3}$$

Thus, Lagrangian equation of motion along r direction is

$$\frac{d}{dt}\left(\frac{\partial L}{\partial \dot{r}}\right) - \frac{\partial L}{\partial r} = 0$$

$$\text{i.e.,}\quad \frac{d}{dt}(m\dot{r}) - \left[mr\dot{\theta}^2 - \frac{\partial V}{\partial r}\right] = 0$$

$$\text{i.e.,}\quad m\ddot{r} - mr\dot{\theta}^2 = -\frac{\partial V}{\partial r} = -\frac{k}{r^2} \tag{4}$$

Equation of motion along θ direction is

$$\frac{d}{dt}\left(\frac{\partial L}{\partial \dot{\theta}}\right) - \frac{\partial L}{\partial \theta} = 0,\ \text{i.e.,}\ \frac{d}{dt}\left(mr^2\dot{\theta}\right) = 0,\ \text{i.e.,}\ mr^2\dot{\theta} = \text{const.} \tag{5}$$

Eqns (4) and (5) determine the motion of the particle under central force.

EXAMPLE 7.10 *A bead of mass M slides on a smooth fixed wire whose inclination to the vertical is α, and has hinged to it a rod of mass m and length $2l$, which can move freely in the vertical plane through the wire. If the system starts from rest with the rod hanging vertically, show that $\{4M + m(1 + 3\cos^2\theta)\}\, l\dot{\theta}^2 = 6(M + m)g\sin\alpha(\sin\theta - \sin\alpha)$ where θ is the angle between the rod and the lower part of the wire.*

Solution. We have shown the arrangement in the FIG. 7.4, where OP is the fixed wire and A is the position of the bead at any time t, so that $OA = x$.

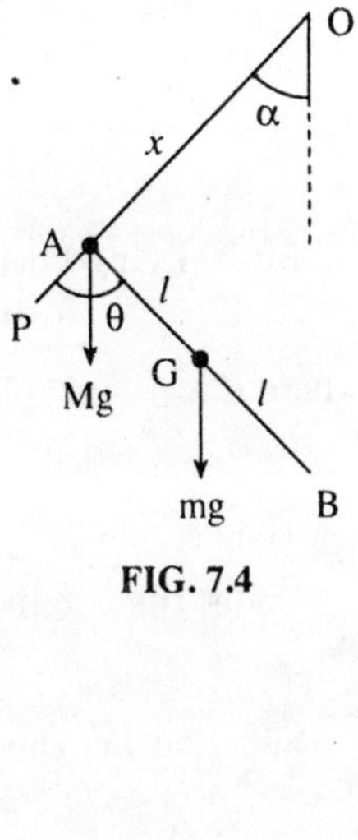

FIG. 7.4

AB represents the rod of length $2l$ with G as centre of mass and

$$\angle PAB = \theta$$

Taking O as origin and OP as x-axis, the coordinates of G are

$$\left.\begin{aligned} X &= x + l\cos\theta \\ Y &= l\sin\theta \end{aligned}\right\} \tag{1}$$

Then velocity of G is

$$v^2 = \dot{X}^2 + \dot{Y}^2 = \left(\dot{x} - l\sin\theta\dot{\theta}\right)^2 + \left(l\cos\theta\dot{\theta}\right)^2 \tag{2}$$

Kinetic energy of the system is

$$T = \text{KE of bead} + \text{KE of the rod}$$

$$T = \frac{1}{2}M\dot{x}^2 + \frac{1}{2}m\left[K^2\dot{\theta}^2 + v^2\right],$$

where K = radius of gyration of the rod about axis through A.

Hence,

$$T = \frac{1}{2}M\dot{x}^2 + \frac{1}{2}m\left[\frac{l^2}{3}\dot{\theta}^2 + \left(\dot{x} - l\sin\theta\dot{\theta}\right)^2 + \left(l\cos\theta\dot{\theta}\right)^2\right]$$

$$\text{i.e., } T = \frac{1}{2}\left[(M+m)\dot{x}^2 - 2ml\dot{x}\dot{\theta}\sin\theta + \frac{4}{3}ml^2\dot{\theta}^2\right] \tag{3}$$

The work function is

$$\begin{aligned} W &= Mgx\cos\alpha + mg\{x\cos\alpha + l\cos(\theta - \alpha)\} \\ &= (M+m)gx\cos\alpha + mgl\cos(\theta - \alpha) \end{aligned} \tag{4}$$

Lagrange's x-equation is

$$\frac{d}{dt}\left(\frac{\partial T}{\partial \dot{x}}\right) - \frac{\partial T}{\partial x} = \frac{\partial W}{\partial x}$$

$$\text{i.e., } \frac{d}{dt}\left[(M+m)\dot{x} - ml\dot{\theta}\sin\theta\right] = (M+m)g\cos\alpha$$

$$\text{or, } (M+m)\ddot{x} - ml\ddot{\theta}\sin\theta - ml\dot{\theta}^2\cos\theta = (M+m)g\cos\alpha \tag{5}$$

Lagrange's θ-equation is

$$\frac{d}{dt}\left(\frac{\partial T}{\partial \dot{\theta}}\right) - \frac{\partial T}{\partial \theta} = \frac{\partial W}{\partial \theta}$$

$$\text{i.e., } \frac{d}{dt}\left[-ml\dot{x}\sin\theta + \frac{4}{3}ml^2\dot{\theta}\right] + ml\dot{x}\dot{\theta}\cos\theta = -mgl\sin(\theta - \alpha)$$

$$\text{or, } ml\ddot{x}\sin\theta + ml\dot{x}\dot{\theta}\cos\theta - \frac{4}{3}ml^2\ddot{\theta} - ml\dot{x}\dot{\theta}\cos\theta = mgl\sin(\theta - \alpha)$$

$$\text{or, } \ddot{x}\sin\theta - \frac{4}{3}l\ddot{\theta} = g\sin(\theta - \alpha) \tag{6}$$

Multiplying Eqn (5) by $\sin\theta$ and Eqn (6) by $(M+m)$ and subtracting one from another, we get

$$\begin{aligned} &\ddot{\theta}\left[-ml\sin^2\theta + \frac{4}{3}(M+m)l\right] - ml\dot{\theta}^2\sin\theta\cos\theta \\ &\qquad = (M+m)[g\cos\alpha\sin\theta - g\sin(\theta - \alpha)] \\ \text{or, } &l\ddot{\theta}\left[(4M+m) + 3m\cos^2\theta\right] - 3ml\dot{\theta}^2\sin\theta\cos\theta \\ &\qquad = 3(M+m)g\cos\theta\sin\alpha \end{aligned} \tag{7}$$

Multiplying both sides of Eqn (7) by $2\dot{\theta}$ and integrating, we get

$$l\dot{\theta}^2\left[(4M+m) + 3m\cos^2\theta\right] = 6(M+m)g\sin\alpha\sin\theta + C \tag{8}$$

where C = constant of integration.

Initially, when

$$\theta = \alpha, \ \dot{\theta} = 0 \tag{9}$$

Using the conditions in (9), we get from (8)

$$C = -6(M+m)g\sin^2\alpha \tag{10}$$

Substituting the value of C from Eqn (10) into (8), we find

$$l\dot{\theta}^2[4M + m + 3m\cos^2\theta] = 6(M+m)g\sin\alpha(\sin\theta - \sin\alpha)$$

which is the required result.

7.6 Rayleigh's Dissipation Function

We consider a system on which act conservative forces $\mathbf{F_i}$ as well as non-conservative forces $\mathbf{F'_i}$. Then the components of generalised force are given by

$$Q_k = \sum_{i=1}^{N} \{\mathbf{F_i} + \mathbf{F'_i}\} \cdot \frac{\partial \mathbf{r_i}}{\partial q_k} \text{ [by Eqn (7.18)]}$$

$$= \sum_{i=1}^{N} \mathbf{F_i} \cdot \frac{\partial \mathbf{r_i}}{\partial q_k} + \sum_{i=1}^{N} \mathbf{F'_i} \cdot \frac{\partial \mathbf{r_i}}{\partial q_k}$$

$$\text{i.e., } Q_k = -\frac{\partial V}{\partial q_k} + Q'_k \tag{7.46}$$

where

$$Q'_k = \sum_{i=1}^{N} \mathbf{F'_i} \cdot \frac{\partial \mathbf{r_i}}{\partial q_k} \tag{7.47}$$

are the components of generalised force due to non-conservative forces. As an example, frictional forces are non-conservative. Many a times we find a situation when the frictional force is proportional to velocity of the particle, so that

$$\mathbf{F'_i} = -\mathbf{k_i v_i} \tag{7.48}$$

Such a force can be derived from a function of the type:

$$R = \frac{1}{2}\sum_{i=1}^{N} \mathbf{k_i v_i^2} = \frac{1}{2}\sum_{i=1}^{N} \left(k_x v_{xi}^2 + k_y v_{yi}^2 + k_z v_{zi}^2\right) \tag{7.49}$$

where k_x, k_y, k_z are respectively components of frictional force/unit velocity along x, y, z directions.

The function R is known as ***Rayleigh's dissipation function***. Then, in terms of this function, we can write non-conservative forces $\mathbf{F'_i}$ in the form:

$$\mathbf{F'_i} = -\nabla_\nu R \tag{7.50}$$

where

$$\nabla_\nu = \mathbf{i}\frac{\partial}{\partial v_x} + \mathbf{j}\frac{\partial}{\partial v_y} + \mathbf{k}\frac{\partial}{\partial v_z} \tag{7.51}$$

Thus, in view of Eqn (7.50), the component of generalised force Q'_k in Eqn (7.47) can be written as

$$Q'_k = \Sigma \mathbf{F'_i} \cdot \frac{\partial \mathbf{r_i}}{\partial q_k} = -\Sigma \nabla_\nu R \cdot \frac{\partial \mathbf{r_i}}{\partial q_k}$$

$$\text{or, } Q'_k = -\Sigma \nabla_\nu R \cdot \frac{\partial \dot{\mathbf{r}}_\mathbf{i}}{\partial \dot{q}_k} \text{ [using Eqn (7.26)]}$$

$$= -\Sigma \nabla_\nu R \cdot \frac{\partial \mathbf{v_i}}{\partial \dot{q}_k}$$

$$= -\Sigma \left[\frac{\partial R}{\partial v_{xi}} \cdot \frac{\partial v_{xi}}{\partial \dot{q}_k} + \frac{\partial R}{\partial v_{yi}} \cdot \frac{\partial v_{yi}}{\partial \dot{q}_k} + \frac{\partial R}{\partial v_{zi}} \cdot \frac{\partial v_{zi}}{\partial \dot{q}_k}\right]$$

$$\text{i.e., } Q'_k = -\frac{\partial R}{\partial \dot{q}_k} \tag{7.52}$$

Using Eqns (7.47) and (7.52), Lagrange's equation is written as

$$\frac{d}{dt}\left(\frac{\partial L}{\partial \dot{q}_k}\right) - \left(\frac{\partial L}{\partial q_k}\right) = Q'_k$$

$$\text{i.e.,} \quad \frac{d}{dt}\left(\frac{\partial L}{\partial \dot{q}_k}\right) - \frac{\partial L}{\partial q_k} = -\frac{\partial R}{\partial \dot{q}_k} \tag{7.53}$$

Again, the work done against frictional force is

$$dW = -\sum \mathbf{F}'_i \cdot d\mathbf{r}_i = -\Sigma \mathbf{F}'_i \cdot \mathbf{v}_i dt$$

$$\text{or, } dW = \Sigma\left(\mathbf{k}_i \mathbf{v}_i^2\right) dt \text{ [using Eqn (7.48)]}$$

Hence,

$$\frac{dW}{dt} = \Sigma \mathbf{k}_i \cdot \mathbf{v}_i^2 = 2R \text{ [by Eqn (7.49)]} \tag{7.54}$$

That is, ***rate of dissipation energy against friction is twice the Rayleigh's dissipation function R***. This is the physical interpretation of ***Rayleigh's dissipation function***.

EXAMPLE 7.11 *Derive the equation of motion of a compound pendulum by Lagrangian formulation and hence find its period of small oscillations in a vertical plane.*

Solution. (See FIG. 5.3, Chapter 5.) The kinetic energy is

$$T = \frac{1}{2}MK^2\dot{\theta}^2 \tag{1}$$

and the potential energy is

$$V = -Mgh\cos\theta \tag{2}$$

Thus, the Lagrangian is

$$L = T - V = \frac{1}{2}MK^2\dot{\theta}^2 + Mgh\cos\theta \tag{3}$$

Hence,

$$\frac{\partial L}{\partial \theta} = -Mgh\sin\theta, \quad \frac{\partial L}{\partial \dot{\theta}} = MK^2\dot{\theta} \tag{4}$$

Lagrange's equation in θ-coordinate is

$$\frac{d}{dt}\left(\frac{\partial L}{\partial \dot{\theta}}\right) - \frac{\partial L}{\partial \theta} = 0 \tag{5}$$

Using (4), we find

$$MK^2\ddot{\theta} + Mgh\sin\theta = 0$$

$$\text{i.e., } \ddot{\theta} = -\frac{gh}{K^2}\sin\theta = -\frac{gh}{K^2}\theta \text{ (for small } \theta)$$

Hence, the period of small oscillation

$$P = 2\pi\sqrt{\frac{K^2}{gh}}.$$

EXAMPLE 7.12 *A cylinder of radius R and mass M rolls down an inclined plane making an angle α with horizon. Set up the Lagrangian and find the equation of motion.*

Solution. Let the cylinder start rolling from the point O. Then, from FIG. 7.5, we have

$$x = R\theta, \text{ so that, } \dot{x} = R\dot{\theta} \quad (1)$$

Kinetic energy

$$T = \frac{1}{2}M\dot{x}^2 + \frac{1}{2}MK^2\dot{\theta}^2 \quad (2)$$

Since, for a cylinder, $K^2 = \frac{1}{2}R^2$, we get from Eqns (1) and (2)

$$T = \frac{1}{2}M\dot{x}^2 + M\frac{1}{4}R^2\frac{\dot{x}^2}{R^2} = \frac{3}{4}M\dot{x}^2 \quad (3)$$

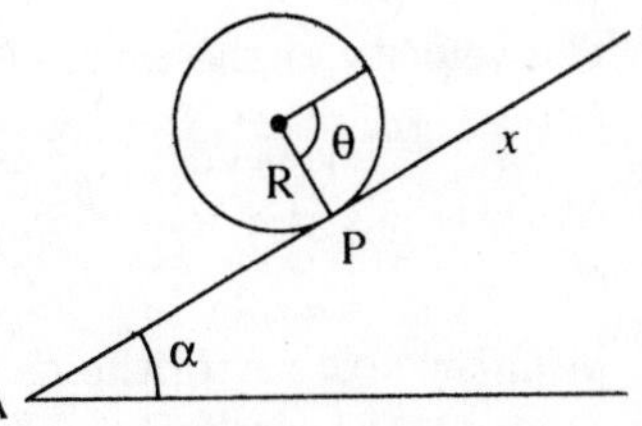

FIG. 7.5

Potential energy is (taking $AO = l$)

$$V = Mg(l - x)\sin\alpha + MgR\cos\alpha$$

Hence, the Lagrangian is

$$L = T - V = \frac{3}{4}M\dot{x}^2 - Mg(l - x)\sin\alpha - MgR\cos\alpha. \quad (4)$$

Lagrangian equation of motion gives

$$\frac{d}{dt}\left(\frac{\partial L}{\partial \dot{x}}\right) - \frac{\partial L}{\partial x} = 0 \text{ or, } \frac{d}{dt}\left(\frac{3}{2}M\dot{x}\right) - Mg\sin\alpha = 0 \text{ or, } \frac{3}{2}\ddot{x} = g\sin\alpha \quad (5)$$

Using Eqn (1),

$$\ddot{\theta} = \frac{2g}{3R}\sin\alpha. \quad (6)$$

EXAMPLE 7.13 *A perfectly rough sphere lying inside a hollow cylinder which rests on a perfectly rough horizontal plane, is slightly displaced from its position of equilibrium. Show that, the time of a small oscillation is*

$$2\pi\left[\frac{a-b}{g}\cdot\frac{14M}{10M+7m}\right]^{1/2},$$

where a = radius of the cylinder, b = radius of the sphere, M = mass of the cylinder, m = mass of the sphere.

Solution. The arrangement is shown in FIG. 7.6. Here O is the initial point of contact of the cylinder with horizontal plane. After time t it has moved to B.

Then

$$OB = \text{arc } AB = a\psi \quad (1)$$

where ψ = angular displacement of the cylinder. $C'E$ is a line fixed in the sphere; it was initially vertical and makes an angle ϕ with vertical after time t. Let θ be the angle between the vertical and the line joining the centres C, C'. Then

$$\text{arc } AD = \text{arc } ED, \text{ i.e., } a(\theta + \psi) = b(\theta + \phi)$$

which gives

$$b\dot{\phi} = (a - b)\dot{\theta} + a\dot{\psi} \quad (2)$$

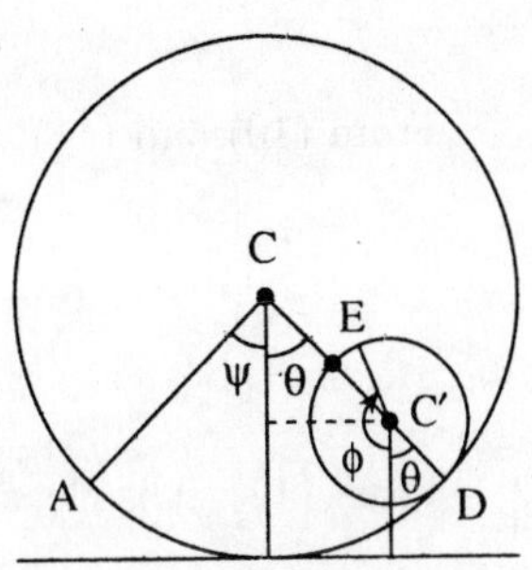

FIG. 7.6

Let the coordinates of C' be (x, y) with origin at O. Then

$$x = OP = OB + BP = a\psi + (a-b)\sin\theta \tag{3}$$

$$y = PC' = a - (a-b)\cos\theta \tag{4}$$

The velocity of the centre C' is given by

$$V_1^2 = \dot{x}^2 + \dot{y}^2 = \left\{a\dot{\psi} + (a-b)\cos\theta\dot{\theta}\right\}^2 + \left\{(a-b)\sin\theta\dot{\theta}\right\}^2$$

$$= a^2\dot{\psi}^2 + (a-b)^2\dot{\theta}^2 + 2a(a-b)\dot{\psi}\dot{\theta} \ [\because \theta \text{ is small}]. \tag{5}$$

Also, the velocity of the centre C is given by

$$V_2^2 = \left(a\dot{\psi}\right)^2 = a^2\dot{\psi}^2 \tag{6}$$

Thus, kinetic energy of the cylinder is

$$\frac{1}{2}MK^2\dot{\psi}^2 = \frac{1}{2}M(a^2 + a^2)\dot{\psi}^2 = Ma^2\dot{\psi}^2 \tag{7}$$

Total kinetic energy of the system is

$$T = Ma^2\dot{\psi}^2 + \frac{1}{2}m\left[\frac{2b^2}{5}\dot{\phi}^2 + a^2\dot{\psi}^2 + (a-b)^2\dot{\theta}^2 + 2a(a-b)\dot{\theta}\dot{\psi}\right]$$

$$= Ma^2\dot{\psi}^2 + \frac{1}{2}m\left[\frac{2}{5}\left\{(a-b)\dot{\theta} + a\dot{\psi}\right\}^2 + a^2\dot{\psi}^2 + (a-b)^2\dot{\theta}^2 + 2a(a-b)\dot{\theta}\dot{\psi}\right]$$

[using Eqn (2)]

$$= \frac{10M + 7m}{10}a^2\dot{\psi}^2 + \frac{7m}{10}\left[(a-b)^2\dot{\theta}^2 + 2a(a-b)\dot{\theta}\dot{\psi}\right] \tag{8}$$

The work function is

$$W = (a-b)mg\cos\theta + \text{const.} \tag{9}$$

Lagrange's θ-equation is

$$\frac{d}{dt}\left[\frac{7m}{10}\left\{(a-b)^2 2\dot{\theta} + 2a(a-b)\dot{\psi}\right\}\right] = -(a-b)mg\sin\theta$$

$$\text{or, } 7(a-b)\ddot{\theta} + 7a\ddot{\psi} = -5g\theta \ [\because \ \theta \text{ is small}]. \tag{10}$$

Again, Lagrange's ψ-equation is

$$\frac{d}{dt}\left[\frac{10M + 7m}{10}\cdot 2a^2\dot{\psi} + \frac{7m}{10}\cdot 2a(a-b)\dot{\theta}\right] = 0$$

$$\text{or, } (10M + 7m)a\ddot{\psi} + 7m(a-b)\ddot{\theta} = 0 \tag{11}$$

From (10) and (11), we eliminate $\ddot{\psi}$ and get

$$70M(a-b)\ddot{\theta} = -5(10M + 7m)g\theta, \ \text{or, } \ddot{\theta} = -\mu\theta \tag{12}$$

where

$$\mu = \frac{10M + 7m}{14M}\frac{g}{a-b} \tag{13}$$

From (12), it is clear that the motion is simple harmonic and its time period is

$$P = \frac{2\pi}{\sqrt{\mu}} = 2\pi\left[\frac{a-b}{g}\cdot\frac{14M}{10M + 7m}\right]^{1/2} \qquad \textbf{(Proved)}$$

7.7 Cyclic or Ignorable Coordinates

The Lagrangian L is a function of q_k and $\dot{q}_k$. Let the coordinate q_k be absent in the expression for the Lagrangian L. Then

$$\frac{\partial L}{\partial q_k} = 0 \tag{7.55}$$

In this case, the equation of motion for q_k-direction is

$$\frac{d}{dt}\left(\frac{\partial L}{\partial \dot{q}_k}\right) = 0 \tag{7.56}$$

Integrating this equation, we get

$$\frac{\partial L}{\partial \dot{q}_k} = \text{constant} = p_k \tag{7.57}$$

This implies that whenever the coordinate q_k is absent in the Lagrangian L, the corresponding linear momentum p_k is a constant of the motion. Such a coordinate q_k is said to be ***cyclic*** or ***ignorable***.

EXERCISE

1. Derive the Lagrangian and obtain the equation of motion in the following cases:
   (a) a simple pendulum
   (b) a harmonic oscillator (two-dimensional)
   (c) a projectile launched with speed u at an angle α with the horizon.
2. A particle of mass m moves on a plane in the field of force given in polar coordinates by
   $$F = -kr\cos\theta, K = \text{constant}.$$
   Set up the Lagrangian and obtain the equation of motion.
3. Two masses connected by inextensible and massless strings passing over a smooth pulley moves under gravity. Set up the Lagrangian and find the common acceleration.
4. A rod of mass m with one end on a smooth floor and the other against a smooth vertical wall slides. Set up the Lagrangian and derive the equation of motion.
5. A solid ball of radius r rolls without slipping on the inside of a cylinder of radius R. Derive the equation of small oscillations.
6. A bead of mass M slides on a smooth uniform circular wire of radius R which is rotating with a constant angular velocity ω about a fixed vertical diameter. Find the equation of motion of the bead.
7. A particle P moves on a smooth horizontal circular wire of radius R which is free to rotate about a vertical axis through a point O at a distance b from the centre C. If $\angle PCO = \theta$, set up the Lagrangian and hence show that
   $$R\ddot{\theta} + \dot{\omega}(R - b\cos\theta) = b\omega^2\sin\theta$$
8. A bead slides on a smooth wire in the form of a cycloid given by
   $$x = a(\theta - \sin\theta)$$
   $$y = a(1 + \cos\theta), \ (0 \leq \theta \leq 2\pi)$$
   Derive the equation of motion.

9. A particle of mass m moves under gravity on the inner smooth surface of a paraboloid of revolution given by $x^2 + y^2 = az$. Find the equations of motion.

10. A spring of mass M and constant k is hung from a roof ceiling. A mass m is placed on the spring and set into motion. Set up the Lagrangian and prove that the system oscillates with a time period

$$2\pi\sqrt{\frac{M+3m}{3k}}$$

CHAPTER 8

Rotation of Rigid Bodies about a Fixed Point

When a rigid body rotates about a fixed point, there is no translational motion. Only three coordinates are necessary for describing such motions. We usually take three angles as the coordinates.

8.1 Angular Momentum

We consider a rigid body rotating about a point O which is fixed in both body and space. Let OX, OY, OZ be mutually perpendicular axes fixed in the body and $\mathbf{i}, \mathbf{j}, \mathbf{k}$ are unit vectors along them respectively. The angular velocity $\boldsymbol{\omega}$ of the body rotating about a line through O can be expressed as

$$\boldsymbol{\omega} = \mathbf{i}\omega_1 + \mathbf{j}\omega_2 + \mathbf{k}\omega_3 \tag{8.1}$$

where $\omega_1, \omega_2, \omega_3$ are components along X-, Y-, Z-axes respectively. Similarly, if J_1, J_2, J_3 are components of angular momentum along the axes, we have

$$\mathbf{J} = \mathbf{i}J_1 + \mathbf{j}J_2 + \mathbf{k}J_3 \tag{8.2}$$

Now, we take a particle of mass m at a point $P(x, y, z)$ so that its position vector is

$$\mathbf{r} = \mathbf{i}x + \mathbf{j}y + \mathbf{k}z \tag{8.3}$$

Then, the angular momentum can be written as

$$\begin{aligned} \mathbf{J} &= \Sigma \mathbf{r} \times m(\boldsymbol{\omega} \times \mathbf{r}) \\ &= \Sigma m \left[(\mathbf{r} \cdot \mathbf{r})\boldsymbol{\omega} - (\mathbf{r} \cdot \boldsymbol{\omega})\mathbf{r}\right] \\ &= \Sigma m \left[\left(x^2 + y^2 + z^2\right)\boldsymbol{\omega} - (x\omega_1 + y\omega_2 + z\omega_3)\,\mathbf{r}\right] \end{aligned} \tag{8.4}$$

Taking component along $\mathbf{i}$-direction,

$$\begin{aligned} J_1 &= \Sigma m \left[\left(x^2 + y^2 + z^2\right)\omega_1 - (x\omega_1 + y\omega_2 + z\omega_3)\,x\right] \\ &= \omega_1 \Sigma m(y^2 + z^2) - \omega_2 \Sigma mxy - \omega_3 \Sigma mzx \\ &= I_{xx}\omega_1 + I_{xy}\omega_2 + I_{xz}\omega_3 \end{aligned} \tag{8.5}$$

Similarly, we get expressions for J_2 and J_3, so that, we write:

$$\left.\begin{aligned} J_1 &= I_{xx}\omega_1 + I_{xy}\omega_2 + I_{xz}\omega_3 \\ J_2 &= I_{yx}\omega_1 + I_{yy}\omega_2 + I_{yz}\omega_3 \\ J_3 &= I_{zx}\omega_1 + I_{zy}\omega_2 + I_{zz}\omega_3 \end{aligned}\right\} \tag{8.6}$$

where $I_{xx}, \cdots,$ etc. are moments and products of inertia of the rigid body with respect to coordinate system in the body. If the axes of this coordinate system coincide with the principal axes of inertia of the body at O, then, $I_{xy} = I_{yz} = I_{zx} = 0$. Then Eqns (8.6) assume the form:

$$\left.\begin{aligned} J_1 &= I_{xx}\omega_1 \\ J_2 &= I_{yy}\omega_2 \\ J_3 &= I_{zz}\omega_3 \end{aligned}\right\} \tag{8.7}$$

Thus, if a rigid body is rotating about a principal axis, the angular momentum J and the angular velocity ω are directed along any one of principal axes. Hence, we write:

$$\mathbf{J} = I\boldsymbol{\omega} \tag{8.8}$$

8.2 Inertia Tensor

Eqn (8.6) can be written in matrix form as

$$\begin{bmatrix} J_1 \\ J_2 \\ J_3 \end{bmatrix} = \begin{bmatrix} I_{xx} & I_{xy} & I_{xz} \\ I_{yx} & I_{yy} & I_{yz} \\ I_{zx} & I_{zy} & I_{zz} \end{bmatrix} \begin{bmatrix} \omega_1 \\ \omega_2 \\ \omega_3 \end{bmatrix} \quad \text{or, } \mathbf{J} = \mathbf{I}\omega \tag{8.9}$$

where we have denoted

$$\mathbf{I} = \begin{bmatrix} I_{xx} & I_{xy} & I_{xz} \\ I_{yx} & I_{yy} & I_{yz} \\ I_{zx} & I_{zy} & I_{zz} \end{bmatrix} \tag{8.10}$$

The matrix in the right-hand side of Eqn (8.10) is called the ***inertia tensor***. Since there are nine quantities called ***elements***, it is a tensor of rank two. Also we have

$$I_{xy} = I_{yx}, \quad I_{xz} = I_{zx}, \quad I_{yz} = I_{zy} \tag{8.11}$$

Hence, the ***inertia tensor is symmetric***. If the coordinate axes happen to be the principal axes, then the inertia tensor becomes

$$\mathbf{I} = \begin{bmatrix} A & 0 & 0 \\ 0 & B & 0 \\ 0 & 0 & C \end{bmatrix} \tag{8.12}$$

where we have denoted

$$A = I_{xx}, \quad B = I_{yy}, \quad C = I_{zz} \tag{8.13}$$

In view of Eqn (8.13), we write Eqn (8.7) in the form:

$$J_1 = A\omega_1, \quad J_2 = B\omega_2, \quad J_3 = C\omega_3 \tag{8.14}$$

8.3 Kinetic Energy of Rotation

The total kinetic energy of the rigid body is given by

$$\begin{aligned} T &= \Sigma\frac{1}{2}mv^2 \\ \text{or, } 2T &= \Sigma m\mathbf{v}\cdot\mathbf{v} = \Sigma(\boldsymbol{\omega}\times\mathbf{r})\cdot m\mathbf{v} \quad [\because\ \mathbf{v} = \boldsymbol{\omega}\times\mathbf{r}] \\ &= \Sigma\boldsymbol{\omega}\cdot(\mathbf{r}\times m\mathbf{v}) \quad [\because\ (\mathbf{a}\times\mathbf{b})\cdot\mathbf{c} = \mathbf{a}\cdot(\mathbf{b}\times\mathbf{c})] \\ &= \boldsymbol{\omega}\cdot\Sigma(\mathbf{r}\times m\mathbf{v}) = \boldsymbol{\omega}\cdot\mathbf{J} \end{aligned}$$

Hence,

$$T = \frac{1}{2}\boldsymbol{\omega} \cdot \mathbf{J}. \tag{8.15}$$

Eqn (8.15) in expanded form becomes

$$T = \frac{1}{2}(\omega_1 J_1 + \omega_2 J_2 + \omega_3 J_3) \tag{8.16}$$

Substituting here the values of J_1, J_2 and J_3 from Eqn (8.6), we get after simplification

$$T = \frac{1}{2}\left(I_{xx}\omega_1^2 + I_{yy}\omega_2^2 + I_{zz}\omega_3^2 + 2I_{xy}\omega_1\omega_2 + 2I_{yz}\omega_2\omega_3 + 2I_{zx}\omega_3\omega_1\right) \tag{8.17}$$

Eqn (8.17) assumes a simple form if the axes of reference are principal axes. In this case,

$$T = \frac{1}{2}\left(A\omega_1^2 + B\omega_2^2 + C\omega_3^2\right) \tag{8.18}$$

8.4 Euler's Equations of Motion

Consider a rigid body rotating about a fixed point under the influence of an external torque $\mathbf{N}$. Then the equation of rotational motion of the rigid body is

$$\mathbf{N} = \frac{d\mathbf{J}}{dt} \tag{8.19}$$

where the time derivative is to be taken under fixed or inertial frame of reference. The body coordinate system is rotating with an instanteneous angular velocity $\boldsymbol{\omega}$. Hence, the time derivatives in the two systems are related as

$$\frac{d}{dt} = \frac{d'}{dt} + \boldsymbol{\omega} \times \tag{8.20}$$

where the derivative d'/dt corresponds to rotating frame of reference.

Thus, Eqn (8.19) assumes the form:

$$\mathbf{N} = \frac{d'\mathbf{J}}{dt} + \boldsymbol{\omega} \times \mathbf{J} \tag{8.21}$$

We take the body axes to be the principal axes. Then

$$\mathbf{J} = \mathbf{i}A\omega_1 + \mathbf{j}B\omega_2 + \mathbf{k}C\omega_3 \tag{8.22}$$

and the angular velocity $\boldsymbol{\omega}$ is written as

$$\boldsymbol{\omega} = \mathbf{i}\omega_1 + \mathbf{j}\omega_2 + \mathbf{k}\omega_3 \tag{8.23}$$

The principal moments of inertia A, B, C as well as the unit vectors $\mathbf{i}, \mathbf{j}, \mathbf{k}$ remain constants in the body coordinate system.

Hence,

$$\frac{d'\mathbf{J}}{dt} = \frac{d\mathbf{J}}{dt} = \mathbf{i}A\dot{\omega}_1 + \mathbf{j}B\dot{\omega}_2 + \mathbf{k}C\dot{\omega}_3 \tag{8.24}$$

Writing

$$\mathbf{N} = N_1\mathbf{i} + N_2\mathbf{j} + N_3\mathbf{k} \tag{8.25}$$

we get from Eqns (8.21) and (8.24)

$$\begin{aligned} N_1\mathbf{i} + N_2\mathbf{j} + N_3\mathbf{k} = A\dot{\omega}_1\mathbf{i} + B\dot{\omega}_2\mathbf{j} + C\dot{\omega}_3\mathbf{k} + (\omega_1\mathbf{i} + \omega_2\mathbf{j} + \omega_3\mathbf{k}) \\ \times(A\omega_1\mathbf{i} + B\omega_2\mathbf{j} + C\omega_3\mathbf{k}) \end{aligned}$$

This gives

$$\left.\begin{aligned} N_1 &= A\dot{\omega}_1 - (B - C)\omega_2\omega_3 \\ N_2 &= B\dot{\omega}_2 - (C - A)\omega_3\omega_1 \\ N_3 &= C\dot{\omega}_3 - (A - B)\omega_1\omega_2 \end{aligned}\right\} \tag{8.26}$$

These are known as ***Euler's equations*** for the motion of a rigid body about a fixed point under the action of a torque.

8.5 Force-free Motion of a Symmetric Top

We consider a symmetric top for which $A = B$, moving under no forces. Then Eqns (8.26) assume the form:

$$A\dot{\omega}_1 = (A - C)\omega_2\omega_3 \tag{8.27}$$

$$A\dot{\omega}_2 = -(A - C)\omega_3\omega_1 \tag{8.28}$$

$$C\dot{\omega}_3 = 0 \tag{8.29}$$

Integrating Eqn (8.29), we get

$$\omega_3 = \text{const.} = n \text{ (say)}. \tag{8.30}$$

Thus, the component of angular velocity about the symmetry axis remains constant throughout the motion of the rigid body. We substitute

$$\Omega = \frac{C - A}{A}\omega_3 = \text{constant}. \tag{8.31}$$

Then Eqns (8.27) and (8.28) can be written as

$$\dot{\omega}_1 = -\Omega\omega_2 \tag{8.32}$$

$$\dot{\omega}_2 = \Omega\omega_1 \tag{8.33}$$

Differentiating Eqn (8.32) w.r.t. time

$$\ddot{\omega}_1 = -\Omega\dot{\omega}_2 = -\Omega^2\omega_1 \text{ [using Eqn (8.33)]}$$

$$\text{or, } \ddot{\omega}_1 + \Omega^2\omega_1 = 0 \tag{8.34}$$

Its solution is

$$\omega_1 = a\sin\Omega t \tag{8.35}$$

where a = constant. We have assumed $\omega_1 = 0$ at $t = 0$ so that phase constant is zero.

From Eqns (8.32) and (8.35), we have

$$\omega_2 = -\frac{\dot{\omega}_1}{\Omega} = -a\cos\Omega t \tag{8.36}$$

Squaring and adding Eqns (8.35) and (8.36),

$$\omega_c^2 = \omega_1^2 + \omega_2^2 = a^2 = \text{constant}. \tag{8.37}$$

Thus, the resultant of ω_1 and ω_2 is always in the xy-plane and rotates with angular frequency Ω (FIG. 8.1).

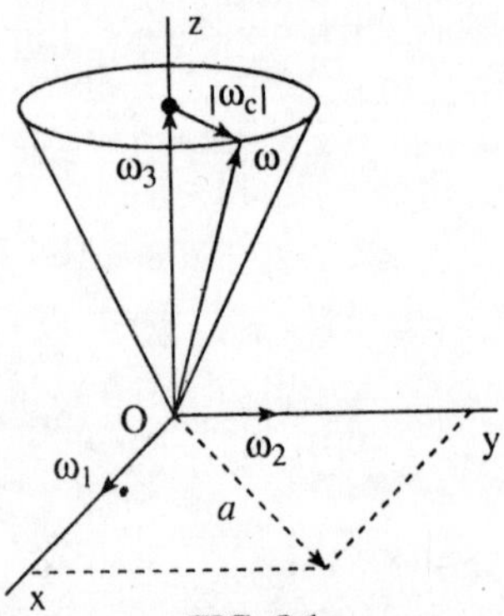

FIG. 8.1

Again

$$\begin{aligned} \boldsymbol{\omega} &= \mathbf{i}\omega_1 + \mathbf{j}\omega_2 + \mathbf{k}\omega_3 \\ &= \mathbf{i}a\sin\Omega t - \mathbf{j}a\cos\Omega t + \mathbf{k}n \end{aligned} \tag{8.38}$$

Hence,

$$|\omega| = \sqrt{a^2 + n^2} = \text{constant}. \tag{8.39}$$

Thus, the vector $\boldsymbol{\omega}$ describes a cone with vertex at the origin. This motion is called ***precession***.

The kinetic energy T and the angular momentum J are the two constants of the motion. The constants a and n can be determined in terms of the kinetic energy T and the magnitude of the angular momentum J.

In view of Eqns (8.18), (8.22) and (8.37), we write:

$$T = \frac{1}{2}Aa^2 + \frac{1}{2}Cn^2 \tag{8.40}$$

$$\text{and } J^2 = A^2a^2 + C^2n^2 \tag{8.41}$$

These equations can be solved to find

$$\omega_3^2 = n^2 = \frac{J^2 - 2AT}{C(C-A)} \tag{8.42}$$

$$\text{and } a^2 = \frac{J^2 - 2CT}{A(A-C)} \tag{8.43}$$

The frequency and period of precession of the angular velocity ω about z-axis are:

$$f = \frac{\Omega}{2\pi} = \frac{\omega_3}{2\pi}\frac{C-A}{A} \tag{8.44}$$

$$\text{and } P = \frac{2\pi}{\omega_3}\frac{A}{C-A} \tag{8.45}$$

8.6 Motion of a Heavy Symmetric Top

We consider the motion of a symmetric top of mass M spinning about its axis of symmetry. The centre of mass of the top remains at a distance h from the tip which is fixed in body as well in space during motion (FIG. 8.2).

We take Euler angles θ, ϕ, ψ (ref. Chapter 6) as generalised coordinates of the motion. Let $\omega_1, \omega_2, \omega_3$ be the components of the angular velocity of the top about principal axes of the body frame of reference. Then the components of angular velocity are, by Eqn (6.77),

$$\left.\begin{aligned} \omega_1 &= \dot\theta\cos\psi + \dot\phi\sin\theta\sin\psi \\ \omega_2 &= -\dot\theta\sin\psi + \dot\phi\sin\theta\cos\psi \\ \omega_3 &= \dot\psi + \dot\phi\cos\theta \end{aligned}\right\} \tag{8.46}$$

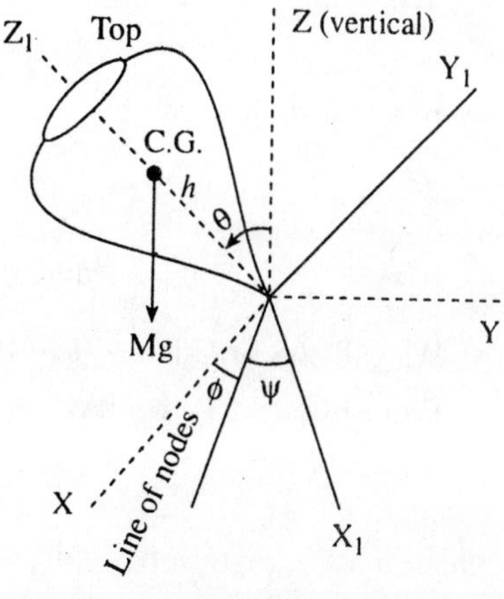

FIG. 8.2

The kinetic energy of the top is

$$T = \frac{1}{2}\left(A\omega_1^2 + B\omega_2^2 + C\omega_3^2\right) \tag{8.47}$$

For the symmetric top, $A = B$; hence,

$$T = \frac{1}{2}A\left(\omega_1^2 + \omega_2^2\right) + \frac{1}{2}C\omega_3^2 = \frac{1}{2}A\left(\dot\theta^2 + \dot\phi^2\sin^2\theta\right) + \frac{1}{2}C\left(\dot\psi + \dot\phi\cos\theta\right)^2 \tag{8.48}$$

The potential energy of the top is (FIG. 8.2)

$$V = Mgh\cos\theta \tag{8.49}$$

The Lagrangian function is

$$L = \frac{1}{2}A\left(\dot{\theta}^2 + \dot{\phi}^2\sin^2\theta\right) + \frac{1}{2}C\left(\dot{\psi} + \dot{\phi}\cos\theta\right)^2 - Mgh\cos\theta \tag{8.50}$$

Lagrange's ψ-equation gives

$$\frac{d}{dt}\left(\frac{\partial L}{\partial\dot{\psi}}\right) - \frac{\partial L}{\partial\psi} = 0, \text{ i.e., } \frac{\partial L}{\partial\dot{\psi}} = \text{const.}$$

Hence,

$$p_\psi = \frac{\partial L}{\partial\dot{\psi}} = C\left(\dot{\psi} + \dot{\phi}\cos\theta\right) = C\omega_3 = \text{const.} \tag{8.51}$$

Similarly, from Lagrange's ϕ-equation, we get

$$p_\phi = \frac{\partial L}{\partial\dot{\phi}} = A\dot{\phi}\sin^2\theta + C\left(\dot{\psi} + \dot{\phi}\cos\theta\right)\cos\theta = \text{const.}$$

$$\text{or, } p_\phi = A\dot{\phi}\sin^2\theta + p_\psi\cos\theta = \text{const.} \tag{8.52}$$

These are two first integrals of the motion. Again, the system being conservative, the total energy E should remain constant. Now,

$$E = T + V = \frac{1}{2}A\left(\dot{\theta}^2 + \dot{\phi}^2\sin^2\theta\right) + \frac{1}{2}C\left(\dot{\psi} + \dot{\phi}\cos\theta\right)^2 + Mgh\cos\theta \tag{8.53}$$

From Eqn (8.52), we have

$$\dot{\phi} = \frac{p_\phi - p_\psi\cos\theta}{A\sin^2\theta} \tag{8.54}$$

In view of this equation, the total energy E is

$$E = \frac{1}{2}A\dot{\theta}^2 + \frac{(p_\phi - p_\psi\cos\theta)^2}{2A\sin^2\theta} + \frac{p_\psi^2}{2C} + Mgh\cos\theta$$

$$\text{or, } E' = \frac{1}{2}A\dot{\theta}^2 + V(\theta) \tag{8.55}$$

where we have defined

$$E' = E - \frac{p_\psi^2}{2C} = \text{constant} \tag{8.56}$$

$$\text{and } V(\theta) = \frac{(p_\phi - p_\psi\cos\theta)^2}{2A\sin^2\theta} + Mgh\cos\theta \tag{8.57}$$

We call the function $V(\theta)$, which depends on θ alone, the ***effective potential energy***.

From Eqn (8.55), we have

$$\dot{\theta} = \frac{d\theta}{dt} = \sqrt{\frac{2}{A}[E' - V(\theta)]}$$

which upon integration yields

$$t(\theta) = \int \frac{d\theta}{\sqrt{\frac{2}{A}[E' - V(\theta)]}} \tag{8.58}$$

Thus, in principle, the problem of the motion of the top is solved. The integral in Eqn (8.58) will in general be elliptic integral.

However, we like to have some qualitative features of the motion. From a glance at $V(\theta)$ in Eqn (8.57), we find that (FIG. 8.3) it tends to infinity when $\theta \to 0$ and again when $\theta \to \pi$. Thus, a physically acceptable range for θ is $0 < \theta < \pi$. For any particular constant value of E', we will get two values of θ, i.e., θ_1 and θ_2 within which the motion is limited. That is, the inclination θ of the top from vertical axis is such that $\theta_1 \le \theta \le \theta_2$. This is called ***nutation***.

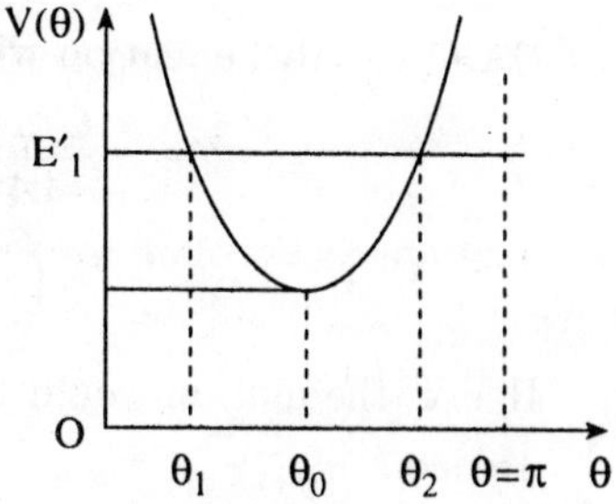

FIG. 8.3

In a special case, when the value of E' is such that we get only one value of $\theta = \theta_0$, which remains constant, we say that the ***precession*** is ***steady***. That is, there is precession without nutation.

This value of $\theta = \theta_0$ is given by

$$\left.\frac{dV}{d\theta}\right|_{\theta=\theta_0} = 0 \text{ or, } (p_\phi - p_\psi \cos\theta_0)(p_\psi - p_\phi \cos\theta_0) - AMgh\sin^4\theta_0 = 0 \tag{8.59}$$

We substitute

$$p_\phi - p_\psi \cos\theta_0 = \xi \tag{8.60}$$

Then

$$p_\psi - p_\phi \cos\theta_0 = p_\psi \sin^2\theta_0 - \xi\cos\theta_0 \tag{8.61}$$

In view of Eqns (8.60) and (8.61), Eqn (8.59) assumes the form:

$$\xi^2 \cos\theta_0 - \xi p_\psi \sin^2\theta_0 + MghA\sin^4\theta_0 = 0$$

which is a quadratic equation in ξ. Since ξ as defined by Eqn (8.60) is real, we must have

$$p_\psi^2 \sin^4\theta_0 - 4\cos\theta_0 MghA\sin^4\theta_0 \ge 0$$

$$\text{i.e., } p_\psi^2 \ge 4MghA\cos\theta_0, \text{ for } \theta_0 < \frac{\pi}{2}$$

$$\text{i.e., } \omega_3 \ge \frac{2}{C}\sqrt{MghA\cos\theta_0} \tag{8.62}$$

Thus, for a steady precession, the value of ω_3 must satisfy the condition in Eqn (8.62).

8.7 Stability of Motion

We consider the steady motion of the top when its axis of rotation is vertical. Thus, initially

$$\theta = 0, \dot{\theta} = 0$$

We intend to see whether the motion would stable if the top is slightly disturbed. Now, initially when $\theta = 0$, we get from Eqn (8.52)

$$p_\phi = p_\psi = C\omega_3 \tag{8.63}$$

For a small disturbance when $\theta = \varepsilon$, we have from Eqn (8.52)

$$A\dot{\phi}\sin^2\varepsilon + C\omega_3\cos\varepsilon = C\omega_3$$

$$\text{or, } \dot{\phi} = \frac{C\omega_3}{A}\frac{(1-\cos\varepsilon)}{1-\cos^2\varepsilon} = \frac{C\omega_3}{A}\frac{1}{1+\cos\varepsilon} = \frac{C\omega_3}{2A} \tag{8.64}$$

since $\cos\varepsilon \simeq 1$.

Now, Lagrange's θ-equation is

$$A\ddot{\theta} = A\dot{\phi}^2 \sin\theta\cos\theta - C\omega_3\dot{\phi}\sin\theta + Mgh\sin\theta \tag{8.65}$$

For $\theta = \varepsilon$, this equation with the use of Eqn (8.64) yields

$$\begin{aligned} A\ddot{\varepsilon} &= \frac{C^2\omega_3^2}{4A}\varepsilon - \frac{C^2\omega_3^2}{2A}\varepsilon + Mgh\varepsilon \quad [\because \ \sin\varepsilon = \varepsilon, \cos\varepsilon = 1] \\ &= -\left(\frac{C^2\omega_3^2}{4A} - Mgh\right)\varepsilon \end{aligned} \tag{8.66}$$

Hence, the motion would be stable, if

$$\frac{C^2\omega_3^2}{4A} - Mgh > 0$$

which yields

$$\omega_3 > \frac{2}{C}\sqrt{MghA} \tag{8.67}$$

The time of nutation is

$$4\pi A\left(C^2\omega_3^2 - 4AMgh\right)^{-1/2} \tag{8.68}$$

It may be observed that Eqn (8.67) is same as Eqn (8.62) for $\theta_0 = 0$, which is expected.

EXAMPLE 8.1 *A body moves under no forces about a point O, the principal moments of inertia at O being $6A, 3A, A$. Initially, the angular velocity of the body has components $\omega_1 = n$, $\omega_2 = 0$, $\omega_3 = 3n$ about the principal axes. Show that, at any later time*

$$\omega_2 = -\sqrt{5}\,n\tanh\left(\sqrt{5}\,nt\right)$$

Solution. Euler's equation of motion under no forces are:

$$A\dot{\omega}_1 - (B - C)\omega_2\omega_3 = 0 \tag{1}$$

$$B\dot{\omega}_2 - (C - A)\omega_3\omega_1 = 0 \tag{2}$$

$$C\dot{\omega}_3 - (A - B)\omega_1\omega_2 = 0 \tag{3}$$

Putting here the values of principal moments of inertia $6A, 3A, A$ as given, we get

$$3\dot{\omega}_1 = \omega_2\omega_3 \tag{4}$$

$$3\dot{\omega}_2 = -5\omega_3\omega_1 \tag{5}$$

$$\dot{\omega}_3 = 3\omega_1\omega_2 \tag{6}$$

From Eqns (4) and (6), we get by eliminating ω_2, $9\omega_1\dot{\omega}_1 - \omega_3\dot{\omega}_3 = 0$.
Upon integration, this yields

$$9\omega_1^2 - \omega_3^2 = \alpha \text{ (say)} \tag{7}$$

Initially, $\omega_1 = n$ and $\omega_3 = 3n$, hence, $\alpha = 0$. Thus, from Eqn (7),

$$9\omega_1^2 = \omega_3^2, \text{ i.e., } 3\omega_1 = \omega_3 \tag{8}$$

Again, from Eqns (5) and (6), we get by eliminating ω_1

$$9\omega_2\dot{\omega}_2 + 5\omega_3\dot{\omega}_3 = 0$$

Integrating this, we have

$$9\omega_2^2 + 5\omega_3^2 = \beta \text{ (say).} \tag{9}$$

Initially, $\omega_2 = 0$, $\omega_3 = 3n$, so that $\beta = 45n^2$.

Hence, Eqn (9) becomes

$$9\omega_2^2 + 5\omega_3^2 = 45n^2 \tag{10}$$

Now, using Eqn (8) in Eqn (5), we get

$$3\dot{\omega}_2 = -5\omega_3\left(\frac{\omega_3}{3}\right) = -\frac{5}{3}\omega_3^2$$

$$\text{or, } 9\dot{\omega}_2 = -5\omega_3^2 = 9\omega_2^2 - 45n^2 \text{ [using Eqn (10)]}$$

$$\text{or, } \dot{\omega}_2 = \omega_2^2 - 5n^2 = -(5n^2 - \omega_2^2)$$

Thus,

$$-dt = \frac{d\omega_2}{5n^2 - \omega_2^2}$$

Integrating,

$$-t = \frac{1}{\sqrt{5}\,n}\tanh^{-1}\left(\frac{\omega_2}{\sqrt{5}\,n}\right) + \gamma \tag{11}$$

where γ is the constant of integration. Initially, when $t = 0$, $\omega_2 = 0$. This gives $\gamma = 0$.

Thus, from Eqn (11),

$$t = -\frac{1}{\sqrt{5}\,n}\tanh^{-1}\left(\frac{\omega_2}{\sqrt{5}\,n}\right)$$

$$\text{or, } \omega_2 = -\sqrt{5}\,n\tanh\left(\sqrt{5}\,nt\right) \quad \textbf{(Proved)}$$

EXAMPLE 8.2 *A circular disc of radius R has a thin rod pushed through its centre perpendicular to its plane, the length of the rod being equal to the radius of the disc. Show that the system cannot spin with the rod vertical unless the angular velocity is greater than*

$$\sqrt{\frac{20g}{R}}$$

Solution. The system behaves like a symmetric top. The moment of inertia about the vertical axis is

$$C = \frac{1}{2}MR^2, \text{ where } M = \text{mass.} \tag{1}$$

The moment of inertia about a line parallel to diameter of the disc passing through the tip is

$$A = \frac{MR^2}{4} + MR^2 = \frac{5}{4}MR^2 \tag{2}$$

For steady motion with axis of rotation vertical, the angular velocity ω should satisfy the condition in Eqn (8.67), i.e.,

$$\omega > \frac{2}{C}\sqrt{MghA} = \frac{4}{MR^2}\sqrt{MgR\frac{5}{4}MR^2} \; [\because h = R]$$

$$= \sqrt{\frac{20g}{R}}.$$

EXAMPLE 8.3 *Show that in a force-free motion of a symmetric top, the angle between the vectors* $\boldsymbol{\omega}$ *and* **J** *is constant. Discuss this situation.*

Solution. Let α be the angle between the vectors $\boldsymbol{\omega}$ and $\mathbf{J}$. Then

$$\cos\alpha = \frac{\boldsymbol{\omega}\cdot\mathbf{J}}{|\boldsymbol{\omega}\cdot\mathbf{J}|} \tag{1}$$

From Eqn (8.15), we have

$$\boldsymbol{\omega}\cdot\mathbf{J} = 2T \tag{2}$$

From (1) and (2),

$$\cos\alpha = \frac{2T}{|\boldsymbol{\omega}\cdot\mathbf{J}|} \tag{3}$$

For the symmetric top, the kinetic energy

$$T = \frac{1}{2}(Aa^2 + Cn^2) = \text{constant.} \tag{4}$$

Hence, from (3) and (4), we conclude that

$$\alpha = \cos^{-1}\left[\frac{2T}{|\omega\cdot J|}\right] = \text{constant.} \tag{5}$$

Thus, $\boldsymbol{\omega}$ rotates about the vector $\mathbf{j}$, which is fixed relative to space axes. It describes a cone of vertical angle 2α. This cone is called a ***space cone***.

The instantaneous axis of rotation is the line of contact between the space cone and the ***body cone***, which is created by the top about Z-axis. Thus, it looks as if the body cone slides on the space cone without any slip. This is shown in FIG. 8.4.

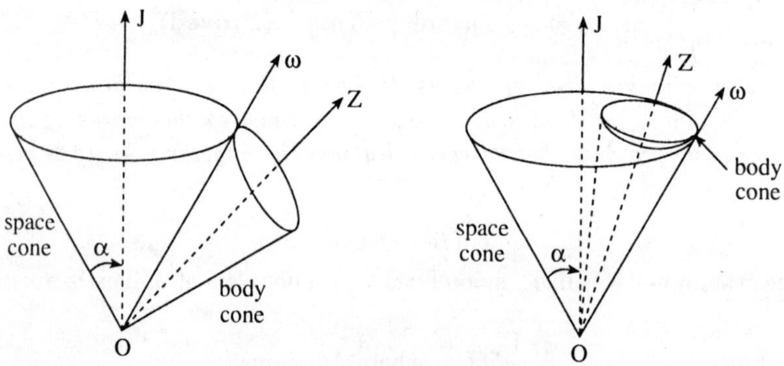

FIG. 8.4

EXAMPLE 8.4 *In the motion of a heavy symmetric top, the initial conditions are given as* $\theta = \theta_0$, $\dot\theta = 0$, $\dot\phi = 0$ *and* $\dot\psi = \omega_3$ *at* $t = 0$.

Deduce the relation $\dot\theta^2 = a(\cos\theta_0 - \cos\theta) - \dfrac{b^2}{\sin^2\theta}(\cos\theta_0 - \cos\theta)^2$, *where* a, b *are some constants.*

Solution. We know that in the motion of the heavy symmetric top, the quantities p_ϕ, p_ψ and E are the constants of motion.

From Eqn (8.51),

$$(p_\psi)_{t=0} = C\omega_3 \tag{1}$$

From Eqn (8.52),

$$(p_\phi)_{t=0} = p_\psi\cos\theta_0 = C\omega_3\cos\theta_0 \tag{2}$$

From Eqn (8.55)

$$(E)_{t=0} = \frac{p_\psi^2}{2C} + Mgh\cos\theta_0 \tag{3}$$

Since energy is conserved, we must have

$$\frac{p_\psi^2}{2C} + Mgh\cos\theta_0 = \frac{1}{2}A\dot{\theta}^2 + \frac{(p_\phi - p_\psi\cos\theta)^2}{2A\sin^2\theta} + \frac{p_\psi^2}{2C} + Mgh\cos\theta$$

$$\begin{aligned}
\text{i.e., } \dot{\theta}^2 &= \frac{2Mgh}{A}(\cos\theta_0 - \cos\theta) - \left(\frac{p_\phi - p_\psi\cos\theta}{A}\right)^2 \frac{1}{\sin^2\theta} \\
&= \frac{2Mgh}{A}(\cos\theta_0 - \cos\theta) - \frac{1}{\sin^2\theta}\left(\frac{C\omega_3\cos\theta_0 - C\omega_3\cos\theta}{A}\right)^2 \\
&= a(\cos\theta_0 - \cos\theta) - \frac{b^2}{\sin^2\theta}(\cos\theta_0 - \cos\theta)^2
\end{aligned}$$

where we have denoted

$$\left.\begin{aligned} a &= \frac{2Mgh}{A} \\ b &= \frac{C\omega_3}{A} \end{aligned}\right\} \tag{4}$$

Hence proved.

EXAMPLE 8.5 *Show that the kinetic energy and angular momentum of torque-free motion of a rigid body are constants.*

Solution. For torque-free motion, $\mathbf{N} = 0$. That is,

$$N_1 = N_2 = N_3 = 0 \tag{1}$$

Then Euler's dynamical equations are:

$$A\dot{\omega}_1 - (B - C)\omega_2\omega_3 = 0 \tag{2}$$

$$B\dot{\omega}_2 - (C - A)\omega_3\omega_1 = 0 \tag{3}$$

$$C\dot{\omega}_3 - (A - B)\omega_1\omega_2 = 0 \tag{4}$$

Multiplying (2) by ω_1, (3) by ω_2, (4) by ω_3 and adding the results, we get

$$A\dot{\omega}_1\omega_1 + B\dot{\omega}_2\omega_2 + C\dot{\omega}_3\omega_3 = 0$$

$$\text{i.e., } \frac{1}{2}\frac{d}{dt}\left(A\omega_1^2 + B\omega_2^2 + C\omega_3^2\right) = 0$$

$$\text{i.e., } \frac{dT}{dt} = 0 \tag{5}$$

Integrating, we get

$$T = \text{constant.} \tag{6}$$

Hence, the kinetic energy is constant.

Again, multiplying (2) by $A\omega_1$, (3) by $B\omega_2$, (4) by $C\omega_3$ and adding the results, we get

$$A^2\dot{\omega}_1\omega_1 + B^2\dot{\omega}_2\omega_2 + C^2\dot{\omega}_3\omega_3 = 0$$

$$\text{i.e., } \frac{d}{dt}\left[(A\omega_1)^2 + (B\omega_2)^2 + (C\omega_3)^2\right] = 0 \tag{7}$$

Integrating, we get

$$(A\omega_1)^2 + (B\omega_2)^2 + (C\omega_3)^2 = \text{constant.} \tag{8}$$

The components of angular momentum **J** are

$$J_1 = A\omega_1,\ J_2 = B\omega_2,\ J_3 = C\omega_3 \tag{9}$$

From (8) and (9),

$$J_1^2 + J_2^2 + J_3^2 = \text{constant, i.e., } |\mathbf{J}|^2 = \text{constant.} \tag{10}$$

Thus, the angular momentum is constant.

EXAMPLE 8.6 *An asymmetric top moves under no external torque such that the components ω_1 and ω_2 of its angular velocity are very small compared to ω_3. Discuss its motion.*

Solution. Euler's equations of motion are

$$A\dot{\omega}_1 = (B - C)\omega_2\omega_3 \tag{1}$$

$$B\dot{\omega}_2 = (C - A)\omega_3\omega_1 \tag{2}$$

$$C\dot{\omega}_3 = (A - B)\omega_1\omega_2 \tag{3}$$

Since ω_1, ω_2 are very small, the top moves nearly with angular velocity ω_3, so that we can accept

$$\omega_1\omega_2 = 0 \tag{4}$$

Then (3) becomes

$$C\dot{\omega}_3 = 0 \tag{5}$$

Hence, ω_3 = constant up to the first order of ω_1 and ω_2.

From (1), we get

$$\dot{\omega}_1 = \left[\frac{B-C}{A}\omega_3\right]\omega_2 \tag{6}$$

From (2), we get

$$\dot{\omega}_2 = \left[\frac{C-A}{B}\omega_3\right]\omega_1 \tag{7}$$

From (6) and (7), we find

$$\ddot{\omega}_1 = \left(\frac{B-C}{A}\omega_3\right)\dot{\omega}_2 = \left(\frac{B-C}{A}\omega_3\right)\left(\frac{C-A}{B}\omega_3\right)\omega_1$$

$$\text{i.e., } \ddot{\omega}_1 = -\Omega_1^2\omega_1 \tag{8}$$

where

$$\Omega_1^2 = \frac{(B-C)(A-C)}{AB}\omega_3^2 \tag{9}$$

Equation (8) represents a simple harmonic motion. Its solution is

$$\omega_1 = P\sin(\Omega_1 t + \lambda) \tag{10}$$

where P and λ are constants.

Inserting (10) in (7), we have

$$\dot{\omega}_2 = P\left(\frac{C-A}{B}\omega_3\right)\sin(\Omega_1 t + \lambda) \tag{11}$$

Integrating,

$$\omega_2 = -\frac{P}{\Omega_1}\left(\frac{C-A}{B}\omega_3\right)\cos(\Omega_1 t + \lambda)$$

$$\text{i.e., } \omega_2 = P\sqrt{\frac{A(A-C)}{B(B-C)}}\cos(\Omega_1 t + \lambda) \text{ [using (9)].} \tag{12}$$

Thus we have found solutions for ω_1 and ω_2 in (10) and (12) respectively. But these solutions are valid only if Ω_1 is real.

From (9), it is obvious that Ω_1 is real if

$$\left.\begin{array}{c} A > C \text{ and } B > C \\ \text{or, if } A < C \text{ and } B < C \end{array}\right\} \tag{13}$$

This means that when C is either greatest or least moment of inertia, Ω_1 is real and ω_1 oscillates harmonically with bounded amplitude.

EXAMPLE 8.7 *Solve the equation of the problem in Example* 8.4 *assuming that*

$$\theta = \theta_0 + \epsilon, (\epsilon \ll \theta_0) \text{ and } \dot{\theta} = \dot{\epsilon}$$

and discuss your results.

Solution. Since $\theta = \theta_0 + \epsilon, \epsilon \ll \theta_0$, we have

$$\cos\theta = \cos(\theta_0 + \epsilon) = \cos\theta_0\cos\epsilon - \sin\theta_0\sin\epsilon \simeq \cos\theta_0 - \epsilon\sin\theta_0 \tag{1}$$

Similarly,

$$\sin\theta = \sin(\theta_0 + \epsilon) = \sin\theta_0\cos\epsilon + \cos\theta_0\sin\epsilon = \sin\theta_0 + \epsilon\cos\theta_0 \tag{2}$$

The given equation is

$$\dot{\theta}^2 = a(\cos\theta_0 - \cos\theta) - b^2\frac{(\cos\theta_0 - \cos\theta)^2}{\sin^2\theta} \tag{3}$$

Using relations (1) and (2),

$$\dot{\theta}^2 = a\epsilon\sin\theta_0 - b^2\epsilon^2$$

$$\therefore \dot{\epsilon}^2 = a\epsilon\sin\theta_0 - b^2\epsilon^2 \quad [\text{since } \dot{\theta} = \dot{\epsilon}]. \tag{4}$$

Equation (4) can be recast as

$$\dot{\epsilon}^2 = b^2(2\epsilon\xi - \epsilon^2) \tag{5}$$

where

$$\xi = \frac{a\sin\theta_0}{2b^2} \tag{6}$$

From (5), we get

$$\frac{d\epsilon}{dt} = b\sqrt{2\xi\epsilon - \epsilon^2}, \text{ i.e., } \frac{d\epsilon}{\sqrt{2\xi\epsilon - \epsilon^2}} = b\,dt \tag{7}$$

Integrating,

$$\epsilon = \xi(1 - \cos bt) \tag{8}$$

Hence,

$$\theta = \theta_0 + \epsilon = \theta_0 + \xi(1 - \cos bt) \tag{9}$$

From (9) it is observed that θ varies from

$$\left.\begin{array}{l}\theta = \theta_0, \text{ when } \cos bt = 1 \\ \theta = \theta_0 + 2\xi, \text{ when } \cos bt = -1\end{array}\right\} \tag{10}$$

and that the nature of change is simple harmonic.

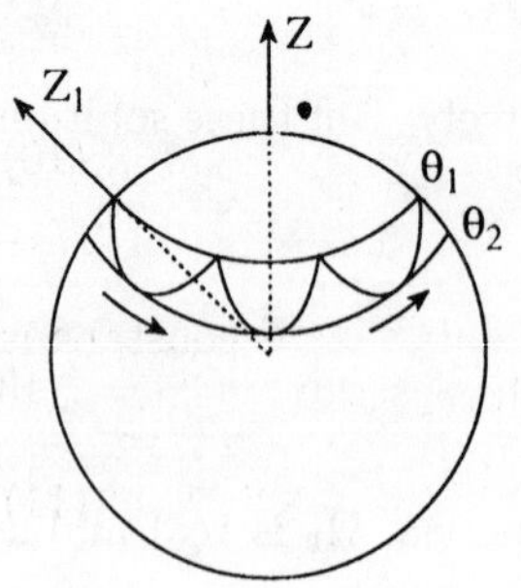

FIG. 8.5

The amplitude ξ is given by (6) and angular velocity is

$$b = \frac{C}{A}\omega_3 \tag{11}$$

This variation of angle θ, as we know, is called ***nutation***. This is shown in FIG. 8.5.

Inserting (11) into (6), we get

$$\xi = \frac{a \sin\theta_0}{b^2} = \frac{a \sin\theta A^2}{C^2\omega_3^2} \propto \frac{1}{\omega_3^2} \tag{12}$$

Thus, the amplitude of nutation ξ is inversely proportional to the square of ω_3.

EXAMPLE 8.8 *Find an expression for the angle of rotation ϕ of the symmetry axis about the Z-axis in the motion of the heavy symmetric top under the initial conditions*

$$\phi = 0, \theta = \theta_0, \dot{\theta} = 0, \dot{\phi} = 0, \dot{\psi} = \omega_3, \text{ at } t = 0.$$

Discuss your results.

Solution. We have from Eqn (8.54)

$$\dot{\phi} = \frac{p_\phi - p_\psi \cos\theta}{A\sin^2\theta} \tag{1}$$

where p_ϕ, p_ψ are constants of motion given by

$$\left.\begin{array}{l}p_\phi = A\dot{\phi}\sin^2\theta + p_\psi\cos\theta \\ p_\psi = C\omega_3\end{array}\right\} \tag{2}$$

Using initial conditions in (2), we have

$$p_\phi = (p_\phi)_{t=0} = C\omega_3 \cos\theta_0 \tag{3}$$

Hence, equation (1) becomes

$$\dot{\phi} = -\frac{C\omega_3}{A}\frac{\cos\theta - \cos\theta_0}{\sin^2\theta} \tag{4}$$

Assuming $\theta = \theta_0 + \epsilon$, $\epsilon \ll \theta_0$, and $\dot{\theta} = \dot{\epsilon}$ as in the previous example, we write (4) in the form:

$$\dot{\phi} = \frac{b\epsilon\sin\theta_0}{\sin^2\theta} = \frac{b\epsilon}{\sin\theta_0} \tag{5}$$

where

$$b = \frac{C\omega_3}{A} \tag{6}$$

$$\epsilon = \xi(1 - \cos bt) \tag{7}$$

as obtained earlier.

From (5) and (7), we have

$$\frac{d\phi}{dt} = \frac{b\xi}{\sin\theta_0}(1 - \cos bt) \tag{8}$$

Integrating,

$$\phi = \frac{\xi b}{\sin\theta_0}t - \frac{\xi}{\sin\theta_0}\sin bt \tag{9}$$

We can ignore the last term in the RHS of (9), since the oscillations of $\sin bt$ are quite fast and is $\sim \omega_3$.

Then

$$\phi = \frac{\xi b}{\sin\theta_0}t \tag{10}$$

which shows that the angle ϕ changes linearly with time.

This kind of rotational motion of the symmetry axis of the top maintaining θ constant is called ***precession***. The ***angular speed*** of the precession without rotation is

$$\omega^* = \frac{\xi b}{\sin\theta_0} \tag{11}$$

Inserting the values of ξ and b, we get

$$\omega^* = \frac{ba\sin\theta_0}{b^2\sin\theta_0} = \frac{a}{b} = \frac{2Mgh}{A}\cdot\frac{A}{C\omega_3} = \frac{2Mgh}{C\omega_3} \tag{12}$$

This shows that the angular speed of precession is inversely proportional to ω_3.

EXERCISE

1. Set up the Euler's dynamical equations of motion. Show that for force-free rotational motion of a rigid body, total angular momentum remains constant. **[C.U. (HONS) 2006]**
2. For rotational motion of rigid bodies, derive an expression for kinetic energy in terms of moment of inertia and angular velocity. **[C.U. (HONS) 2005]**
3. Find the kinetic energy of rotation of a rigid body with respect to principal axes in terms of Euler's angles.

   Show that the kinetic energy is constant when there is no external torque.
4. A spherical top is rotating about a fixed point under no external forces. Show that its angular velocity remains constant.
5. A spherical top of radius R is spinning about a vertical axis supported at its lowest point. Show that its angular velocity ω should satisfy the condition

$$\omega > \sqrt{\frac{35g}{R}}$$

6. Use Euler's equations of motion to prove that

$$\frac{dT}{dt} = \mathbf{N}\cdot\omega$$

CHAPTER 9

Hamiltonian Dynamics

In Lagrangian formulation, the dynamics of a system are governed by a set of second-order differential equations. In Hamiltonian dynamics, the set of second-order differential equations are reduced to a set of first-order differential equations by introducing the concept of generalised momenta.

9.1 Generalised Momentum

We consider a holonomic system of independent coordinates $q_1, q_2, \cdots, q_n$. We define the ***generalised momentum*** corresponding to the generalised coordinate q_k as

$$p_k = \frac{\partial L}{\partial \dot{q}_k} \tag{9.1}$$

where L denotes the Lagrangian function. For a conservative system, we know that

$$\frac{d}{dt}\left(\frac{\partial L}{\partial \dot{q}_k}\right) - \frac{\partial L}{\partial q_k} = 0 \tag{9.2}$$

which upon using Eqn (9.1) assumes the form:

$$\frac{d}{dt}(p_k) - \frac{\partial L}{\partial q_k} = 0 \text{ or, } \dot{p}_k = \frac{\partial L}{\partial q_k} \tag{9.3}$$

It is to be noted that the generalised momentum p_k may not have the dimension of momentum.

9.2 Cyclic Coordinates

We know that the Lagrangian L is a function of generalised coordinates q_k and generalised velocities $\dot{q}_k$. If any one coordinate, say q_k, does not appear explicitly in the Lagrangian L, then

$$\frac{\partial L}{\partial q_k} = 0 \tag{9.4}$$

In such a situation, we find from Eqn (9.3), that

$$\dot{p}_k = 0 \tag{9.5}$$

which gives on integration

$$p_k = \text{constant.} \tag{9.6}$$

Thus, whenever the coordinate q_k is absent in the Lagrangian function L, the corresponding generalised momentum p_k is a constant of the motion. Such a coordinate q_k is said to be ***cyclic*** or ***ignorable***, and p_k is said to be the first integral of motion.

As an example, the Lagrangian L for a particle moving under central force does not contain the coordinate θ. Hence, θ is the cyclic coordinate. Thus,

$$p_\theta = \frac{\partial L}{\partial \dot{\theta}} = mr^2\dot{\theta} = \text{constant.} \tag{9.7}$$

Here, p_θ represents the angular momentum, which is conserved in the system.

9.3 Hamilton's Equations of Motion

We define the function

$$H = -L + \sum_{k=1}^{n} p_k \dot{q}_k \tag{9.8}$$

The function H is called the ***Hamiltonian function***, which in general, is a function of generalised coordinates, generalised momenta and time. Hence, we can write:

$$H = H(p_1, p_2, \dots, p_n, q_1, q_2, \dots, q_n, t) \text{ or, } H = H(p_k, q_k, t) \tag{9.9}$$

Then

$$dH = \sum_k \frac{\partial H}{\partial p_k} dp_k + \sum_k \frac{\partial H}{\partial q_k} dq_k + \frac{\partial H}{\partial t} dt \tag{9.10}$$

Also, from Eqn (9.8), we have

$$dH = -dL + \Sigma p_k d\dot{q}_k + \Sigma \dot{q}_k dp_k \tag{9.11}$$

The Lagrangian is a function of generalised coordinates, generalised velocities and time, so that, we write:

$$L = L(q_k, \dot{q}_k, t) \tag{9.12}$$

Hence,

$$dL = \Sigma \frac{\partial L}{\partial q_k} dq_k + \Sigma \frac{\partial L}{\partial \dot{q}_k} d\dot{q}_k + \frac{\partial L}{\partial t} dt \tag{9.13}$$

But, by Eqns (9.1) and (9.3), we have

$$p_k = \frac{\partial L}{\partial \dot{q}_k} \text{ and } \dot{p}_k = \frac{\partial L}{\partial q_k} \tag{9.14}$$

Using these relations in Eqn (9.13), we get

$$dL = \Sigma \dot{p}_k dq_k + \Sigma p_k d\dot{q}_k + \frac{\partial L}{\partial t} dt \tag{9.15}$$

Substituting from this equation the expression for dL into Eqn (9.11), we get

$$dH = -\Sigma \dot{p}_k dq_k + \Sigma \dot{q}_k dp_k - \frac{\partial L}{\partial t} dt \tag{9.16}$$

Comparing Eqns (9.10) and (9.16), we have

$$\dot{q}_k = \frac{\partial H}{\partial p_k} \tag{9.17}$$

$$\dot{p}_k = -\frac{\partial H}{\partial q_k} \tag{9.18}$$

$$\text{and } \frac{\partial L}{\partial t} = -\frac{\partial H}{\partial t} \tag{9.19}$$

These are known as ***Hamilton's canonical equations of motion***.

If the Lagrangian L does not contain time explicitly, then

$$\frac{\partial L}{\partial t} = 0, \text{ i.e., } \frac{\partial H}{\partial t} = 0 \tag{9.20}$$

Hence, from Eqn (9.10), we have

$$\frac{dH}{dt} = \sum_k \left(\frac{\partial H}{\partial p_k} \dot{p}_k + \frac{\partial H}{\partial q_k} \dot{q}_k \right) \tag{9.21}$$

By virtue of Eqns (9.17) and (9.18), the RHS of Eqn (9.21) vanishes. Hence,

$$\frac{dH}{dt} = 0, \text{ i.e., } H = \text{const.} \tag{9.22}$$

If the system is conservative, the potential energy V does not depend on velocity coordinates, so that

$$\frac{\partial V}{\partial \dot{q}_k} = 0 \tag{9.23}$$

From Eqn (9.14), we have

$$p_k = \frac{\partial L}{\partial \dot{q}_k} = \frac{\partial}{\partial \dot{q}_k}(T - V) = \frac{\partial T}{\partial \dot{q}_k} \tag{9.24}$$

Hence,

$$H = \sum_k p_k \dot{q}_k - L = \sum_k \frac{\partial T}{\partial \dot{q}_k} \dot{q}_k - L \tag{9.25}$$

If the position vector $\mathbf{r}_i$ does not depend on time explicitly, then T is a homogeneous quadratic function of coordinates $\dot{q}_k$. By Euler's theorem for homogeneous functions, we can write:

$$\sum_k \frac{\partial T}{\partial \dot{q}_k} \dot{q}_k = 2T \tag{9.26}$$

Hence, from Eqn (9.25), we can write:

$$H = -L + 2T = T + V = \text{const.} \tag{9.27}$$

Thus, Hamiltonian equation gives an equation of energy.

EXAMPLE 9.1 *Find Hamilton's equation of motion for a particle under central forces.*

Solution. If m be the mass of the particle, the Lagrangian in polar coordinates is

$$L = T - V = \frac{1}{2} m \left(\dot{r}^2 + r^2 \dot{\theta}^2 \right) - V \tag{1}$$

where

$$-\frac{dV}{dr} = f(r), \text{ the central force.}$$

Now

$$p_r = \frac{\partial L}{\partial \dot{r}} = m\dot{r}; \text{ hence, } \dot{r} = \frac{p_r}{m} \tag{2}$$

Similarly,

$$p_\theta = \frac{\partial L}{\partial \dot{\theta}} = mr^2 \dot{\theta}; \text{ hence, } \dot{\theta} = \frac{p_\theta}{mr^2} \tag{3}$$

Hence, the Hamiltonian function is

$$H = T + V = \frac{1}{2}m\left[\left(\frac{p_r}{m}\right)^2 + r^2\left(\frac{p_\theta}{mr^2}\right)^2\right] + V(r)$$

$$\text{i.e., } H = \frac{1}{2m}\left[p_r^2 + \frac{p_\theta^2}{r^2}\right] + V(r) \tag{4}$$

We have generalised coordinates $q_1 = r$, $q_2 = \theta$. Hence, Hamilton's equations are:

For $q_1 = r$,

$$\dot{r} = \frac{\partial H}{\partial p_r} = \frac{p_r}{m} \tag{5}$$

$$-\dot{p}_r = \frac{\partial H}{\partial r} = -\frac{p_\theta^2}{mr^3} + \frac{\partial V}{\partial r} \tag{6}$$

For $q_2 = \theta$,

$$\dot{\theta} = \frac{\partial H}{\partial p_\theta} = \frac{p_\theta}{mr^2} \tag{7}$$

$$-\dot{p}_\theta = \frac{\partial H}{\partial \theta} = 0 \tag{8}$$

From Eqns (7) and (8), we get

$$p_\theta = mr^2\dot{\theta} = \text{const.} \tag{9}$$

which states that the angular momentum of the particle under central force is conserved.

From Eqn (5), we get

$$m\dot{r} = p_r, \text{ i.e., } m\ddot{r} = \dot{p}_r \tag{10}$$

From Eqns (6) and (10), we can write:

$$-m\ddot{r} = -\frac{p_\theta^2}{mr^3} + \frac{\partial V}{\partial r}$$

$$\text{i.e., } m\ddot{r} - \frac{p_\theta^2}{mr^3} + \frac{\partial V}{\partial r} = 0 \tag{11}$$

which is the required equation.

EXAMPLE 9.2 *Find Hamilton's equation of motion in polar cylindrical coordinates.*

Solution. In polar cylindrical coordinates, we have

$$L = T - V = \frac{1}{2}m\left(\dot{r}^2 + r^2\dot{\theta}^2 + \dot{z}^2\right) - V(r, \theta, z)$$

Then, the generalised momenta are:

$$p_r = \frac{\partial L}{\partial \dot{r}} = m\dot{r}, \text{ i.e., } \dot{r} = \frac{p_r}{m} \tag{1}$$

$$p_\theta = \frac{\partial L}{\partial \dot{\theta}} = mr^2\dot{\theta}, \text{ i.e., } \dot{\theta} = \frac{p_\theta}{mr^2} \tag{2}$$

$$p_z = \frac{\partial L}{\partial \dot{z}} = m\dot{z}, \text{ i.e., } \dot{z} = \frac{p_z}{m} \tag{3}$$

Hence, Hamilton function is

$$H = \Sigma p_k\dot{q}_k - L = \left(p_r\dot{r} + p_\theta\dot{\theta} + p_z\dot{z}\right) - \frac{1}{2}m\left(\dot{r}^2 + r^2\dot{\theta}^2 + \dot{z}^2\right) + V$$

$$\text{i.e., } H = \frac{1}{2m}\left(p_r^2 + \frac{p_\theta^2}{r^2} + p_z^2\right) + V \tag{4}$$

Hence, Hamilton's equations are:

$$\text{For } q_1 = r, \begin{cases} \dot{r} = \dfrac{\partial H}{\partial p_r} = \dfrac{p_r}{m} & (5) \\ -\dot{p}_r = \dfrac{\partial H}{\partial r} = -\dfrac{p_\theta^2}{mr^3} + \dfrac{\partial V}{\partial r} & (6) \end{cases}$$

$$\text{For } q_2 = \theta, \begin{cases} \dot{\theta} = \dfrac{\partial H}{\partial p_\theta} = \dfrac{p_\theta}{mr^2} & (7) \\ -\dot{p}_\theta = \dfrac{\partial H}{\partial \theta} = \dfrac{\partial V}{\partial \theta} & (8) \end{cases}$$

$$\text{For } q_3 = z, \begin{cases} \dot{z} = \dfrac{\partial H}{\partial p_z} = \dfrac{p_z}{m} & (9) \\ -\dot{p}_z = \dfrac{\partial H}{\partial z} = \dfrac{\partial V}{\partial z} & (10) \end{cases}$$

EXAMPLE 9.3 *Derive the Hamiltonian equations of a simple pendulum.*

Solution. The Lagrangian for the motion of a simple pendulum is

$$L = T - V = \frac{1}{2}ml^2\dot{\theta}^2 - mgl(1 - \cos\theta) \tag{1}$$

Hence,

$$p_\theta = \frac{\partial L}{\partial \dot{\theta}} = ml^2\dot{\theta}, \text{ i.e., } \dot{\theta} = \frac{p_\theta}{ml^2} \tag{2}$$

Now, Hamilton's function is

$$\begin{aligned} H &= \Sigma p_k \dot{q}_k - L = p_\theta\dot{\theta} - \left[\frac{1}{2}ml^2\dot{\theta}^2 - mgl(1 - \cos\theta)\right] \\ &= ml^2\dot{\theta}^2 - \frac{1}{2}ml^2\dot{\theta}^2 + mgl(1 - \cos\theta) = \frac{1}{2}ml^2\dot{\theta}^2 + mgl(1 - \cos\theta) \\ &= \frac{1}{2}ml^2\left(\frac{p_\theta}{ml^2}\right)^2 + mgl(1 - \cos\theta) \text{ [using Eqn (2)]} \end{aligned}$$

$$\text{or, } H = \frac{p_\theta^2}{2ml^2} + mgl(1 - \cos\theta) \tag{3}$$

Hence, Hamilton's equations are:

$$\dot{\theta} = \frac{\partial H}{\partial p_\theta} = \frac{p_\theta}{ml^2} \tag{4}$$

$$\text{and } -\dot{p}_\theta = \frac{\partial H}{\partial \theta} = mgl\sin\theta \tag{5}$$

From Eqn (4), we have

$$\dot{p}_\theta = ml^2\ddot{\theta} \tag{6}$$

From Eqns (5) and (6), we get

$$ml^2\ddot{\theta} = -mgl\sin\theta \text{ or, } l\ddot{\theta} + g\sin\theta = 0 \tag{7}$$

For small θ, $\sin\theta = \theta$; hence, we get from (7)

$$\ddot{\theta} + \frac{g}{l}\theta = 0 \tag{8}$$

which is the equation of motion of a simple pendulum.

From Eqn (3), we find that $H = T + V = E$, where E is the total energy.

EXAMPLE 9.4 *Derive Hamilton's equation of motion in spherical polar coordinates.*

Solution. In spherical polar coordinates

$$\left.\begin{aligned} x &= r\sin\theta\cos\phi \\ y &= r\sin\theta\sin\phi \\ z &= r\cos\theta \end{aligned}\right\} \tag{1}$$

Hence,

$$\dot{x} = \dot{r}\sin\theta\cos\phi + r\dot{\theta}\cos\theta\cos\phi - r\dot{\phi}\sin\theta\sin\phi \tag{2}$$

$$\dot{y} = \dot{r}\sin\theta\sin\phi + r\dot{\theta}\cos\theta\sin\phi + r\dot{\phi}\sin\theta\cos\phi \tag{3}$$

$$\dot{z} = \dot{r}\cos\theta - r\dot{\theta}\sin\theta \tag{4}$$

Thus,

$$T = \frac{1}{2}m\left(\dot{x}^2 + \dot{y}^2 + \dot{z}^2\right) = \frac{1}{2}m\left(\dot{r}^2 + r^2\dot{\theta}^2 + r^2\sin^2\theta\dot{\phi}^2\right) \tag{5}$$

Hence,

$$L = T - V = \frac{1}{2}m\left(\dot{r}^2 + r^2\dot{\theta}^2 + r^2\sin^2\theta\dot{\phi}^2\right) - V(r,\theta,\phi) \tag{6}$$

Now

$$p_r = \frac{\partial L}{\partial \dot{r}} = m\dot{r}, \text{ i.e., } \dot{r} = \frac{p_r}{m} \tag{7}$$

$$p_\theta = \frac{\partial L}{\partial \dot{\theta}} = mr^2\dot{\theta}, \text{ i.e., } \dot{\theta} = \frac{p_\theta}{mr^2} \tag{8}$$

$$p_\phi = \frac{\partial L}{\partial \dot{\phi}} = mr^2\sin^2\theta\dot{\phi}, \text{ i.e., } \dot{\phi} = \frac{p_\phi}{mr^2\sin^2\theta} \tag{9}$$

Hence,

$$\begin{aligned} H &= \Sigma p_k\dot{q}_k - L \\ &= p_r\dot{r} + p_\theta\dot{\theta} + p_\phi\dot{\phi} - \frac{1}{2}m\left(\dot{r}^2 + r^2\dot{\theta}^2 + r^2\dot{\phi}^2\sin^2\theta\right) + V(r,\theta,\phi) \end{aligned}$$

$$\text{or, } H = \frac{1}{2m}\left(p_r^2 + \frac{p_\theta^2}{r^2} + \frac{p_\phi^2}{r^2\sin^2\theta}\right) + V \text{ [using Eqns (7)-(9)]} \tag{10}$$

Hence, Hamiltonian equations are:

$$\dot{r} = \frac{\partial H}{\partial p_r} = \frac{p_r}{m} \tag{11}$$

$$-\dot{p}_r = \frac{\partial H}{\partial r} = -\frac{p_\theta^2}{mr^3} - \frac{p_\phi^2}{mr^3\sin^2\phi} + \frac{\partial V}{\partial r} \tag{12}$$

$$\dot{\theta} = \frac{\partial H}{\partial p_\theta} = \frac{p_\theta}{mr^2} \tag{13}$$

$$-\dot{p}_\theta = \frac{\partial H}{\partial \theta} = -\frac{p_\phi^2\cos\theta}{mr^2\sin^3\theta} + \frac{\partial V}{\partial \theta} \tag{14}$$

$$\dot{\phi} = \frac{\partial H}{\partial p_\phi} = \frac{p_\phi}{mr^2\sin^2\theta} \tag{15}$$

$$-\dot{p}_\phi = \frac{\partial H}{\partial \phi} = \frac{\partial V}{\partial \phi} \tag{16}$$

EXAMPLE 9.5 *Derive Hamilton's equation of motion of a projectile in space, neglecting the earth's rotation.*

Solution. If m be the mass of the projectile, the kinetic energy at any time is

$$K = \frac{1}{2}m\left(\dot{x}^2 + \dot{y}^2 + \dot{z}^2\right) \tag{1}$$

Potential energy relative to surface of the earth is

$$V = mgz \tag{2}$$

Hence,

$$L = T - V = \frac{1}{2}m\left(\dot{x}^2 + \dot{y}^2 + \dot{z}^2\right) - mgz \tag{3}$$

Since L does not depend on time t explicitly, the Hamiltonian is

$$H = T + V = \frac{1}{2}m\left(\dot{x}^2 + \dot{y}^2 + \dot{z}^2\right) + mgz \tag{4}$$

Hence,

$$p_x = \frac{\partial H}{\partial \dot{x}} = m\dot{x}, \text{ i.e., } \dot{x} = \frac{p_x}{m} \tag{5}$$

$$p_y = \frac{\partial H}{\partial \dot{y}} = m\dot{y}, \text{ i.e., } \dot{y} = \frac{p_y}{m} \tag{6}$$

$$p_z = \frac{\partial H}{\partial \dot{z}} = m\dot{z}, \text{ i.e., } \dot{z} = \frac{p_z}{m} \tag{7}$$

Using Eqns (5)-(7), we get from (4)

$$H = \frac{1}{2m}\left(p_x^2 + p_y^2 + p_z^2\right) + mgz \tag{8}$$

Hamilton's equations are:

$$\dot{x} = \frac{\partial H}{\partial p_x} = \frac{p_x}{m}; \; \dot{p}_x = -\frac{\partial H}{\partial x} = 0 \tag{9}$$

$$\dot{y} = \frac{\partial H}{\partial p_y} = \frac{p_y}{m}; \; \dot{p}_y = -\frac{\partial H}{\partial y} = 0 \tag{10}$$

$$\dot{z} = \frac{\partial H}{\partial p_z} = \frac{p_z}{m}; \; \dot{p}_z = -\frac{\partial H}{\partial z} = -mg \tag{11}$$

Equations of motion of the projectile is obtained from (9)–(11) as

$$\ddot{x} = \frac{\dot{p}_x}{m} = 0 \tag{12}$$

$$\ddot{y} = \frac{\dot{p}_y}{m} = 0 \tag{13}$$

$$\ddot{z} = \frac{\dot{p}_z}{m} = -g \tag{14}$$

EXAMPLE 9.6 *A particle of mass m is free to slide on a smooth helical wire whose position in cylindrical coordinates is represented as $\rho = a$ and $z = b\phi$. The particle is released from rest at $\rho = a, \phi = 0$ and $z = 0$. Discuss the motion of the particle.*

Solution. In cylindrical coordinates,

$$T = \frac{1}{2}m\left(\dot{\rho}^2 + \rho^2\dot{\phi}^2 + \dot{z}^2\right) \tag{1}$$

For helical wire

$$\rho = a, \text{ hence, } \dot{\rho} = 0 \tag{2}$$

and, $b\phi = z$, so that

$$\dot{\phi} = \frac{\dot{z}}{b} \tag{3}$$

Hence,

$$T = \frac{1}{2}m\left(\frac{a^2}{b^2}\dot{z}^2 + \dot{z}^2\right) = \frac{m}{2b^2}(a^2 + b^2)\dot{z}^2 \tag{4}$$

Potential energy measured from $z = 0$ is

$$V = -mgz \tag{5}$$

Hence, the Lagrangian is

$$L = T - V = \frac{m}{2b^2}(a^2 + b^2)\dot{z}^2 + mgz \tag{6}$$

Hence,

$$p_z = \frac{\partial L}{\partial \dot{z}} = \frac{m}{b^2}(a^2 + b^2)\dot{z} \tag{7}$$

From (4) and (7),

$$T = \frac{1}{2m}\frac{b^2}{a^2 + b^2}p_z^2 \tag{8}$$

Hence, Hamiltonian is

$$H = T + V = \frac{1}{2m}\frac{b^2}{a^2 + b^2}p_z^2 - mgz \tag{9}$$

Thus, Hamiltonian equations are:

$$\dot{z} = \frac{\partial H}{\partial p_z} = \frac{b^2}{m(a^2 + b^2)}p_z \tag{10}$$

$$-\dot{p}_z = \frac{\partial H}{\partial z} = -mg \tag{11}$$

From (10) and (11), we get

$$\ddot{z} = \frac{b^2}{a^2 + b^2}g \tag{12}$$

On integration,

$$z = \frac{b^2 g}{a^2 + b^2}\left(\frac{t^2}{2} + At + B\right) \tag{13}$$

Initially, at $t = 0, z = 0$ and $\dot{z} = 0$. Hence, $A = 0$ and $B = 0$. Thus,

$$z = \frac{b^2}{2(a^2 + b^2)}gt^2. \tag{14}$$

EXAMPLE 9.7 *The point of suspension of a simple pendulum moves harmonically along the vertical upward direction. Discuss the motion by Hamiltonian method.*

Solution. Let the point of suspension move according to the law, $y = a\sin\omega t$. Let O be the initial position of the point of suspension which moves to O' after time t so that $OO' = a\sin\omega t$.

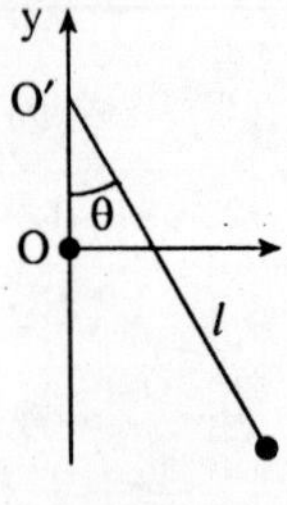

FIG. 9.1

Then the coordinates of the bob of the pendulum at any instant of time are (FIG. 9.1):

$$\left.\begin{aligned} x &= l\sin\theta \\ y &= -l\cos\theta + a\sin\omega t \end{aligned}\right\} \tag{1}$$

Hence,

$$\left.\begin{aligned} \dot{x} &= l\dot{\theta}\cos\theta \\ \dot{y} &= l\dot{\theta}\sin\theta + a\omega\cos\omega t \end{aligned}\right\} \tag{2}$$

Thus, kinetic energy is

$$T = \frac{1}{2}m\left(\dot{x}^2 + \dot{y}^2\right) = \frac{1}{2}m\left(l^2\dot{\theta}^2 + a^2\omega^2\cos^2\omega t + 2a\omega l\dot{\theta}\sin\theta\cos\omega t\right) \tag{3}$$

Potential energy is

$$V = mg(l - l\cos\theta + a\sin\omega t) \tag{4}$$

The Hamiltonian function is

$$\begin{aligned} H &= T + V \\ &= \frac{1}{2}m\left(l^2\dot{\theta}^2 + a^2\omega^2\cos^2\omega t + 2a\omega l\dot{\theta}\sin\theta\cos\omega t\right) + mg(l - l\cos\theta + a\sin\omega t) \end{aligned} \tag{5}$$

From the Lagrangian $L = T - V$, we find

$$p_\theta = \frac{\partial L}{\partial\dot{\theta}} = m\left(l^2\dot{\theta} + a\omega l\sin\theta\cos\omega t\right)$$

$$\text{i.e., } \dot{\theta} = \frac{1}{ml^2}(p_\theta - ma\omega l\sin\theta\cos\omega t) \tag{6}$$

Substituting this value for $\dot{\theta}$ into Eqn (5), we get

$$\begin{aligned} H = \frac{1}{2}m\Bigg[&l^2\left(\frac{p_\theta}{ml^2} - \frac{a\omega}{l}\sin\theta\cos\omega t\right)^2 + a^2\omega^2\cos^2\omega t \\ &+ 2a\omega l\sin\theta\cos\omega t\left(\frac{p_\theta}{ml^2} - \frac{a\omega}{l}\sin\theta\cos\omega t\right)\Bigg] \\ &+ mg\,(l - l\cos\theta + a\sin\omega t) \end{aligned} \tag{7}$$

Hamilton's equation for $\dot{\theta}$ is

$$\dot{\theta} = \frac{\partial H}{\partial p_\theta} = \frac{1}{2}m\left[2l^2\left(\frac{p_\theta}{ml^2} - \frac{a\omega}{l}\sin\theta\cos\omega t\right)\cdot\frac{1}{ml^2} + 2a\omega l\sin\theta\cos\omega t\cdot\frac{1}{ml^2}\right]$$

$$\text{i.e., } \dot{\theta} = \frac{p_\theta}{ml^2} \tag{8}$$

The other Hamilton's equation is

$$\dot{p}_\theta = -\frac{\partial H}{\partial\theta} = -\frac{1}{2}m\left[2a\omega\dot{\theta}l\cos\theta\cos\omega t\right] - mgl\sin\theta$$

$$\text{or, } ml^2\ddot{\theta} = -ml\left[a\omega\dot{\theta}\cos\theta\cos\omega t + g\sin\theta\right] \quad \text{[using (8)]}$$

$$\text{or, } l\ddot{\theta} = -\left(a\omega\dot{\theta}\cos\theta\cos\omega t + g\sin\theta\right) \tag{9}$$

This is the equation of motion of the simple pendulum in this case. If, in particular, θ is small, then $\cos\theta \simeq 1$ and $\sin\theta \simeq \theta$, so that the Eqn (9) assumes the form:

$$l\ddot{\theta} + a\omega\dot{\theta}\cos\omega t + g\theta = 0. \tag{10}$$

9.4 Calculus of Variation

We have obtained earlier the Lagrange's equations from D'Alembert's principle. Now we intend to obtain Lagrange's equations from the variational principle where we consider the entire motion of the system between two instants of time t_1 and t_2.

For this purpose, we like to discuss some aspects of calculus of variation. We consider a function $y(x)$ in one dimension between two points x_1 and x_2. Let the integral J be defined as

$$J = \int_{x_1}^{x_2} F(y, y', x)dx \tag{9.28}$$

where $y' = \dfrac{dy}{dx}$. Here, $F(y, y', x)$ is called a ***functional***. The fundamental problem of the calculus of variation lies in finding a function $y(x)$ such that the integral J is extremum for small variation in $y(x)$.

We consider a curve $y(x, \alpha)$ very close to $y(x)$ between the points x_1, x_2, and write:

$$y(x, \alpha) = y(x, 0) + \alpha\eta(x) \tag{9.29}$$

where α is a small scale factor and $\eta(x)$ is an arbitrary differentiable function of x, such that (see FIG. 9.2)

$$\eta(x_1) = \eta(x_2) = 0 \tag{9.30}$$

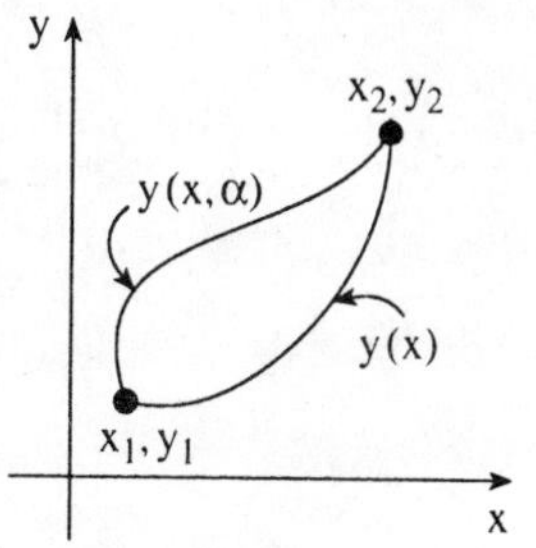

FIG. 9.2

Then Eqn (9.28) becomes

$$J(\alpha) = \int_{x_1}^{x_2} F\left[y(x, \alpha), y'(x, \alpha), x\right] dx \tag{9.31}$$

The condition for extremum value of J is

$$\left.\frac{dJ}{d\alpha}\right|_{\alpha=0} = 0, \text{ for all } \eta(x) \tag{9.32}$$

Differentiating Eqn (9.31) under integral sign w.r.t. α, we get

$$\frac{dJ}{d\alpha} = \int_{x_1}^{x_2} \left[\frac{\partial F}{\partial y}\frac{\partial y}{\partial \alpha} + \frac{\partial F}{\partial y'}\frac{\partial y'}{\partial \alpha}\right] dx \tag{9.33}$$

But from Eqn (9.29), we have

$$\frac{\partial y}{\partial \alpha} = \eta(x) \text{ and } \frac{\partial y'}{\partial \alpha} = \frac{d\eta}{d\alpha} \tag{9.34}$$

In view of these relations, Eqn (9.33) assumes the form:

$$\frac{dJ}{d\alpha} = \int_{x_1}^{x_2} \left[\frac{\partial F}{\partial y}\eta(x) + \frac{\partial F}{\partial y'}\frac{d\eta}{dx}\right] dx \tag{9.35}$$

Now

$$\begin{aligned}\int_{x_1}^{x_2} \frac{\partial F}{\partial y'}\frac{d\eta}{dx}dx &= \left.\frac{\partial F}{\partial y'}\eta(x)\right|_{x_1}^{x_2} - \int_{x_1}^{x_2} \frac{d}{dx}\left(\frac{\partial F}{\partial y'}\right)\eta(x)dx \\ &= -\int_{x_1}^{x_2} \frac{d}{dx}\left(\frac{\partial F}{\partial y'}\right)\eta(x)dx \end{aligned} \tag{9.36}$$

since $\eta(x_1) = \eta(x_2) = 0$.

Substituting Eqn (9.36) into (9.35), we get

$$\frac{dJ}{d\alpha} = \int_{x_1}^{x_2} \left[\frac{\partial F}{\partial y} - \frac{d}{dx}\left(\frac{\partial F}{\partial y'}\right)\right] \eta(x)dx \tag{9.37}$$

Since $\eta(x)$ is an arbitrary function, the condition in Eqn (9.32) will be satisfied when the integrand in Eqn (9.37) vanishes. Thus, we have

$$\frac{\partial F}{\partial y} - \frac{d}{dx}\left(\frac{\partial F}{\partial y'}\right) = 0 \tag{9.38}$$

This equation is known as ***Euler's equation.***

EXAMPLE 9.8 *Show that the shortest distance between two points in a plane is a straight line.*

Solution. The elemental length in xy-plane is $ds = (dx^2 + dy^2)^{1/2} = dx\left[1 + \left(\frac{dy}{dx}\right)^2\right]^{1/2}$.

Thus, the distance between two points (x_1, y_1) and (x_2, y_2) is

$$J = \int_{x_1}^{x_2} ds = \int_{x_1}^{x_2} \left[1 + \left(\frac{dy}{dx}\right)^2\right]^{1/2} dx \tag{1}$$

This integral J should be minimum for the path to be the shortest. Thus, here we have

$$F = \sqrt{1 + y'^2},\ y' = \frac{dy}{dx} \tag{2}$$

Hence,

$$\frac{\partial F}{\partial y} = 0 \text{ and } \frac{\partial F}{\partial y'} = \frac{y'}{\sqrt{1 + y'^2}} \tag{3}$$

Substituting these values in Euler's Eqn (9.38), we get

$$\frac{d}{dx}\left[\frac{y'}{\sqrt{1 + y'^2}}\right] = 0 \text{ or, } \frac{y'}{\sqrt{1 + y'^2}} = \text{const.} = C \text{ (say)} \tag{4}$$

Simplifying this equation, we get $y' = \frac{C}{\sqrt{1 - C^2}} = \text{const.} = A$ (say)

Integrating this, we have $y = Ax + B$, which represents a straight line.

Hence proved.

EXAMPLE 9.9 *Find the equation to the curve passing through two fixed points such that the area of the surface obtained by revolving the curve about y-axis is minimum.*

Solution. In FIG. 9.3, we have shown the curve AB passing through two fixed points (x_1, y_1), (x_2, y_2), and the surface obtained by revolving it about y-axis.

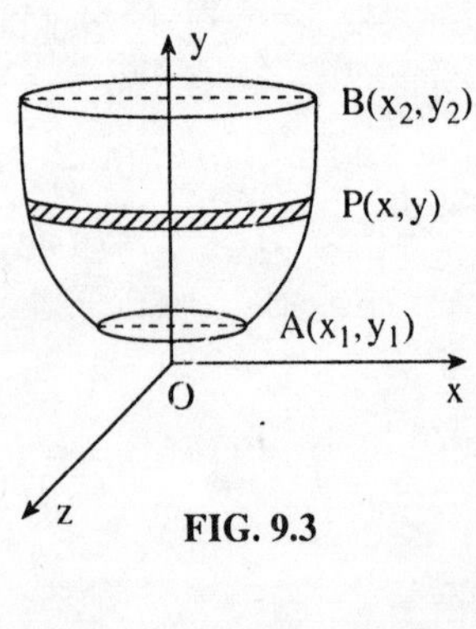

FIG. 9.3

We take an element of ring-like strip of radius x whose area is

$$dA = 2\pi x ds = 2\pi x\sqrt{1 + y'^2}dx \tag{1}$$

Hence, the total area of surface is

$$A = \int_{x_1}^{x_2} 2\pi x\sqrt{1 + y'^2}dx \tag{2}$$

Thus, here $F = x\sqrt{1 + y'^2}$.

Hence,

$$\frac{\partial F}{\partial y} = 0 \text{ and } \frac{\partial F}{\partial y'} = \frac{xy'}{\sqrt{1 + y'^2}} \tag{3}$$

For minimum value of A, we get by using Euler's Eqn (9.38)

$$\frac{d}{dx}\left[\frac{xy'}{\sqrt{1+y'^2}}\right] = 0, \text{ i.e., } \frac{xy'}{\sqrt{1+y'^2}} = \text{const.} = a \text{ (say).} \tag{4}$$

Squaring this, we get

$$x^2 y'^2 = a^2\left(1+y'^2\right) \text{ or, } y' = \frac{dy}{dx} = \frac{a}{\sqrt{x^2-a^2}}.$$

Hence,

$$y = \int \frac{a}{\sqrt{x^2-a^2}}dx = a\cosh^{-1}\left(\frac{x}{a}\right) + b \tag{5}$$

where b is another constant.

Eqn (5) can be written as

$$x = a\cosh\left(\frac{y-b}{a}\right) \tag{6}$$

which represents a catenary.

EXAMPLE 9.10 *A particle slides from rest from a point along a smooth frictionless wire under gravity to some other point. If the time of travel is minimum, show that, the path followed by the particle is a cycloid.*

Solution. Taking origin at the point of start and v as the speed of the particle, time of travel is

$$t = \int_0^{x_1} \frac{ds}{v} \tag{1}$$

Now, at any point of the path, the kinetic energy is

$$T = \frac{1}{2}mv^2 \tag{2}$$

and the potential energy is

$$V = -mgx \text{ (taking zero at } x = 0) \tag{3}$$

From conservation of energy, initial total energy (which is zero) = final total energy.

Thus,

$$\frac{1}{2}mv^2 - mgx = 0 \text{ or, } v = \sqrt{2gx} \tag{4}$$

Hence, from (1) and (4), the time of travel is

$$t = \int \frac{ds}{\sqrt{2gx}} = \int_0^{x_1}\sqrt{\frac{1+y'^2}{2gx}} \tag{5}$$

Thus, here

$$F = \sqrt{\frac{1+y'^2}{2gx}} \tag{6}$$

Hence,

$$\frac{\partial F}{\partial y} = 0, \quad \frac{\partial F}{\partial y'} = \frac{y'}{\sqrt{2gx(1+y'^2)}} \tag{7}$$

Thus, by Euler's equation, for minimum value of t,

$$\frac{d}{dx}\left(\frac{\partial F}{\partial y'}\right) = 0, \text{ i.e., } \frac{\partial F}{\partial y'} = \text{const.} = A \text{ (say).} \tag{8}$$

Hence, from (7) and (8),

$$\frac{y'^2}{2gx\,(1+y'^2)} = A^2 \text{ or, } \frac{y'^2}{x\,(1+y'^2)} = \frac{1}{2a} \text{ (say) or, } y' = \frac{dy}{dx} = \frac{\sqrt{x}}{\sqrt{2a-x}}$$

Hence,

$$y = \int \frac{\sqrt{x}}{\sqrt{2a-x}}dx \tag{9}$$

We substitute

$$x = a(1-\cos\phi) \tag{10}$$

so that

$$dx = a\sin\phi d\phi \text{ and } \sqrt{\frac{x}{2a-x}} = \tan\frac{\phi}{2} \tag{11}$$

Hence, from (9) and (11), we get

$$y = \int \tan\frac{\phi}{2}a\sin\phi d\phi = \int a(1-\cos\phi)d\phi = a(\phi-\sin\phi) + \text{const.} \tag{12}$$

Initially, when $x = 0, y = 0$, then $\phi = 0$. Thus, the constant term vanishes. Hence, the equation of the path of the particle is

$$x = a(1-\cos\phi)$$
$$y = a(\phi-\sin\phi)$$

which represents a cycloid passing through the origin. This problem is known as ***brachistochrone*** or ***shortest time problem***.

EXAMPLE 9.11 *Show that, the geodesics of a spherical surface are great circles, i.e., the circles having centres at the centre of the sphere.*

Solution. Taking spherical coordinates, the elemental arc-length on the surface of the sphere is given by

$$ds^2 = R^2(d\theta^2 + \sin^2\theta d\phi^2), \text{ i.e., } ds = Rd\theta\left[1+\sin^2\theta\left(\frac{d\phi}{d\theta}\right)^2\right]^{1/2} \tag{1}$$

Hence, the total distance between two points $A(R, \theta_1, \phi_1)$ and $B(R, \theta_2, \phi_2)$ is

$$J = \int_A^B ds = \int_{\theta_1}^{\theta_2} R\left[1+\sin^2\theta\left(\frac{d\phi}{d\theta}\right)^2\right]^{1/2} d\theta \tag{2}$$

Thus, here the functional is

$$F = R\left[1+\sin^2\theta\left(\frac{d\phi}{d\theta}\right)^2\right]^{1/2}$$

Hence,

$$\frac{\partial F}{\partial\phi} = 0 \text{ and } \frac{\partial F}{\partial\phi'} = \frac{\phi'\sin^2\theta}{(1+\sin^2\theta\phi'^2)^{1/2}} \tag{3}$$

From Euler's equation, we get

$$\frac{d}{d\theta}\left(\frac{\partial F}{\partial \phi'}\right) = 0, \text{ i.e., } \frac{\partial F}{\partial \phi'} = \text{const.} = c \tag{4}$$

Hence, from (3) and (4),

$$\frac{\phi' \sin^2\theta}{\left(1+\phi'^2 \sin^2\theta\right)^{1/2}} = c \text{ or, } \phi' = \frac{d\phi}{d\theta} = \frac{c}{\sin\theta\left(\sin^2\theta - c^2\right)^{1/2}}$$

$$\text{or, } \frac{d\phi}{d\theta} = \frac{c \operatorname{cosec}^2\theta}{\left(1-c^2-c^2\cot^2\theta\right)^{1/2}} \tag{5}$$

On integration, we get

$$\phi = \xi - \sin^{-1}(\eta \cot\theta) \tag{6}$$

where ξ and $\eta = \dfrac{c}{\sqrt{1-c^2}}$ are constants.

From (6), we have $\sin(\xi - \phi) = \eta\cot\theta$, which gives

$$R\eta\cos\theta = R\sin(\xi-\phi)\sin\theta = R(\sin\xi\cos\phi\sin\theta - \cos\xi\sin\phi\sin\theta) \tag{7}$$

Reverting to Cartesian coordinates, we get from (7)

$$\eta z = x\sin\xi - y\cos\xi \tag{8}$$

which represents a plane passing through the origin and hence cutting the surface of the sphere in a great circle. Its centre will lie at the centre of the sphere.

NOTE 1. A ***geodesics*** is a line restricted to be on some surface which represents the shortest path between any two points.

NOTE 2. To integrate (5), substitute $p = c\cos\theta$, so that $dp = -c\operatorname{cosec}^2\theta d\theta$.

Hence, $I = -\displaystyle\int \frac{dp}{\left[\left(\sqrt{1-c^2}\right)^2 - p^2\right]^{1/2}}$, which is a standard integral.

9.5 Hamilton's Principle

Hamilton's principle is a more general foundation for the formulation of Lagrangian mechanics. The principle states: ***During the motion of a dynamical system from one point to another within a given time interval, the actual path followed is that which minimizes the time integral of the Lagrangian.***

Thus, the motion of a dynamical system from time t_1 to t_2 is such that

$$J = \int_{t_1}^{t_2} L dt \tag{9.39}$$

is an extremum for the actual path of motion. Here $L = T - V$, the Lagrangian. Thus, in terms of calculus of variation, we can write the principle as

$$\delta J = \delta \int_{t_1}^{t_2} L dt = 0 \tag{9.40}$$

Now, the Lagrangian in general is a function of generalised coordinates, generalised velocities and time, so that

$$L = L(q_k, \dot{q}_k, t) \tag{9.41}$$

Hence, Eqn (9.40) can be written as

$$\delta \int_{t_1}^{t_2} L(q_k, \dot{q}_k, t)\, dt = 0 \tag{9.42}$$

This is similar to the Eqn (9.28), so that we can write the Euler's equations corresponding to Eqn (9.42) as

$$\frac{\partial L}{\partial q_k} - \frac{d}{dt}\left(\frac{\partial L}{\partial \dot{q}_k}\right) = 0, \quad (k = 1, 2, \dots, n)$$

$$\text{or,} \quad \frac{d}{dt}\left(\frac{\partial L}{\partial \dot{q}_k}\right) - \frac{\partial L}{\partial q_k} = 0, \quad (k = 1, 2, \dots, n) \tag{9.43}$$

Thus, we have obtained the Lagrange's equations of motion for a system of particles.

9.6 Modified Hamilton's Principle

We have known the Hamilton's principle as

$$\delta \int_{t_1}^{t_2} L(q_k, \dot{q}_k, t)\, dt = 0 \tag{9.44}$$

This can be written in terms of Hamiltonian H using the relation (see Eqn 9.8)

$$H(p_k, q_k, t) = \sum_k p_k \dot{q}_k - L(q_k, \dot{q}_k, t)$$

Thus, Eqn (9.44) is written as

$$\delta \int_{t_1}^{t_2} \left(\sum_k p_k \dot{q}_k - H\right) dt = 0 \tag{9.45}$$

This is known as ***modified Hamilton's principle***. Here the δ-variation means the independent variations of both generalised coordinates q_k and generalised momenta p_k.

9.7 Hamilton's Principle from D'Alembert's Principle

We have from D'Alembert's principle

$$\sum_k \dot{p}_k \cdot \delta r_k = \sum_k F_k \cdot \delta r_k \tag{9.46}$$

Now

$$\dot{p}_k \cdot \delta r_k = m_k \ddot{r}_k \cdot \delta r_k = \frac{d}{dt}(m_k \dot{r}_k) \cdot \delta r_k$$

$$\text{or,} \quad \dot{p}_k \cdot \delta r_k = \frac{d}{dt}(m_k \dot{r}_k \cdot \delta r_k) - m_k \dot{r}_k \cdot \frac{d}{dt}(\delta r_k)$$

$$= \frac{d}{dt}(m_k \dot{r}_k \cdot \delta r_k) - m_k \dot{r}_k \cdot \delta\left(\frac{dr_k}{dt}\right) \tag{9.47}$$

In view of Eqn (9.47), Eqn (9.46) assumes the form:

$$\frac{d}{dt}\left[\sum_k m_k \dot{r}_k \cdot \delta r_k\right] - \delta\left(\sum_k \frac{1}{2} m_k \dot{r}_k^2\right) = \sum_k F_k \cdot \delta r_k \tag{9.48}$$

If the forces are conservative, then

$$\sum_k F_k \cdot \delta r_k = -\delta V \tag{9.49}$$

Also

$$\sum_k \frac{1}{2} m_k \dot{r}_k^2 = \text{kinetic energy} = T \tag{9.50}$$

Thus, from Eqns (9.48)-(9.50), we get

$$\frac{d}{dt}\left[\sum_k m_k \dot{r}_k \cdot \delta r_k\right] = \delta T - \delta V = \delta(T - V) \tag{9.51}$$

Integrating this from t_1 to t_2 with the condition that $\delta r_k = 0$ at ends of the path, we get

$$\left[\sum_k m_k \dot{r}_k \cdot \delta r_k\right]_{t_1}^{t_2} = \int_{t_1}^{t_2} \delta(T - V)dt, \text{ i.e., } 0 = \int_{t_1}^{t_2} \delta(T - V)dt$$

Hence,

$$\delta \int_{t_1}^{t_2} (T - V)dt = 0 \; [\because \delta(dt) = 0] \text{ or, } \delta \int_{t_1}^{t_2} L dt = 0 \tag{9.52}$$

where $L = T - V$.

9.8 Lagrange's Equation from Hamilton's Principle

We have the Lagrangian L as

$$L = L(q_k, \dot{q}_k, t) \tag{9.53}$$

If L does not depend on time t explicitly, then

$$\delta L = \sum_k \frac{\partial L}{\partial q_k}\delta q_k + \sum_k \frac{\partial L}{\partial \dot{q}_k}\delta \dot{q}_k \tag{9.54}$$

Integrating both sides from time t_1 to t_2,

$$\int_{t_1}^{t_2} \delta L dt = \int_{t_1}^{t_2} \sum_k \frac{\partial L}{\partial q_k}\delta q_k dt + \int_{t_1}^{t_2} \sum_k \frac{\partial L}{\partial \dot{q}_k}\delta \dot{q}_k dt \tag{9.55}$$

But, according to the Hamilton's principle,

$$\delta \int_{t_1}^{t_2} L dt = 0 \tag{9.56}$$

Hence, from Eqn (9.55), we get

$$\int_{t_1}^{t_2} \sum \frac{\partial L}{\partial q_k}\delta q_k dt + \int_{t_1}^{t_2} \sum_k \frac{\partial L}{\partial \dot{q}_k}\delta \dot{q}_k dt = 0 \tag{9.57}$$

We consider the second term of the LHS. Thus,

$$\int_{t_1}^{t_2} \frac{\partial L}{\partial \dot{q}_k}\delta \dot{q}_k dt = \left[\frac{\partial L}{\partial \dot{q}_k}\delta q_k\right]_{t_1}^{t_2} - \int_{t_1}^{t_2} \frac{d}{dt}\left(\frac{\partial L}{\partial \dot{q}_k}\right)\delta q_k dt \tag{9.58}$$

But, at the end points of the path, q_k at times t_1 and t_2 does not vary, i.e.,

$$\delta q_k = 0 \text{ for } t = t_1 \text{ and } t = t_2$$

Hence, the first term in Eqn (9.58) vanishes, so that

$$\int_{t_1}^{t_2} \frac{\partial L}{\partial \dot{q}_k} \delta \dot{q}_k dt = -\int_{t_1}^{t_2} \frac{d}{dt}\left(\frac{\partial L}{\partial \dot{q}_k}\right) \delta q_k dt \tag{9.59}$$

Thus, from Eqns (9.57) and (9.59), we get

$$\int_{t_1}^{t_2} \sum_k \frac{\partial L}{\partial q_k} \delta q_k dt - \int_{t_1}^{t_2} \sum_k \frac{d}{dt}\left(\frac{\partial L}{\partial \dot{q}_k}\right) \delta q_k dt = 0$$

$$\text{or,} \quad \sum_k \int_{t_1}^{t_2} \left[\frac{d}{dt}\left(\frac{\partial L}{\partial \dot{q}_k}\right) - \frac{\partial L}{\partial q_k}\right] \delta q_k dt = 0 \tag{9.60}$$

For a holonomic system, the generalised coordinates δq_k are all independent of each other. Hence, the coefficients of each δq_k should vanish. Thus,

$$\frac{d}{dt}\left(\frac{\partial L}{\partial \dot{q}_k}\right) - \frac{\partial L}{\partial q_k} = 0, \quad (k = 1, 2, \ldots, n) \tag{9.61}$$

which are Lagrange's equations of motion.

9.9 Principles of Least Action

According to the principle of least action, for a conservative system

$$\Delta \int_{t_1}^{t_2} \sum_k p_k \dot{q}_k dt = 0 \tag{9.62}$$

where the quantity

$$W = \int_{t_1}^{t_2} \sum_k p_k \dot{q}_k dt \tag{9.63}$$

is called ***action***, and the Δ-operation is

$$\Delta = \delta + \Delta t \frac{d}{dt} \tag{9.64}$$

Now, for a conservative system

$$\sum p_k \dot{q}_k = 2T \text{ (see Eqn 9.26).}$$

Hence, the principle of least action, i.e., Eqn (9.62) assumes the form:

$$\Delta \int_{t_1}^{t_2} 2T dt = 0 \tag{9.65}$$

That is, action on the actual path is a minimum when compared with some neighbouring paths. If the system consists of only one particle of mass m, then its kinetic energy is

$$T = \frac{1}{2} m v^2 = \frac{1}{2} m \left(\frac{ds}{dt}\right)^2 \tag{9.66}$$

where ds = element of arc traversed by the particle in time dt.

Eqn (9.66) yields

$$dt = \sqrt{\frac{m}{2T}} ds \tag{9.67}$$

Thus, Eqn (9.65) assumes the form:

$$\Delta \int 2T\sqrt{\frac{m}{2T}}ds = 0$$

$$\text{or, } \Delta \int \sqrt{2mT}ds = 0 \text{ or, } \Delta \int \sqrt{2m(H-V)}ds = 0 \tag{9.68}$$

which represents the principle of least action in terms of arc length of the trajectory of the particle.

If T = constant, Eqn (9.65) gives

$$\Delta \int_{t_1}^{t_2} dt = 0, \text{ i.e., } \Delta(t_2 - t_1) = 0 \tag{9.69}$$

which implies that, out of all possible paths between two points, the system for which T is conserved moves along that path for which time of transit is minimum. In geometrical optics, Fermat's principle is stated in this form.

9.10 Jacobi's Form of Principle of Least Action

The transformation equation in general is of the form:

$$\mathbf{r_k} = \mathbf{r_k}\,(q_k, \dot{q}_k, t) \tag{9.70}$$

When this equation does not contain time t explicitly, the kinetic energy of the system is expressed as

$$T = \frac{1}{2}\sum_{ik} m_{ik}\frac{dq_i}{dt}\frac{dq_k}{dt} \tag{9.71}$$

Let

$$d\rho^2 = \sum_{ik} m_{ik}dq_idq_k \tag{9.72}$$

where $d\rho$ represents the element of path length in configuration space coordinates $q_1, q_2, \ldots, q_n$.

Then, from Eqn (9.71),

$$T = \frac{1}{2}\left(\frac{d\rho}{dt}\right)^2 \text{ or, } dt = \frac{d\rho}{\sqrt{2T}} \tag{9.73}$$

Hence, the principle of least action becomes

$$\Delta \int 2Tdt = \Delta \int 2T\frac{d\rho}{\sqrt{2T}} = 0$$

$$\text{or, } \Delta \int \sqrt{T}d\rho = 0 \text{ or, } \Delta \int \sqrt{H-V}d\rho = 0 \tag{9.74}$$

which is the Jacobi's form of the principle of least action.

EXAMPLE 9.12 *A particle falls a given distance y_0 in time $t_0 = \sqrt{2y_0/g}$ and distance travelled in time t is given by $y = at + bt^2$, where the constants a and b are so adjusted that the time to fall y_0 is always correctly given by t_0. Show that the integral $\int_0^t Ldt$ is an extremum for real values of coefficients, only when $a = 0$ and $b = g/2$.*

Solution. Here kinetic energy $T = \frac{1}{2}m\dot{y}^2$; potential energy $V = -mgy$. Hence, the Lagrangian

$$L = \frac{1}{2}m\dot{y}^2 + mgy.$$

Lagrange's equation of motion is

$$\frac{d}{dt}(m\dot{y}) - mg = 0, \text{ i.e., } \ddot{y} = g \tag{1}$$

But, it is given that

$$y = at + bt^2 \tag{2}$$

Hence,

$$\ddot{y} = 2b \tag{3}$$

From Eqns (1) and (3),

$$2b = g, \text{ i.e., } b = \frac{g}{2} \tag{4}$$

Again,

$$y_0 = at_0 + bt_0^2 \text{ or, } \frac{gt_0^2}{2} = at_0 + \frac{g}{2}t_0^2 \quad \left[\because\ t_0 = \sqrt{\frac{2y_0}{g}} \text{ given}\right] \tag{5}$$

Hence, $a = 0$.

Thus, $\int_0^{t_0} L dt$ is extremum for $a = 0$ and $b = g/2$.

EXAMPLE 9.13 *Given the following Lagrangian for an anharmonic oscillator, find its corresponding Hamiltonian:*

$$L(x, \dot{x}) = \frac{1}{2}\dot{x}^2 - \frac{1}{2}\omega^2 x^2 - \alpha x^3 + \beta x \dot{x}^2,$$

where α, β and ω are constants. [C.U. (HONS.)]

Solution. We have

$$L(x, \dot{x}) = \frac{1}{2}\dot{x}^2 - \frac{1}{2}\omega^2 x^2 - \alpha x^3 + \beta x \dot{x}^2 \tag{1}$$

Hence,

$$p_x = \frac{\partial L}{\partial \dot{x}} = \dot{x} + 2\beta x \dot{x} = (1 + 2\beta x)\dot{x} \tag{2}$$

The Hamiltonian of this single-degree-of-freedom-system is

$$\begin{aligned}
H(x, p_x) &= p_x \dot{x} - L(x, \dot{x}) = (1 + 2\beta x)\dot{x}^2 - \frac{1}{2}\dot{x}^2 + \frac{1}{2}\omega^2 x^2 + \alpha x^3 - \beta x \dot{x}^2 \\
&= \left(1 + 2\beta x - \frac{1}{2} - \beta x\right)\dot{x}^2 + \frac{1}{2}\omega^2 x^2 + \alpha x^3 \\
&= \left(\frac{1}{2} + \beta x\right)\left[\frac{p_x}{1 + 2\beta x}\right]^2 + \frac{1}{2}\omega^2 x^2 + \alpha x^3 \quad \text{[using (2)]} \\
&= \frac{p_x^2}{2(1 + 2\beta x)} + \frac{1}{2}\omega^2 x^2 + \alpha x^3
\end{aligned}$$

EXAMPLE 9.14 *From the Lagrangian $L(q, \dot{q}, t) = \frac{1}{2}G(q, t)\dot{q}^2 + F(q, t)\dot{q} - V(q, t)$, obtain the Hamiltonian.*

Solution. We have

$$L(q, \dot{q}, t) = \frac{1}{2}G(q, t)\dot{q}^2 + F(q, t)\dot{q} - V(q, t) \tag{1}$$

The degree of freedom of the system is one. Hence, the Hamiltonian is

$$H = p\dot{q} - L \tag{2}$$

From (1), we have

$$p = \frac{\partial L}{\partial \dot{q}} = G(q,t)\dot{q} + F(q,t) \tag{3}$$

Then,

$$\begin{aligned} H &= G(q,t)\dot{q}^2 + F(q,t)\dot{q} - \frac{1}{2}G(q,t)\dot{q}^2 - F(q,t)\dot{q} + V(q,t) \\ &= \frac{1}{2}G(q,t)\dot{q}^2 + V(q,t) \end{aligned} \tag{4}$$

From (3), we get

$$\dot{q} = \frac{p - F(q,t)}{G(q,t)} \tag{5}$$

Inserting this in (4), we get

$$H = \frac{1}{2}G(q,t)\left[\frac{p - F(q,t)}{G(q,t)}\right]^2 + V(q,t) = \frac{[p - F(q,t)]^2}{2G(q,t)} + V(q,t)$$

EXAMPLE 9.15 *Set up the Hamilton's equations of motion for the following Lagrangian:*

$$L = L(q, \dot{q}, t) = \frac{1}{2}m(\dot{q}^2 sin^2\omega t + q\dot{q}\omega \sin 2\omega t + q^2\omega^2)$$

Solution. We have

$$L(q, \dot{q}, t) = \frac{1}{2}m\left(\dot{q}^2 \sin^2 \omega t + q\dot{q}\omega \sin 2\omega t + q^2\omega^2\right) \tag{1}$$

Then

$$p = \frac{\partial L}{\partial \dot{q}} = m\dot{q}\sin^2 \omega t \tag{2}$$

The Hamiltonian is

$$H = p\dot{q} - L = p\dot{q} - \frac{1}{2}m\left(\dot{q}^2 \sin^2 \omega t + q\dot{q}\omega \sin 2\omega t + q^2\omega^2\right) \tag{3}$$

From (2), we have

$$\dot{q} = \frac{p}{m \sin^2 \omega t} \tag{4}$$

Inserting this in (3), we get

$$\begin{aligned} H &= \frac{p^2}{m \sin^2 \omega t} - \frac{1}{2}m\left[\frac{p^2 \sin^2 \omega t}{m^2 \sin^4 \omega t} + \frac{pq\omega \sin 2\omega t}{m \sin^2 \omega t} + q^2\omega^2\right] \\ &= \left[\frac{p^2}{2m \sin^2 \omega t} - pq\omega \cot \omega t - \frac{1}{2}mq^2\omega^2\right] \end{aligned} \tag{5}$$

Hamilton's equations of motion are:

$$\left.\begin{aligned} \dot{q} &= \frac{\partial H}{\partial p} = \frac{p}{m \sin^2 \omega t} - q\omega \cos \omega t \\ \text{and } \dot{p} &= -\frac{\partial H}{\partial q} = p\omega \cos \omega t + mq\omega^2 \end{aligned}\right\}$$

EXAMPLE 9.16 *A bead of mass m slides along a frictionless straight rod. If the rod rotates at a constant angular speed ω in a horizontal plane, find the radial position r of the bead at any instant of time. Assume the initial conditions:*

At $t = 0, r = R, \dot{r} = v.$

Solution. Since angular speed ω is constant, we have in polar coordinates,

$$\dot{\theta} = \omega = \text{constant.} \tag{1}$$

The kinetic energy is

$$T = \frac{1}{2}m\left(\dot{r}^2 + r^2\dot{\theta}^2\right) = \frac{1}{2}m\left(\dot{r}^2 + r^2\omega^2\right) \tag{2}$$

The potential energy V measured from the horizontal plane is zero.

Hence, the Lagrangian is

$$L = T - V = \frac{1}{2}m\left(\dot{r}^2 + r^2\omega^2\right) \tag{3}$$

Then we have

$$\left.\begin{aligned} \frac{\partial L}{\partial \dot{r}} &= m\dot{r} \\ \frac{\partial L}{\partial r} &= mr\omega^2 \end{aligned}\right\} \tag{4}$$

The equation of motion is

$$\frac{d}{dt}\left(\frac{\partial L}{\partial \dot{r}}\right) - \frac{\partial L}{\partial r} = 0$$

$$\therefore m\ddot{r} - mr\omega^2 = 0$$

$$\text{i.e., } \ddot{r} = \omega^2 r \tag{5}$$

We assume its solution as

$$r(t) = e^{\lambda t} \quad (\lambda = \text{constant}). \tag{6}$$

We find the characteristic equation

$$\lambda^2 = \omega^2 \tag{7}$$

which yields

$$\lambda = \pm\omega \tag{8}$$

Hence, the solution of equation (5) can be written as

$$r(t) = Ae^{\omega t} + Be^{-\omega t} \tag{9}$$

where A and B are constants of integration.

Then

$$\dot{r}(t) = A\omega e^{\omega t} - B\omega e^{-\omega t} \tag{10}$$

Using the given initial conditions in (9) and (10), we get

$$\left.\begin{aligned} R &= A + B \\ v &= (A - B)\omega \end{aligned}\right\} \tag{11}$$

Solving these, we find

$$\left.\begin{aligned} A &= \frac{1}{2}\left(R + \frac{v}{\omega}\right) \\ B &= \frac{1}{2}\left(R - \frac{v}{\omega}\right) \end{aligned}\right\} \tag{12}$$

Inserting these values in (9), we get

$$r(t) = \frac{1}{2}\left(R + \frac{v}{\omega}\right)e^{\omega t} + \frac{1}{2}\left(R - \frac{v}{\omega}\right)e^{-\omega t}$$

EXAMPLE 9.17 *A particle of mass m moves on the inner side of frictionless cone*

$$x^2 + y^2 = z^2 \tan^2\alpha, \ \alpha = \textit{semi-vertical angle.}$$

Determine the Hamilton's equations of motion.

Solution. We take cylindrical coordinates (r, θ, z) of any point on the cone.

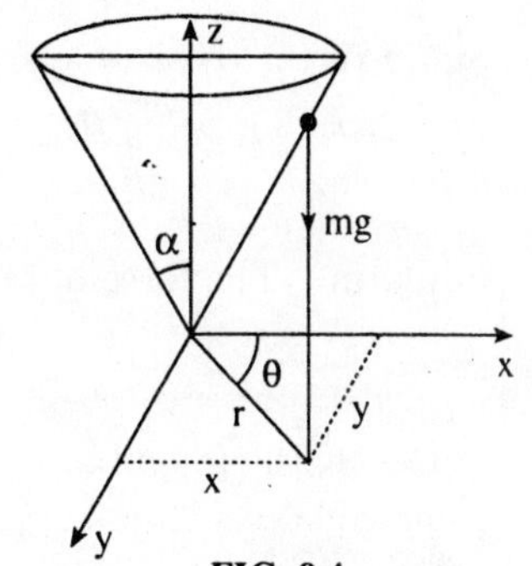

FIG. 9.4

Then from FIG. 9.4,

$$\left.\begin{aligned} x &= r\cos\theta \\ y &= r\sin\theta \\ z &= r\cot\alpha \end{aligned}\right\} \tag{1}$$

Hence,

$$\left.\begin{aligned} \dot{x} &= \dot{r}\cos\theta - r\dot{\theta}\sin\theta \\ \dot{y} &= \dot{r}\sin\theta + r\dot{\theta}\cos\theta \\ \dot{z} &= \dot{r}\cot\alpha \end{aligned}\right\} \tag{2}$$

The kinetic energy is

$$\begin{aligned} T &= \frac{1}{2}m\left(\dot{x}^2 + \dot{y}^2 + \dot{z}^2\right) = \frac{1}{2}m\left(\dot{r}^2 + r^2\dot{\theta}^2 + \dot{r}^2\cot^2\alpha\right) \\ &= \frac{1}{2}m\left(\dot{r}^2\operatorname{cosec}^2\alpha + r^2\dot{\theta}^2\right) \end{aligned} \tag{3}$$

The potential energy is

$$V = mgz = mgr\cot\alpha \tag{4}$$

Hence, the Lagrangian is

$$L = T - V = \frac{1}{2}m\left(\dot{r}^2\operatorname{cosec}^2\alpha + r^2\dot{\theta}^2\right) - mgr\cot\alpha \tag{5}$$

Then the generalised momenta are:

$$\left.\begin{aligned} p_r &= \frac{\partial L}{\partial \dot{r}} = m\dot{r}\operatorname{cosec}^2\alpha \\ p_\theta &= \frac{\partial L}{\partial \dot{\theta}} = mr^2\dot{\theta} \end{aligned}\right\} \tag{6}$$

The Hamiltonian is given by

$$H = T + V = \frac{1}{2}m\left(\dot{r}^2\operatorname{cosec}^2\alpha + r^2\dot{\theta}^2\right) + mgr\cot\alpha \tag{7}$$

From (6), we have

$$\dot{r} = \frac{p_r}{m\operatorname{cosec}^2\alpha}; \ \dot{\theta} = \frac{p_\theta}{mr^2} \tag{8}$$

Inserting these relations in (7), we get

$$H = \frac{1}{2}m\left(\frac{p_r^2}{m^2 \operatorname{cosec}^2 \alpha} + \frac{p_\theta^2}{m^2r^2}\right) + mgr\cot\alpha \tag{9}$$

Hence, Hamilton's equations are:

$$\left.\begin{aligned} \dot{r} &= \frac{\partial H}{\partial p_r} = \frac{p_r}{m \operatorname{cosec}^2 \alpha} \\ \dot{\theta} &= \frac{\partial H}{\partial p_\theta} = \frac{p_\theta}{mr^2} \end{aligned}\right\} \tag{10}$$

$$\left.\begin{aligned} \dot{p}_r &= -\frac{\partial H}{\partial r} = \frac{p_\theta^2}{mr^3} - mg\cot\alpha \\ \dot{p}_\theta &= -\frac{\partial H}{\partial \theta} = 0 \end{aligned}\right\} \tag{11}$$

EXAMPLE 9.18 *A particle of mass m is attached to a fixed point by an inverse square force $F = -\mu/r^2$, μ being the gravitational constant. Determine the Hamilton's equation of motion in plane polar coordinates.*

Solution. The force of attraction is conservative. Hence,

$$F(r) = -\frac{\mu}{r^2} = -\frac{\partial V}{\partial r} \tag{1}$$

Hence, the potential energy

$$V = -\frac{\mu}{r} \tag{2}$$

The kinetic energy of the particle

$$T = \frac{1}{2}m\left(\dot{r}^2 + r^2\dot{\theta}^2\right) \tag{3}$$

Hence, the Lagrangian

$$L = T - V = \frac{1}{2}m\left(\dot{r}^2 + r^2\dot{\theta}^2\right) + \frac{\mu}{r} \tag{4}$$

Then

$$\left.\begin{aligned} p_r &= \frac{\partial L}{\partial \dot{r}} = m\dot{r} \\ p_\theta &= \frac{\partial L}{\partial \dot{\theta}} = mr^2\dot{\theta} \end{aligned}\right\} \tag{5}$$

These yield the relations

$$\left.\begin{aligned} \dot{r} &= \frac{p_r}{m} \\ \dot{\theta} &= \frac{p_\theta}{mr^2} \end{aligned}\right\} \tag{6}$$

The Hamiltonian is

$$H = T + V = \frac{1}{2}m\left(\dot{r}^2 + r^2\dot{\theta}^2\right) - \frac{\mu}{r} \tag{7}$$

Inserting relations (6) in (7), we write:

$$H = \frac{1}{2}m\left[\frac{p_r^2}{m^2} + \frac{p_\theta^2}{m^2r^2}\right] - \frac{\mu}{r} \tag{8}$$

Hence, Hamilton's equation in r-direction is:

$$\dot{p}_r = -\frac{\partial H}{\partial r} = -\left[-\frac{p_\theta^2}{mr^3} + \frac{\mu}{r^2}\right] \tag{9}$$

From(5) and (9),

$$m\ddot{r} = \frac{p_\theta^2}{mr^3} - \frac{\mu}{r^2} \tag{10}$$

This is the equation of motion in r-direction.

Again,

$$\dot{p}_\theta = \frac{\partial H}{\partial \theta} = 0 \tag{11}$$

From (5) and (11), we get

$$mr^2\ddot{\theta} = 0, \text{ i.e., } \ddot{\theta} = 0 \tag{12}$$

This is the equation of motion in θ-direction.

Eqn (12) yields, $\theta = At + B$, where A and B are constants.

EXAMPLE 9.19 *A particle of mass m is constrained to move on the surface of the cylinder*

$$x^2 + y^2 = R^2, \ R = \text{radius of circular base.}$$

It is subjected to a force $F = -\mu r$, proportional to its distance from origin and directed to it. Write Hamilton's equations and show that the motion in the z-direction is simple harmonic.

Solution. If V be the potential, then

$$F(r) = -\frac{\partial V}{\partial r} = -\mu r \tag{1}$$

Integrating, we get

$$V = \frac{1}{2}\mu r^2 = \frac{1}{2}\mu(x^2 + y^2 + z^2) \tag{2}$$

Equation of the cylinder being

$$x^2 + y^2 = R^2 \tag{3}$$

we write (2) in the form:

$$V = \frac{1}{2}\mu(R^2 + z^2) \tag{4}$$

According to the given constraint, the particle is to remain on the surface at every moment. So it cannot have radial velocity. Hence,

$$\dot{r} = 0 \tag{5}$$

The kinetic energy is

$$T = \frac{1}{2}m\left(\dot{r}^2 + r^2\dot{\theta}^2 + \dot{z}^2\right) = \frac{1}{2}m\left(R^2\dot{\theta}^2 + \dot{z}^2\right) \tag{6}$$

The Lagrangian is

$$L = T - V = \frac{1}{2}m\left(R^2\dot{\theta}^2 + \dot{z}^2\right) - \frac{1}{2}\mu(R^2 + z^2) \tag{7}$$

The generalised momenta are:

$$\left.\begin{aligned} p_\theta &= \frac{\partial L}{\partial \dot\theta} = mR^2\dot\theta, \text{ i.e., } \dot\theta = \frac{p_\theta}{mR^2} \\ p_z &= \frac{\partial L}{\partial \dot z} = m\dot z, \text{ i.e., } \dot z = \frac{p_z}{m} \end{aligned}\right\} \tag{8}$$

The Hamiltonian is

$$\begin{aligned} H &= p_\theta\dot\theta + p_z\dot z - L \\ &= mR^2\dot\theta^2 + m\dot z^2 - \frac{1}{2}m\left(R^2\dot\theta^2 + \dot z^2\right) + \frac{1}{2}\mu(R^2 + z^2) \\ &= \frac{1}{2}mR^2\dot\theta^2 + \frac{1}{2}m\dot z^2 + \frac{1}{2}\mu(R^2 + z^2) \\ &= \frac{p_\theta^2}{2mR^2} + \frac{p_z^2}{2m} + \frac{1}{2}\mu\left(R^2 + z^2\right) \quad \text{[using (8)]} \end{aligned} \tag{9}$$

Hence, Hamilton's equations are:

$$\left.\begin{aligned} \dot\theta &= \frac{\partial H}{\partial p_\theta} = \frac{p_\theta}{mR^2}; \quad \dot p_\theta = -\frac{\partial H}{\partial \theta} = 0 \\ \dot z &= \frac{\partial H}{\partial p_z} = \frac{p_z}{m}; \quad \dot p_z = -\frac{\partial H}{\partial z} = -\mu z \end{aligned}\right\} \tag{10}$$

From (10), we find $p_\theta = mR^2\dot\theta =$ constant, which means the angular momentum along z-axis is constant. From last equations, we get

$$\ddot z = \frac{\dot p_z}{m} = -\frac{\mu}{m}z, \text{ i.e., } \ddot z + \omega^2 z = 0, \text{ where } \omega^2 = \frac{\mu}{m} \tag{11}$$

This shows that the motion along z-axis is simple harmonic.

EXAMPLE 9.20 *Show that the extremal of the functional* $\int_{x_0}^{x_1} \frac{\sqrt{1+y'^2}}{x}dx$ *is a circle with its centre on y-axis.*

Solution. The given functional depends on only x and y'. Thus,

$$f = \frac{\sqrt{1+y'^2}}{x} \tag{1}$$

From Euler's equation, we get

$$\frac{d}{dx}\left(\frac{\partial f}{\partial y'}\right) = 0 \tag{2}$$

This yields

$$\frac{\partial f}{\partial y'} = \text{constant, i.e., } \frac{\partial}{\partial y'}\left(\frac{\sqrt{1+y'^2}}{x}\right) = \text{constant.}$$

Hence,

$$\frac{y'}{x\sqrt{1+y'^2}} = \text{constant} = \frac{1}{A} \text{ (say)} \tag{3}$$

Let $y' = \tan\alpha$, i.e., $\sqrt{1+y'^2} = \sec\alpha$.

From (3), we get

$$x = \frac{Ay'}{\sqrt{1+y'^2}} = \frac{A\tan\alpha}{\sec\alpha} = A\sin\alpha \tag{4}$$

Also,

$$dy = \tan \alpha dx = \tan \alpha \left(\frac{dx}{d\alpha}\right) d\alpha = \tan \alpha (A \cos \alpha) d\alpha = A \sin \alpha d\alpha$$

Hence,

$$y = -A \cos \alpha + B \tag{5}$$

Eliminating α from (4) and (5), we get

$$x^2 + (y - B)^2 = A^2 \left(\sin^2 \alpha + \cos^2 \alpha\right) = A^2 \tag{6}$$

This represents a circle with its centre on y-axis. Hence proved.

EXAMPLE 9.21 *Using Hamilton's principle, find the equations of motion of a particle over a plane moving in a conservative field of force.*

Solution. If V be the potential at any point (x, y) of the position of the particle, then the components of forces along axes are:

$$X = -\frac{\partial V}{\partial x};\ Y = -\frac{\partial V}{\partial y} \tag{1}$$

The kinetic energy of the particle of mass m is

$$T = \frac{1}{2} m \left(\dot{x}^2 + \dot{y}^2\right) \tag{2}$$

Hence, the Lagrangian is

$$L = T - V = \frac{1}{2} m \left(\dot{x}^2 + \dot{y}^2\right) - V \tag{3}$$

According to Hamilton's principle, we have

$$\delta \int_{t_1}^{t_2} L dt = 0, \text{ i.e., } \int_{t_1}^{t_2} \delta L dt = 0 \tag{4}$$

From (3) and (4), we have

$$\int_{t_1}^{t_2} \left[m \left(\dot{x} \delta \dot{x} + \dot{y} \delta \dot{y}\right) - \delta V\right] dt = 0$$

$$\text{i.e., } \int_{t_1}^{t_2} \left[m \left(\dot{x} \delta \dot{x} + \dot{y} \delta \dot{y}\right) - \frac{\partial V}{\partial x} \delta x - \frac{\partial V}{\partial y} \delta y\right] dt = 0 \tag{5}$$

But

$$\begin{aligned} \int_{t_1}^{t_2} (\dot{x} \delta \dot{x})\, dt &= \int_{t_1}^{t_2} \dot{x} \frac{d}{dt} (\delta x) dt \\ &= \dot{x} \delta x \Big|_{t_1}^{t_2} - \int_{t_1}^{t_2} (\ddot{x} \delta x)\, dt \quad \text{[integrating by parts]} \\ &= -\int_{t_1}^{t_2} (\ddot{x} \delta x)\, dt \quad \text{[since } \delta x = 0 \text{ at end points } t_1 \text{ and } t_2\text{].} \end{aligned} \tag{6}$$

Similarly, we have

$$\int_{t_1}^{t_2} (\dot{y} \delta \dot{y})\, dt = -\int_{t_1}^{t_2} (\ddot{y} \delta y)\, dt \tag{7}$$

From (5), (6) and (7), we write:

$$\int_{t_1}^{t_2} \left[\left(m\ddot{x} + \frac{\partial V}{\partial x} \right) \delta x + \left(m\ddot{y} + \frac{\partial V}{\partial y} \right) \delta y \right] dt = 0 \tag{8}$$

Since δx and δy are independent and arbitrary, we have

$$\left. \begin{aligned} m\ddot{x} &= -\frac{\partial V}{\partial x} = X \\ m\ddot{y} &= -\frac{\partial V}{\partial y} = Y \end{aligned} \right\} \tag{9}$$

which are the required equations of motion.

EXERCISE

1. Set up the Hamiltonian for a compound pendulum and hence, find its equation of motion.
2. Set up the Hamiltonian for a two-dimensional harmonic oscillator in polar coordinates and derive its equation of motion.
3. A particle of mass m is falling freely under gravity. Neglecting the earth's rotation, set up the Hamiltonian. Hence, find the equation of motion.
4. Deduce the Hamilton's function for the motion of a particle under inverse square law of central force. Hence, find the equation of motion.
5. A particle of mass m moves in a force-field of potential V. Set up the Hamiltonian in rectangular coordinates and hence, write the Hamilton's equations.
6. The Lagrangian for anharmonic oscillator is given by

$$L(x, \dot{x}) = \frac{1}{2}\dot{x}^2 - \frac{1}{2}\omega^2 x^2 - \lambda x^3 + \mu x \dot{x}^2$$

   Write the Hamiltonian.
7. Prove that if the function F in the integral $\int F(x, y, y')dx$ is independent of x, then the integral is an extremum, if

$$F - y'\frac{\partial F}{\partial y'} = \text{constant.}$$

8. A solid cylinder of mass M and radius R rolls down an inclined plane of angle α. Find its equations of motion by using Hamilton's principle.
9. A particle of mass m moves on the surface of a cylinder of radius R under a central attractive force towards the origin proportional to the distance of the particle. Find the Hamiltonian. Also show that the motion of the particle along z-direction is simple harmonic with a period

$$P = 2\pi\sqrt{\frac{m}{k}}$$

   where k = constant of proportionality of the central force.
10. The Hamiltonian of a dynamical system is given by $H = p_1 q_1, -p_2 q_2 - \alpha q_1^2 + \beta q_2^2$, where α, β are constants. Solve the problem.

CHAPTER 10

Canonical Transformations and Brackets

10.1 Canonical Transformations

Suppose, we have a set of coordinates p_k and q_k. From this set of coordinates we change to a new set of coordinates P_k and Q_k such that

$$P_k = P_k\left(p_k, q_k, t\right) \text{ and } Q_k = Q_k\left(p_k, q_k, t\right) \tag{10.1}$$

Then, if there exists a Hamiltonian

$$H^* = H^*\left(P_k, Q_k, t\right) \tag{10.2}$$

in the new coordinates, such that

$$\dot{P}_k = \frac{\partial H^*}{\partial Q_k} \text{ and } \dot{Q}_k \doteq \frac{\partial H^*}{\partial P_k} \tag{10.3}$$

then the transformations given by Eqn (10.1) are called ***canonical transformations*** or ***contact transformations***.

10.2 Generating Functions

Let L and L^* be the Lagrangian in the old and new sets of position and momentum coordinates respectively. Then by Hamilton's principle

$$\delta \int_{t_1}^{t_2} L dt = 0 \text{ and } \delta \int_{t_1}^{t_2} L^* dt = 0 \tag{10.4}$$

are valid simultaneously. Upon subtraction, we write Eqn (10.4) as

$$\delta \int_{t_1}^{t_2} (L - L^*) dt = 0 \tag{10.5}$$

Eqn (10.5) can be satisfied if there exists a function G such that

$$L - L^* = \frac{dG}{dt} \tag{10.6}$$

since in such case,

$$\delta \int_{t_1}^{t_2} \frac{dG}{dt} dt = \delta\left[G(t_2) - G(t_1)\right] = 0 \tag{10.7}$$

The function G is called ***generating function***. This function can be chosen in the following four possible forms:

$$G_1 = G_1(q_k, Q_k, t) \tag{10.8}$$
$$G_2 = G_2(q_k, P_k, t) \tag{10.9}$$
$$G_3 = G_3(p_k, Q_k, t) \tag{10.10}$$
$$G_4 = G_4(p_k, P_k, t) \tag{10.11}$$

• Case I

Let us choose the generating function

$$G_1 = G_1(q_k, Q_k, t) \tag{10.12}$$

Then, by Eqn (10.6),

$$\frac{dG_1}{dt} = L - L^* = \left(\sum p_k \dot{q}_k - H\right) - \left(\sum P_k \dot{Q}_k - H^*\right) \tag{10.13}$$

Eqn (10.13) yields

$$dG_1 = \sum p_k dq_k - \sum P_k dQ_k + (H^* - H)dt \tag{10.14}$$

But from Eqn (10.12), we have

$$dG_1 = \sum \frac{\partial G_1}{\partial q_k} dq_k + \sum \frac{\partial G_1}{\partial Q_k} dQ_k + \frac{\partial G_1}{\partial t} dt \tag{10.15}$$

Comparing Eqns (10.14) and (10.15), we have

$$p_k = \frac{\partial G_1}{\partial q_k}, \quad P_k = -\frac{\partial G_1}{\partial Q_k} \text{ and } H^* - H = \frac{\partial G_1}{\partial t} \tag{10.16}$$

• Case II

Let the generating function be of the type

$$G_2 = G_2(q_k, P_k, t) \tag{10.17}$$

Then

$$dG_2 = \sum \frac{\partial G_2}{\partial q_k} dq_k + \sum \frac{\partial G_2}{\partial P_k} dP_k + \frac{\partial G_2}{\partial t} dt \tag{10.18}$$

Again, by Eqn (10.14),

$$\begin{aligned} dG_1 &= \sum p_k dq_k - \sum P_k dQ_k + (H^* - H)dt \\ &= \sum p_k dq_k - d\left[\sum P_k Q_k\right] + \sum Q_k dP_k + (H^* - H)dt \end{aligned}$$

$$\text{or, } d\left(G_1 + \sum P_k Q_k\right) = \sum p_k dq_k + \sum Q_k dP_k + (H^* - H)dt \tag{10.19}$$

Comparing Eqns (10.18) and (10.19), we get

$$p_k = \frac{\partial G_2}{\partial q_k}, \quad Q_k = \frac{\partial G_2}{\partial P_k} \text{ and } H^* - H = \frac{\partial G_2}{\partial t} \tag{10.20}$$

where

$$G_2 = G_1 + \sum P_k Q_k \tag{10.21}$$

• Case III

Let us choose the generating function

$$G_3 = G_3\,(p_k, Q_k, t) \tag{10.22}$$

Then

$$dG_3 = \sum \frac{\partial G_3}{\partial p_k} dp_k + \sum \frac{\partial G_3}{\partial Q_k} dQ_k + \frac{\partial G_3}{\partial t} dt \tag{10.23}$$

By Eqn (10.14), we have

$$\begin{aligned} dG_1 &= \sum p_k dq_k - \sum P_k dQ_k + (H^* - H)dt \\ &= d\left[\sum p_k q_k\right] - \sum q_k dp_k - \sum P_k dQ_k + (H^* - H)dt \end{aligned}$$

$$\text{or, } d\left[G_1 - \sum p_k q_k\right] = -\sum q_k dp_k - \sum P_k dQ_k + (H^* - H)dt \tag{10.24}$$

Comparing Eqns (10.23) and (10.24), we find

$$q_k = -\frac{\partial G_3}{\partial p_k},\ P_k = -\frac{\partial G_3}{\partial Q_k} \text{ and } H^* - H = \frac{\partial G_3}{\partial t} \tag{10.25}$$

where

$$G_3 = G_1 - \sum p_k q_k \tag{10.26}$$

• Case IV

Let the generating function be of the form:

$$G_4 = G_4\,(p_k, P_k, t) \tag{10.27}$$

Then

$$dG_4 = \sum \frac{\partial G_4}{\partial p_k} dp_k + \sum \frac{\partial G_4}{\partial P_k} dP_k + \frac{\partial G_3}{\partial t} dt \tag{10.28}$$

By Eqn (10.14), we have

$$\begin{aligned} dG_1 &= \sum p_k dq_k - \sum P_k dQ_k + (H^* - H)dt \\ &= d\left(\sum p_k q_k\right) - \sum q_k dp_k - d\left(\sum P_k Q_k\right) + \sum Q_k dP_k + (H^* - H)dt \end{aligned}$$

$$\text{or, } d\left(G_1 - \sum p_k q_k + \sum P_k Q_k\right) = -\sum q_k dp_k + \sum Q_k dP_k + (H^* - H)dt \tag{10.29}$$

Comparing Eqns (10.28) and (10.29), we find

$$q_k = -\frac{\partial G_4}{\partial p_k},\ Q_k = \frac{\partial G_4}{\partial P_k} \text{ and } H^* - H = \frac{\partial G_4}{\partial t} \tag{10.30}$$

where

$$G_4 = G_1 - \sum p_k q_k + \sum P_k Q_k \tag{10.31}$$

From the above discussions we note that:

(1) the functions G_2, G_3 and G_4 are expressed in terms of the function G_1, and

(2) in all the cases,

$$H^* - H = \frac{\partial G}{\partial t} \tag{10.32}$$

If the function G does not depend explicitly on time so that $\partial G/\partial t = 0$, then we get

$$H^*(P_k, Q_k) = H(p_k, q_k) \tag{10.33}$$

10.3 Condition for Canonical Transformations

Consider the generating function

$$G = G(q_k, Q_k) \tag{10.34}$$

when time is held fixed, so that

$$\frac{\partial G}{\partial t} = 0 \tag{10.35}$$

Then from Eqns (10.16), we have

$$p_k = \frac{\partial G}{\partial q_k} \text{ and } P_k = -\frac{\partial G}{\partial Q_k} \tag{10.36}$$

We also have, by Eqn (10.34),

$$dG = \sum \frac{\partial G}{\partial q_k} dq_k + \sum \frac{\partial G}{\partial Q_k} dQ_k = \sum p_k dq_k - \sum P_k dQ_k \tag{10.37}$$

where we have used Eqns (10.36). But dG is an exact differential. Hence, the expression

$$\sum p_k dq_k - \sum P_k dQ_k \tag{10.38}$$

must also be an exact differential. We can obtain this result by using other types of generating functions also. Thus, the transformation given by Eqn (10.1) is canonical only if the expression in Eqn (10.38) is an exact differential.

EXAMPLE 10.1 *Show that the generating function* $G = \sum q_k P_k$ *generates the identity transformation.*

Solution. The given generating function is of the type two. From Eqns (10.20), we have

$$p_k = \frac{\partial G}{\partial q_k} = P_k \tag{1}$$

$$Q_k = \frac{\partial G}{\partial P_k} = q_k \tag{2}$$

$$\text{and } H^* = H \tag{3}$$

Thus, the old and new variables are equal. Hence, the function G generates identity transformation.

EXAMPLE 10.2 *Show that the generating function* $G = \sum q_k Q_k$ *generates the exchange transformation.*

Solution. The given generating function is of type one. From Eqns (10.16), we have

$$p_k = \frac{\partial G}{\partial q_k} = Q_k \tag{1}$$

$$P_k = -\frac{\partial G}{\partial Q_k} = -q_k \tag{2}$$

$$\text{and } H^* = H \tag{3}$$

Thus, the coordinates and momenta are interchanged by the transformation.

EXAMPLE 10.3 *Show that the transformation*

$$Q = q \tan p, \; P = \log(\sin p)$$

is canonical.

Solution. We are given

$$Q = q \tan p \tag{1}$$

$$P = \log(\sin p) \tag{2}$$

The expression

$$\begin{aligned} pdq - PdQ &= pdq - \log(\sin p)[q \sec^2 p\, dp + \tan p\, dq] \\ &= [p - \tan p \log(\sin p)]dq - q \sec^2 p \log(\sin p) dp \end{aligned} \tag{3}$$

Let

$$F(p) = p - \tan p \log(\sin p)$$

Then

$$dF = 1 - \tan p \cdot \frac{\cos p}{\sin p} - \log(\sin p) \sec^2 p = -\log(\sin p) \sec^2 p \tag{4}$$

Thus, Eqn (3) can be written as

$$pdq - PdQ = Fdq + qdF = d(Fq)$$

Hence, the expression in the LHS is an exact differential. Thus, the transformation is canonical.

EXAMPLE 10.4 *For what values of α and β do the equations*

$$Q = q^\alpha \cos \beta p, \quad P = q^\alpha \sin \beta p$$

represent a canonical transformation? Find the generating function G_3 in this case.

Solution. We are given

$$Q = q^\alpha \cos \beta p$$

$$\therefore\ dQ = \alpha \cos \beta p (q^{\alpha-1}) dq - \beta q^\alpha \sin \beta p\, dp \tag{1}$$

$$\text{and } P = q^\alpha \sin \beta p$$

Then the expression

$$\begin{aligned} pdq - PdQ &= pdq - q^\alpha \sin \beta p \left[\alpha \cos \beta p (q^{\alpha-1}) dq - \beta q^\alpha \sin \beta p\, dp\right] \\ &= \left(p - \alpha q^{2\alpha-1} \sin \beta p \cos \beta p\right) dq + \beta q^{2\alpha} \sin^2 \beta p\, dp \end{aligned} \tag{2}$$

Let the expression (2) be a total differential of a function $F(p, q)$. Then

$$dF = \frac{\partial F}{\partial p} dp + \frac{\partial F}{\partial q} dq \tag{3}$$

Comparing Eqns (2) and (3), we get

$$\frac{\partial F}{\partial p} = \beta q^{2\alpha} \sin^2 \beta p = \frac{1}{2} \beta q^{2\alpha} (1 - \cos 2\beta p) \tag{4}$$

$$\frac{\partial F}{\partial q} = p - \alpha q^{2\alpha-1} \sin \beta p \cos \beta p \tag{5}$$

Integrating Eqn (4) w.r.t. p and Eqn (5) w.r.t. q, we find

$$F = \frac{1}{2} \beta q^{2\alpha} \left(p - \frac{\sin 2\beta p}{2\beta}\right) + f_1(q) \tag{6}$$

$$\text{and } F = pq - \alpha \frac{q^{2\alpha}}{2\alpha} \sin \beta p \cos \beta p + f_2(p) \tag{7}$$

Comparing Eqns (6) and (7), we get

$$f_1(q) = f_2(p) = 0 \text{ and } \frac{1}{2}\beta q^{2\alpha} p = pq$$

Hence,

$$\beta = 2 \text{ and } \alpha = \frac{1}{2} \tag{8}$$

Thus, the given transformations are canonical for values of α, β as in Eqn (8).

In view of this, the given transformations are

$$Q = \sqrt{q}\cos 2p, \quad P = \sqrt{q}\sin 2p$$

Hence,

$$Q\tan 2p = P \tag{9}$$

Let us find the generating function G_3 of type three.

By Eqn (10.25),

$$P = -\frac{\partial G_3}{\partial Q} \tag{10}$$

From Eqns (9) and (10), we get

$$-\frac{\partial G_3}{\partial Q} = Q\tan 2p$$

which upon integration yields

$$G_3 = -\frac{1}{2}Q^2 \tan 2p$$

We have taken the constant of integration equal to zero.

EXAMPLE 10.5 *The transformation equations between two sets of coordinates are*

$$P = 2\,(1 + \sqrt{q}\cos p)\,\sqrt{q}\sin p \text{ and } Q = \log\,(1 + \sqrt{q}\cos p)$$

Show that (i) *the transformation is canonical and* (ii) *the generating function of this transformation is*

$$G_3 = -(e^Q - 1)^2 \tan p$$

Solution. We are given

$$P = 2\,(1 + \sqrt{q}\cos p)\,\sqrt{q}\sin p \tag{1}$$

$$\text{and } Q = \log\,(1 + \sqrt{q}\cos p) \tag{2}$$

$$\therefore\ dQ = \frac{-\sqrt{q}\sin p\, dp + \frac{1}{2}\cos p/\sqrt{q}\, dq}{1 + \sqrt{q}\cos p} \tag{3}$$

Hence, the expression

$$pdq - PdQ = pdq - 2\,(1 + \sqrt{q}\cos p)\,\sqrt{q}\sin p \times \frac{\left[-\sqrt{q}\sin p\, dp + \frac{1}{2}\frac{\cos p}{\sqrt{q}}dq\right]}{1 + \sqrt{q}\cos p}$$

$$= pdq + 2q\sin^2 p\, dp - \sin p\cos p\, dq$$

$$= \left(p - \frac{1}{2}\sin 2p\right) dq + q(1 - \cos 2p)dp$$

$$= d\left[q\left(p - \frac{1}{2}\sin 2p\right)\right]$$

Hence, the expression $(pdq - PdQ)$ is an exact differential.

So the transformation is canonical.

Again, from Eqn (2),

$$e^Q = 1 + \sqrt{q}\cos p \tag{4}$$

which yields

$$q = \frac{(e^Q - 1)^2}{\cos^2 p} \tag{5}$$

For this transformation, let the generating function be of type three. That is,

$$G_3 = G_3(p, Q) \tag{6}$$

By Eqn (10.25), we have

$$q = -\frac{\partial G_3}{\partial p} \tag{7}$$

Hence, from Eqns (5) and (7), we get

$$-\frac{\partial G_3}{\partial p} = \frac{(e^Q - 1)^2}{\cos^2 p} = (e^Q - 1)^2 \sec^2 p$$

which upon integration yields

$$G_3 = -(e^Q - 1)^2 \tan p$$

the constant of integration taken to be zero.

EXAMPLE 10.6 *Determine the canonical transformations defined by the generating function*

$$G(q, Q, t) = \frac{1}{2} m\omega q^2 \cot Q$$

Hence, obtain an expression for the displacement of a harmonic oscillator.

Solution. The given generating function

$$G(q, Q, t) = \frac{1}{2} m\omega q^2 \cot Q \tag{1}$$

is of type one. By Eqns (10.16), we have

$$p = \frac{\partial G}{\partial q} = m\omega q \cot Q \tag{2}$$

$$P = -\frac{\partial G}{\partial Q} = \frac{m\omega q^2}{2\sin^2 Q} \tag{3}$$

$$\text{and } H^* = H \tag{4}$$

From Eqn (3),

$$q = \sqrt{\frac{2P}{m\omega}} \sin Q \tag{5}$$

Substituting q from Eqn (5) into Eqn (2), we get

$$p = \sqrt{2m\omega P} \cos Q \tag{6}$$

Eqns (5) and (6) give the required transformation.

The Hamiltonian H for the harmonic oscillator is

$$H = \frac{1}{2m} (p^2 + m^2\omega^2 q^2) \tag{7}$$

Using Eqns (5) and (6), we get

$$H^* = H = \frac{1}{2m}(2m\omega P\cos^2 Q + 2m\omega P\sin^2 Q) = \omega P \tag{8}$$

Then equation of motion for Q is

$$\dot{Q} = \frac{\partial H^*}{\partial P} = \omega \tag{9}$$

Upon integration we get

$$Q = \omega t + \phi \tag{10}$$

where ϕ = constant of integration.

Hence, from Eqns (5) and (10), we have

$$q = \sqrt{\frac{2P}{m\omega}}\sin(\omega t + \phi) \tag{11}$$

Since by Eqn (8), the Hamiltonian is cyclic in Q, the conjugate momentum P is constant. The Hamiltonian in this case is actually the total energy E, so that

$$E = \omega P, \text{ i.e., } P = \frac{E}{\omega} \tag{12}$$

Using this value of P in Eqn (11), we write:

$$q = \sqrt{\frac{2E}{m\omega^2}}\sin(\omega t + \phi)$$

which is the required expression for the displacement.

10.4 Poisson Brackets

Consider two functions F and G which depend on the position coordinates q_k, the momentum coordinates p_k and time t. The Poisson bracket of F and G is defined by the equation

$$[F, G]_{q,p} = \sum\left[\frac{\partial F}{\partial q_k}\frac{\partial G}{\partial p_k} - \frac{\partial F}{\partial p_k}\frac{\partial G}{\partial q_k}\right] \tag{10.39}$$

We drop the subscripts q, p and denote the Poisson bracket as $[F, G]$. From the definition of the Poisson bracket we note the following properties:

PROPERTY 1. If we put $F = G$ in Eqn (10.39), then we find that

$$[F, F] = 0 \tag{10.40}$$

That is, the Poisson bracket of a function with itself is zero.

PROPERTY 2. If G = constant = C (say), then

$$[F, C] = 0 \tag{10.41}$$

PROPERTY 3. If there are three functions F_1, F_2 and G, then

$$[F_1 + F_2, G] = [F_1, G] + [F_2, G] \tag{10.42}$$

That is, the Poisson bracket obeys the distributive law.

Similarly,

$$[F_1, F_2 G] = [F_1, F_2]G + F_2[F_1, G] \tag{10.43}$$

PROPERTY 4. It can be verified that

$$[F, G] = -[G, F] \tag{10.44}$$

That is, the Poisson bracket does not obey commutative law.

PROPERTY 5. A very important property of Poisson bracket is that

$$[q_k, p_j] = \delta_{kj} \tag{10.45}$$

where δ_{kj} is the ***Kronecker delta*** with the property

$$\delta_{kj} = 0 \text{ for } k \neq j \text{ and } \delta_{kj} = 1 \text{ for } k = j.$$

To prove this property, we write from definition:

$$[q_k, p_j] = \sum_i \left(\frac{\partial q_k}{\partial q_i} \frac{\partial p_j}{\partial p_i} - \frac{\partial q_k}{\partial p_i} \frac{\partial p_j}{\partial q_i} \right) \tag{10.46}$$

But we have

$$\left. \begin{aligned} \frac{\partial q_k}{\partial q_i} = \delta_{ki}, \quad \frac{\partial p_j}{\partial p_i} = \delta_{ji} \\ \frac{\partial q_k}{\partial p_i} = 0, \quad \frac{\partial p_j}{\partial q_i} = 0 \end{aligned} \right\} \tag{10.47}$$

$$\text{and} \quad \sum_i \delta_{ki} \delta_{ji} = \delta_{kj} \tag{10.48}$$

In view of Eqns (10.47) and (10.48), we find

$$[q_k, p_j] = \delta_{kj} \tag{10.49}$$

PROPERTY 6. It can also be proved that

$$[q_k, q_j] = [p_k, p_j] = 0 \tag{10.50}$$

Eqns (10.49) and (10.50) are called ***fundamental Poisson brackets***.

PROPERTY 7. For any three functions F, G and K of q_k and p_k,

$$[F, [G, K]] + [G, [F, K]] + [K, [F, G]] = 0 \tag{10.51}$$

Eqn (10.51) is known as ***Jacobi's identity***.

EXAMPLE 10.7 *Prove the distributive law*

$$[F + G, K] = [F, K] + [G, K]$$

of Poisson bracket for three functions F, G and K.

Solution. We write from definition:

$$\begin{aligned}
[F + G, K] &= \sum_k \left[\frac{\partial (F+G)}{\partial q_k} \frac{\partial K}{\partial p_k} - \frac{\partial (F+G)}{\partial p_k} \frac{\partial K}{\partial q_k} \right] \\
&= \sum_k \left[\left(\frac{\partial F}{\partial q_k} \frac{\partial K}{\partial p_k} + \frac{\partial G}{\partial q_k} \frac{\partial K}{\partial p_k} \right) - \left(\frac{\partial F}{\partial p_k} \frac{\partial K}{\partial q_k} + \frac{\partial G}{\partial p_k} \frac{\partial K}{\partial q_k} \right) \right] \\
&= \sum_k \left(\frac{\partial F}{\partial q_k} \frac{\partial K}{\partial p_k} - \frac{\partial F}{\partial p_k} \frac{\partial K}{\partial q_k} \right) + \sum_k \left(\frac{\partial G}{\partial q_k} \frac{\partial K}{\partial p_k} - \frac{\partial G}{\partial p_k} \frac{\partial K}{\partial q_k} \right) \\
&= [F, K] + [G, K]
\end{aligned}$$

Hence proved.

EXAMPLE 10.8 *If H is the Hamiltonian and f is any function depending on position, momenta and time, then show that*

$$\frac{df}{dt} = \frac{\partial f}{\partial t} - [H, f]$$

Solution. Since H is the Hamiltonian, we have

$$\dot{p}_k = -\frac{\partial H}{\partial q_k}, \; \dot{q}_k = \frac{\partial H}{\partial p_k} \tag{1}$$

Since f is a function depending on q_k, p_k and t, we have

$$\frac{df}{dt} = \frac{\partial f}{\partial t} + \sum_k \left(\frac{\partial f}{\partial p_k}\dot{p}_k + \frac{\partial f}{\partial q_k}\dot{q}_k \right) \tag{2}$$

Using Eqns (1) in (2), we get

$$\begin{aligned} \frac{df}{dt} &= \frac{\partial f}{\partial t} + \sum_k \left(\frac{\partial f}{\partial q_k}\frac{\partial H}{\partial p_k} - \frac{\partial f}{\partial p_k}\frac{\partial H}{\partial q_k} \right) \\ &= \frac{\partial f}{\partial t} + [f, H] \\ &= \frac{\partial f}{\partial t} - [H, f] \end{aligned} \tag{3}$$

10.5 Poisson's Theorem

Consider a function $F(q_k, p_k, t)$. Then we have (see Example 10.8)

$$\frac{dF}{dt} = \frac{\partial F}{\partial t} + [F, H] \tag{10.52}$$

where H is the Hamiltonian. If the function F does not depend on time explicitly, then Eqn (10.52) yields

$$[F, H] = 0 \tag{10.53}$$

Thus, if a function F not depending on time becomes a constant of motion, then its Poisson bracket with the Hamiltonian vanishes. Conversely, a function whose Poisson bracket with the Hamiltonian vanishes is a constant of motion.

Now, let the function K in Jacobi's identity be replaced by the Hamiltonian H. Then

$$[F, [G, H]] + [G, [F, H]] + [H, [F, G]] = 0 \tag{10.54}$$

If now the functions F, G are constants of motion, then by Eqn (10.53)

$$[F, H] = 0, \; [G, H] = 0 \tag{10.55}$$

In view of this, Eqn (10.54) reduces to

$$[H, [F, G]] = 0 \tag{10.56}$$

Eqn (10.56) implies that $[F, G]$ must be a constant of motion. Thus,

The Poisson bracket of two constants of motion is a constant of motion. This is Poisson's theorem.

10.6 Canonical Transformation in Terms of Poisson Bracket

Let us suppose that we have the transformation from a set of coordinates p_k and q_k to a new set of coordinates P_k and Q_k, such that

$$P_k = P_k(p_k, q_k); Q_k = Q_k(p_k, q_k) \tag{10.57}$$

Then

$$[P_k, P_k]_{q,p} = \sum \left[\frac{\partial P_k}{\partial q_k} \frac{\partial P_k}{\partial p_k} - \frac{\partial P_k}{\partial p_k} \frac{\partial P_k}{\partial q_k} \right] = 0 \tag{10.58}$$

Similarly, we have

$$[Q_k, Q_k]_{q,p} = 0 \tag{10.59}$$

Now,

$$[Q_k, P_j]_{q,p} = \sum_i \left[\frac{\partial Q_k}{\partial q_i} \frac{\partial P_j}{\partial p_i} - \frac{\partial Q_k}{\partial p_i} \frac{\partial P_j}{\partial q_i} \right] \tag{10.60}$$

We can have some relations for the four conjugate variables as follows:
For the generating function G_1 in Eqn (10.12), we have, keeping t fixed, the relations

$$p_k = \frac{\partial G_1}{\partial q_k}; P_k = -\frac{\partial G_1}{\partial Q_k} \tag{10.61}$$

Hence,

$$\frac{\partial p_k}{\partial Q_k} = \frac{\partial^2 G_1}{\partial Q_k \partial q_k} = \frac{\partial}{\partial q_k} \left(\frac{\partial G_1}{\partial Q_k} \right) = -\frac{\partial P_k}{\partial q_k} \tag{10.62}$$

Similarly, considering other generating functions, we can prove the following relations:

$$\frac{\partial p_k}{\partial P_k} = \frac{\partial Q_k}{\partial q_k} \tag{10.63}$$

$$\frac{\partial q_k}{\partial Q_k} = \frac{\partial P_k}{\partial p_k} \tag{10.64}$$

$$\frac{\partial q_k}{\partial P_k} = -\frac{\partial Q_k}{\partial p_k} \tag{10.65}$$

By relation (10.64), we have

$$\frac{\partial P_j}{\partial p_i} = \frac{\partial q_j}{\partial Q_i} \tag{10.66}$$

Again, by relation (10.62), we have

$$\frac{\partial P_j}{\partial q_i} = -\frac{\partial p_j}{\partial Q_i} \tag{10.67}$$

Using relations (10.66) and (10.67) in Eqn (10.60), we get

$$[Q_k, P_j]_{q,p} = \sum_i \left(\frac{\partial Q_k}{\partial q_i} \frac{\partial q_j}{\partial Q_i} + \frac{\partial Q_k}{\partial p_i} \frac{\partial p_j}{\partial Q_i} \right) = \frac{\partial Q_k}{\partial Q_i} = \delta_{ki} = 1 \tag{10.68}$$

Thus, for the transformation in (10.57) to be canonical, we must have

$$[Q_k, P_j]_{q,p} = 1 \tag{10.69}$$

This gives the condition for a canonical transformation in terms of Poisson bracket.

10.7 Lagrange Bracket

The Lagrange bracket of two functions u and v with respect to the basis (q_k, p_k) is defined as

$$\{u, v\}_{q,p} = \sum \left(\frac{\partial q_k}{\partial u} \frac{\partial p_k}{\partial v} - \frac{\partial q_k}{\partial v} \frac{\partial p_k}{\partial u} \right) \tag{10.70}$$

We can drop the subscripts q, p and denote the Lagrange bracket as $\{u, v\}$.

From definition, we note the following properties of the Lagrange bracket:

PROPERTY 8.

$$\{q_k, q_j\} = 0; \{p_k, p_j\} = 0 \text{ and } \{q_j, p_j\} = \delta_{ij} \tag{10.71}$$

Proof. From definition, we have

$$\{q_k, q_j\} = \sum_i \left(\frac{\partial q_k}{\partial q_k} \frac{\partial p_i}{\partial q_j} - \frac{\partial q_i}{\partial q_j} \frac{\partial p_i}{\partial q_k} \right) = \sum_i (\delta_{ik} \cdot 0 - \delta_{ij} \cdot 0) = 0$$

Similarly, $\{p_k, p_j\} = 0$.

Again,

$$\begin{aligned} \{q_k, p_j\} &= \sum_i \left(\frac{\partial q_i}{\partial q_k} \frac{\partial p_i}{\partial p_j} - \frac{\partial q_i}{\partial p_j} \frac{\partial p_i}{\partial q_k} \right) \\ &= \sum_i (\delta_{ik}\delta_{ij} - 0) \left[\text{since } \frac{\partial p_i}{\partial q_k} = 0, \frac{\partial q_i}{\partial q_k} = \delta_{ik} \right] \\ &= \delta_{ij} \end{aligned}$$

PROPERTY 9. The Lagrange bracket is not commutative. That is,

$$\{u, v\} \neq \{v, u\} \tag{10.72}$$

Proof. From definition, we have

$$\{u, v\} = \sum_i \left(\frac{\partial q_i}{\partial u} \frac{\partial p_i}{\partial v} - \frac{\partial q_i}{\partial v} \frac{\partial p_i}{\partial u} \right) = -\sum_i \left(\frac{\partial q_i}{\partial v} \frac{\partial p_i}{\partial u} - \frac{\partial q_i}{\partial u} \frac{\partial p_i}{\partial v} \right) = -\{v, u\}$$

PROPERTY 10. The Lagrange bracket is invariant under canonical transformation. Thus,

$$\{u, v\}_{q,p} = \{u, v\}_{Q,P} \tag{10.73}$$

10.8 Canonical Transformation Test

The Lagrange bracket may also be used to test whether a given transformation is canonical. Let us have the transformations:

$$\left. \begin{aligned} Q_k &= Q_k(q, p, t) \\ P_k &= P_k(q, p, t) \end{aligned} \right\} \tag{10.74}$$

It can be proved that the above transformations will be canonical, if

$$\left. \begin{aligned} \{Q_j, Q_m\}_{q,p} &= 0 \\ \{P_j, P_m\}_{q,p} &= 0 \\ \text{and } \{Q_i, P_m\} &= \delta_{im} \end{aligned} \right\} \tag{10.75}$$

where $i, m = 1, 2, \ldots, n$.

EXAMPLE 10.9 *If ϕ and ψ are two functions, then prove that*

$$\frac{\partial}{\partial t}[\phi, \psi] = \left[\frac{\partial \phi}{\partial t}, \psi\right] + \left[\phi, \frac{\partial \psi}{\partial t}\right].$$

Solution. From definition of Poisson bracket, we have

$$[\phi, \psi] = \sum_k \left(\frac{\partial \phi}{\partial q_k}\frac{\partial \psi}{\partial p_k} - \frac{\partial \phi}{\partial p_k}\frac{\partial \psi}{\partial q_k}\right) \tag{1}$$

Hence,

$$\begin{aligned}\frac{\partial}{\partial t}[\phi, \psi] &= \sum_k \left[\frac{\partial \phi}{\partial q_k}\frac{\partial}{\partial p_k}\left(\frac{\partial \psi}{\partial t}\right) + \frac{\partial}{\partial q_k}\left(\frac{\partial \phi}{\partial t}\right)\frac{\partial \psi}{\partial p_k}\right.\\ &\qquad \left. - \frac{\partial}{\partial p_k}\left(\frac{\partial \phi}{\partial t}\right)\frac{\partial \psi}{\partial q_k} - \frac{\partial \phi}{\partial p_k}\frac{\partial}{\partial q_k}\left(\frac{\partial \psi}{\partial t}\right)\right]\\ &= \sum_k \left[\frac{\partial}{\partial q_k}\left(\frac{\partial \phi}{\partial t}\right)\frac{\partial \psi}{\partial p_k} - \frac{\partial}{\partial p_k}\left(\frac{\partial \phi}{\partial t}\right)\frac{\partial \psi}{\partial q_k}\right]\\ &\qquad + \sum_k \left[\frac{\partial}{\partial p_k}\left(\frac{\partial \psi}{\partial t}\right)\frac{\partial \phi}{\partial q_k} - \frac{\partial}{\partial q_k}\left(\frac{\partial \psi}{\partial t}\right)\frac{\partial \phi}{\partial p_k}\right]\\ &= \left[\frac{\partial \phi}{\partial t}, \psi\right] + \left[\phi, \frac{\partial \psi}{\partial t}\right] \qquad \textbf{(Proved)}\end{aligned}$$

EXAMPLE 10.10 *Deduce Hamilton's equations in terms of Poisson bracket.*

Solution. The Poisson bracket of functions q_k and H is given by

$$[q_k, H] = \sum_j \left(\frac{\partial q_k}{\partial q_j}\frac{\partial H}{\partial p_j} - \frac{\partial q_k}{\partial p_j}\frac{\partial H}{\partial q_j}\right) \tag{1}$$

But

$$\frac{\partial q_k}{\partial p_j} = 0 \tag{2}$$

In view of this, we write (1) as

$$[q_k, H] = \sum_j \left(\frac{\partial q_k}{\partial q_j}\frac{\partial H}{\partial p_j}\right) = \sum_j \frac{\partial H}{\partial p_j}\delta_{kj} = \frac{\partial H}{\partial p_k} \tag{3}$$

Here δ_{kj} is the Kronecker delta defined by $\delta_{kj} = 1, \delta_{kk} = 0$.
Similarly, we have

$$[p_k, H] = -\frac{\partial H}{\partial q_k} \tag{4}$$

Hamilton's equations are:

$$\left.\begin{aligned}\dot{q}_k &= \frac{\partial H}{\partial p_k}\\ \dot{p}_k &= -\frac{\partial H}{\partial q_k}\end{aligned}\right\} \tag{5}$$

Using relations (3) and (4) in (5),

$$\left.\begin{aligned}\dot{q}_k &= [q_k, H]\\ \dot{p}_k &= [p_k, H]\end{aligned}\right\} \tag{6}$$

Thus, Hamilton's equations are expressed in terms of Poisson bracket.

EXAMPLE 10.11 *For what values of the parameter λ, the following transformation would be canonical?*

$$Q = q\cos\lambda - p\sin\lambda$$
$$P = q\sin\lambda + p\cos\lambda$$

Solution. From the given transformations, we find,

$$\frac{\partial Q}{\partial q} = \cos\lambda; \frac{\partial Q}{\partial p} = -\sin\lambda \tag{1}$$

$$\text{and } \frac{\partial P}{\partial q} = \sin\lambda; \frac{\partial P}{\partial p} = \cos\lambda \tag{2}$$

For the transformation to be canonical, we must have

$$[Q, P] = 1 \tag{3}$$

Now,

$$[Q, P] = \frac{\partial Q}{\partial q}\frac{\partial P}{\partial p} - \frac{\partial Q}{\partial p}\frac{\partial P}{\partial q}$$
$$= \cos\lambda \cdot \cos\lambda - (-\sin\lambda)\cdot\sin\lambda = \cos^2\lambda + \sin^2\lambda = 1 \tag{4}$$

Thus, the given transformation satisfies the condition (3) for any value of the parameter λ.

EXAMPLE 10.12 *Show, by using Poisson bracket, that the transformation defined by*

$$\left.\begin{aligned} q &= \sqrt{2P}\sin Q \\ p &= \sqrt{2P}\cos Q \end{aligned}\right\}$$

is canonical.

Solution. We are given

$$\left.\begin{aligned} q &= \sqrt{2P}\sin Q \\ p &= \sqrt{2P}\cos Q \end{aligned}\right\} \tag{1}$$

Hence,

$$\tan Q = \frac{q}{p} \tag{2}$$

$$\text{and } P = \frac{1}{2}(p^2 + q^2) \tag{3}$$

For the transformation to be canonical, we must have

$$[P, P] = 0,\ [Q, Q] = 0 \text{ and } [Q, P] = 1 \tag{4}$$

Now

$$[P, P] = \frac{\partial P}{\partial q}\frac{\partial P}{\partial p} - \frac{\partial P}{\partial p}\frac{\partial P}{\partial q} = 0 \tag{5}$$

Similarly,

$$[Q, Q] = 0 \tag{6}$$

Also,

$$[Q, P] = \frac{\partial Q}{\partial q}\frac{\partial P}{\partial p} - \frac{\partial Q}{\partial p}\frac{\partial P}{\partial q} \tag{7}$$

From (2) and (3), wel find

$$\sec^2 Q \frac{\partial Q}{\partial q} = \frac{1}{p}; \text{ i.e., } \frac{\partial Q}{\partial q} = \frac{\cos^2 Q}{p} \tag{8}$$

$$\frac{\partial P}{\partial p} = p; \frac{\partial P}{\partial q} = q \tag{9}$$

$$\text{and } \sec^2 Q \frac{\partial Q}{\partial p} = -\frac{q}{p^2}; \text{ i.e., } \frac{\partial Q}{\partial p} = -\frac{q\cos^2 Q}{p^2} \tag{10}$$

Using relations (8), (9) and (10) in (7), we get

$$\begin{aligned}[Q, P] &= \frac{\cos^2 Q}{p} \cdot p + \frac{q\cos^2 Q}{p^2} \cdot q \\ &= \cos^2 Q \left(1 + \frac{q^2}{p^2}\right) = \cos^2 Q(1 + \tan^2 Q), \quad \text{[using (2)]} \\ &= \cos^2 Q \cdot \sec^2 Q = 1\end{aligned}$$

Hence, the transformation is canonical.

EXAMPLE 10.13 *Find Poisson brackets for angular momentum of a particle.*

Solution. Let the particle have the position vector $\mathbf{r}$ and momentum vector $\mathbf{p}$ at any instant. Then

$$\left.\begin{aligned}\mathbf{r} &= \mathbf{i}x + \mathbf{j}y + \mathbf{k}z \\ \text{and } \mathbf{p} &= \mathbf{i}p_x + \mathbf{j}p_y + \mathbf{k}p_z\end{aligned}\right\} \tag{1}$$

Hence, the angular momentum

$$\begin{aligned}\mathbf{J} &= \mathbf{r} \times \mathbf{p} \\ &= (\mathbf{i}x + \mathbf{j}y + \mathbf{k}z) \times (\mathbf{i}p_x + \mathbf{j}p_y + \mathbf{k}p_z) \\ &= \mathbf{i}(yp_z - zp_y) + \mathbf{j}(zp_x - xp_z) + \mathbf{k}(xp_y - yp_x)\end{aligned} \tag{2}$$

Thus, the components of angular momentum are:

$$\left.\begin{aligned}J_x &= yp_z - zp_y \\ J_y &= zp_x - xp_z \\ J_z &= xp_y - yp_x\end{aligned}\right\} \tag{3}$$

Then

$$\begin{aligned}[J_x, J_y] &= [(yp_z - zp_y), (zp_x - xp_z)] \\ &= [yp_z, zp_x] - [yp_z, xp_z] - [zp_y, zp_x] + [zp_y, xp_z]\end{aligned}$$

Now,

$$\begin{aligned}[yp_z, zp_x] &= y[p_z, zp_x] + p_z[y, zp_x] + yp_x[p_z, z] + zp_z[y, p_x] + p_z p_x[y, z] \\ &= -yp_x\end{aligned} \tag{4}$$

Similarly, we can have

$$\left.\begin{aligned}[yp_z, xp_z] &= 0 \\ [zp_y, zp_x] &= 0 \\ [zp_y, xp_z] &= xp_y\end{aligned}\right\} \tag{5}$$

Thus,

$$[J_x, J_y] = xp_y - yp_x = J_z \tag{6}$$

Similarly, we can find

$$\left.\begin{aligned} [J_y, J_z] &= J_x \\ \text{and } [J_z, J_x] &= J_y \end{aligned}\right\} \tag{7}$$

EXAMPLE 10.14 *Using Poisson bracket, show that the transformation given by*

$$Q = \left(e^{-2q} - p^2\right)^{1/2}$$
$$P = \cos^{-1}(pe^q)$$

is canonical.

Solution. We have

$$Q = \left(e^{-2q} - p^2\right)^{1/2} \tag{1}$$

$$\therefore \frac{\partial Q}{\partial q} = -\frac{e^{-2q}}{\left(e^{-2q} - p^2\right)^{1/2}} = -\frac{e^{-2q}}{Q} \tag{2}$$

$$\frac{\partial Q}{\partial p} = -\frac{p}{\left(e^{-2q} - p^2\right)^{1/2}} = -\frac{p}{Q} \tag{3}$$

Similarly, we have

$$P = \cos^{-1}(pe^q) \tag{4}$$

$$\therefore \frac{\partial P}{\partial q} = -\frac{p}{\left(e^{-2q} - p^2\right)^{1/2}} = -\frac{p}{Q} \tag{5}$$

$$\text{and } \frac{\partial P}{\partial p} = -\frac{1}{\left(e^{-2q} - p^2\right)^{1/2}} = -\frac{1}{Q} \tag{6}$$

Now,

$$[Q, P] = \frac{\partial Q}{\partial q}\frac{\partial P}{\partial p} - \frac{\partial Q}{\partial p}\frac{\partial P}{\partial q} \tag{7}$$

Using the relations (2), (3), (5) and (6) in (7), we get

$$[Q, P] = \left(-\frac{e^{-2q}}{Q}\right)\left(-\frac{1}{Q}\right) - \left(-\frac{p}{Q}\right)\left(-\frac{p}{Q}\right)$$
$$= \frac{1}{Q^2}\left(e^{-2q} - p^2\right) = \frac{1}{Q^2}\cdot Q^2 = 1 \tag{8}$$

Also, we have

$$\left.\begin{aligned} [Q, Q] &= 0 \\ \text{and } [P, P] &= 0 \end{aligned}\right\} \tag{9}$$

Thus, all the conditions for a canonical transformation are satisfied. Hence, the given transformation is necessarily canonical.

EXAMPLE 10.15 *Show, by the use of Lagrange brackets, that the transformation*

$$Q = \log\left(\frac{\sin p}{q}\right);\ P = q\cot p$$

is canonical.

Solution. We are given

$$Q = \log\left(\frac{\sin p}{q}\right)$$

$$\therefore \frac{\partial Q}{\partial q} = \frac{q}{\sin p}\left(-\frac{\sin p}{q^2}\right) = -\frac{1}{q} \quad (1)$$

$$\frac{\partial Q}{\partial p} = \frac{q}{\sin p}\left(\frac{\cos p}{q}\right) = \cot p \quad (2)$$

Also

$$P = q \cot p$$

$$\therefore \frac{\partial P}{\partial q} = \cot p \quad (3)$$

$$\frac{\partial P}{\partial p} = -q \operatorname{cosec}^2 p \quad (4)$$

Hence,

$$\{q, q\} = \frac{\partial Q}{\partial q}\frac{\partial P}{\partial q} - \frac{\partial Q}{\partial q}\frac{\partial P}{\partial q} = 0 \quad (5)$$

$$\{p, p\} = \frac{\partial P}{\partial p}\frac{\partial Q}{\partial p} - \frac{\partial Q}{\partial p}\frac{\partial P}{\partial p} = 0 \quad (6)$$

and $\{q, p\} = \dfrac{\partial Q}{\partial q}\dfrac{\partial P}{\partial p} - \dfrac{\partial Q}{\partial p}\dfrac{\partial P}{\partial q} = \left(-\dfrac{1}{q}\right)(-q \operatorname{cosec}^2 p) - (\cot p)(\cot p)$

$$= \operatorname{cosec}^2 p - \cot^2 p = 1 \quad (7)$$

Hence the given transformation is canonical.

EXERCISE

1. Show that the function $G = \sum Q_k p_k$ generates the identity transformation.
2. Show that the transformation $P = \frac{1}{2}(p^2 + q^2)$, $Q = \tan^{-1}\left(\frac{p}{q}\right)$ is canonical.
3. Show that the transformations:
   (a) $q = \sqrt{2P} \sin Q$ and $P = \sqrt{2q} \cos Q$
   (b) $Q = \log\left(\frac{1}{q}\sin p\right)$ and $P = q \cot p$
   are canonical.
4. Find the generating function for the transformation $Q = \sqrt{q} \cos 2p$ and $P = \sqrt{q} \sin 2p$
5. Prove the following relations for canonical transformations:
   (a) $\dfrac{\partial q_j}{\partial Q_k} = \dfrac{\partial P_k}{\partial p_j}$ (b) $\dfrac{\partial q_j}{\partial P_k} = -\dfrac{\partial Q_k}{\partial p_j}$
6. Find the canonical transformations defined by the generating functions:
   (a) $G_1(q, Q) = qQ - \frac{1}{2}m\omega q^2 - \dfrac{Q^2}{4m\omega}$
   (b) $G_3(Q, p) = -\frac{1}{2}Q^2 \tan 2p$

7. Prove that, if a function F does not depend explicitly on time but is a constant of motion, then its Poisson bracket with the Hamiltonian vanishes.

8. Prove the identities:

   (a) $[F, q_k] = -\dfrac{\partial F}{\partial p_k}$ (b) $[F, p_k] = \dfrac{\partial F}{\partial q_k}$

9. Verify the following properties of Poisson brackets:

   (a) $\dot{q}_k = [q_k, H]$ (b) $\dot{q}_k = [p_k, H]$

   where H is the Hamiltonian.

10. If $[\phi, \psi]$ be the Poisson bracket, then prove that $\dfrac{\partial}{\partial t}[\phi, \psi] = \left[\dfrac{\partial \phi}{\partial t}, \psi\right] + \left[\phi, \dfrac{\partial \psi}{\partial t}\right]$

CHAPTER 11

Small Oscillations

Small oscillations occur in a physical system when it is given a small displacement from its equilibrium position. Small oscillations of a pendulum, motion of charges in electrical circuits having capacitance and inductance are examples of small harmonic oscillations.

11.1 Simple Harmonic Oscillator

Consider a body of mass m attached to one end of a spring of force constant k hanging vertically from a roof ceiling. Now the body is pushed downward a little and released immediately. It will then oscillate vertically (FIG. 11.1) about its position of equilibrium.

The force acting on the body will be k-times the displacement and is called ***restoring force***.

Hence, the equation of motion is

$$m\frac{d^2x}{dt^2} = -kx$$

$$\text{or,} \quad \frac{d^2x}{dt^2} + \xi^2 x = 0 \tag{11.1}$$

where

$$\xi^2 = \frac{k}{m} \tag{11.2}$$

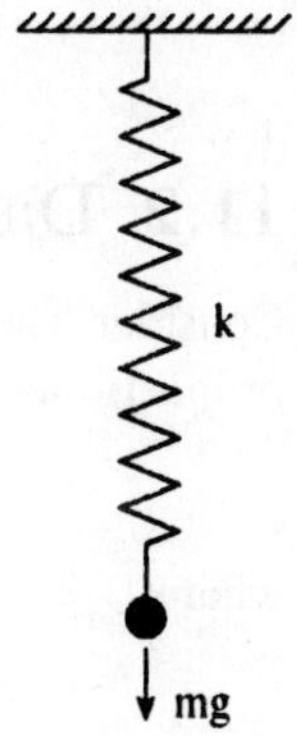

FIG. 11.1

Eqn (11.1) is a standard form for simple harmonic motion and the system is called ***simple harmonic oscillator***.

The general solution of Eqn (11.1) is

$$x = A\sin(\xi t + \phi) \tag{11.3}$$

where A and ϕ are constants of integration to be determined from initial conditions. Let, initially at $t = 0$,

$$x = x_0 \text{ and } \dot{x} = v = v_0 \tag{11.4}$$

Then we can find

$$A = \sqrt{x_0^2 + \frac{v_0^2}{\xi^2}} \text{ and } \phi = \tan^{-1}\left(\frac{\xi x_0}{v_0}\right) \tag{11.5}$$

If P be the period of oscillations, then

$$\xi t + \phi + 2\pi = \xi(t + P) + \phi \text{ or, } P = \frac{2\pi}{\xi} \tag{11.6}$$

Frequency of vibration is

$$n = \frac{1}{P} = \frac{\xi}{2\pi} \tag{11.7}$$

so that

$$\xi = 2n\pi \tag{11.8}$$

Eqn (11.8) shows that the constant $\xi = \sqrt{k/m}$ is actually the number of oscillations in 2π seconds.

The velocity of the body at any instant of time is

$$v = \frac{dx}{dt} = A\xi \cos(\xi t + \phi) \quad \text{or,} \; v = \xi\sqrt{A^2 - x^2} \tag{11.9}$$

The kinetic energy of the particle at any instant of time is

$$T = \frac{1}{2}mv^2 = \frac{1}{2}m\xi^2(A^2 - x^2) \tag{11.10}$$

The potential energy of the body at any instant of time is

$$V = \int_0^x m\xi^2 x\, dx = \frac{1}{2}m\xi^2 x^2 \tag{11.11}$$

Thus, the total energy of the body at any instant of time is

$$E = T + V = \frac{1}{2}m^2\xi^2 A^2 = \text{constant.} \tag{11.12}$$

11.2 Damped Harmonic Oscillator

Consider the harmonic oscillator in a medium of viscous damping coefficient c which is proportional to the velocity. Then the equation of motion is

$$m\frac{d^2x}{dt^2} = -c\frac{dx}{dt} - kx \text{ or, } \frac{d^2x}{dt^2} + 2\alpha\frac{dx}{dt} + \xi^2 x = 0 \tag{11.13}$$

where

$$\alpha = \frac{c}{2m}, \; \xi^2 = \frac{k}{m} \tag{11.14}$$

We assume the solution of Eqn (11.13) in the form:

$$x = e^{\lambda t} \tag{11.15}$$

Feeding this into Eqn (11.13), we get the characteristic equation

$$\lambda^2 + 2\alpha\lambda + \xi^2 = 0 \tag{11.16}$$

Roots of this equation are:

$$\left.\begin{aligned} \lambda_1 &= -\alpha + \sqrt{\alpha^2 - \xi^2} \\ \lambda_2 &= -\alpha - \sqrt{\alpha^2 - \xi^2} \end{aligned}\right\} \tag{11.17}$$

Thus, the solution of Eqn (11.13) is

$$x(t) = e^{-\alpha t}\left(Ae^{\beta t} + Be^{-\beta t}\right) \tag{11.18}$$

where we have denoted

$$\beta = \sqrt{\alpha^2 - \xi^2} \tag{11.19}$$

The nature of the solution function given by Eqn (11.18) depends on the value of β.

If

$$\beta^2 > 0, \text{ i.e., } \alpha^2 - \xi^2 > 0 \tag{11.20}$$

the roots λ_1, λ_2 are real and the solution is given by Eqn (11.18). In this case the damping is very large.

If

$$\beta^2 = 0, \text{ i.e., } \alpha^2 - \xi^2 = 0 \tag{11.21}$$

the roots λ_1, λ_2 are real and equal, and the solution of Eqn (11.13) is

$$x(t) = e^{-\alpha t}(A + Bt) \tag{11.22}$$

In this case the damping is said to be ***critical***.

If

$$\beta^2 < 0, \text{ i.e., } \alpha^2 - \xi^2 < 0 \tag{11.23}$$

the solution of Eqn (11.13) is given by

$$x(t) = Ae^{-\alpha t} \sin\left(\sqrt{\xi^2 - \alpha^2}\, t + \phi\right) \tag{11.24}$$

In this case, the damping is very small and the motion is called ***underdamped***. This is the most common type of oscillatory motion and hence, we consider this in detail. Let, initially at $t = 0$, the conditions for the underdamped motion be given as

$$x = x_0 \text{ and } v = \frac{dx}{dt} = v_0 \tag{11.25}$$

Then the constants of integration A and ϕ of the Eqn (11.24) are obtained as

$$A = \sqrt{x_0^2 + \frac{(v_0 + \alpha x_0)^2}{\xi^2 - \alpha^2}} \tag{11.26}$$

$$\text{and } \phi = \tan^{-1}\left(\frac{x_0\sqrt{\xi^2 - \alpha^2}}{v_0 + \alpha x_0}\right) \tag{11.27}$$

The amplitude of oscillation will gradually decrease due to the damping effect. The maximum amplitude will occur at those points where the velocity vanishes. This condition yields

$$\frac{dx}{dt} \equiv e^{-\alpha t}\left[\sqrt{\xi^2 - \alpha^2} \cos\left(\sqrt{\xi^2 - \alpha^2}\, t + \phi\right) - \alpha \sin\left(\sqrt{\xi^2 - \alpha^2}\, t + \phi\right)\right] = 0$$

$$\text{or, } \tan\left(\sqrt{\xi^2 - \alpha^2}\, t + \phi\right) = \frac{\sqrt{\xi^2 - \alpha^2}}{\alpha} \tag{11.28}$$

Hence, in general, we have

$$\sqrt{\xi^2 - \alpha^2}\, t + \phi = \tan^{-1}\left(\frac{\sqrt{\xi^2 - \alpha^2}}{\alpha}\right) + n\pi \tag{11.29}$$

where $n = 0, 1, 2, \ldots$

From Eqn (11.29) we find the times corresponding to maximum amplitude as

$$t = \frac{1}{\sqrt{\xi^2 - \alpha^2}}\left[\tan^{-1}\left(\frac{\sqrt{\xi^2 - \alpha^2}}{\alpha}\right) + n\pi - \phi\right] \tag{11.30}$$

Eqn (11.24) describes a sinusoidal oscillation bounded by the curves $Ae^{-\alpha t}$ and $-Ae^{-\alpha t}$. This is shown in (FIG. 11.2) taking the phase angle ϕ equal to $\pi/2$.

Let t_1 and t_2 be the times for two successive maximum amplitudes. Then from Eqn (11.24) we write:

$$\frac{x_n}{x_{n+2}} = e^{-\alpha(t_1-t_2)}\frac{\sin(\omega t_1+\phi)}{\sin(\omega t_2+\phi)} \tag{11.31}$$

where

$$\omega = \sqrt{\xi^2-\alpha^2} \tag{11.32}$$

With the use of Eqn (11.30), Eqn (11.31) is written as

$$\frac{x_n}{x_{n+2}} = e^{\frac{2\pi\alpha}{\omega}}\frac{\sin\left[\tan^{-1}\left(\frac{\omega}{\alpha}\right)+n\pi\right]}{\sin\left[\tan^{-1}\left(\frac{\omega}{\alpha}\right)+(n+2)\pi\right]}$$

$$= e^{\frac{2\pi\alpha}{\omega}}\frac{\sin\left\{\tan^{-1}\left(\frac{\omega}{\alpha}\right)\right\}\cos n\pi}{\sin\left\{\tan^{-1}\left(\frac{\omega}{\alpha}\right)\right\}\cos(n+2)\pi} = e^{\frac{2\pi\alpha}{\omega}} \tag{11.33}$$

which is independent of time.

The natural logarithm of this ratio is called the ***logarithmic decrement*** of the motion and is denoted by δ. Thus,

$$\delta = \ln\left(\frac{x_n}{x_{n+2}}\right) = \frac{2\pi\alpha}{\omega} \tag{11.34}$$

Substituting the values of α and ω, we find

$$\delta = 2\pi\left(\frac{c}{2m}\right)\frac{1}{\sqrt{\left(\frac{k}{m}\right)^2-\left(\frac{c}{2m}\right)^2}}$$

$$\text{or, } \delta = \frac{2\pi c}{\sqrt{4km-c^2}}. \tag{11.35}$$

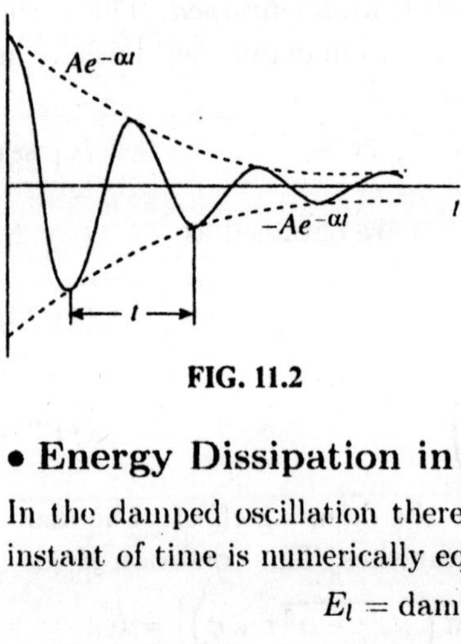

FIG. 11.2

• Energy Dissipation in Damped Oscillations

In the damped oscillation there will be a gradual loss of energy. This loss of energy at any instant of time is numerically equal to

$$E_l = \text{damping force} \times \text{velocity} = (c\dot{x})\times\dot{x} = c(\dot{x})^2$$

Now, from the equation of motion, we have

$$c\left(\frac{dx}{dt}\right)^2 = -\left[m\frac{dx}{dt}\cdot\frac{d^2x}{dt^2}+kx\frac{dx}{dt}\right] = -\left[\frac{1}{2}m\frac{d}{dt}(\dot{x})^2+\frac{k}{2}\frac{d}{dt}(x^2)\right]$$

$$= -\frac{d}{dt}(T+V) = -\frac{dE}{dt}$$

$$\text{or, } \frac{dE}{dt} = -c\left(\frac{dx}{dt}\right)^2 \tag{11.36}$$

where E is the total energy of motion.

Thus, the total energy of the particle goes on decreasing due to damping. When the damping force is proportional to velocity, the rate of decrease in energy is proportional to square of the velocity. In particular case, if $c = 0$, i.e., there is no damping, we have

$$\frac{dE}{dt} = 0, \text{ i.e., } E = T+V = \text{constant}. \tag{11.37}$$

11.3 Forced Oscillations

We have seen that in a damped system, the amplitude of oscillation goes on decreasing due to energy dissipation which causes the oscillations discontinued. Hence, in order to maintain the oscillations, an external energy has to be supplied to the system to compensate the energy loss. This can be done by providing an external force, which depends on time. The oscillations are then called ***forced oscillations***. Forced oscillations can also occur when there is no damping.

11.4 Forced Oscillation with Damping

Let us now consider the presence of damping in the forced oscillation under a sinusoidal force function. The equation of motion in this case is

$$m\frac{d^2x}{dt^2} + c\frac{dx}{dt} + kx = P_0 \sin\omega t$$

$$\left.\begin{aligned} &\text{or,} \quad \frac{d^2x}{dt^2} + 2\alpha\frac{dx}{dt} + \xi^2 x = \frac{P_0}{m}\sin\omega t \\ &\text{where as before,} \quad \alpha = \frac{c}{2m}, \ \xi^2 = \frac{k}{m} \end{aligned}\right\} \tag{11.38}$$

This is a linear non-homogeneous differential equation having complementary solution and the particular integral. The complementary solution, as we have seen, describes the damped oscillation. We are interested in a steady-state oscillation given by the particular integral of Eqn (11.38) which does not die out due to external driving force. Thus, for the study of forced vibration, we ignore the complementary solution and seek the particular integral in the form:

$$x = A\sin(\omega t - \phi) \tag{11.39}$$

Substituting this into Eqn (11.38), writing

$$\frac{P_0}{m}\sin\omega t = \frac{P_0}{m}\sin(\omega t - \phi + \phi) \tag{11.40}$$

and comparing coefficients of sine and cosine terms, we find

$$\left.\begin{aligned} -A\omega^2 + \xi^2 A &= \frac{P_0}{m}\cos\phi \\ \text{and} \quad 2\alpha A\omega &= \frac{P_0}{m}\sin\phi \end{aligned}\right\} \tag{11.41}$$

Solving these equations, we get

$$A = \frac{P_0/m}{\sqrt{(\xi^2 - \omega^2)^2 + 4\alpha^2\omega^2}} \tag{11.42}$$

$$\text{and} \quad \tan\phi = \frac{2\alpha\omega}{\xi^2 - \omega^2} \tag{11.43}$$

Thus, the solution given by Eqn (11.39) is written as

$$x = \frac{P_0/m}{\sqrt{(\xi^2 - \omega^2)^2 + 4\alpha^2\omega^2}}\sin(\omega t - \phi) \tag{11.44}$$

From this equation, we find that the amplitude A of the oscillation depends on the frequency of the applied load. Hence, for some particular value of this frequency, the amplitude will be maximum. This is called ***amplitude resonance***. Let us now determine this maximum value for which

$$\frac{dA}{d\omega} = 0 \text{ and } \frac{d^2A}{d\omega^2} < 0$$

This gives $-2(\xi^2 - \omega^2)2\omega + 8\alpha^2\omega = 0$, i.e., $\omega^2 = \xi^2 - 2\alpha^2$.

Hence, the resonant frequency is

$$\omega' = \sqrt{\xi^2 - 2\alpha^2},\ \xi^2 > 2\alpha^2 \tag{11.45}$$

With this frequency, the maximum amplitude is

$$A_{\max} = \frac{P_0/m}{\sqrt{4\alpha^4 + 4\alpha^2(\xi^2 - 2\alpha^2)}} = \frac{P_0/m}{2\alpha\sqrt{\xi^2 - \alpha^2}} \tag{11.46}$$

Thus, $A_{\max}$ depends upon damping factor α. This is shown in the FIG. 11.3. It is seen from the figure that smaller the value of the damping factor, greater is the amplitude $A_{\max}$ and steeper is the resonance curve.

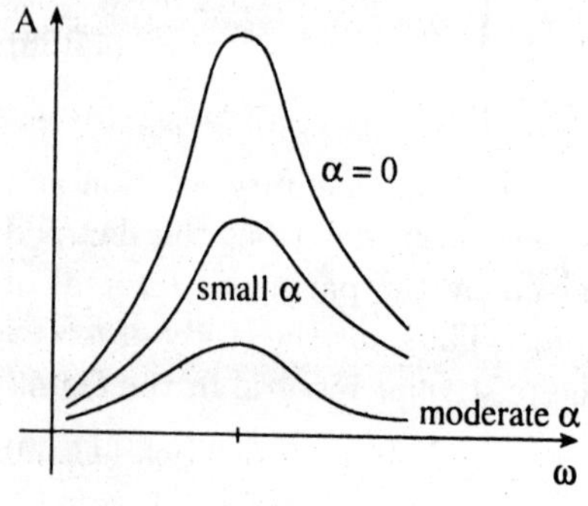

FIG. 11.3

Again, the velocity at any instant of the particle is given by

$$\frac{dx}{dt} = A\omega\cos(\omega t - \phi) \tag{11.47}$$

so that the velocity amplitude is

$$A\omega = \frac{P_0/m}{\sqrt{f(\omega)}} \tag{11.48}$$

where

$$f(\omega) = \left(\frac{\xi^2 - \omega^2}{\omega}\right)^2 + 4\alpha^2 \tag{11.49}$$

Thus, the velocity amplitude will be maximum when $f(\omega)$ is minimum. Now, $f(\omega)$ will be minimum when

$$\left(\frac{\xi^2 - \omega^2}{\omega}\right)^2 = 0, \text{ i.e., } \omega = \xi \tag{11.50}$$

Hence, the velocity amplitude is maximum when frequency of the driving force is equal to the natural frequency of the oscillator. At this moment resonance occurs and we call it ***velocity resonance***. Now, the maximum value of velocity amplitude is obtained equal to

$$(A\omega)_{\max} = \frac{P_0}{2m\alpha} = \frac{P_0}{c} \tag{11.51}$$

which shows that it also depends on the damping coefficient. In the absence of damping, i.e., when $c = 0$, the maximum value of velocity amplitude becomes infinite.

Electric current being analogous to velocity, velocity resonance has got much importance in alternating currents.

11.5 Sharpness of Resonance

The sharpness of resonance is measured in terms of a quantity Q, called the ***Q-factor*** of the system. It is defined as 2π times the ratio of the average energy stored in the system to the average energy dissipated per cycle.

Now, the energy stored in the system is

$$E = \frac{1}{2}m\dot{x}^2 + \frac{1}{2}m\xi^2 x^2 = \frac{A^2}{2}m\left[\omega^2\cos^2(\omega t - \phi) + \xi^2\sin^2(\omega t - \phi)\right] \tag{11.52}$$

Time average of this expression over a cycle is

$$\langle E\rangle = \frac{A^2}{4}m(\omega^2 + \xi^2) \tag{11.53}$$

since time average of $\cos^2(\omega t - \phi)$ and $\sin^2(\omega t - \phi)$ is $\frac{1}{2}$.

Similarly, the energy dissipated per cycle is

$$R = (\text{damping force} \times \text{velocity})\frac{2\pi}{\omega}$$

$$\text{i.e.,}\quad R = (2m\alpha\dot{x}^2)\,\frac{2\pi}{\omega} = \frac{4m\pi}{\omega}\alpha\dot{x}^2 \tag{11.54}$$

Time average of this expression per cycle is

$$\langle R\rangle = \frac{4m\pi}{\omega}\alpha\left(\frac{1}{2}A^2\omega^2\right) = 2m\pi\alpha A^2\omega \tag{11.55}$$

Hence, the Q-factor is

$$Q = \frac{2\pi\,\langle E\rangle}{\langle R\rangle} = \frac{\omega^2+\xi^2}{4\alpha\omega} \tag{11.56}$$

Resonance occurs when $\omega \simeq \xi$. In this situation, we get

$$Q = \frac{\xi}{2\alpha} \tag{11.57}$$

which shows that the Q-factor is inversely proportional to the damping factor. Thus, smaller the value of α, greater is the value of Q. But, for smaller values of α, sharpness is greater. Hence, greater the value of Q, greater is the sharpness of resonance.

EXAMPLE 11.1 *A simple pendulum of length l hangs vertically from a fixed point. The bob of mass m is given an initial horizontal velocity v_0. Prove that the arc through which the bob swings up has a length equal to*

$$l\cos^{-1}\left(1 - \frac{v_0^2}{2gl}\right)$$

Solution. The system is shown in FIG. 11.4. The height risen by the bob is

$$BC = l - l\cos\theta = l(1-\cos\theta) \tag{1}$$

From energy considerations, $\frac{1}{2}mv_0^2 = mgl(1-\cos\theta)$, which yields

$$\frac{v_0^2}{2gl} = 1-\cos\theta,\ \text{i.e.,}\ \cos\theta = 1 - \frac{v_0^2}{2gl} \tag{2}$$

The length of the arc AC is

$$l\theta = l\cos^{-1}\left(1 - \frac{v_0^2}{2gl}\right).$$

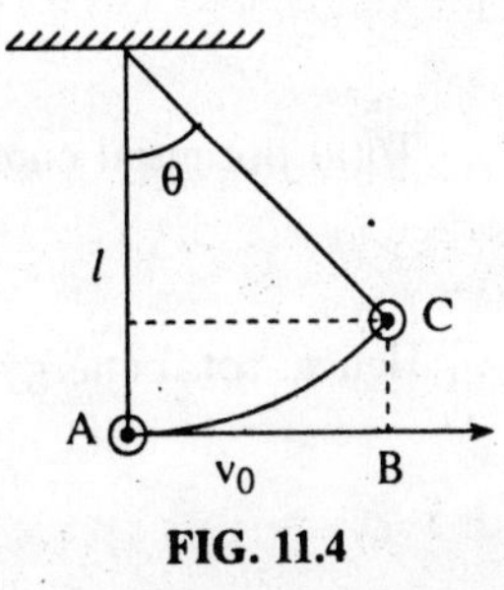

FIG. 11.4

EXAMPLE 11.2 *A right circular cylinder of mass M and radius r floats in a liquid of density ρ with its axis vertical. It is pushed slightly down and then released. Calculate the time period of vertical oscillations of the cylinder.* **[C.U. (HONS) 2007]**

Solution. Let $ABCD$ be the cylinder in equilibrium position (FIG. 11.5). At any instant of time, let z be the depression from bottom face CD. Then the restoring force by Archimedes' principle is

$$(\pi r^2) z \rho g \tag{1}$$

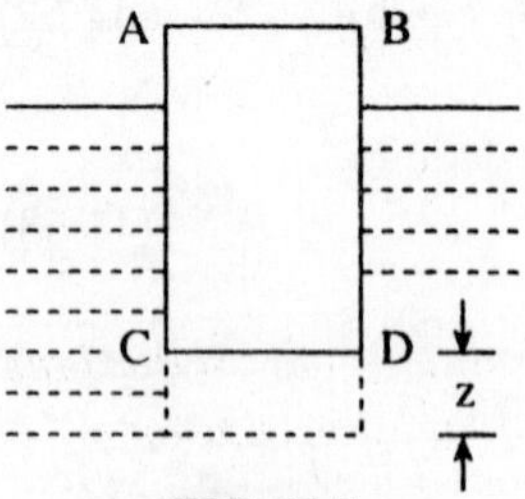

FIG. 11.5

Hence, by Newton's second law,

$$M\frac{d^2 z}{dt^2} = -\pi r^2 \rho g z \text{ or, } \frac{d^2 z}{dt^2} = -\xi^2 z \tag{2}$$

where we have denoted

$$\xi^2 = \frac{\pi r^2 \rho g}{M} \tag{3}$$

Eqn (2) represents a simple harmonic motion with period of oscillation

$$P = \frac{2\pi}{\xi} = 2\pi\sqrt{\frac{M}{\pi r^2 \rho g}}.$$

EXAMPLE 11.3 *A solid cylinder of mass M remaining on a smooth horizontal table is connected by a massless spring of force constant k, the other end of the spring being fixed with a vertical wall. The cylinder is slightly pushed away from the wall and then released. Assuming that the cylinder can roll without sliding on the table, find the period of small oscillations of the cylinder.*

Solution. Let R be the radius of the cylinder (FIG. 11.6). At any instant of time, let x be the displacement of its centre of mass. Then linear velocity of the centre of mass is

$$\frac{dx}{dt} \tag{1}$$

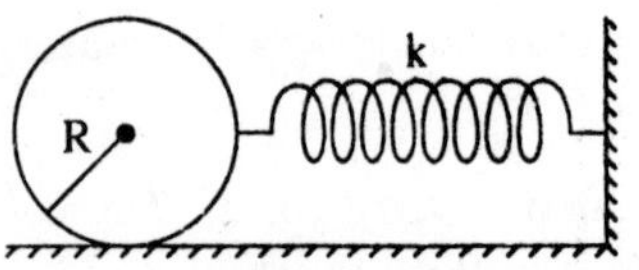

FIG. 11.6

Angular velocity of the cylinder is

$$\omega = \frac{1}{R}\frac{dx}{dt} \tag{2}$$

Hence, total kinetic energy of the system

$$\begin{aligned} T &= \frac{1}{2}M\left(\frac{dx}{dt}\right)^2 + \frac{1}{2}I\omega^2 \\ &= \frac{1}{2}M\left(\frac{dx}{dt}\right)^2 + \frac{1}{4}MR^2\left(\frac{1}{R}\frac{dx}{dt}\right)^2 = \frac{3}{4}M\left(\frac{dx}{dt}\right)^2 \end{aligned} \tag{3}$$

Total potential energy of the system

$$V = \frac{1}{2}kx^2 \tag{4}$$

Hence, total energy of the system

$$E = T + V = \frac{3}{4}M\left(\frac{dx}{dt}\right)^2 + \frac{1}{2}kx^2 \tag{5}$$

But the total energy E should be constant for a conservative system. Hence,

$$\frac{dE}{dt} \equiv \frac{3}{2}M\left(\frac{dx}{dt}\right)\frac{d^2x}{dt^2} + kx\frac{dx}{dt} = 0 \tag{6}$$

which yields

$$\frac{d^2x}{dt^2} = -\frac{2k}{3M}x \quad \left[\text{since } \frac{dx}{dt} \neq 0\right] \tag{7}$$

Eqn (7) represents a simple harmonic motion with period

$$P = 2\pi\sqrt{\frac{3M}{2k}}.$$

EXAMPLE 11.4 *A solid hemisphere of radius R and mass M is performing small oscillations on a horizontal floor rough enough to prevent sliding. Find the period of small oscillations.*

Solution. Let C be the centre and G be the centre of mass of the hemisphere (FIG. 11.7). Then $CG = \frac{3}{8}R$. Let N be the point of contact with the floor in the vertical plane through C. Since the floor is perfectly rough, the point N will be the instanteneous centre of rotation. Let $NG = a$ and $\angle NCG = \theta$. Hence, taking moments about N,

$$M\,(K^2 + a^2)\,\ddot{\theta} = -Mg\left(\frac{3R}{8}\right)\sin\theta \tag{1}$$

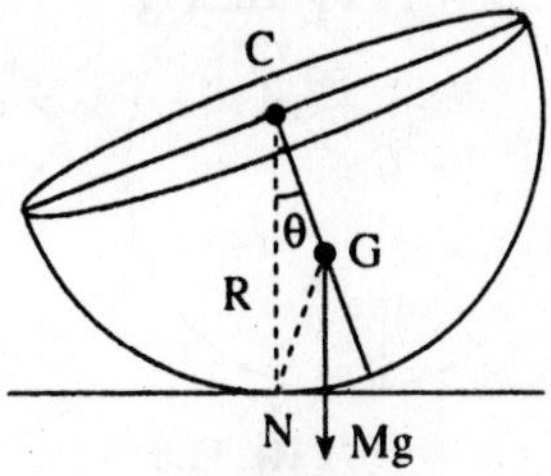

FIG. 11.7

where K is the radius of gyration about G.

Now, from the $\triangle CNG$, we have

$$a^2 = R^2 + \left(\frac{3}{8}R\right)^2 - 2R\left(\frac{3}{8}R\right)\cos\theta \tag{2}$$

For small θ, Eqns (1) and (2) assume the forms:

$$\left.\begin{aligned} M\,(K^2 + a^2)\,\ddot{\theta} &= -Mg\left(\frac{3}{8}R\right)\theta \\ \text{and } a^2 = R^2 + \frac{9}{64}R^2 - \frac{3}{4}R^2 &= \frac{R^2}{4} + \frac{9}{64}R^2 \end{aligned}\right\} \tag{3}$$

From these equations, we find

$$M\left(K^2 + \frac{9}{64}R^2 + \frac{R^2}{4}\right)\ddot{\theta} = -\frac{3MgR}{8}\theta \tag{4}$$

But

$$M\left(K^2 + \frac{9}{64}R^2\right) = \text{moment of inertia of the hemisphere about an axis through } C \text{ is}$$

$$= \frac{2}{5}MR^2$$

Hence, from Eqn (4), we get

$$\left(\frac{2}{5}R^2 + \frac{1}{4}R^2\right)\ddot{\theta} = -\frac{3gR}{8}\theta \ \text{ or, } \ddot{\theta} = -\frac{3g}{8}\left(\frac{20}{13R}\right)\theta = -\frac{15}{26}\frac{g}{R}\theta \tag{5}$$

Eqn (5) represents a simple harmonic motion with period

$$P = 2\pi\sqrt{\frac{26R}{15g}}.$$

EXAMPLE 11.5 *A uniform bar of length 2l and depth 2d performs small oscillation on a horizontal cylinder, axis of which is perpendicular to the axis of the bar. Determine the period of oscillations.*

Solution. Let A be the point on the bar in contact with the cylinder at O in equilibrium position. After a small displacement of the bar, B is the point of contact. G is the centre of mass of the bar (FIG. 11.8). GL is perpendicular drawn on the horizontal from B. If M be the mass of the bar, then, taking moment about B, we get

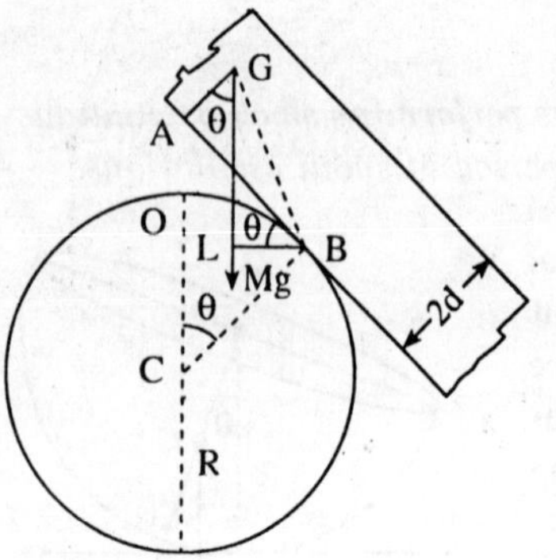

FIG. 11.8

$$M\left(K^2 + GB^2\right)\ddot{\theta} = -Mg \cdot LB \tag{1}$$

where K is the radius of gyration of the bar about the axis through G parallel to the axis of the cylinder.

Now

$$\begin{aligned} LB &= AB\cos\theta - AG\sin\theta \\ &= R\theta\cos\theta - d\sin\theta \text{ [since } AB = R\theta] \\ &= (R-d)\theta \text{ [since } \theta \text{ is small]} \end{aligned} \tag{2}$$

$$\text{and } GB^2 = GA^2 + AB^2 = d^2 + R^2\theta^2 \simeq d^2 \tag{3}$$

Then Eqn (1) becomes

$$M\left(K^2 + d^2\right)\ddot{\theta} = -Mg(R-d)\theta$$

$$\text{or, } \ddot{\theta} = -\frac{g(R-d)}{K^2+d^2}\theta \tag{4}$$

Eqn (4) represents a simple harmonic motion with period

$$P = 2\pi\sqrt{\frac{K^2+d^2}{g(R-d)}} \tag{5}$$

For the uniform bar,

$$K^2 = \frac{1}{3}\left(l^2 + d^2\right) \tag{6}$$

Substituting this value in Eqn (5), we get

$$P = 2\pi\sqrt{\frac{l^2+4d^2}{3g(R-d)}}.$$

11.6 Coupled Oscillations

Two oscillating systems in which the motion of one system gets influenced by that of the other are said to be coupled to each other. The resulting oscillations are called ***coupled oscillations***.

Consider two identical simple pendulums of length l each having a bob of mass m. The bobs are connected with a horizontal spring of force constant k but of negligible mass (FIG. 11.9).

Assuming oscillations to be small, the equations of motion of the two pendulums are written as

$$m\ddot{x}_1 = -\frac{mg}{l}x_1 + k(x_2 - x_1) \tag{11.58}$$

$$\text{and } m\ddot{x}_2 = -\frac{mg}{l}x_2 - k(x_2 - x_1) \tag{11.59}$$

The first terms in the RHS of these equations represent the restoring force due to mass of the bob and the second terms represent the restoring force due to stretching of the spring by an amount $(x_2 - x_1)$.

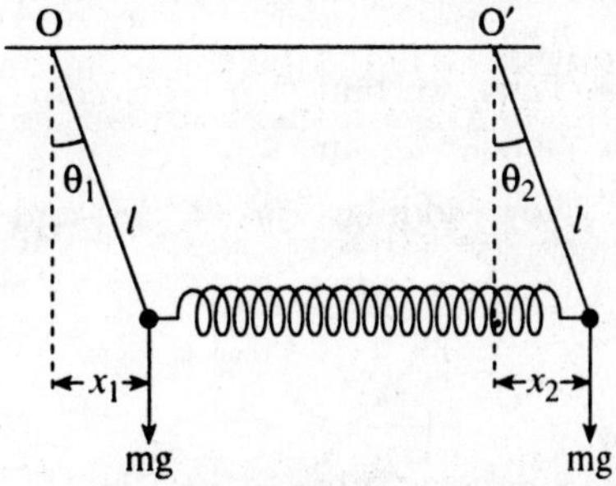

FIG. 11.9

We assume the solutions of Eqns (11.58) and (11.59) in the form:

$$x_1 = A_1 e^{i\omega t} \text{ and } x_2 = A_2 e^{i\omega t} \tag{11.60}$$

Substituting these values of x_1 and x_2 in Eqns (11.58) and (11.59), we get

$$\left.\begin{aligned} -m\omega^2 A_1 + \frac{mg}{l}A_1 - k(A_2 - A_1) = 0 \\ \text{and } -m\omega^2 A_2 + \frac{mg}{l}A_2 + k(A_2 - A_1) = 0 \end{aligned}\right\} \tag{11.61}$$

These equations can be written in the form:

$$\left.\begin{aligned} \left(\omega^2 - \frac{g}{l} - \frac{k}{m}\right) A_1 + \frac{k}{m}A_2 = 0 \\ \text{and } \frac{k}{m}A_1 + \left(\omega^2 - \frac{g}{l} - \frac{k}{m}\right) A_2 = 0 \end{aligned}\right\} \tag{11.62}$$

For non-trivial solutions of these two simultaneous equations in A_1 and A_2, we must have

$$\begin{vmatrix} \omega^2 - \frac{g}{l} - \frac{k}{m} & \frac{k}{m} \\ \frac{k}{m} & \omega^2 - \frac{g}{l} - \frac{k}{m} \end{vmatrix} = 0 \tag{11.63}$$

This equation yields

$$\omega^2 - \frac{g}{l} - \frac{k}{m} = \pm\frac{k}{m} \tag{11.64}$$

If ω_1 and ω_2 be two values of ω, then from Eqn (11.64), we get

$$\omega_1 = \pm\sqrt{\frac{g}{l}} \text{ and } \omega_2 = \pm\sqrt{\frac{g}{l} + \frac{2k}{m}} \tag{11.65}$$

Hence, the general solution of Eqns (11.58) and (11.59) are:

$$\left.\begin{aligned} x_1 = A_1 e^{i\omega_1 t} + A_1' e^{-i\omega_1 t} + B_1 e^{i\omega_2 t} + B_1' e^{-i\omega_2 t} \\ x_2 = A_2 e^{i\omega_1 t} + A_2' e^{-i\omega_1 t} + B_2 e^{i\omega_2 t} + B_2' e^{-i\omega_2 t} \end{aligned}\right\} \tag{11.66}$$

where the constants of integration are arbitrary but they are not all independent.

From Eqn (11.62), we find that

$$\left.\begin{aligned} \text{for } \omega = \pm\omega_1 = \pm\sqrt{\frac{g}{l}},\ A_1 = A_2 \\ \text{and, for } \omega = \pm\omega_2 = \pm\sqrt{\frac{g}{l} + \frac{2k}{m}},\ A_1 = -A_2 \end{aligned}\right\} \tag{11.67}$$

Hence, the general solution is

$$\left.\begin{aligned} x_1 = A_1 e^{i\omega_1 t} + A_1' e^{-i\omega_1 t} + B_1 e^{i\omega_2 t} + B_1' e^{-i\omega_2 t} \\ x_2 = A_1 e^{i\omega_1 t} + A_1' e^{-i\omega_1 t} - B_1 e^{i\omega_2 t} - B_1' e^{-i\omega_2 t} \end{aligned}\right\} \tag{11.68}$$

These equations can be written as

$$\left.\begin{aligned} x_1 &= A\cos(\omega_1 t + \phi_1) + B\cos(\omega_2 t + \phi_2) \\ x_2 &= A\cos(\omega_1 t + \phi_1) - B\cos(\omega_2 t + \phi_2) \end{aligned}\right\} \tag{11.69}$$

Thus, we find that each coordinate is a superposition of two simple harmonic motions of frequencies ω_1 and ω_2.

Now, adding Eqns (11.58) and (11.59), we get

$$m\,(\ddot{x}_1 + \ddot{x}_2) = -\frac{mg}{l}(x_1 + x_2)$$

$$\text{or,}\quad \frac{d^2}{dt^2}(x_1 + x_2) + \frac{g}{l}(x_1 + x_2) = 0 \tag{11.70}$$

Similarly, subtracting Eqn (11.59) from Eqn (11.58), we obtain

$$\frac{d^2}{dt^2}(x_1 - x_2) + \left(\frac{g}{l} + \frac{2k}{m}\right)(x_1 - x_2) = 0 \tag{11.71}$$

We define new variables

$$\xi = x_1 + x_2 \text{ and } \eta = x_1 - x_2 \tag{11.72}$$

Then Eqns (11.70) and (11.71) assume the forms:

$$\frac{d^2\xi}{dt^2} + \omega_1^2\xi = 0 \text{ and } \frac{d^2\eta}{dt^2} + \omega_2^2\eta = 0 \tag{11.73}$$

Each of these equations describes a simple harmonic motion of single frequency ω_1 or ω_2. Thus, the coupled system of differential equations of motion has been decoupled. The solutions of Eqn (11.73) are

$$\xi = A\cos(\omega_1 t + \phi_1) \text{ and } \eta = B\cos(\omega_2 t + \phi_2) \tag{11.74}$$

These two simple harmonic motions involving only one frequency ω_1 or ω_2 are called ***normal modes*** and the frequency of vibrations of each mode is the ***normal frequency***. The coordinates ξ or η which describes the normal mode of oscillations are called ***normal coordinates***.

Let us now find the kinetic and potential energies of the coupled oscillations under discussion. The kinetic energy of the system is given by

$$T = \frac{1}{2}m\dot{x}_1^2 + \frac{1}{2}m\dot{x}_2^2 \tag{11.75}$$

The potential energy of the pendulum bobs is

$$\int_0^{x_1} \frac{mgx}{l}dx + \int_0^{x_2} \frac{mgx}{l}dx = \frac{mg}{2l}\left(x_1^2 + x_2^2\right) \tag{11.76}$$

Potential energy due to the extension of spring by an amount $(x_2 - x_1)$ is

$$\frac{1}{2}k(x_2 - x_1)^2 \tag{11.77}$$

Hence, the total potential energy is

$$V = \frac{mg}{2l}(x_1^2 + x_2^2) + \frac{k}{2}(x_2 - x_1)^2 \tag{11.78}$$

In terms of normal coordinates ξ and η, the kinetic energy is

$$T = \frac{1}{2}m\frac{1}{4}\left(\dot{\xi} + \dot{\eta}\right)^2 + \frac{1}{2}m\frac{1}{4}\left(\dot{\xi} - \dot{\eta}\right)^2 = \frac{1}{4}m\left(\dot{\xi}^2 + \dot{\eta}^2\right) \tag{11.79}$$

The total potential energy in terms of normal coordinates is

$$V = \frac{mg}{2l}\left[\frac{1}{4}(\xi+\eta)^2 + \frac{1}{4}(\xi-\eta)^2\right] + \frac{k}{2}\eta^2$$
$$= \frac{mg}{4l}\xi^2 + \left(\frac{mg}{4l} + \frac{k}{2}\right)\eta^2 \qquad (11.80)$$

Thus, both the kinetic and potential energies in terms of normal coordinates are homogeneous quadratic functions.

The normal modes of oscillations may be symmetric and antisymmetric.

Suppose that the normal coordinate η is initially zero, i.e., $x_1 = x_2$. Then there is no oscillation of frequency ω_2 described by the second of differential Eqns (11.73). There is normal modes of oscillation of frequency $\omega_1 = \sqrt{g/l}$ only. The value of ω_1 does not depend on the spring constant k so that the spring remains absolutely free during the process of oscillation. This normal mode of oscillation is called ***symmetric mode***. All the particles in motion remain in same phase (FIG. 11.10).

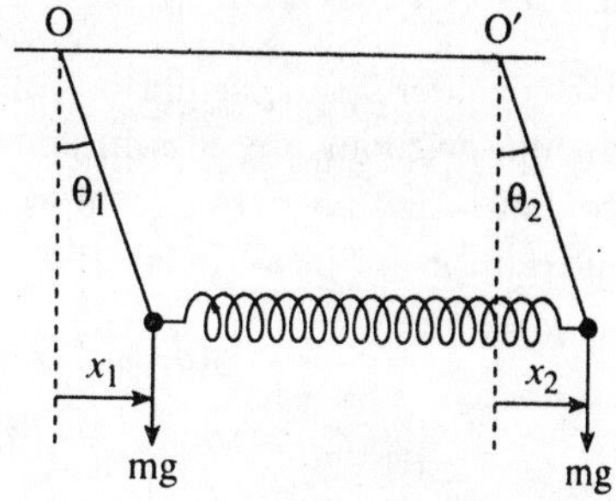

FIG. 11.10

Now suppose that the normal coordinate ξ is initially zero, i.e., $x_1 = -x_2$. In this case, there will be no oscillation of frequency ω_1, oscillations of frequency ω_2 alone will exist. Moreover, the pendulums oscillate with a phase difference of π. Since ω_2 by Eqn (11.67) depends on the spring constant k, the spring remains active in exerting a restoring force during oscillations. These oscillations are called ***antisymmetric mode*** of oscillations (FIG. 11.11).

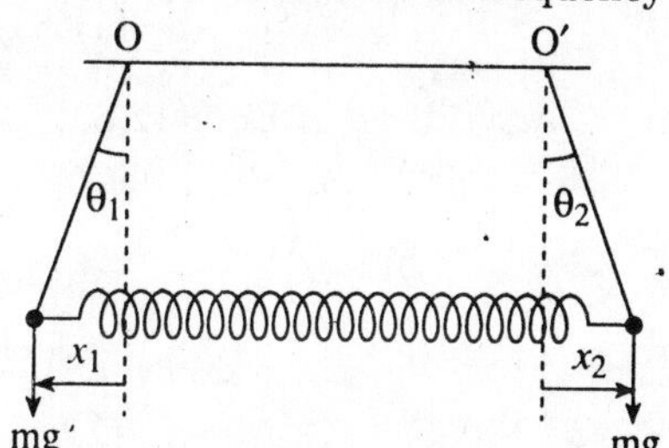

FIG. 11.11

We also note that $\omega_1 < \omega_2$. That is, the frequency of oscillations in symmetric mode is lower than that of oscillations in antisymmetric mode.

11.7 Stable and Unstable Equilibrium

We consider a conservative system having many degrees of freedom and assume that the generalised coordinates q_k do not depend on time explicitly. Then the potential energy of the system is given by

$$V = V(q_1, q_2, \ldots, q_n) \qquad (11.81)$$

and the generalised forces are given by

$$Q_k = -\frac{\partial V}{\partial q_k} \qquad (11.82)$$

The system is said to be in equilibrium if the generalised forces Q_k acting on it vanish. Thus, for the system to be in equilibrium,

$$Q_k = -\left(\frac{\partial V}{\partial q_k}\right)_0 = 0 \qquad (11.83)$$

where by the suffix zero we have indicated the equilibrium configuration.

Eqn (11.83) shows that the potential energy has an extremum value at the equilibrium configuration of the system which we shall represent by the generalised coordinates $q_{01}, q_{02}, \ldots, q_{0n}$.

The equilibrium is said to be ***stable***, if a small displacement from the equilibrium position results in a small bounded motion about that position. If, on the other hand, the system performs an unbounded motion when disturbed slightly from it, it is in an ***unstable equilibrium***. The pendulum in its rest position is an example of a stable equilibrium, while an egg standing on its tips is in unstable equilibrium.

11.8 General Theory of Small Oscillations

We consider small oscillations of a system about its position of stable equilibrium. Moreover, we assume without any loss in generality, that the generalised coordinates q_{0k} at the equilibrium configuration are zero. Then we can expand the potential energy function V about the position of equilibrium in a Taylor's series in the form:

$$V(q_1, q_2, \ldots, q_n) = V(q_{01}, q_{02}, \ldots, q_{0n}) + \sum_i \left(\frac{\partial V}{\partial q_i}\right)_0 q_i + \frac{1}{2}\sum_i \sum_j \left(\frac{\partial^2 V}{\partial q_i \partial q_j}\right)_0 q_i q_j \tag{11.84}$$

We assume that the potential energy in the equilibrium position is zero. That is,

$$V(q_{01}, q_{02}, \ldots, q_{0n}) = 0 \tag{11.85}$$

Moreover, the second term in the RHS of Eqn (11.84) vanishes by virtue of Eqn (11.83).

Thus, we are left with

$$V = \frac{1}{2}\sum_i \sum_j \left(\frac{\partial^2 V}{\partial q_i \partial q_j}\right)_0 q_i q_j = \frac{1}{2}\sum_i \sum_j V_{ij} q_i q_j \tag{11.86}$$

where

$$V_{ij} = \left(\frac{\partial^2 V}{\partial q_i \partial q_j}\right)_0 \tag{11.87}$$

is a constant for $q_i = q_{0i}$ and $q_j = q_{0j}$.

It is obvious that these constants are symmetric. That is

$$V_{ij} = V_{ji} \tag{11.88}$$

The kinetic energy of the system is given by

$$T = \frac{1}{2}\sum_i \sum_j m_{ij} \dot{q}_i \dot{q}_j \tag{11.89}$$

where the coefficients m_{ij} are, in general, functions of generalised coordinates q_k. These coefficients can also be expanded in Taylor's series, so that

$$m_{ij}(q_1, q_2, \ldots, q_n) = m_{ij}(q_{01}, \ldots, q_{0n}) + \sum_k \left(\frac{\partial m_{ij}}{\partial q_k}\right)_0 q_k + \cdots \tag{11.90}$$

Now, Eqn (11.89) is already quadratic in $\dot{q}_i$'s. Hence, the lowest non-vanishing approximation to T is obtained by considering only the first term in the expansion (11.90). We denote it by T_{ij}. Then the kinetic energy of the system is

$$T = \frac{1}{2}\sum_i\sum_j T_{ij}\dot{q}_i\dot{q}_j \tag{11.91}$$

Obviously, the constants T_{ij} are elements of a symmetric matrix. That is,

$$T_{ij} = T_{ji} \tag{11.92}$$

Lagrangian L of the system can now be written as

$$L = T - V = \frac{1}{2}\sum_i\sum_j \left(T_{ij}\dot{q}_i\dot{q}_j - V_{ij}q_iq_j\right) \tag{11.93}$$

Hence, considering q_i's as the generalised coordinates, the Lagrange's equations

$$\frac{d}{dt}\left(\frac{\partial L}{\partial \dot{q}_i}\right) - \frac{\partial L}{\partial q_i} = 0$$

in this case yield

$$\sum_{j=1}^{n}\left(T_{ij}\ddot{q}_j + V_{ij}q_j\right) = 0 \tag{11.94}$$

These are n equations for $i = 1, 2, \ldots, n$. We can describe the motion of the system near its equilibrium position if we can solve these equations.

• Eigenfrequencies and Eigenvectors

Each of the Eqns (11.94) is a second order differential equation so that there will be two constants of integration. Hence, we assume for the solution of oscillatory motion

$$q_j = a_j \exp[i(\omega t + \phi_j)] \tag{11.95}$$

where a_j are real amplitudes of oscillations corresponding to coordinates q_j, and ϕ_j represent the phase. Actual motion will be represented by the real part of the solution in Eqn (11.95).

Substituting this solution function into Eqn (11.94), we get

$$\sum_{j=1}^{n}\left(V_{ij} - \omega^2 T_{ij}\right)a_j = 0 \tag{11.96}$$

This is a set of n linear homogeneous algebraic equations in ω and a_j which in expanded form is

$$\left.\begin{array}{l}\left(V_{11} - \omega^2 T_{11}\right)a_1 + \left(V_{12} - \omega^2 T_{12}\right)a_2 + \cdots + \left(V_{1n} - \omega^2 T_{1n}\right)a_n = 0\\ \quad\cdots\qquad\qquad\quad\cdots\qquad\qquad\cdots\qquad\qquad\cdots\\ \quad\cdots\qquad\qquad\quad\cdots\qquad\qquad\cdots\qquad\qquad\cdots\\ \left(V_{n1} - \omega^2 T_{n1}\right)a_1 + \left(V_{n2} - \omega^2 T_{n2}\right)a_2 + \cdots + \left(V_{nn} - \omega^2 T_{nn}\right)a_n = 0\end{array}\right\} \tag{11.97}$$

This set of equations can have a non-trivial solution only when the determinant of the coefficients of a_j vanishes.

Thus,

$$\left|V_{ij} - \omega^2 T_{ij}\right| = 0 \tag{11.98}$$

or,
$$\begin{vmatrix} V_{11} - \omega^2 T_{11} & V_{12} - \omega^2 T_{12} & \cdots & V_{1n} - \omega^2 T_{1n} \\ V_{12} - \omega^2 T_{21} & V_{22} - \omega^2 T_{22} & \cdots & V_{2n} - \omega^2 T_{2n} \\ \cdots & \cdots & & \cdots \\ \cdots & \cdots & & \cdots \\ V_{n1} - \omega^2 T_{n1} & V_{n2} - \omega^2 T_{n2} & \cdots & V_{nn} - \omega^2 T_{nn} \end{vmatrix} = 0 \tag{11.99}$$

Eqn (11.99) is called the ***characteristic*** or ***secular equation*** of the system. It is an equation of degree n in ω^2 and its solution will provide the roots ω_k^2 $(k = 1, 2, \ldots, n)$. These values are called ***eigenfrequencies*** of the system.

The motion is oscillatory about the position of stable equilibrium only for $\omega_k^2 > 0$. If two or more of the values of ω_k are equal, the phenomenon is called ***degeneracy***.

For each value of ω_k^2, we can solve the system of Eqns (11.97) to determine the $(n-1)$ values of amplitudes a_j in terms of the remaining values. We denote it by a_{ik} $(i = 1, 2, \ldots, n)$. The symbol a_{ik} represents the amplitude of the i-th coordinate in the k-th mode of vibration.

Thus, corresponding to one value of ω_k, we get n values of a_j. Since there are n values of ω_k, we get n such sets. Each set can be defined to be the component of an n-dimensional vector. This vector is called the ***eigenvector*** of the system.

Thus, the general solution for q_j is written as

$$q_j(t) = \sum_k a_{jk} \exp[i(\omega_k t + \phi_k)] \tag{11.100}$$

Taking the real part, we get the general solution by the principle of superposition as

$$q_j(t) = \sum_k a_{jk} \cos(\omega_k t + \phi_k) \tag{11.101}$$

11.9 Orthogonality of Eigenvectors

Let us consider the k-th and m-th mode of oscillations. Then, from Eqn (11.96), we can write:

$$\sum_j (V_{ij} - \omega_k^2 T_{ij}) a_{jk} = 0 \tag{11.102}$$

$$\text{and } \sum_i \left(V_{ij} - \omega_m^2 T_{ij}\right) a_{im} = 0 \tag{11.103}$$

Multiplying Eqn (11.102) by a_{im} and summing over i, we get

$$\sum_i \sum_j V_{ij} a_{jk} a_{im} = \sum_i \sum_j \omega_k^2 T_{ij} a_{jk} a_{im} \tag{11.104}$$

Similarly, multiplying Eqn (11.103) by a_{jk} and summing over j, we get

$$\sum_i \sum_j V_{ij} a_{im} a_{jk} = \sum_i \sum_j \omega_m^2 T_{ij} a_{im} a_{jk} \tag{11.105}$$

Subtracting these equations, we find

$$\left(\omega_k^2 - \omega_m^2\right) \sum_i \sum_j T_{ij} a_{im} a_{jk} = 0 \tag{11.106}$$

Since $\omega_k \neq \omega_m$, we must have

$$\sum_i \sum_j T_{ij} a_{im} a_{jk} = 0 \tag{11.107}$$

This shows that the eigenvectors are ***orthogonal***.

Again, as we have seen earlier, the coefficients a_{jk} in Eqn (11.102) cannot be completely determined. This indeterminacy is removed by accepting an additional condition like

$$\sum_i \sum_j T_{ij} a_{ik} a_{jk} = 1 \tag{11.108}$$

Using the symbol of Kronecker delta, Eqns (11.107) and (11.108) can be written as

$$\sum_i \sum_j T_{ij} a_{im} a_{jk} = \delta_{km} \tag{11.109}$$

11.10 Matrix Notations

We denote the matrices

$$[V] = \begin{bmatrix} V_{11} & V_{12} & \cdots & V_{1n} \\ V_{21} & V_{22} & \cdots & V_{2n} \\ \cdots & \cdots & \cdots & \cdots \\ V_{n1} & V_{n2} & \cdots & V_{nn} \end{bmatrix} \tag{11.110}$$

$$[T] = \begin{bmatrix} T_{11} & T_{12} & \cdots & T_{1n} \\ T_{21} & T_{22} & \cdots & T_{2n} \\ \cdots & \cdots & \cdots & \cdots \\ T_{n1} & T_{n2} & \cdots & T_{nn} \end{bmatrix} \tag{11.111}$$

$$\text{and } [A] = \begin{bmatrix} a_{11} & a_{12} & \cdots & a_{1n} \\ a_{21} & a_{22} & \cdots & a_{2n} \\ \cdots & \cdots & \cdots & \cdots \\ a_{n1} & a_{n2} & \cdots & a_{nn} \end{bmatrix} \tag{11.112}$$

We recall that for k-th mode of oscillation, the eigenvector a_k will have n components $a_{1k}, a_{2k}, \ldots$ so that there will be $n \times n$ eigenvector components. Then Eqn (11.109) can be expressed as

$$[A]^T \; [T] \; [A] = I \tag{11.113}$$

where I represents the identity matrix. We can express the potential and kinetic energy in matrix notation in the following way:

The potential energy

$$V = \frac{1}{2} \sum_i \sum_j V_{ij} q_i q_j = \frac{1}{2} \sum_i \sum_j q_i V_{ij} q_j = \frac{1}{2} \{q\}^T [V] \{q\} \tag{11.114}$$

where $\{q\}$ is a single-column matrix defined by

$$\{q\} = \begin{bmatrix} q_1 \\ q_2 \\ \vdots \\ q_n \end{bmatrix} \tag{11.115}$$

Similarly, the kinetic energy of the system is

$$T = \frac{1}{2}\sum_i\sum_j T_{ij}\dot{q}_i\dot{q}_j = \frac{1}{2}\{\dot{q}\}^T[T]\{\dot{q}\} \tag{11.116}$$

where

$$\{\dot{q}\} = \frac{d}{dt}\{q\} \tag{11.117}$$

11.11 Normal Coordinates

Let the generalised coordinates q_j be expressed as

$$q_j = \sum_k a_{jk}Q_k \tag{11.118}$$

We intend to show that Q_k are the normal coordinates.

In matrix notation, we write Eqn (11.118) as

$$\{q\} = [A]\ \{Q\} \tag{11.119}$$

where

$$\{q\} = \begin{bmatrix} q_1 \\ q_2 \\ \vdots \\ q_n \end{bmatrix} \text{ and } \{Q\} = \begin{bmatrix} Q_1 \\ Q_2 \\ \vdots \\ Q_n \end{bmatrix} \tag{11.120}$$

are single-column matrices.

Then

$$\{q\}^T = \{Q\}^T\ [A]^T \tag{11.121}$$

Hence, the potential energy V from Eqn (11.114) can be expressed as

$$V = \frac{1}{2}\{Q\}^T[A]^T\ [V]\ [A]\ \{Q\} \tag{11.122}$$

For $\omega^2 = \omega_k^2$, Eqn (11.102) can be written in matrix notation in the form:

$$[V]\ [A] = [T]\ [A]\ [\omega^2] \tag{11.123}$$

where

$$[\omega^2] = \begin{bmatrix} \omega_1^2 & 0 & \cdots & 0 \\ 0 & \omega_2^2 & \cdots & 0 \\ \cdots & \cdots & \cdots & \cdots \\ 0 & 0 & \cdots & \omega_n^2 \end{bmatrix} \tag{11.124}$$

From Eqns (11.122) and (11.123), we get

$$\begin{aligned} V &= \frac{1}{2}\{Q\}^T\ [A]^T\ [T]\ [A]\ [\omega^2]\ \{Q\} \\ &= \frac{1}{2}\{Q\}^T\ [\omega^2]\ \{Q\} \text{ [by using Eqn (11.113)]} \\ &= \frac{1}{2}\sum_j \omega_j^2 Q_j^2 \end{aligned} \tag{11.125}$$

Similarly, the kinetic energy of the system is

$$T = \frac{1}{2}\sum_i\sum_j T_{ij}\,\dot{q}_i\,\dot{q}_j = \frac{1}{2}\{\dot{q}\}^T\,[T]\,\{\dot{q}\}$$

$$= \frac{1}{2}\{\dot{Q}\}^T[A]^T\,[T]\,[A]\,\{\dot{Q}\},\ \text{[by using Eqn (11.121)]}$$

$$= \frac{1}{2}\{\dot{Q}\}^T\,I\{\dot{Q}\},\ \text{[by using Eqn (11.113)]}$$

$$= \frac{1}{2}\sum_j \dot{Q}_j^2 \tag{11.126}$$

Hence, the Lagrangian is

$$L = T - V = \frac{1}{2}\sum \dot{Q}_j^2 - \frac{1}{2}\sum \omega_j^2 Q_j^2 \tag{11.127}$$

Lagrange's equations of motion are

$$\frac{d}{dt}\left(\frac{\partial L}{\partial \dot{Q}_j}\right) - \frac{\partial L}{\partial Q_j} = 0 \text{ or, } \ddot{Q}_j + \omega_j^2 Q_j = 0,\ (j = 1, 2, \ldots, n) \tag{11.128}$$

Eqn (11.128) shows that each of the coordinates Q_j executes a simple harmonic motion with a single frequency ω_j. Hence, the coordinates Q_j $(j = 1, 2, \ldots, n)$ are ***normal coordinates*** of the system. The corresponding frequencies ω_j $(j = 1, 2, \ldots, n)$ are the ***normal frequencies***. The component oscillation corresponding to the particular frequency is the normal mode of vibration.

EXAMPLE 11.6 *A double pendulum consists of a mass m_1 and length l_1 to which a second pendulum of mass m_2 and length l_2 is suspended. Discuss its small oscillations in a vertical plane.*

Solution. The system has two degrees of freedom. Let (x_1, y_1) and (x_2, y_2) be the coordinates of the masses m_1 and m_2 respectively at an instant of time (FIG. 11.12).

Then we have

$$\left.\begin{aligned} x_1 &= l_1\sin\theta,\ x_2 = l_1\sin\theta + l_2\sin\phi \\ y_1 &= l_1\cos\theta,\ y_2 = i_1\cos\theta + l_2\cos\phi \end{aligned}\right\} \tag{1}$$

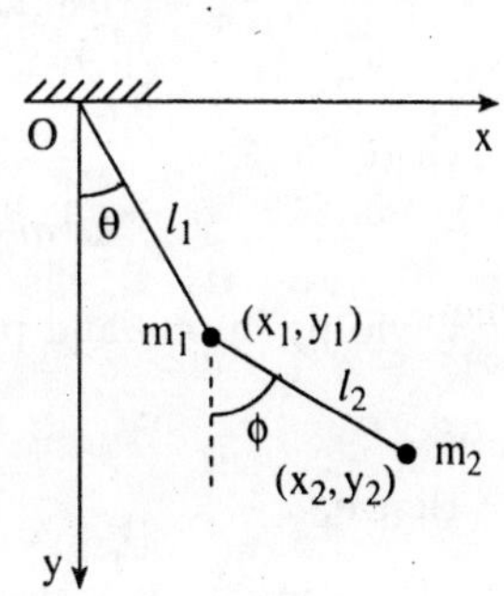

FIG. 11.12

where θ and ϕ are the angles made by the pendulums with the vertical. These are taken as the generalised coordinates of the system.

Taking the potential energy at the level Ox to be zero, we get the total potential energy of the system as

$$V = -m_1 g y_1 - m_2 g y_2$$

$$= -m_1 g l_1 \cos\theta - m_2 g(l_1\cos\theta + l_2\cos\phi) \tag{2}$$

For small values of θ and ϕ,

$$\cos\theta = 1 - \frac{1}{2}\theta^2;\ \cos\phi = 1 - \frac{1}{2}\phi^2 \tag{3}$$

Then the potential energy becomes

$$V = -m_1 g l_1 - m_2 g(l_1 + l_2) + \frac{1}{2}m_1 g l_1\theta^2 + \frac{1}{2}m_2 g\left(l_1\theta^2 + l_2\phi^2\right) \tag{4}$$

Hence, by Eqn (11.87),

$$\left.\begin{aligned} V_{11} &= \left(\frac{\partial^2 V}{\partial \theta^2}\right)_0 = (m_1 + m_2)gl_1 \\ V_{21} = V_{12} &= \left(\frac{\partial^2 V}{\partial \theta \partial \phi}\right)_0 = 0 \\ V_{22} &= \left(\frac{\partial^2 V}{\partial \phi^2}\right)_0 = m_2 g l_2 \end{aligned}\right\} \tag{5}$$

The kinetic energy of the system is

$$T = \frac{1}{2}\left(\dot{x}_1^2 + \dot{y}_1^2\right) + \frac{1}{2}m_2\left(\dot{x}_2^2 + \dot{y}_2^2\right)$$

which, by using Eqn (1), assumes the form:

$$T = \frac{1}{2}m_1 l_1^2 \dot{\theta}^2 + \frac{1}{2}m_2\left[l_1^2\dot{\theta}^2 + l_2^2\dot{\phi}^2 + 2l_1 l_2 \dot{\theta}\dot{\phi}\cos(\theta - \phi)\right]$$

Since θ and ϕ are small, $\cos(\theta - \phi) \simeq 1$. Hence,

$$T = \frac{1}{2}(m_1 + m_2)l_1^2\dot{\theta}^2 + m_2 l_1 l_2 \dot{\theta}\dot{\phi} + \frac{1}{2}m_2 l_2^2 \dot{\phi}^2 \tag{6}$$

Comparing this expression with Eqn (11.116), we find

$$\left.\begin{aligned} T_{11} &= (m_1 + m_2)l_1^2 \\ T_{12} = T_{21} &= m_2 l_1 l_2 \\ T_{22} &= m_2 l_2^2 \end{aligned}\right\} \tag{7}$$

The eigenfrequencies are given by Eqn (11.99) which, in this case, takes the form:

$$\begin{vmatrix} V_{11} - \omega^2 T_{11} & V_{12} - \omega^2 T_{12} \\ V_{21} - \omega^2 T_{21} & V_{22} - \omega^2 T_{22} \end{vmatrix} = 0 \tag{8}$$

Using Eqns (5) and (7), we get from Eqn (8)

$$\begin{vmatrix} (m_1 + m_2)gl_1 - \omega^2(m_1 + m_2)l_1^2 & -\omega^2 m_2 l_1 l_2 \\ -\omega^2 m_2 l_1 l_2 & m_2 g l_2 - \omega^2 m_2 l_2^2 \end{vmatrix} = 0$$

which reduces to

$$\omega^4 m_2 l_1 l_2 - \omega^2(m_1 + m_2)(l_1 + l_2)g + (m_1 + m_2)g^2 = 0 \tag{9}$$

Dividing by m_2 and putting $\lambda = m_1/m_2$, we get

$$\omega^4 l_1 l_2 - \omega^2(1 + \lambda)(l_1 + l_2)g + (1 + \lambda)g^2 = 0 \tag{10}$$

Hence,

$$\omega^2 = \frac{1}{2l_1 l_2}\left[(1 + \lambda)(l_1 + l_2)g \pm \sqrt{g^2(1 + \lambda)^2(l_1 + l_2)^2 - 4g^2(1 + \lambda)l_1 l_2}\right] \tag{11}$$

Eqn (11) shows that the values of ω^2 are real.

• Particular Case

Let $l_1 = l_2 = l$ (say) and $m_1 = m_2$ so that $\lambda = 1$. Then we get

$$\omega^2 = \frac{1}{2l^2}\left[4gl \pm \sqrt{16g^2l^2 - 8g^2l^2}\right] = \frac{g}{l}\left(2 \pm \sqrt{2}\right)$$

Thus,

$$\omega_1^2 = \frac{g}{l}\left(2+\sqrt{2}\right),\ \omega_2^2 = \frac{g}{l}\left(2-\sqrt{2}\right) \tag{12}$$

The eigenvectors are determined by Eqn (11.96) which, in this case, is

$$\begin{bmatrix} 2g-2\omega^2 l & -\omega^2 l \\ -\omega^2 l & g-\omega^2 l \end{bmatrix} \begin{bmatrix} a_{1k} \\ a_{2k} \end{bmatrix} = 0 \tag{13}$$

For $\omega^2 = \omega_1^2 = \frac{g}{l}\left(2+\sqrt{2}\right)$, we get from Eqn (13)

$$\begin{bmatrix} -2g-2\sqrt{2}g & -(2+\sqrt{2})g \\ -(2+\sqrt{2})g & -g-\sqrt{2}g \end{bmatrix} \begin{bmatrix} a_{11} \\ a_{21} \end{bmatrix} = 0$$

This yields

$$\frac{a_{21}}{a_{11}} = -\frac{2+\sqrt{2}}{1+\sqrt{2}}$$

We can take

$$a_{11} = \left(1+\sqrt{2}\right)\lambda \text{ and } a_{21} = -\left(2+\sqrt{2}\right)\lambda \tag{14}$$

Similarly, for $\omega^2 = \omega_2^2 = \frac{g}{l}\left(2-\sqrt{2}\right)$, we have

$$\begin{bmatrix} -2g+2\sqrt{2}g & -(2-\sqrt{2})g \\ -(2-\sqrt{2})g & -g+\sqrt{2}g \end{bmatrix} \begin{bmatrix} a_{12} \\ a_{22} \end{bmatrix} = 0$$

This yields

$$\frac{a_{22}}{a_{12}} = -\frac{2-\sqrt{2}}{1-\sqrt{2}}$$

So we can take

$$a_{12} = \left(1-\sqrt{2}\right)\beta \text{ and } a_{22} = -\left(2-\sqrt{2}\right)\beta \tag{15}$$

Then the matrix $[A]$ is

$$[A] = \begin{bmatrix} a_{11} & a_{12} \\ a_{21} & a_{22} \end{bmatrix} = \begin{bmatrix} (1+\sqrt{2})\,\lambda & (1-\sqrt{2})\,\beta \\ -(2+\sqrt{2})\,\lambda & -(2-\sqrt{2})\,\beta \end{bmatrix} \tag{16}$$

Hence, its transpose is

$$[A]^T = \begin{bmatrix} (1+\sqrt{2})\,\lambda & -(2+\sqrt{2})\,\lambda \\ (1-\sqrt{2})\,\beta & -(2-\sqrt{2})\,\beta \end{bmatrix} \tag{17}$$

For $m_1 = m_2 = m$ and $l_1 = l_2 = l$, the $[T]$ matrix is

$$[T] = \begin{bmatrix} 2ml^2 & ml^2 \\ ml^2 & ml^2 \end{bmatrix} = ml^2 \begin{bmatrix} 2 & 1 \\ 1 & 2 \end{bmatrix} \tag{18}$$

We now use the condition in Eqn (11.113) and get

$$ml^2 \begin{bmatrix} (1+\sqrt{2})\,\lambda & -(2+\sqrt{2})\,\lambda \\ (1-\sqrt{2})\,\beta & -(2-\sqrt{2})\,\beta \end{bmatrix} \begin{bmatrix} 2 & 1 \\ 1 & 1 \end{bmatrix} \begin{bmatrix} (1+\sqrt{2})\,\lambda & (1-\sqrt{2})\,\beta \\ -(2+\sqrt{2})\,\lambda & -(2-\sqrt{2})\,\beta \end{bmatrix} = \begin{bmatrix} 1 & 0 \\ 0 & 1 \end{bmatrix}$$

This simplifies to

$$ml^2 \begin{bmatrix} (4+2\sqrt{2})\,\lambda^2 & 0 \\ 0 & (4-2\sqrt{2})\,\beta^2 \end{bmatrix} = \begin{bmatrix} 1 & 0 \\ 0 & 1 \end{bmatrix}$$

which yields

$$\lambda = \frac{1}{l\sqrt{(4+2\sqrt{2})\,m}} \text{ and } \beta = \frac{1}{l\sqrt{(4-2\sqrt{2})\,m}} \tag{19}$$

From Eqns (14), (15) and (19) we find the eigenvectors as

$$a_1 = \begin{Bmatrix} a_{11} \\ a_{21} \end{Bmatrix} = \frac{1}{l\sqrt{(4+2\sqrt{2})\,m}} \begin{Bmatrix} 1+\sqrt{2} \\ -(2+\sqrt{2}) \end{Bmatrix} \tag{20}$$

$$\text{and } a_2 = \begin{Bmatrix} a_{21} \\ a_{22} \end{Bmatrix} = \frac{1}{l\sqrt{(4-2\sqrt{2})\,m}} \begin{Bmatrix} 1-\sqrt{2} \\ -(2-\sqrt{2}) \end{Bmatrix} \tag{21}$$

11.12 Oscillations of a Linear Triatomic Molecule

We consider a linear triatomic molecule in which two atoms P each of mass m_1 are placed at the same distances on either side of the third atom Q of mass m_2. We make the following assumptions:

(1) The oscillations are along the line joining the atoms.

(2) The interaction force between the atoms P and Q is equal to the elastic force of a spring having force constant k.

(3) There is no interaction between the P atoms.

Let q_1, q_2 and q_3 be the displacements of the atoms from the equilibrium position. Then the kinetic energy of the system is

$$T = \frac{1}{2} m_1 \left(\dot{q}_1^2 + \dot{q}_3^2\right) + \frac{1}{2} m_2 \dot{q}_2^2 \tag{11.129}$$

Hence, $T_{11} = T_{33} = m_1$, $T_{22} = m_2$ and other elements are zero.

Thus,

$$[T] = \begin{bmatrix} m_1 & 0 & 0 \\ 0 & m_2 & 0 \\ 0 & 0 & m_1 \end{bmatrix} \tag{11.130}$$

The potential energy of the system is

$$V = \frac{1}{2} k (q_2 - q_1)^2 + \frac{1}{2} k (q_2 - q_3)^2 = \frac{1}{2} k \left(q_1^2 + 2q_2^2 + q_3^2 - 2q_1 q_2 - 2q_2 q_3\right) \tag{11.131}$$

Hence,

$$\begin{aligned} V_{11} &= k, & V_{12} &= -k, & V_{13} &= 0 \\ V_{21} &= -k, & V_{22} &= 2k, & V_{23} &= -k \\ V_{31} &= 0, & V_{32} &= -k, & V_{33} &= k \end{aligned}$$

Thus,

$$[V] = \begin{bmatrix} k & -k & 0 \\ -k & 2k & -k \\ 0 & -k & k \end{bmatrix} \tag{11.132}$$

The characteristic Eqn (11.99) in this case assumes the form:

$$\begin{vmatrix} k-\omega^2 m_1 & -k & 0 \\ -k & 2k-\omega^2 m_2 & -k \\ 0 & -k & k-\omega^2 m_1 \end{vmatrix} = 0 \tag{11.133}$$

$$\text{or,}\quad (k-\omega^2 m_1)\left(\omega^4 m_1 m_2 - k\omega^2 m_2 - 2k\omega^2 m_1\right) = 0$$

$$\text{or,}\quad \omega^2 (k-\omega^2 m_1)\left\{\omega^2 m_1 m_2 - k(m_2+2m_1)\right\} = 0 \tag{11.134}$$

Eqn (11.134) yields three roots of ω like

$$\left.\begin{aligned} \omega_1 &= 0 \\ \omega_2 &= \sqrt{\frac{k}{m_1}} \\ \text{and}\quad \omega_3 &= \sqrt{\frac{k}{m_1}\left(1+\frac{2m_1}{m_2}\right)} \end{aligned}\right\} \tag{11.135}$$

The frequency $\omega_1 = 0$ corresponds to uniform translatory motion of the molecule as a whole.

• Determination of Eigenvectors

CASE 1. $\omega = \omega_1 = 0$

Let the eigenvectors be (a_{11}, a_{21}, a_{31}). Then

$$\begin{bmatrix} k & -k & 0 \\ -k & 2k & -k \\ 0 & -k & k \end{bmatrix}\begin{bmatrix} a_{11} \\ a_{21} \\ a_{31} \end{bmatrix} = 0 \tag{11.136}$$

This yields three equations like

$$\left.\begin{aligned} ka_{11} - ka_{21} &= 0 \\ -ka_{11} + 2ka_{21} - ka_{31} &= 0 \\ \text{and}\quad -ka_{21} + ka_{31} &= 0 \end{aligned}\right\} \tag{11.137}$$

Solving these equations we get

$$a_{11} = a_{21} = a_{31} = \lambda \text{ (say)}.$$

Hence, the eigenvector corresponding to $\omega = \omega_1 = 0$ is given by

$$[a_1] = \begin{bmatrix} \lambda \\ \lambda \\ \lambda \end{bmatrix} \tag{11.138}$$

This shows that each atom is equally displaced in the same direction.

CASE 2. $\omega = \omega_2 = \sqrt{\dfrac{k}{m_1}}$

Let the eigenvectors be (a_{12}, a_{22}, a_{32}). Then we have

$$\begin{bmatrix} 0 & -k & 0 \\ -k & 2k - \dfrac{km_2}{m_1} & -k \\ 0 & -k & 0 \end{bmatrix}\begin{bmatrix} a_{12} \\ a_{22} \\ a_{32} \end{bmatrix} = 0 \tag{11.139}$$

We get two equations like

$$\left.\begin{aligned} -ka_{22} &= 0 \\ -ka_{12} + k\left(2 - \frac{m_2}{m_1}\right)a_{22} - ka_{32} &= 0 \end{aligned}\right\} \tag{11.140}$$

Solving we get

$$a_{22} = 0,\ a_{12} = -a_{32} = \mu \text{ (say)}.$$

Hence, the eigenvector corresponding to $\omega = \omega_2$ is

$$\{a_2\} = \begin{bmatrix} \mu \\ 0 \\ -\mu \end{bmatrix} \tag{11.141}$$

Eqn (11.141) shows that the central atom remains at rest while the other two atoms are equally displaced in opposite directions.

CASE 3. $\omega = \omega_3 = \sqrt{\dfrac{k}{m_1}\left(1 + \dfrac{2m_1}{m_2}\right)}$

Let the eigenvectors be (a_{13}, a_{23}, a_{33}). Then we have

$$\begin{bmatrix} -2k\dfrac{m_1}{m_2} & -k & 0 \\ -k & -k\dfrac{m_2}{m_1} & -k \\ 0 & -k & -2k\dfrac{m_1}{m_2} \end{bmatrix} \begin{bmatrix} a_{13} \\ a_{23} \\ a_{33} \end{bmatrix} = 0 \tag{11.142}$$

This yields three equations like

$$\left.\begin{aligned} -2k\frac{m_1}{m_2}a_{13} - ka_{23} &= 0 \\ -ka_{13} - k\frac{m_2}{m_1}a_{23} - ka_{33} &= 0 \\ \text{and} \quad -ka_{23} - 2k\frac{m_1}{m_2}a_{33} &= 0 \end{aligned}\right\} \tag{11.143}$$

Solving, we get

$$a_{13} = a_{33} = \nu \text{ (say) and } a_{23} = -\frac{2m_1}{m_2}\nu$$

Thus, the eigenvector for $\omega = \omega_3$ is

$$\{a_3\} = \begin{bmatrix} \nu \\ -\dfrac{2m_1}{m_2}\nu \\ \nu \end{bmatrix} \tag{11.144}$$

Eqn (11.144) shows that the displacements of end atoms are equal in magnitude as well as in direction while the displacement of the central atom is different in magnitude and opposite in direction to the displacements of end atoms.

In order to find the unknown constants λ, μ, ν, we use the condition given in Eqn (11.113).

Then we get

$$\begin{bmatrix} \lambda & \lambda & \lambda \\ \mu & 0 & -\mu \\ \nu & -\dfrac{2m_1}{m_2}\nu & \nu \end{bmatrix} \begin{bmatrix} m_1 & 0 & 0 \\ 0 & m_2 & 0 \\ 0 & 0 & m_1 \end{bmatrix} \begin{bmatrix} \lambda & \mu & \nu \\ \lambda & 0 & -\dfrac{2m_1}{m_2}\nu \\ \lambda & -\mu & \nu \end{bmatrix} = \begin{bmatrix} 1 & 0 & 0 \\ 0 & 1 & 0 \\ 0 & 0 & 1 \end{bmatrix}$$

$$\text{or,} \quad \begin{bmatrix} \lambda^2(2m_1+m_2) & 0 & 0 \\ 0 & 2m_1\mu^2 & 0 \\ 0 & 0 & 2m_1\left(1+\dfrac{2m_1}{m_2}\right)\nu^2 \end{bmatrix} = \begin{bmatrix} 1 & 0 & 0 \\ 0 & 1 & 0 \\ 0 & 0 & 1 \end{bmatrix}$$

which yields

$$\left.\begin{aligned} \lambda &= \frac{1}{\sqrt{2m_1+m_2}} \\ \mu &= \frac{1}{\sqrt{2m_1}} \\ \nu &= \frac{1}{\sqrt{2m_1\left(1+\dfrac{2m_1}{m_2}\right)}} \end{aligned}\right\} \tag{11.145}$$

Thus, we get the eigenvectors

$$\{a_1\} = \frac{1}{\sqrt{2m_1+m_2}} \begin{bmatrix} 1 \\ 1 \\ 1 \end{bmatrix} \tag{11.146}$$

$$\{a_2\} = \frac{1}{\sqrt{2m_1}} \begin{bmatrix} 1 \\ 0 \\ -1 \end{bmatrix} \tag{11.147}$$

$$\text{and} \quad \{a_3\} = \frac{1}{\sqrt{2m_1\left(1+\dfrac{2m_1}{m_2}\right)}} \begin{bmatrix} 1 \\ -2m_1/m_2 \\ 1 \end{bmatrix} \tag{11.148}$$

EXAMPLE 11.7 *In a damped vibration, it is observed that during one period of vibration, the amplitude has reduced to half.*

Determine logarithmic decrement and change in natural frequency.

Solution. Logarithmic decrement

$$\delta = \ln\left[\frac{x(t)}{x(t+T)}\right] = \ln 2 = 0.693 \tag{1}$$

Again, by Eqn (11.34),

$$\delta = \frac{2\pi\alpha}{\omega} = \alpha T, \text{ i.e., } \alpha = \frac{\delta}{T} \tag{2}$$

From (1) and (2),

$$\alpha = \frac{0.693}{T} = \frac{0.693}{2\pi}\sqrt{\xi^2-\alpha^2}, \text{ i.e., } \alpha^2 = 0.012\,(\xi^2-\alpha^2)$$

$$\therefore \alpha^2 \simeq 0.012\xi^2 \tag{3}$$

Frequency of damped vibration

$$\omega = \sqrt{\xi^2 - \alpha^2} = \sqrt{\xi^2 - 0.012\xi^2} = 0.994\xi \tag{4}$$

Hence, change in natural frequency

$$= \frac{1 - 0.994}{0.994} \times 100\% = 0.6\%$$

EXAMPLE 11.8 *In an under-damped system, the maximum overshoot is 5%. What is its minimum damping ratio?*

Solution. Overshoot in an under-damped vibration is defined as the maximum displacement at the end of its half-cycle, i.e., at $t = T/2$.

The displacement function for this system is

$$x(t) = Ae^{-\alpha t}\sin(\omega t + \phi) \tag{1}$$

At $t = 0$,

$$x(0) = A\sin\phi \tag{2}$$

At $t = T/2$,

$$x(T/2) = Ae^{-\alpha T/2}\sin\left(\frac{\omega T}{2} + \phi\right) = -Ae^{-\alpha T/2}\sin\phi \quad \left[\because T = \frac{2\pi}{\omega}\right]$$

$$= -x(0)e^{-\alpha T/2} \tag{3}$$

By condition given,

$$x(T/2) = -5\% \text{ of } x(0) = -0.05x(0) \tag{4}$$

From (3) and (4),

$$e^{-\alpha T/2} = 0.05 = \frac{1}{20}, \text{ i.e., } \frac{\alpha T}{2} = \ln 20 = 2.9957$$

$$\therefore T = 2 \times \frac{2.9957}{\alpha} = \frac{5.9915}{\alpha} \tag{5}$$

But

$$T = \frac{2\pi}{\sqrt{\xi^2 - \alpha^2}} = \frac{2\pi}{\xi\sqrt{1 - \eta^2}} \tag{6}$$

where $\eta = \alpha/\xi$ = damping ratio.

Hence, from (5) and (6), we get

$$\frac{2\pi}{\xi\sqrt{1-\eta^2}} = \frac{5.9915}{\alpha} = \frac{5.9915}{\xi\eta}, \text{ i.e., } \eta^2 = (1 - \eta^2) \times 0.9093$$

$$\therefore \eta = 0.723$$

EXAMPLE 11.9 *An automobile of mass* 250 kg *has a shock absorber such that the initial amplitude gets reduced to* $\frac{1}{10}$*th in one cycle. If its natural frequency is* 7.0 rad/s, *find the stiffness and the damping coefficients of the shock absorber.*

Solution. The logarithmic decrement is

$$\delta = \ln\left(\frac{x_1}{x_2}\right) = \ln\left(\frac{x_1}{x_1/10}\right) = \ln 10 = 2.30 \tag{1}$$

But

$$\delta = \frac{2\pi\alpha}{\omega} = \frac{2\pi\alpha}{\sqrt{\xi^2 - \alpha^2}} = \frac{2\pi\eta}{\sqrt{1-\eta^2}} \tag{2}$$

where $\eta = \alpha/\xi$ = damping ratio.

From (1) and (2),

$$\frac{2\pi\eta}{\sqrt{1-\eta^2}} = 2.30, \text{ i.e., } 39.44\eta^2 = 5.29(1-\eta^2)$$

$$\therefore \eta = 0.344 \tag{3}$$

Damping coefficient is

$$\begin{aligned} c &= 2m\alpha = 2m\xi\eta \\ &= 2 \times 250 \times 7.0 \times 0.344 \quad [\text{since } \xi = 7.0 \text{ rad/s}] \\ &= 1204\,\text{Ns/m}. \end{aligned}$$

The stiffness coefficient is

$$k = m\xi^2 = 250 \times 7.0 \times 7.0 = 1.225 \times 10^4\,\text{N/m}.$$

EXAMPLE 11.10 *The amplitudes of forced vibration at frequencies ω_1 and ω_2 are found to be equal. What is the value of resonant frequency?*

Solution. The amplitude of forced vibration is

$$A = \frac{P_0/m}{\sqrt{\left(\xi^2 - \omega^2\right)^2 + 4\alpha^2\omega^2}} \tag{1}$$

Hence, for $\omega = \omega_1$,

$$A_1 = \frac{P_0/m}{\sqrt{\left(\xi^2 - \omega_1^2\right)^2 + 4\alpha^2\omega_1^2}} \tag{2}$$

Similarly, for $\omega = \omega_2$,

$$A_2 = \frac{P_0/m}{\sqrt{\left(\xi^2 - \omega_2^2\right)^2 + 4\alpha^2\omega_2^2}} \tag{3}$$

By condition given, $A_1 = A_2$. Hence, from (2) and (3), we have

$$\begin{aligned} &\left(\xi^2 - \omega_1^2\right)^2 + 4\alpha^2\omega_1^2 = \left(\xi^2 - \omega_2^2\right)^2 + 4\alpha^2\omega_2^2 \\ \text{i.e., } &\left(\xi^2 - \omega_1^2\right)^2 - \left(\xi^2 - \omega_2^2\right)^2 = 4\alpha^2\left(\omega_2^2 - \omega_1^2\right) \\ \text{i.e., } &\left(2\xi^2 - \omega_1^2 - \omega_2^2\right)\left(\omega_2^2 - \omega_1^2\right) = 4\alpha^2\left(\omega_2^2 - \omega_1^2\right) \\ \therefore\ &2\xi^2 - \omega_1^2 - \omega_2^2 = 4\alpha^2 \ (\text{since } \omega_1 \neq \omega_2) \end{aligned} \tag{4}$$

The resonant frequency, by Eqn (11.45), is

$$\omega' = \sqrt{\xi^2 - 2\alpha^2} = \sqrt{\xi^2 - \frac{1}{2}\left(2\xi^2 - \omega_1^2 - \omega_2^2\right)} = \left[\frac{1}{2}\left(\omega_1^2 + \omega_2^2\right)\right]^{1/4}$$

EXAMPLE 11.11 *Determine the steady state response of a system, given:*

$$m = 4.0\,\text{kg},\ P_0 = 12.0\,\text{kg},\ \omega = 15\,\text{rad/s}$$
$$k = 4500\,\text{N/m and } c = 12\,\text{Ns/m}.$$

Solution. Natural frequency

$$\xi = \sqrt{\frac{k}{m}} = \sqrt{\frac{4500}{4.0}} = 33.5\,\text{rad/s} \tag{1}$$

Damping ratio

$$\eta = \frac{\alpha}{\xi} = \frac{c}{2m\xi} = \frac{12.0}{2 \times 4.0 \times 33.5} = 0.044 \tag{2}$$

Hence, the vibration is under-damped. By Eqn (11.43), the phase angle is

$$\phi = \tan^{-1}\left(\frac{2\alpha\omega}{\xi^2 - \omega^2}\right)$$

where

$$\alpha = \frac{c}{2m} = \frac{12.0}{2 \times 4.0} = \frac{3}{2} = 1.5$$

Hence,

$$\phi = \tan^{-1}\left(\frac{2 \times 1.5 \times 15}{33.5 \times 33.5 - 15.0 \times 15.0}\right) = \tan^{-1}(0.0492) = 2.8^\circ = 0.0489\,\text{rad}$$

The amplitude is

$$A = \frac{P_0/m}{\sqrt{(\xi^2-\omega^2)^2 + 4\alpha^2\omega^2}} = \frac{12.0/4.0}{\sqrt{(33.5^2 - 15^2)^2 + (3.0)^2 \cdot (15)^2}}$$

$$\approx \frac{3.0}{(48.5)(18.5)} = 3{,}325 \times 10^{-3}\,\text{m} \tag{3}$$

Hence, the steady state response is $x(t) = A\sin(\omega t - \phi)$

$$\therefore x(t) = 3.325 \times 10^{-3} \sin(15t - 0.049)$$

EXAMPLE 11.12 *Prove that, in the steady state of forced vibration, the average power supply by the external harmonic excitation is equal to the average rate of work done by the damping force.*

Solution. The instantenous power supply by the exciting force is

$$P = \text{applied force} \times \text{velocity at time } t$$

$$= (P_0 \sin\omega t)[A\cos(\omega t - \phi)] = \frac{1}{2}A\omega P_0\,[\sin\phi + \sin(2\omega t - \phi)] \tag{1}$$

Hence, average power supply in one cycle

$$P_{\text{av}} = \frac{1}{T}\int_0^T P dt,\ T = \frac{2\pi}{\omega}$$

$$= \frac{\omega}{2\pi}\int_0^T \frac{1}{2}A\omega P_0\,[\sin\phi + \sin(2\omega t - \phi)]\,dt$$

$$= \frac{\omega}{2\pi}\cdot\frac{1}{2}A\omega P_0\left(\frac{2\pi}{\omega}\right)\sin\phi = \frac{1}{2}A\omega P_0\sin\phi \tag{2}$$

Putting here the values of A and $\sin\phi$ from Eqns (11.42) and (11.43),

$$P_{\text{av}} = \frac{1}{2}\omega P_0 \frac{P_0/m}{\sqrt{(\xi^2-\omega^2)^2 + 4\alpha^2\omega^2}} \cdot \frac{2\alpha\omega}{\sqrt{(\xi^2-\omega^2)^2 + 4\alpha^2\omega^2}}$$

$$= \frac{\alpha\omega^2 P_0^2}{m}\left(\frac{1}{(\xi^2-\omega^2)^2 + 4\alpha^2\omega^2}\right) \tag{3}$$

Now, the rate of work done by the damping force is

$$W = c\,(\dot{x})^2 = cA^2\omega^2\cos^2(\omega t - \phi) \tag{4}$$

Then average rate of work done in one cycle

$$\begin{aligned} W_{\text{av}} &= \frac{1}{T}\int_0^T W\,dt \\ &= \frac{1}{T}\int_0^T cA^2\omega^2\cos^2(\omega t-\phi)\,dt \\ &= \frac{1}{2}\cdot\frac{cA^2\omega^2}{T}\int_0^T [1+\cos\{2(\omega t-\phi)\}]\,dt \\ &= \frac{1}{2}\cdot c\omega^2A^2 \\ &= m\alpha\omega^2A^2\ [\text{since } c = 2m\alpha] \\ &= \frac{\alpha\omega^2P_0^2}{m}\cdot\frac{1}{(\xi^2-\omega^2)^2+4\alpha^2\omega^2} \end{aligned} \tag{5}$$

From (3) and (5), $P_{\text{av}} = W_{\text{av}}$. **(Proved)**

EXAMPLE 11.13 *Two similar springs each of spring constant k, hang vertically from a rigid support with two equal masses m attached to them as shown in* FIG. 11.13. *Find the normal mode of frequencies along vertical direction only.*

Solution. Let x_1, x_2 be the vertical displacements of the masses from their equilibrium positions. Then kinetic energy of the system

$$T = \frac{1}{2}m\left(\dot{x}_1^2 + \dot{x}_2^2\right) \tag{1}$$

$$\therefore 2T = \{\dot{x}_1, \dot{x}_2\}\begin{bmatrix} m & 0 \\ 0 & m\end{bmatrix}\begin{Bmatrix}\dot{x}_1 \\ \dot{x}_2\end{Bmatrix} \tag{2}$$

so that

$$[T] = \begin{bmatrix} m & 0 \\ 0 & m\end{bmatrix} \tag{3}$$

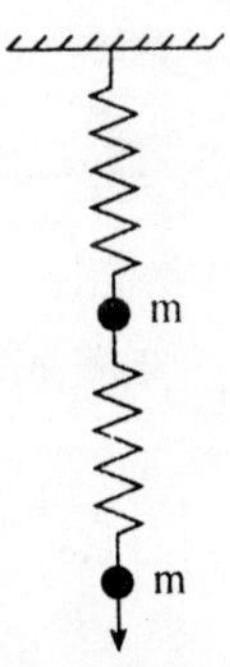

FIG. 11.13

The potential energy is

$$V = \frac{1}{2}k\left\{x_1^2 + (x_2 - x_1)^2\right\} = \frac{1}{2}k\left(2x_1^2 + x_2^2 - 2x_1x_2\right).$$

Hence,

$$\left.\begin{aligned} V_{11} &= \frac{\partial^2 V}{\partial x_1^2} = 2k \\ V_{12} &= \frac{\partial^2 V}{\partial x_1 \partial x_2} = -k \\ V_{22} &= \frac{\partial V}{\partial x_2^2} = k \end{aligned}\right\} \tag{4}$$

Hence,

$$[V] = \begin{bmatrix} 2k & -k \\ -k & k\end{bmatrix} \tag{5}$$

The characteristic equation is

$$|V-\omega^2T| = \begin{bmatrix} 2k-\omega^2 m & -k \\ -k & k-\omega^2 m \end{bmatrix} = 0$$

$$\text{i.e.,}\ (2k-\omega^2 m)\,(k-\omega^2 m) - k^2 = 0$$

$$\text{i.e.,}\ \omega^4 m^2 - 3\omega^2 mk - k^2 = 0$$

$$\therefore \omega^2 = \frac{1}{2m^2}\left[3mk \pm \sqrt{9m^2k^2 - 4m^2k^2}\right] = \frac{3k}{2m} \pm \frac{\sqrt{5}k}{2m} \tag{6}$$

Thus, the normal mode of frequencies are:

$$\left.\begin{aligned} \omega_1 &= \left[\frac{k}{2m}\left(3+\sqrt{5}\right)\right]^{1/2} \\ \omega_2 &= \left[\frac{k}{2m}\left(3-\sqrt{5}\right)\right]^{1/2} \end{aligned}\right\}$$

EXAMPLE 11.14 *Determine the fundamental frequencies and the mode shapes for the system shown in* FIG. 11.14. *Neglect friction between the wheels and floor.*

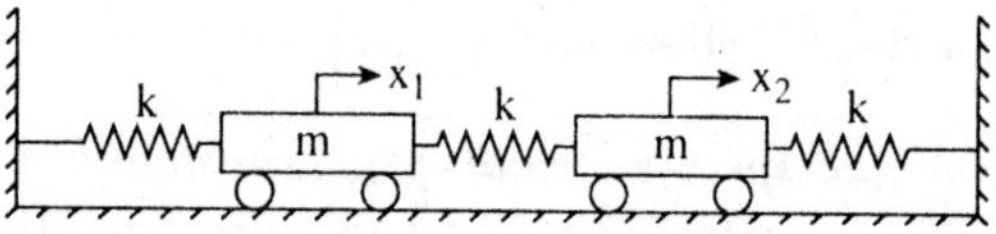

FIG. 11.14

Solution. Kinetic energy of the system

$$T = \frac{1}{2}m\dot{x}_1^2 + \frac{1}{2}m\dot{x}_2^2 \tag{1}$$

Potential energy of the system

$$V = \frac{1}{2}kx_1^2 + \frac{1}{2}k(x_2-x_1)^2 + \frac{1}{2}kx_2^2 \tag{2}$$

Lagrange's equations of motion are:

$$\frac{d}{dt}\left(\frac{\partial T}{\partial \dot{x}_1}\right) - \frac{\partial T}{\partial x_1} = -\frac{\partial V}{\partial x_1} \tag{3}$$

$$\frac{d}{dt}\left(\frac{\partial T}{\partial \dot{x}_2}\right) - \frac{\partial T}{\partial x_2} = -\frac{\partial V}{\partial x_2} \tag{4}$$

Using (1) and (2) in (3) and (4), we get equations of motion

$$m\ddot{x}_1 + 2kx_1 - kx_2 = 0 \tag{5}$$

$$m\ddot{x}_2 - kx_1 + 2kx_2 = 0 \tag{6}$$

We assume

$$x_1 = A_1 \sin(\omega t + \phi) \tag{7}$$

$$x_2 = A_2 \sin(\omega t + \phi) \tag{8}$$

Inserting these functions in (5) and (6),

$$\left.\begin{aligned} (2k-m\omega^2)\,A_1 - kA_2 &= 0 \\ -kA_1 + (2k-m\omega^2)\,A_2 &= 0 \end{aligned}\right\} \tag{9}$$

For non-zero values of A_1, A_2, we must have

$$\begin{vmatrix} 2k - m\omega^2 & -k \\ -k & 2k - m\omega^2 \end{vmatrix} = 0 \tag{10}$$

This yields

$$\left(2k - m\omega^2\right)^2 - k^2 = 0, \text{ i.e., } m^2\omega^4 - 4mk\omega^2 + 3k^2 = 0 \tag{11}$$

Its solutions are

$$\omega_{1,2}^2 = \frac{1}{2m^2}\left(4mk \pm \sqrt{16m^2k^2 - 12m^2k^2}\right) = \frac{k}{m}(2 \pm 1)$$

$$\therefore \omega_1 = \sqrt{\frac{k}{m}}, \omega_2 = \sqrt{\frac{3k}{m}} \tag{12}$$

These are the fundamental frequencies.

• Mode shapes

For $\omega = \omega_1$, the responses of the masses are assumed as

$$x_1 = A_{11} \sin(\omega_1 t + \phi) \tag{13}$$

$$x_2 = A_{21} \sin(\omega_2 t + \phi) \tag{14}$$

where

$$A_{11} = \text{amplitude of mass } m_1 \text{ due to } \omega_1$$

$$A_{21} = \text{amplitude of mass } m_2 \text{ due to } \omega_1.$$

Inserting (13) and (14) into (5) and (6),

$$\left(2k - m\omega_1^2\right) A_{11} - kA_{21} = 0 \tag{15}$$

$$-kA_{11} + \left(2k - m\omega_1^2\right) A_{21} = 0 \tag{16}$$

$$\therefore \mu_1 = \frac{A_{21}}{A_{11}} = \frac{2k - m\omega_1^2}{k} \text{ [from (15)]} \tag{17}$$

$$= \frac{k}{2k - m\omega_1^2} \text{ [from (16)]} \tag{18}$$

Inserting $\omega_1^2 = k/m$, we get $\mu_1 = 1$.
Taking for simplicity $A_{11} = 1, A_{21} = \mu_1 A_{11} = 1$.
Thus, the first modal vector

$$[A]_1 = \begin{Bmatrix} A_{11} \\ \mu A_{21} \end{Bmatrix} = \begin{Bmatrix} 1 \\ 1 \end{Bmatrix} \tag{19}$$

Similarly, for $\omega = \omega_2$, the responses of the system are assumed as

$$x_1 = A_{12} \sin\left(\omega_2 t + \phi\right) \tag{20}$$

$$x_2 = A_{22} \sin\left(\omega_2 t + \phi\right) \tag{21}$$

Inserting these functions into (5) and (6),

$$\left(2k - m\omega_2^2\right) A_{12} - kA_{22} = 0 \tag{22}$$

$$-kA_{12} + \left(2k - m\omega_2^2\right) A_{22} = 0 \tag{23}$$

Then amplitude ratio

$$\mu_2 = \frac{A_{22}}{A_{12}} = \frac{2k - m\omega_2^2}{k} \text{ [from (22)]} \tag{24}$$

$$= \frac{k}{2k - m\omega_2^2} \text{ [from (23)].} \tag{25}$$

Putting the value of $\omega_2 = \sqrt{3k/m}$, we get from any of (24) and (25),

$$\mu_2 = \frac{2k - m(3k/m)}{k} = -1 \tag{26}$$

Taking for simplicity $A_{12} = 1$, the modal vector is

$$[A]_2 = \begin{Bmatrix} A_{12} \\ A_{22} \end{Bmatrix} = \begin{Bmatrix} A_{12} \\ \mu_2 A_{12} \end{Bmatrix} = \begin{Bmatrix} 1 \\ -1 \end{Bmatrix} \tag{27}$$

Thus,

$$[A] = \begin{bmatrix} A_{11} & A_{12} \\ A_{21} & A_{22} \end{bmatrix} = \begin{bmatrix} 1 & 1 \\ 1 & -1 \end{bmatrix} \tag{28}$$

The mode shapes are shown in FIG. 11.15:

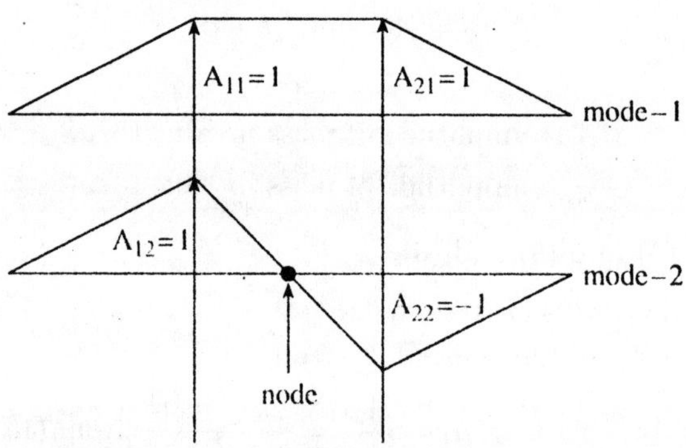

FIG. 11.15

EXAMPLE 11.15 *Write the differential equation of motion of the system in* FIG. 11.16 *in matrix form.*

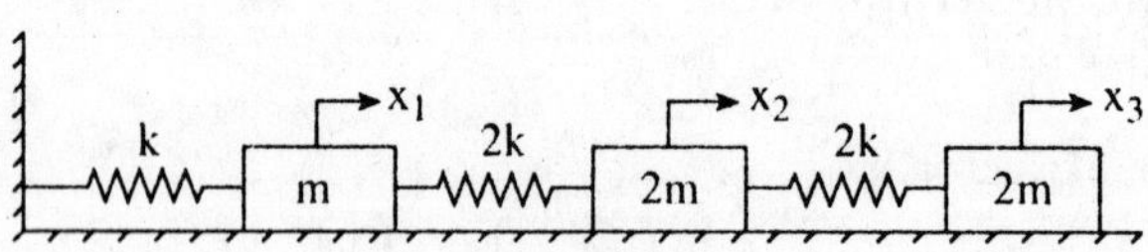

FIG. 11.16

Solution. Let x_1, x_2, x_3 be the displacements of the masses from equilibrium positions at any instant.

Then kinetic energy of the system is

$$T = \frac{1}{2}m\dot{x}_1^2 + \frac{1}{2}(2m)\dot{x}_2^2 + \frac{1}{2}(2m)\dot{x}_3^2 \tag{1}$$

The potential energy of the system is

$$V = \frac{1}{2}kx_1^2 + \frac{1}{2}(2k)(x_2 - x_1)^2 + \frac{1}{2}k(x_3 - x_2)^2 \tag{2}$$

The Lagrangian is

$$L = T - V$$
$$= \frac{1}{2}m\dot{x}_1^2 + m\dot{x}_2^2 + m\dot{x}_3^2 - \frac{1}{2}kx_1^2 - k(x_2 - x_1)^2 - \frac{1}{2}k(x_3 - x_2)^2 \tag{3}$$

Lagrange's equations of motion are

$$\frac{d}{dt}\left(\frac{\partial L}{\partial \dot{x}_i}\right) - \frac{\partial L}{\partial x_i} = 0, i = 1, 2, 3 \tag{4}$$

Thus, for $i = 1$,

$$m\ddot{x}_1 + kx_1 - 2k(x_2 - x_1) = 0$$
$$\text{i.e., } m\ddot{x}_1 + 3kx_1 - 2kx_2 = 0 \tag{5}$$

Similarly,

$$\text{for } i = 2,\ 2m\ddot{x}_2 - 2kx_1 + 3kx_2 - kx_3 = 0 \tag{6}$$
$$\text{for } i = 3,\ 2m\ddot{x}_3 - kx_2 + kx_3 = 0 \tag{7}$$

Thus, in matrix form, Eqns (5), (6) and (7) are:

$$\begin{bmatrix} m & 0 & 0 \\ 0 & 2m & 0 \\ 0 & 0 & 2m \end{bmatrix}\begin{bmatrix} \ddot{x}_1 \\ \ddot{x}_2 \\ \ddot{x}_3 \end{bmatrix} + \begin{bmatrix} 3k & -2k & 0 \\ -2k & 3k & -k \\ 0 & -k & k \end{bmatrix}\begin{bmatrix} x_1 \\ x_2 \\ x_3 \end{bmatrix} = \begin{bmatrix} 0 \\ 0 \\ 0 \end{bmatrix}$$

EXERCISE

1. A uniformly wound spring of length l and mass m hangs vertically from a roof-ceiling. A mass M is attached at its free end, pushed down a little and released immediately. Show that the period of its small oscillation is

$$2\pi\sqrt{\frac{1}{k}\left(M + \frac{m}{3}\right)}$$

where k is the spring constant.

2. In a damped harmonic oscillation the damping is proportional to the velocity. Show that the rate of energy dissipation is proportional to the square of the velocity.

3. In a forced oscillation there is no damping, but the frequency of the impressed force is equal to the natural frequency of oscillation. Show that oscillations will be growing along with time.

4. A particle of mass m oscillates under a conservative force between $x = x_1$ and $x = x_2$. Show that the period of oscillation is

$$P = \int_{x_1}^{x_2}\left[\frac{m}{2\{V(x_2) - V(x_1)\}}\right]^{1/2} dx$$

5. Discuss the problem of two coupled simple pendulums by the general theory.

6. Two identical masses m are connected to two fixed walls by two massless springs of force constant k. The masses are also connected by a massless spring of force constant k_0. The whole system remains in equilibrium on a smooth horizontal table. If the system is slightly displaced, show that the frequencies of oscillation are given by

$$\omega_1 = \pm\sqrt{\frac{k}{m}} \text{ and } \omega_2 = \pm\sqrt{\frac{k}{m} + \frac{2k_0}{m}}$$

7. The Lagrangian of a system is described by

$$L = \frac{1}{2}\left(\dot{x}^2 + \dot{y}^2\right) - \frac{1}{2}\left(\omega_1^2 x^2 + \omega_2^2 y^2\right) + 2xy$$

Find the normal mode frequencies.

8. A triatomic molecule of CO_2 consists of atoms of masses m at equal distances on opposite sides from atom of mass M. Determine the normal coordinates of oscillations.

CHAPTER 12

Anharmonic Oscillations

12.1 Equation of Motion

When the restoring force is a non-linear function of displacement and the damping force is a function of velocity, the motion of the oscillator systems are no more simple harmonic. We call these ***anharmonic oscillations***.

Let m be the mass of the oscillator and

$g(\dot{x})$ — non-linear damping force function

$f(x)$ — non-linear restoring force function

$F(t)$ — the driving force applied to the anharmonic oscillator.

Then the general equation of motion is

$$m\ddot{x} + g(\dot{x}) + f(x) = F(t) \tag{12.1}$$

For the undriven anharmonic oscillator, $F(t) = 0$, so that the equation of motion is

$$m\ddot{x} + g(\dot{x}) + f(x) = 0 \tag{12.2}$$

All these are non-linear equations which cannot be solved exactly. However, ***in general, the oscillation period of non-linear systems is a function of the amplitude of the oscillations***.

Let us show this in the case of the anharmonic oscillator when the damping and the driving terms are absent. In this case, the equation of motion is

$$m\ddot{x} + P(x) = 0 \tag{12.3}$$

This equation can be written as

$$\ddot{x} + f(x) = 0 \tag{12.4}$$

where we have denoted

$$f(x) = \frac{P(x)}{m} \tag{12.5}$$

If the velocity at any instant of time is denoted by v, then $v = dx/dt$, so that

$$\ddot{x} = \frac{dv}{dt} = \frac{dv}{dx} \cdot \frac{dx}{dt} = \frac{dv}{dx}v$$

In view of this, Eqn (12.4) becomes

$$v\,dv = -f(x)dx \tag{12.6}$$

Let a = amplitude of oscillation. Then $v = 0$, for $x = a$.

Integrating Eqn (12.6),

$$\int_0^v v\,dv = -\int_a^x f(x)dx, \text{ i.e., } \frac{1}{2}v^2 = -\int_a^x f(x)dx = \int_x^a f(x)dx \tag{12.7}$$

This equation represents the principle of conservation of energy. The left-hand side is the gain in kinetic energy due to the motion from $(x = a, v = 0)$ to any moment (x, v) and the right-hand side represents the loss of potential energy of the non-linear restoring force (FIG. 12.1).

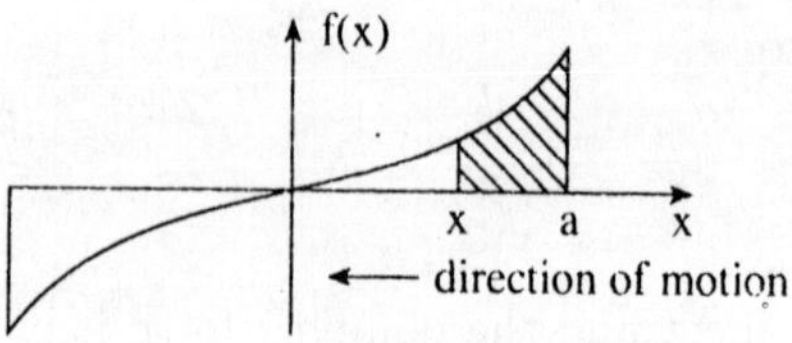

FIG. 12.1

Denoting

$$\int_x^a f(x)dx = F(a) - F(x) \tag{12.8}$$

we get from Eqn (12.7)

$$v = \frac{dx}{dt} = -\sqrt{2}\{F(a) - F(x)\}^{1/2} \tag{12.9}$$

The negative sign has been taken in Eqn (12.9), since the velocity in the interval considered is negative.

Integrating Eqn (12.9) from $x = 0$ to $x = a$, we will get one-fourth of period of oscillation T for a system with symmetric characteristics. Hence,

$$T = 4\sqrt{\frac{1}{2}}\int_0^a \frac{dx}{\{F(a) - F(x)\}^{1/2}} \tag{12.10}$$

Hence, it is obvious that the period of such anharmonic oscillators depends on the amplitude a of vibration.

12.2 Some Examples of Anharmonic Oscillations

(A) Let us consider the anharmonic oscillation when the restoring force function is given by

$$f(x) = Ax^{2n-1}, \ (n = 1, 2, 3, \ldots) \tag{12.11}$$

where A is a known constant.

In this case,

$$\int_x^a f(x)dx = \int_x^a Ax^{2n-1}dx = \frac{A}{2n}\left(a^{2n} - x^{2n}\right) \tag{12.12}$$

Thus,

$$\sqrt{2}\{F(a) - F(x)\}^{1/2} = \sqrt{\frac{A}{n}(a^{2n} - x^{2n})} \tag{12.13}$$

Hence, the period of oscillation is, by Eqn (12.10),

$$T = 4\int_0^a \frac{dx}{\sqrt{\frac{A}{n}\left(a^{2n} - x^{2n}\right)}} = 4\sqrt{\frac{n}{A}} \cdot \frac{1}{a^{n-1}}\int_0^1 \frac{d\xi}{\sqrt{1 - \xi^{2n}}} \tag{12.14}$$

where $\xi = x/a$.

Eqn (12.14) shows that only for $n = 1$, when $f(x) = Ax$, a linear function, the period of oscillation T does not depend on the amplitude. For other values of n, there always exists a relation between the period and amplitude of the oscillation.

In particular, when $n = 2$,

$$f(x) = Ax^3 \tag{12.15}$$

and the period of oscillation is given by

$$T = \frac{4}{a}\sqrt{\frac{2}{A}}\int_0^1 \frac{d\xi}{\sqrt{1-\xi^4}} \tag{12.16}$$

Now

$$\begin{aligned}\int_0^1 \frac{\sqrt{2}d\xi}{\sqrt{1-\xi^4}} &= \int_0^1 \sqrt{2}\,(1-\xi^4)^{-\frac{1}{2}}\,d\xi \\ &= \int_0^1 \sqrt{2}\left(1+\frac{1}{2}\xi^4+\frac{3}{8}\xi^8+\frac{5}{16}\xi^{12}+\frac{35}{128}\xi^{16}+\cdots\right)d\xi \\ &= \sqrt{2}\left(1+\frac{1}{2\cdot 5}+\frac{3}{8\cdot 9}+\frac{5}{16\cdot 13}+\frac{35}{128\cdot 17}+\cdots\right) \\ &= 1.8541\end{aligned}$$

Hence, from Eqn (12.16),

$$T = \frac{4}{a}\sqrt{\frac{1}{A}}\cdot 1.8541 = \frac{7.4164}{a}\sqrt{\frac{1}{A}} \tag{12.17}$$

The frequency of free vibration is

$$p = \frac{2\pi}{T} = 0.8472a\sqrt{A} \tag{12.18}$$

This shows that the frequency of free vibration is directly proportional to the amplitude.

(B) Simple Pendulum

A simple pendulum consists of a heavy particle suspended by a weightless inextensible rod from a rigid support. It can oscillate freely under the action of gravity about an axis perpendicular to the suspended rod.

Let m = mass of the particle called bob,
l = length of the pendulum taken from the point of suspension to the centre of gravity of the particle
and g = acceleration due to gravity.

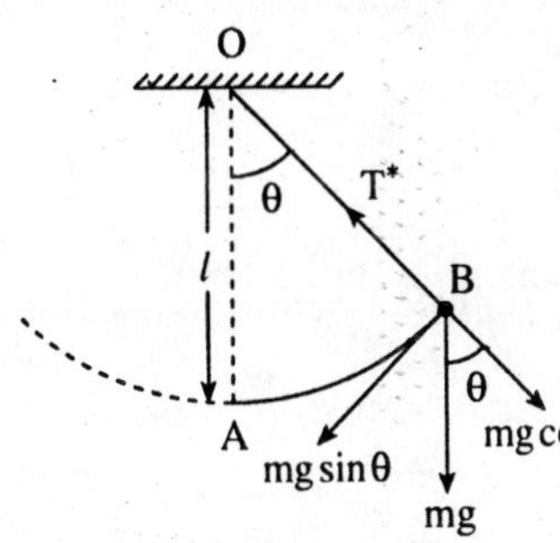

FIG. 12.2

In the figure (FIG. 12.2), OA represents the equilibrium position of the pendulum and OB represents its displaced position.

The forces acting on the bob are:

(1) The weight mg acting vertically downwards and

(2) Tension T^* in the rod acting along BO.

The tangential component $mg\sin\theta$ of the weight mg is responsible for restoring the bob to its original position.

Then the equation of motion is

$$ml\ddot{\theta} = -mg\sin\theta, \text{ i.e., } \ddot{\theta} = -\frac{g}{l}\sin\theta \tag{12.19}$$

The negative sign in the right-hand side is taken since the restoring force is always directed opposite to the direction of motion.

If the oscillation is very small, $\sin\theta \simeq \theta$, so that Eqn (12.19) assumes the form:

$$\ddot{\theta} = -\frac{g}{l}\theta \tag{12.20}$$

As we know, this represents a simple harmonic motion with time period

$$T_0 = 2\pi\sqrt{\frac{l}{g}} \tag{12.21}$$

The situation, however, is totally different when the amplitude of oscillations are quite large. In this case, we have to consider the non-linear second order differential Eqn (12.19). Multiplying both sides of the equation by $\dot{\theta}dt$, we get

$$\ddot{\theta}\dot{\theta}dt = -\frac{g}{l}\sin\theta\dot{\theta}dt = -\frac{g}{l}\sin\theta\, d\theta, \text{ i.e., } \frac{1}{2}d(\dot{\theta}^2) = \frac{g}{l}d(\cos\theta)$$

Integrating,

$$\frac{1}{2}\dot{\theta}^2 = \frac{g}{l}\cos\theta + C \tag{12.22}$$

where C is a constant.

We assume that when the pendulum has got its maximum angular displacement α, it momentarily comes to rest. That is

$$\dot{\theta} = 0, \text{ when } \theta = \alpha. \tag{12.23}$$

Using this condition in Eqn (12.22), we find

$$C = -\frac{g}{l}\cos\alpha \tag{12.24}$$

Hence,

$$\dot{\theta}^2 = \frac{2g}{l}(\cos\theta - \cos\alpha), \text{ i.e., } \sqrt{\frac{2g}{l}}dt = \frac{d\theta}{\sqrt{\cos\theta - \cos\alpha}} \tag{12.25}$$

If T be the period of oscillation, then integrating between the limits $\theta = +\alpha$ to $\theta = -\alpha$, we have

$$\sqrt{\frac{2g}{l}}\cdot\frac{T}{2} = \int_{-\alpha}^{\alpha}\frac{d\theta}{\sqrt{\cos\theta - \cos\alpha}}, \text{ i.e., } \sqrt{\frac{g}{l}}T = \int_{-\alpha}^{\alpha}\frac{d\theta}{\sqrt{\sin^2\frac{\alpha}{2} - \sin^2\frac{\theta}{2}}}. \tag{12.26}$$

where we have used the trigonometrical identities

$$\cos\theta = 1 - 2\sin^2\frac{\theta}{2} \text{ and } \cos\alpha = 1 - 2\sin^2\frac{\alpha}{2} \tag{12.27}$$

To evaluate the integral in the right-hand side, we substitute

$$\sin\frac{\theta}{2} = \sin\frac{\alpha}{2}\sin\phi \tag{12.28}$$

Then

$$\frac{1}{2}\cos\frac{\theta}{2}d\theta = \sin\frac{\alpha}{2}\cos\phi d\phi$$

$$\text{i.e., } d\theta = \frac{2\sin\frac{\alpha}{2}\cos\phi}{\cos\frac{\theta}{2}}d\phi = \frac{2\sin\frac{\alpha}{2}\cos\phi}{\sqrt{1-\sin^2\frac{\theta}{2}}}d\phi = \frac{2\sin\frac{\alpha}{2}\cos\phi}{\sqrt{1-\sin^2\frac{\alpha}{2}\sin^2\phi}}d\phi \tag{12.29}$$

The limits of integration change to:

$$\left.\begin{aligned} \text{for } \theta = -\alpha, \quad \phi = -\frac{\pi}{2} \\ \text{and for } \theta = +\alpha, \quad \phi = +\frac{\pi}{2} \end{aligned}\right\} \tag{12.30}$$

Using the relations (12.28) and (12.29) and the new limits of integration in (12.30), we write the integral as

$$\begin{aligned} \int_{-\pi/2}^{\pi/2} \frac{2d\phi}{\sqrt{1-\sin^2\frac{\alpha}{2}\sin^2\phi}} &= \int_{-\pi/2}^{\pi/2} 2\left(1-\sin^2\frac{\alpha}{2}\sin^2\phi\right)^{-1/2} d\phi \\ &= \int_{-\pi/2}^{\pi/2} 2\left[1+\frac{1}{2}\sin^2\frac{\alpha}{2}\sin^2\phi+\frac{3}{8}\sin^4\frac{\alpha}{2}\sin^4\phi+\cdots\right] \\ &= 2\left[\pi+\frac{1}{2}\sin^2\frac{\alpha}{2}\left(\frac{\pi}{2}\right)+\frac{3}{8}\sin^4\frac{\alpha}{2}\left(\frac{3\pi}{8}\right)+\cdots\right] \\ &= 2\pi\left[1+\frac{1}{4}\sin^2\frac{\alpha}{2}+\frac{9}{64}\sin^4\frac{\alpha}{2}+\cdots\right] \end{aligned} \tag{12.31}$$

Hence, from Eqns (12.26) and (12.31), we get

$$T = T_0\left(1+\frac{1}{4}\sin^2\frac{\alpha}{2}+\frac{9}{64}\sin^4\frac{\alpha}{2}+\cdots\right) \tag{12.32}$$

This gives the period of anharmonic oscillation when the amplitude is large.

For $\alpha = 90°$, we get

$$T = T_0\left[1+\frac{1}{4}\left(\frac{1}{\sqrt{2}}\right)^2+\frac{9}{64}\left(\frac{1}{\sqrt{2}}\right)^4+\cdots\right] = 1.16T_0$$

12.3 Approximate Methods of Solution

(A) Method of Successive Approximations

We consider a driven anharmonic oscillator, the restoring force of which is given by

$$P(x) = Kx + \epsilon x^3 \tag{12.33}$$

where K and ϵ are positive constants and ϵ is sufficiently small.

Let the driving force be given by

$$F(t) = F_0 \cos\omega t \tag{12.34}$$

Then the equation of motion is

$$\begin{aligned} m\ddot{x} + Kx + \epsilon x^3 &= F_0\cos\omega t \\ \text{i.e.,}\ \ddot{x} &= -\frac{K}{m}x - \frac{\epsilon}{m}x^3 + \frac{F_0}{m}\cos\omega t \\ \text{i.e.,}\ \ddot{x} &= -p_0^2 x - \alpha x^3 + P_0\cos\omega t \end{aligned} \tag{12.35}$$

where we have denoted

$$\left.\begin{aligned} p_0^2 &= K/m \\ \alpha &= \epsilon/m \\ P_0 &= F_0/m \end{aligned}\right\} \tag{12.36}$$

As a first approximation of the solution of Eqn (12.35), we assume that

$$x_1 = A\cos\omega t \tag{12.37}$$

We substitute this into the right side of Eqn (12.35) and get the equation for the second approximation in the form:

$$\ddot{x}_2 = (P_0 - p_0^2 A)\cos\omega t - \alpha A^3\cos^3\omega t \tag{12.38}$$

Recalling the trigonometric identity,

$$\cos^3\theta = \frac{3}{4}\cos\theta + \frac{1}{4}\cos 3\theta \tag{12.39}$$

we write Eqn (12.38) as

$$\ddot{x}_2 = \left(P_0 - p_0^2 A - \frac{3\alpha A^3}{4}\right)\cos\omega t - \frac{\alpha A^3}{4}\cos 3\omega t \tag{12.40}$$

Integrating twice, we get

$$x_2 = \frac{1}{\omega^2}\left[\left(p_0^2 A - P_0 + \frac{3\alpha A^3}{4}\right)\cos\omega t + \frac{\alpha A^3}{36}\cos 3\omega t\right] \tag{12.41}$$

Proceeding as above, we can obtain the third approximation to the solution of Eqn (12.35). However, since α is very small, the second approximation given by Eqn (12.41) should be sufficient.

Eqn (12.41) shows that, whatever may be the value of the frequency ω of the driving force, the amplitude cannot increase indefinitely.

(B) Lyapunov's Method

We consider the anharmonic oscillator without driving force for which the equation of motion is

$$\ddot{x} + p_0^2 x + \alpha x^3 = 0 \tag{12.42}$$

where the coefficient α is sufficiently small.

The solution of Eqn (12.42) is assumed in the form of a series containing powers of the small parameter α. Thus, we take

$$x = x_0 + \alpha x_1 + \alpha^2 x_2 + \cdots \tag{12.43}$$

We also take the coefficient p_0^2 in the form of the series

$$p_0^2 = p^2 + \alpha c_1 + \alpha^2 c_2 + \cdots \tag{12.44}$$

where $p^2, c_1, c_2, \ldots$ are unknown constants.

Considering only three terms in Eqns (12.43) and (12.44) and substituting the corresponding expressions in Eqn (12.42), we get

$$\begin{aligned}(\ddot{x}_0 + \alpha\ddot{x}_1 + \alpha^2\ddot{x}_2) + (p^2 + \alpha c_1 + \alpha^2 c_2)(x_0 + \alpha x_1 + \alpha^2 x_2)&\\ + \alpha(x_0 + \alpha x_1 + \alpha^2 x_2)^3 = 0&\end{aligned} \tag{12.45}$$

Retaining terms only up to α^2 and arranging them in powers of α, we get

$$\begin{aligned}(\ddot{x}_0 + p^2 x_0)\alpha^0 + \alpha(\ddot{x}_1 + p^2 x_1 + c_1 x_0 + x_0^3)&\\ + \alpha^2(\ddot{x}_2 + p^2 x_2 + c_2 x_0 + c_1 x_1 + 3x_0^2 x_1) = 0&\end{aligned} \tag{12.46}$$

This equation should hold good for any value of the small parameter α. Hence, the coefficients of $\alpha^0, \alpha^1, \alpha^2$ should vanish separately. Thus,

$$\left.\begin{aligned} \ddot{x}_0 + p^2 x_0 &= 0 \\ \ddot{x}_1 + p^2 x_1 &= -c_1 x_0 - x_0^3 \\ \ddot{x}_2 + p^2 x_2 &= -c_2 x_0 - c_1 x_1 - 3x_0^2 x_1 \end{aligned}\right\} \tag{12.47}$$

The system of equations can be solved like this:

We solve the first equation and determine x_0 consistent with the initial conditions. Then we substitute the known function x_0 to the right side of the second equation and find x_1 consistent with initial condition. Similarly, the function x_2 can be determined. Then the solution of Eqn (12.42) will be given by Eqn (12.43).

Let us now consider the initial conditions

$$\left.\begin{aligned} x &= a, \quad \text{for } t = 0 \\ \text{and } \dot{x} &= 0, \quad \text{for } t = 0 \end{aligned}\right\} \tag{12.48}$$

Using these conditions in Eqn (12.43), we have

$$\left.\begin{aligned} x_0(0) + \alpha x_1(0) + \alpha^2 x_2(0) &= a \\ \dot{x}_0(0) + \alpha \dot{x}_1(0) + \alpha^2 \dot{x}_2(0) &= 0 \end{aligned}\right\} \tag{12.49}$$

These equations can be satisfied, only if

$$\left.\begin{aligned} x_0(0) = a, \quad x_1(0) = 0, \quad x_2(0) = 0 \\ \text{and } \dot{x}_0(0) = 0, \quad \dot{x}_1(0) = 0, \quad \dot{x}_2(0) = 0 \end{aligned}\right\} \tag{12.50}$$

With these initial conditions, the solution of the first equation in (12.47) can be written as

$$x_0 = a \cos pt \tag{12.51}$$

Substituting this function in the right-hand side of the second equation in (12.47), we get

$$\ddot{x}_1 + p^2 x_1 = -c_1(a \cos pt) - a^3 \cos^3 pt \tag{12.52}$$

Using the trigonometric identity (12.39), we write the above equation in the form:

$$\ddot{x}_1 + p^2 x_1 = -\left(c_1 a + \frac{3}{4} a^3\right) \cos pt - \frac{1}{4} a^3 \cos 3pt \tag{12.53}$$

If the coefficient of $\cos pt$ in the right-hand side of Eqn (12.53) is not zero, there will be a term $t \sin pt$ in the solution. This term grows continuously along with time so that the solution function will be unbounded. Hence, in order to have a bounded solution, we must have

$$c_1 a + \frac{3}{4} a^3 = 0, \text{ i.e., } c_1 = -\frac{3}{4} a^2 \tag{12.54}$$

Thus, the constant c_1 is determined. Eqn (12.53) now assumes the form:

$$\ddot{x}_1 + p^2 x_1 = -\frac{1}{4} a^3 \cos 3pt \tag{12.55}$$

Its general solution is

$$x_1 = A \cos pt + B \sin pt + \frac{a^3}{32p^2} \cos 3pt \tag{12.56}$$

where A and B are constants of integration to be determined from initial conditions.

Using the conditions for x_1 as given in Eqn (12.50) into Eqn (12.56), we get

$$\left.\begin{aligned} A + \frac{a^3}{32p^2} &= 0 \\ B &= 0 \end{aligned}\right\} \tag{12.57}$$

Hence, the function x_1 in Eqn (12.56) becomes

$$x_1 = \frac{a^3}{32p^2}(\cos 3pt - \cos pt) \tag{12.58}$$

Thus, the solution of Eqn (12.42) up to two terms is

$$x = x_0 + \alpha x_1 = a\cos pt + \frac{\alpha a^3}{32p^2}(\cos 3pt - \cos pt) \tag{12.59}$$

The value of p in this equation is obtained from Eqns (12.44) and (12.54). Thus,

$$p^2 = p_0^2 - \alpha c_1 = p_0^2 + \frac{3}{4}\alpha a^2 \tag{12.60}$$

In order to determine the function x_2, we substitute x_0 and x_1 from Eqns (12.51) and (12.58) into the right-hand side of the third equation in Eqn (12.47). Then we get

$$\ddot{x}_2 + p^2 x_2 = -c_2 a\cos pt - c_1\frac{a^3}{32p^2}(\cos 3pt - \cos pt)$$
$$-3(a^2\cos^2 pt)\frac{a^3}{32p^2}(\cos 3pt - \cos pt)$$

Since $\cos^2 pt = \frac{1}{2}(1+\cos 2pt)$, we get

$$\ddot{x}_2 + p^2 x_2 = \left[-c_2 a + \left(c_1 + \frac{3}{2}a^2\right)\frac{a^3}{32p^2}\right]\cos pt$$
$$-\left(c_1 + \frac{3}{4}a^2\right)\frac{a^3}{32p^2}\cos 3pt - \frac{3a^2}{4}\cdot\frac{a^3}{32p^2}\cos 5pt \tag{12.61}$$

Reasoning as before, we equate to zero the coefficient of $\cos pt$. Thus,

$$-c_2 a + \left(c_1 + \frac{3}{2}a^2\right)\frac{a^3}{32p^2} = 0$$

$$\text{i.e., } c_2 = \left(c_1 + \frac{3}{2}a^2\right)\frac{a^2}{32p^2} = \left(-\frac{3}{4}a^2 + \frac{3}{2}a^2\right)\frac{a^2}{32p^2} = \frac{3a^4}{128p^2} \tag{12.62}$$

Thus, the value of c_2 is obtained.

Then Eqn (12.61) assumes the form (after using the value of c_1):

$$\ddot{x}_2 + p^2 x_2 = -\frac{3a^5}{128p^2}\cos 5pt \tag{12.63}$$

Its general solution is

$$x_2 = C\cos pt + D\sin pt + \frac{a^5}{1024p^4}\cos 5pt \tag{12.64}$$

Using the conditions for x_2 as given in Eqn (12.50), we have

$$C + \frac{a^5}{1024p^4} = 0 \text{ and } D = 0 \tag{12.65}$$

Hence, the solution in Eqn (12.64) is written as

$$x_2 = \frac{a^5}{1024p^4}(\cos 5pt - \cos pt) \tag{12.66}$$

Thus, the solution of Eqn (12.42) up to three terms is

$$x = x_0 + \alpha x_1 + \alpha^2 x_2$$
$$= a\cos pt + \frac{\alpha a^3}{32p^2}(\cos 3pt - \cos pt) + \frac{\alpha^2 a^5}{1024p^4}(\cos 5pt - \cos pt) \tag{12.67}$$

The value of p in this equation is obtained from Eqn (12.44) by substituting the values of c_1 and c_2. Thus,

$$p^2 = p_0^2 - \alpha c_1 - \alpha^2 c_2 = p_0^2 + \frac{3}{4}\alpha a^2 - \frac{3}{128}\frac{\alpha^2 a^4}{p^2} \tag{12.68}$$

This shows that the frequency of free vibration p is a function of the amplitude a. We also find that the oscillation process in Eqn (12.67) is described by three harmonic functions $\cos pt$, $\cos 3pt$, $\cos 5pt$ with decreasing amplitudes.

(C) Galerkin's Method

We consider the anharmonic oscillations described by the equation of motion

$$\ddot{x} + f(x) = 0 \tag{12.69}$$

We seek its approximate solution in the form:

$$x = a\cos(pt + \epsilon) \tag{12.70}$$

where a, p, ϵ are constants, a representing the amplitude, p the frequency and ϵ the phase.

If we substitute the function in Eqn (12.70) into the Eqn (12.69), it will not be identically zero. The essence of the method is that the integral

$$\int_0^{2\pi/p} [\ddot{x} + f(x)]\, x\, dt$$

taken during one period of oscillation should vanish.

Thus, we have

$$\int_0^{2\pi/p} [\ddot{x} + f(x)]\, x\, dt = 0 \tag{12.71}$$

Substituting the function in Eqn (12.70) into Eqn (12.71), we get

$$\int_0^{2\pi/p} [-ap^2\cos(pt + \epsilon) + f\{a\cos(pt + \epsilon)\}]\, a\cos(pt + \epsilon)dt = 0 \tag{12.72}$$

Now

$$\int_0^{2\pi/p} \cos^2(pt + \epsilon)dt = \frac{1}{2}\int_0^{2\pi/p} [1 + \cos 2(pt + \epsilon)]dt$$

Putting $z = pt + \epsilon$, i.e., $dz = p\,dt$, the integral is

$$= \frac{1}{2p}\int_\epsilon^{2\pi+\epsilon} (1 + \cos 2z)dz = \frac{1}{2p}\left[z + \frac{\sin 2z}{2}\right]_\epsilon^{2\pi+\epsilon}$$

$$= \frac{1}{2p}\left[(2\pi + \epsilon) - \epsilon + \frac{1}{2}\{\sin(4\pi + 2\epsilon) - \sin 2\epsilon\}\right] = \frac{\pi}{p}$$

Hence, Eqn (12.72) becomes

$$-\pi ap + \int_0^{2\pi/p} f[a\cos(pt + \epsilon)]\cos(pt + \epsilon)dt = 0$$

$$\text{i.e.,}\quad -\pi ap + \frac{1}{p}\int_0^{2\pi} f(a\cos z)\cos z dz = 0$$

Hence,

$$p^2 = \frac{1}{\pi a}\int_0^{2\pi} f(a\cos z)\cos z dz \tag{12.73}$$

This determines the frequency of free vibration.

As an example, let us consider the case when

$$f(x) = p_0^2 x + \alpha x^3 \tag{12.74}$$

Then $f(a \cos z) = ap_0^2 \cos z + \alpha a^3 \cos^3 z$.

Hence, the frequency of vibration by formula (12.73) is

$$p^2 = \frac{1}{\pi a} \int_0^{2\pi} (ap_0^2 \cos z + \alpha a^3 \cos^3 z) \cos z \, dz$$

$$\text{i.e., } p^2 = \frac{1}{\pi a} \int_0^{2\pi} (ap_0^2 \cos^2 z + \alpha a^3 \cos^4 z) dz$$

$$\text{i.e., } p^2 = \frac{1}{\pi a} \left[ap_0^2(\pi) + \alpha a^3 \left(\frac{3\pi}{4} \right) \right]$$

$$\text{i.e., } p^2 = p_0^2 + \frac{3}{4} \alpha a^2 \tag{12.75}$$

which coincides exactly with the result obtained by Lyapunov's method in Eqn (12.60).

Again, we can write Eqn (12.69) in the form:

$$\ddot{x} + p^2 x = p^2 x - f(x) \tag{12.76}$$

Assuming the approximate solution (12.70) and substituting it in Eqn (12.76), we get

$$\ddot{x} + p^2 x = F(t) \tag{12.77}$$

where

$$F(t) = ap^2 \cos(pt + \epsilon) - f[a \cos(pt + \epsilon)] \tag{12.78}$$

This function $F(t)$ is a periodical function with period $2\pi/p$.

Hence, we can expand it in Fourier series in the form:

$$F(t) = a_0 + a_1 \cos(pt + \epsilon) + a_2 \cos 2(pt + \epsilon) + \cdots \tag{12.79}$$

The coefficient a_1 is given by

$$a_1 = \frac{2}{T} \int_0^T F(t) \cos(pt + \epsilon) dt, \; T = \frac{2\pi}{p} \tag{12.80}$$

Reasoning as before, we must have $a_1 = 0$ in order that the solution does not increase indefinitely.

Thus,

$$\int_0^{2\pi/p} F(t) \cos(pt + \epsilon) dt = 0 \tag{12.81}$$

Substituting here the function $F(t)$ from Eqn (12.78), we get

$$\int_0^{2\pi/p} \{ap^2 \cos(pt + \epsilon) - f[a \cos(pt + \epsilon)\} \cos(pt + \epsilon) dt = 0 \tag{12.82}$$

which is exactly the Galerkin's Eqn (12.72).

This concept of ***harmonic linearisation*** is due to the Russian scientist N.M. Krylov.

12.4 Duffing's Method

Let the non-linear characteristic function per unit mass be

$$f(x) = p^2x + \beta x^3 \tag{12.83}$$

and the external exciting force/unit mass be $q\sin\omega t$. Then equation of motion is

$$\ddot{x} + p^2x + \beta x^3 = q\sin\omega t \tag{12.84}$$

This may be recast in the form:

$$\ddot{x} + \omega^2 x = (\omega^2 - p^2)x - \beta x^3 + q\sin\omega t \tag{12.85}$$

We assume that the solution of this equation is periodic in nature with period $T = 2\pi/\omega$ and hence take the initial conditions

$$x(t) = A, \dot{x}(t) = 0 \tag{12.86}$$

for $t = T/4$. Here A is the amplitude of vibration.

In the first approximation, we take

$$x(t) = A\sin\omega t \tag{12.87}$$

Inserting this function into the RHS of Eqn (12.84), we get

$$\begin{aligned}\ddot{x} + \omega^2 x &= \left[A(\omega^2 - p^2) + q\right]\sin\omega t - \beta A^3\sin^3\omega t \\ &= \left[A(\omega^2 - p^2) - \frac{3}{4}\beta A^3 + q\right]\sin\omega t + \frac{1}{4}\beta A^3\sin 3\omega t\end{aligned} \tag{12.88}$$

Solution of this equation will contain terms like $t\sin\omega t$ due to the presence of the term $\sin\omega t$ in the RHS. This will cause the displacement to increase indefinitely along with time.

Hence, we take

$$A\left(\omega^2 - p^2\right) - \frac{3}{4}\beta A^3 + q = 0 \tag{12.89}$$

Thus,

$$A = \frac{\frac{3}{4}\beta A^3 - q}{\omega^2 - p^2} \tag{12.90}$$

This determines the value of the amplitude.

Eqn (12.88) now assumes the form:

$$\ddot{x} + \omega^2 x = \frac{1}{4}\beta A^3\sin 3\omega t \tag{12.91}$$

Its general solution is

$$x(t) = C\sin\omega t + D\cos\omega t - \frac{\beta A^3}{32\omega^2}\sin 3\omega t \tag{12.92}$$

Using the initial conditions (12.86) by putting $t = \pi/2\omega$, we get

$$C = A + \frac{\beta A^3}{32\omega^2}, D = 0 \tag{12.93}$$

Hence, the final solution is

$$x(t) = A\sin\omega t + \frac{\beta A^3}{32\omega^2}(\sin\omega t - \sin 3\omega t) \tag{12.94}$$

This shows that the vibration contains harmonics of frequencies ω and 3ω.

12.5 Effect of Non-linear Frictional Force

Let the inelastic force of friction acting on a system be a non-linear function of velocity, so that

$$F = F(\dot{x}) \tag{12.95}$$

We assume that F^* be the ***equivalent constant force of resistance*** due to viscous damping with coefficient c^*.

Then

$$F^* = c^*\dot{x} \tag{12.96}$$

Let the driving force be harmonic in the form:

$$P(t) = P_0 \sin \omega t \tag{12.97}$$

We assume the response of the vibrating system as

$$x(t) = A \sin \omega t \tag{12.98}$$

Work done by the force F^* in one cycle

$$W_1 = \int_0^T (c^*\dot{x})\dot{x}dt = \int_0^T c^* A^2\omega^2 \cos^2 \omega t dt = \pi c^* A^2 \omega \tag{12.99}$$

Work done by the force F in one cycle

$$W_2 = \int_0^T F(\dot{x})\dot{x}dt \tag{12.100}$$

The value of the equivalent damping coefficient is obtained from the condition, $W_1 = W_2$.

This yields

$$c^* = \frac{1}{\pi A^2 \omega} \int_0^T F(\dot{x})\dot{x}dt \tag{12.101}$$

This can be evaluated if the force function $F(\dot{x})$ is given. Knowing c^*, we can determine the amplitude of vibration by putting $c = c^*$ in the well-known equation. Thus,

$$A = \frac{P_0/m}{\sqrt{(\xi^2 - \omega^2) + 4\alpha^2\omega^2}} \tag{12.102}$$

where $2\alpha = c^*/m$.

Obviously, Eqn (12.102) gives a non-linear relation for determining A.

If, in particular,

$$F = \lambda \dot{x} \, |\dot{x}|^{n-1} \tag{12.103}$$

then

$$\int_0^T F(\dot{x})\dot{x}dt = \int_0^T \lambda|\dot{x}|^{n+1} = \lambda(A\omega)^{n+1} \int_0^T \cos^2 \omega t| \cos \omega t|^{n-1} dt$$

$$= \lambda A^{n+1}\omega^n \int_0^{2\pi} |\cos \psi|^{n+1} d\psi = \lambda A^{n+1} \omega^n I \tag{12.104}$$

where we have denoted

$$I = \int_0^{2\pi} |\cos \psi|^{n+1} d\psi \tag{12.105}$$

Then, from Eqns (12.101) and (12.104), we get

$$c^* = \frac{1}{\pi A^2 \omega} \cdot \lambda A^{n+1} \omega^n I = \frac{\lambda I}{\pi} (A\omega)^{n-1} \tag{12.106}$$

The values of I are 3.14 for $n = 1$, 2.67 for $n = 2$ and 2.36 for $n = 3$.

EXAMPLE 12.1 *Find the equation of horizontal motion of the mass m attached at the middle of an elastic string of length 2L as shown in* FIG. 12.3.

Solution. Let N_0 = initial tension in the string

A = cross-sectional area of the string

E = elastic modulus of the material of the string

x = small displacement of the mass m in the horizontal direction.

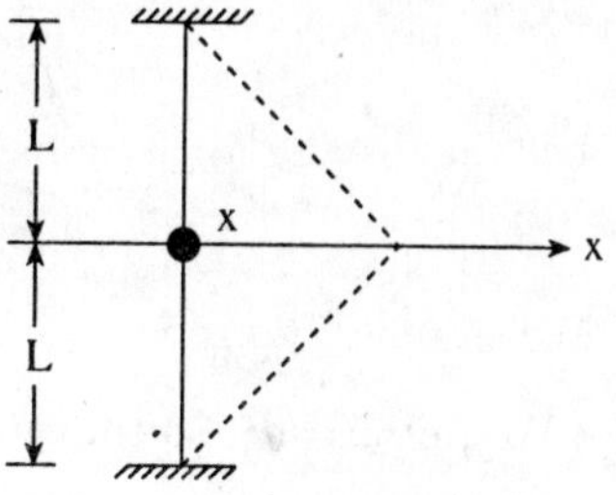

FIG. 12.3

Relative extension of each portion of the string

$$= \frac{\sqrt{L^2 + x^2} - L}{L} = \left(1 + \frac{x^2}{L^2}\right)^{1/2} - 1 \simeq \frac{x^2}{2L^2} \tag{1}$$

Hence, the total tension in the string is

$$N = N_0 + AE\frac{x^2}{2L^2} \tag{2}$$

The horizontal component of this force is

$$P = 2N\frac{x}{\sqrt{x^2 + L^2}} \approx 2\left(N_0 + AE\frac{x^2}{2L^2}\right)\frac{x}{L}\left(1 - \frac{1}{2}\frac{L^2}{x^2}\right) = \frac{2N_0 x}{L} + AE\frac{x^3}{L^3} \tag{3}$$

Hence, the equation of horizontal motion of the mass m is

$$m\ddot{x} + \frac{2N_0 x}{L} + AE\frac{x^3}{L^3} = 0 \tag{4}$$

This can be written as

$$\ddot{x} + \alpha x + \beta x^3 = 0 \tag{5}$$

where we have denoted

$$\alpha = \frac{2N_0}{L}, \; \beta = \frac{AE}{L^3} \tag{6}$$

If the initial tension of the string is absent, then $\alpha = 0$, and the equation of motion assumes the form:

$$\ddot{x} + \beta x^3 = 0 \tag{7}$$

This shows that the equation of motion is non-linear and the response will be anharmonic.

EXAMPLE 12.2 *Determine the time period of vibration of the system shown in* FIG. 12.3 *by exact method.*

Solution. We write the equation of motion in the form:

$$\ddot{x} + f(x) = 0 \tag{1}$$

where

$$f(x) = \alpha(x + \lambda x^3), \lambda = \frac{\beta}{\alpha} \tag{2}$$

Then

$$\int_x^a f(x)dx = \int_x^a \alpha\left(x + \lambda x^3\right) dx$$

$$= \frac{\alpha}{2}\left(a^2 + \frac{\lambda a^4}{2}\right) - \frac{\alpha}{2}\left(x^2 + \frac{\lambda x^4}{2}\right)$$

Here a = amplitude of vibration.

Hence, by Equ (12.10), the time period is

$$T = \frac{4}{\sqrt{\alpha}}\int_0^a \frac{dx}{\sqrt{\left(\frac{\lambda a^4}{2} + a^2\right) - \left(\frac{\lambda x^4}{2} + x^2\right)}}$$

$$= \frac{4}{\sqrt{\alpha}}\int_0^a \frac{dx}{\sqrt{(a^2 - x^2)\left(1 + \frac{\lambda}{2}x^2 + \frac{\lambda}{2}a^2\right)}} \tag{3}$$

We put $x = a\xi$, so that $dx = ad\xi$.

Then

$$1 + \frac{\lambda}{2}x^2 + \frac{\lambda}{2}a^2 = 1 + \frac{\lambda}{2}a^2 + \frac{\lambda}{2}a^2\xi^2 = \frac{2 + \lambda a^2}{2} + \frac{\lambda}{2}a^2\xi^2$$

$$= \frac{\lambda a^2}{2}\left(\frac{2 + \lambda a^2}{\lambda a^2} + \xi^2\right) = \frac{\lambda a^2}{2}(b^2 + \xi^2) \tag{4}$$

where

$$b^2 = \frac{2 + \lambda a^2}{\lambda a^2} \tag{5}$$

Hence, from (3), we get

$$T = \frac{4}{\sqrt{\alpha}}\frac{1}{a}\sqrt{\frac{2}{\lambda}}\int_0^1 \frac{d\xi}{\sqrt{(1 - \xi^2)(b^2 + \xi^2)}} \tag{6}$$

The integral in (6) represents a standard elliptic integral of first kind and its value is given by

$$I = \frac{1}{c}F\left(\frac{1}{c}, \phi\right) \tag{7}$$

where

$$\left.\begin{aligned} c^2 &= 1 + b^2 = 1 + \frac{2 + \lambda a^2}{\lambda a^2} \\ \phi &= \sin^{-1}\left(\frac{c}{\sqrt{1 + b^2}}\right) = \frac{\pi}{2} \end{aligned}\right\} \tag{8}$$

From (6) and (7), the time period is

$$T = \frac{4\sqrt{2}}{a\sqrt{\beta}}\frac{1}{c}F\left(\frac{1}{c}, \frac{\pi}{2}\right) \tag{9}$$

Thus, we find that the time period depends on the amplitude of vibration.

EXAMPLE 12.3 *The characteristic function of a system is given by*

$$f(x) = p^2x + \alpha x^2.$$

Determine by the method of successive approximation the response function.

Solution. The equation of motion in this case is

$$\ddot{x} + p^2 x + \alpha x^2 = 0 \tag{1}$$

We assume the initial conditions

$$x = A, \dot{x} = 0, \text{ at } t = 0 \tag{2}$$

where A is the amplitude of vibration.

In the first approximation, we take the solution of (1) as

$$x_1(t) = A \cos pt \tag{3}$$

Inserting this into (1), we get for the second approximation

$$\ddot{x}_2 + p^2 x_2 + \alpha A^2 \cos^2 pt = 0$$

$$\text{i.e.,} \quad \ddot{x}_2 + p^2 x_2 + \frac{1}{2}\alpha A^2(1 + \cos 2pt) = 0$$

$$\text{i.e.,} \quad \ddot{x}_2 + p^2\left(x_2 + \frac{\alpha A^2}{2p^2}\right) + \frac{\alpha A^2}{2}\cos 2pt = 0$$

$$\text{i.e.,} \quad \ddot{z} + p^2 z = -\frac{\alpha A^2}{2}\cos 2pt \tag{4}$$

where

$$z = x_2 + \frac{\alpha A^2}{2p^2} \tag{5}$$

Solution of (4) is

$$z(t) = C_1 \cos pt + C_2 \sin pt + \frac{\alpha A^2}{6p^2}\cos 2pt$$

$$\text{i.e.,} \quad x_2(t) = -\frac{\alpha A^2}{2p^2} + C_1 \cos pt + C_2 \sin pt + \frac{\alpha A^2}{6p^2}\cos 2pt \tag{6}$$

Using the initial conditions at (2), we find

$$C_1 = \frac{\alpha A^2}{3p^2}, C_2 = 0 \tag{7}$$

Hence,

$$x_2(t) = -\frac{\alpha A^2}{2p^2} + \frac{\alpha A^2}{6p^2}(2\cos pt + \cos 2pt) \tag{8}$$

The complete solution, therefore, is

$$x(t) = x_1(t) + x_2(t)$$

$$= -\frac{\alpha A^2}{2p^2} + A\cos pt\left(1 + \frac{\alpha A}{3p^2}\right) + \frac{\alpha A^2}{6p^2}\cos 2pt$$

EXAMPLE 12.4 *Determine the time period of a simple pendulum for large amplitude vibration by Galerkin's method.*

Solution. The equation of motion of a simple pendulum is

$$\ddot{\theta} + \frac{g}{l}\sin\theta = 0 \tag{1}$$

where l is the length of the pendulum.

We expand $\sin\theta$ in series and retain only two terms, so that

$$\sin\theta = \theta - \frac{\theta^3}{6} \tag{2}$$

Then equation of motion assumes the form:

$$\ddot{\theta} + \frac{g}{l}\theta - \frac{g}{6l}\theta^3 = 0 \tag{3}$$

Comparing this equation with Eqn (12.74), we get

$$p_0^2 = \frac{g}{l}, \alpha = -\frac{g}{6l} \tag{4}$$

Hence, by Eqn (12.75), frequency of vibration p is given by

$$p^2 = p_0^2 + \frac{3}{4}\alpha\theta_0^2 = \frac{g}{l} - \frac{g}{8l}\theta_0^2 \tag{5}$$

where θ_0 is the angular amplitude.

Now, from (5),

$$\frac{1}{p} = \sqrt{\frac{l}{g}}\left(1 - \frac{\theta_0^2}{8}\right)^{-\frac{1}{2}} = \sqrt{\frac{l}{g}}\left(1 + \frac{\theta_0^2}{16}\right) \tag{6}$$

Thus, the time period is

$$T = \frac{2\pi}{p} = 2\pi\sqrt{\frac{l}{g}}\left(1 + \frac{\theta_0^2}{16}\right) = T_0\left(1 + \frac{\theta_0^2}{16}\right) \tag{7}$$

where $T_0 = 2\pi\sqrt{l/g}$ = time period for small oscillations.

For $\theta_0 = \pi/2$, we get from (7)

$$T = T_0\left[1 + \frac{(3.14)^2}{4 \times 16}\right] = 1.154T_0 \tag{8}$$

which compares well with the known result.

EXAMPLE 12.5 *The inelastic force of friction in a non-linear system is given as*

$$F(\dot{x}) = \lambda\dot{x}^2,\ \lambda = constant.$$

Determine an expression for the amplitude of vibration.

Solution. Here $n = 2$ in Eqn (12.105). Then

$$I = \int_0^{2\pi} |\cos\psi|^3 d\psi = 2.67 \tag{1}$$

By Eqn (12.106),

$$c^* = \frac{\lambda I}{\pi}(A\omega) = \frac{2.67}{3.14}(\lambda A\omega) = 0.85\lambda A\omega \tag{2}$$

Then the amplitude is given by

$$A = \frac{P_0/m}{\sqrt{(\xi^2 - \omega^2)^2 + (c^*\omega/m)^2}} = \frac{P_0/m}{\sqrt{(\xi^2 - \omega^2)^2 + (0.85\lambda A\omega^2/m)^2}} \tag{3}$$

Now

$$\left.\begin{aligned} \frac{P_0}{m} &= \frac{P_0}{k}\cdot\frac{k}{m} = \frac{P_0}{k}\xi^2 \\ \text{and } \frac{1}{m} &= \frac{k}{m}\cdot\frac{1}{k} = \frac{\xi^2}{k} \end{aligned}\right\} \tag{4}$$

Using (4), we get from (3)

$$A = \frac{P_0/k}{\sqrt{\left(1-\frac{\omega^2}{\xi^2}\right)^2 + \left(\frac{0.85\lambda A\omega^2}{k}\right)^2}} \tag{5}$$

Squaring and arranging the terms,

$$A^4\left(\frac{0.85\lambda\omega^2}{k}\right)^2 + A^2\left(1-\frac{\omega^2}{\xi^2}\right)^2 - \left(\frac{P_0}{k}\right)^2 = 0 \tag{6}$$

This yields

$$A^2 = \frac{1}{2B}\left[-\left(1-\frac{\omega^2}{\xi^2}\right)^2 + \sqrt{D}\right] \tag{7}$$

where

$$B = \left(\frac{0.85\lambda\omega^2}{k}\right)^2 \tag{8}$$

$$\text{and } D = \left(1-\frac{\omega^2}{\xi^2}\right)^4 + 4\left(\frac{P_0}{k}\right)^2 B \tag{9}$$

The maximum amplitude is obtained when $\omega = \xi$ approximately. Thus,

$$A_{\text{max}}^2 = \frac{1}{2B}\cdot\sqrt{4\left(\frac{P_0}{k}\right)^2 B} = \frac{P_0}{k}\cdot\frac{1}{\sqrt{B}} = \frac{P_0}{k}\cdot\frac{k}{0.85\lambda\omega^2}$$

$$= \frac{k}{0.85\lambda\xi^2}\frac{P_0}{k} = \frac{P_0}{k}\cdot\frac{m\xi^2}{0.85\lambda\xi^2} = \frac{1}{0.85}\frac{mP_0}{\lambda k} \tag{10}$$

Hence,

$$A_{\text{max}} = 1.08\sqrt{\frac{mP_0}{\lambda k}}$$

EXERCISE

1. A system is governed by the equation

$$\ddot{x} + x + \lambda x\cos x = 0, \lambda = \text{constant}.$$

Develop an equation for velocity if $v = v_0$, when $x = 0$.

[**Ans.** $v = \sqrt{v_0^2 - x^2 + 2\lambda[1-\cos x - x\sin x]}$]

2. Determine by the method of successive approximation the response of the system $f(x) = p^2x + \alpha x^3$.

[**Ans.** $x(t) = A\cos pt + \frac{\alpha A^3}{4(9p^2-\omega^2)}[\cos 3pt - \cos pt]$
$\omega^2 = p^2 + \frac{3}{4}\alpha A^2$]

3. Assume an approximate solution of the system

$$\ddot{x} + 2\alpha\dot{x} + p^2(x + \beta x^3) = 0, \beta = \text{small},$$

in the form $x(t) = A\cos\omega t$ with conditions $x = A$ and $\dot{x} = 0$, when $t = 0$. Determine an approximate value of ω.

[**Ans.** $\omega = p\sqrt{1+\beta A^2}$]

4. A mass $m = 45$ kg hung from ceiling by a spring whose stiffness characteristic is given by $K(x) = k_0x + k_1x^3$ executes vibration in vertical direction. Find the time period by Lyapunov's formula, given

$$k_0 = 7200\,\text{kg/cm},\ k_1 = 100\,\text{kg/cm}^3,\ x(0) = 2.5\,\text{cm},\ \dot{x}(0) = 0$$

[**Ans.** 0.48 second]

5. Use Galerkin's method to find an approximate value of amplitude of the system

$$\ddot{x} + x + \alpha(\dot{x}^2 + x^2 - 1)\dot{x} = 0.$$

[**Ans.** $\sqrt{1 + \frac{1}{\alpha}}$]

CHAPTER 13

Gravitational Attraction and Potential

13.1 The Law of Gravitation

The law of gravitation discovered by Newton states:

> ***Every particle in the universe attracts every other particle with a force which is directly proportional to the product of the masses of the particles and inversely proportional to the square of the distance between them.***

Thus, if m_1, m_2 denote the masses of two particles and r their distance apart, then the force of attraction F between them is

$$F = \gamma \frac{m_1 m_2}{r^2} \tag{13.1}$$

where γ is known as the ***constant of gravitation***.

The value of γ as determined experimentally is 6.67×10^{-8} in C.G.S. unit.

In order to avoid a continuous repetition of the constant γ, we sometimes choose a new unit of mass such that $\gamma = 1$. Such units are called ***astronomical units***.

Now, if f be the acceleration produced by the attraction of a particle of mass m_1 on the particle of mass m_2, then, by Eqn (13.1),

$$m_2 f = \gamma \frac{m_1 m_2}{r^2}, \text{ i.e., } f = \frac{\gamma m_1}{r^2} \tag{13.2}$$

Hence, $\gamma = 1$, when f, m_1 and r are all unity.

Thus, the astronomical unit of mass is the mass of the particle which produces unit acceleration at unit distance by virtue of its attraction.

Let us consider a point $P(x, y, z)$ remaining external to a body or a system of particles. The force of attraction which the body or the system of particles would exert on a unit mass placed at the point P determines the attraction of the body or the system at that point. This attraction must have a definite value at every point where the unit mass can be placed. Thus, we have the conception of a ***field of force***.

We define it as follows:

> ***A field of force is a region of space, every point of which is associated with a force definite in magnitude and direction.***

13.2 Attraction of a System of Particles

We consider a system of n particles of masses $m_1, m_2, \ldots, m_n$ situated at points $A_1(a_1, b_1, c_1)$, $A_2(a_2, b_2, c_2), \ldots$ with reference to a Cartesian coordinate system. $P(x, y, z)$ is a point external to them.

Let $r_1, r_2, \ldots, r_n$ denote the distances $PA_1, PA_2, \ldots$ etc. (FIG. 13.1).

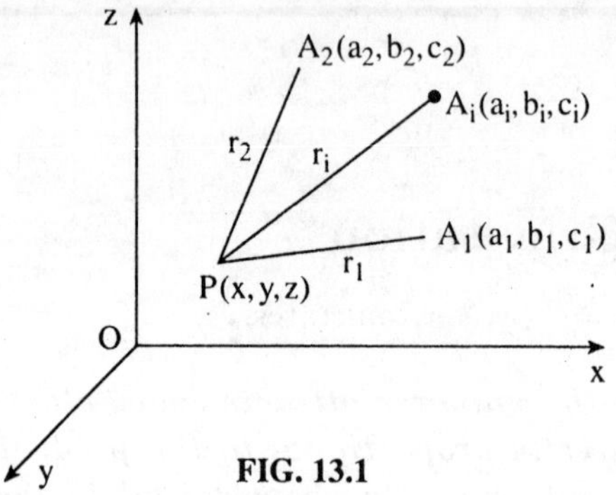

FIG. 13.1

Then

$$r_i^2 = (a_i - x)^2 + (b_i - y)^2 + (c_i - z)^2 \tag{13.3}$$

where $i = 1, 2, \ldots, n$. The attraction at P due to the mass m_i at A_i is

$$F_i = \frac{m_i}{r_i^2} \tag{13.4}$$

The line PA_i has direction cosines

$$\frac{a_i - x}{r_i}, \frac{b_i - y}{r_i}, \frac{c_i - z}{r_i}$$

Hence, the components of attraction force at P due to mass m_i along coordinate axes are

$$\left.\begin{aligned} F_{xi} &= \frac{m_i(a_i - x)}{r_i^3} \\ F_{yi} &= \frac{m_i(b_i - y)}{r_i^3} \\ F_{zi} &= \frac{m_i(c_i - z)}{r_i^3} \end{aligned}\right\} \tag{13.5}$$

By the principle of superposition of fields of force, the components of attraction force at P due to all masses along the coordinate axes are

$$\left.\begin{aligned} X &= \sum_{i=1}^{n} \frac{m_i(a_i - x)}{r_i^3} \\ Y &= \sum_{i=1}^{n} \frac{m_i(b_i - y)}{r_i^3} \\ Z &= \sum_{i=1}^{n} \frac{m_i(c_i - z)}{r_i^3} \end{aligned}\right\} \tag{13.6}$$

13.3 Gravitational Potential

We define a function $V(x, y, z)$ by the formula

$$V = \sum \frac{m_i}{r_i} \tag{13.7}$$

where r_i is as defined by Eqn (13.3). Obviously, the function V is single-valued and is related to a system of attracting particles and the summation does not depend on any particular coordinate system.

By differentiating Eqn (13.7) with respect to x, we get

$$\frac{\partial V}{\partial x} = -\sum \frac{m_i}{r_i^2} \frac{\partial r_i}{\partial x} \tag{13.8}$$

But, from Eqn (13.3) we have

$$r_i \frac{\partial r_i}{\partial x} = -(a_i - x), \text{ i.e., } \frac{\partial r_i}{\partial x} = -\frac{a_i - x}{r_i} \tag{13.9}$$

From Eqns (13.8) and (13.9), we get

$$\frac{\partial V}{\partial x} = \sum \frac{m_i (a_i - x)}{r_i^3} \tag{13.10}$$

Comparing this with Eqn (13.6), we find $X = \dfrac{\partial V}{\partial x}$. Thus, derivative of the function V with respect to x gives the component of attraction at $P(x, y, z)$ in the direction of x-axis. We can get similar other two relations, so that we write

$$\left. \begin{aligned} X &= \frac{\partial V}{\partial x} \\ Y &= \frac{\partial V}{\partial y} \\ Z &= \frac{\partial V}{\partial z} \end{aligned} \right\} \tag{13.11}$$

The function V defined by Eqn (13.7) is called the ***gravitational potential.***

Now, we can choose the direction of coordinate axes arbitrarily. Hence, from Eqn (13.11), we can say that ***the space derivative of V in any direction gives the component of attraction in that direction.*** Thus, the potential function plays a very important role in the theory of attraction.

Let us now consider a path from A to B (FIG. 13.2) and find the difference of potential $(V_B - V_A)$ between these two points. Since

$$\begin{aligned} dV &= \frac{\partial V}{\partial x} dx + \frac{\partial V}{\partial y} dy + \frac{\partial V}{\partial z} dz \\ &= X\,dx + Y\,dy + Z\,dz \end{aligned} \tag{13.12}$$

according to Eqn (13.11), we find

$$V_B - V_A = \int_A^B (X\,dx + Y\,dy + Z\,dz) \tag{13.13}$$

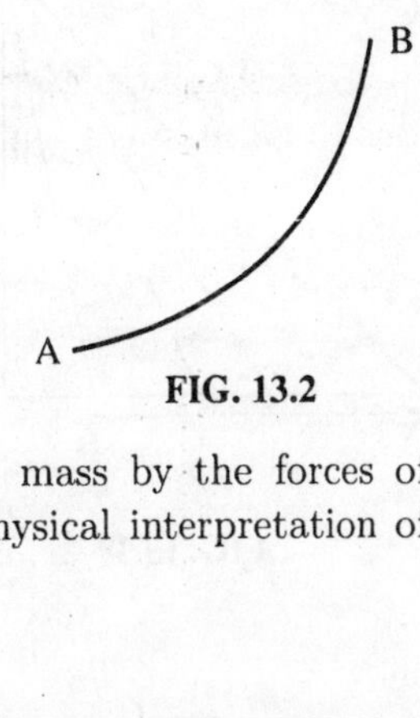

FIG. 13.2

The integral in Eqn (13.13) represents the work done on a unit mass by the forces of attraction as it moves from A to B along any path. This gives a physical interpretation of the gravitational potential.

If the point A be at infinity where the potential is zero, i.e., $V_A = 0$, then Eqn (13.13) determines the potential at B. Thus, ***the potential at a given point due to a given attracting system is the work done by the attractions of the system on a particle of unit mass when it is moved along any path from infinity up to the point considered.***

Considering a gravitational field due to a single mass m, the force of attraction on a unit mass at a distance r measured from origin of coordinates will be $F = \frac{m}{r^2}$.

Hence, work done in moving a unit mass by a small distance dr in the direction of attraction force is $dW = \frac{m}{r^2}dr$.

Then the total work done in moving the unit mass from infinity up to a point distance x from the mass m is

$$W = \int_{\infty}^{x} \frac{m}{r^2}dr = -\frac{m}{x}$$

Thus, the potential at A is

$$V_A = -\frac{m}{x} \tag{13.14}$$

Since the particle is moved in the direction of the attraction force, work is done on the unit mass and hence the negative sign appears in Eqn (13.14).

The potential V of a given attracting system is a function of the coordinates x, y, z of the point considered. Hence, the equation

$$V(x, y, z) = \text{constant}, \tag{13.15}$$

represents a surface over which the potential remains constant. Such surfaces are called ***equipotential surfaces***. It is obvious from definition of potential that only one such surface passes through any point of space. It means that ***no two equipotential surfaces can intersect.***

Again, there is no difference of potentials between any two points on the equipotential surface. This means that no work is done against gravitational force in moving unit mass along any path between two points on the equipotential surface. Hence, the resultant attraction on any point of this surface is in the direction perpendicular to the surface through the point.

13.4 Attraction of Some Solids

(A) Uniform Straight Rods

Let AB be the rod of mass per unit length m, P is the point where attraction due to the rod is to be determined, and $PM = h$. Considering a small element of length dx at a distance x from M, attraction at P is

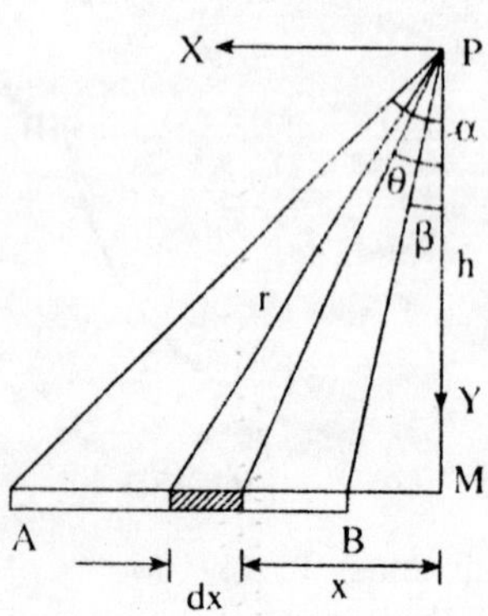

FIG. 13.3

$$dF = \gamma \frac{m}{r^2}dx \tag{13.16}$$

From the figure (FIG. 13.3),

$$x = h \tan \theta, \text{ i.e., } dx = h \sec^2 \theta d\theta.$$

Hence,

$$\begin{aligned} dF &= \gamma \frac{m}{r^2} h \sec^2 \theta d\theta \\ &= \gamma \frac{m}{r^2} h \cdot \frac{r^2}{h^2} d\theta \\ &= \frac{\gamma m}{h} d\theta \end{aligned} \tag{13.17}$$

Let X and Y be the components of attraction of the rod parallel and perpendicular to the rod. Also, let $\angle MPA = \alpha$, $\angle MPB = \beta$.

Then

$$X = \int_{\beta}^{\alpha} \frac{\gamma m}{h} \sin\theta d\theta$$

$$\text{i.e., } X = \frac{\gamma m}{h}(\cos\beta - \cos\alpha) = \frac{2\gamma m}{h} \sin\frac{1}{2}(\alpha+\beta)\sin\frac{1}{2}(\alpha-\beta) \tag{13.18}$$

Similarly,

$$Y = \int_{\beta}^{\alpha} \frac{\gamma m}{h} \cos\theta d\theta = \frac{\gamma m}{h}(\sin\alpha - \sin\beta)$$

$$= \frac{2\gamma m}{h} \cos\frac{1}{2}(\alpha+\beta)\sin\frac{1}{2}(\alpha-\beta) \tag{13.19}$$

The resultant attraction is

$$R = \sqrt{X^2+Y^2} = \frac{2\gamma m}{h} \sin\frac{1}{2}(\alpha-\beta) \tag{13.20}$$

which makes an angle of

$$\phi = \tan^{-1}\left(\frac{X}{Y}\right) = \frac{1}{2}(\alpha+\beta) \tag{13.21}$$

with the vertical line MP.

Thus, the resultant attraction acts along the bisector of the $\angle APB$.

If, in particular, the rod is infinite in length in both directions, then $\alpha = 90°$ and $\beta = -90°$.

Hence, $X = 0$, $Y = \dfrac{\gamma m}{h}[1-(-1)] = \dfrac{2\gamma m}{h}$, $R = \dfrac{2\gamma m}{h}$ and $\phi = 0$.

The resultant R acts along PM.

(B) Uniform Circular Disc

We consider a circular strip of thickness dx at a distance x from the centre (FIG. 13.4). Let m be the mass per unit area of the disc. Area of the elementary strip is $\pi(x+dx)^2 - \pi x^2 = 2\pi x\, dx$ so that its mass $= 2\pi m x\, dx$.

Let $OP = h$, $\angle OPA = \theta$ and $\angle OPB = \alpha$.

Then attraction at P due to the elementary mass is

$$dF = \frac{\gamma}{PA^2}(2\pi x m\, dx)\cos\theta \text{ along } PO.$$

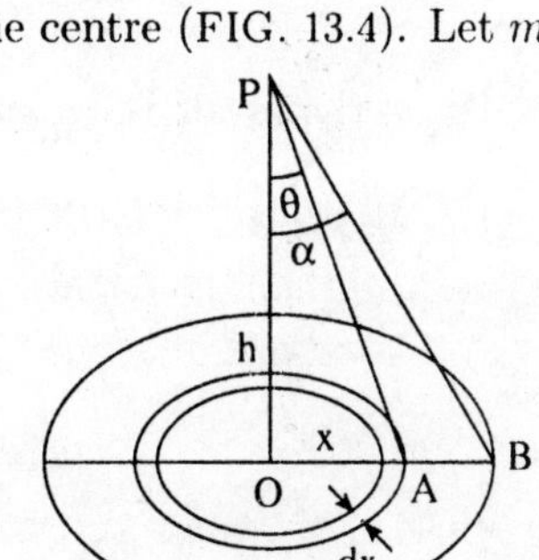

FIG. 13.4

Since $x = h\tan\theta$, $dx = h\sec^2\theta d\theta$.

$$\text{Thus, } dF = \frac{\gamma m}{h^2\sec^2\theta}(2\pi h\tan\theta)\cdot h\sec^2\theta\, d\theta \cdot \cos\theta$$

$$= 2\pi\gamma m \sin\theta\, d\theta.$$

Hence, the total attraction at P due to the whole disc is

$$F = 2\pi\gamma m \int_0^{\alpha} \sin\theta\, d\theta = 2\pi\gamma m(1-\cos\alpha) \tag{13.22}$$

If the radius of the disc is very large, then $\alpha = \pi/2$. In this case,

$$F = 2\pi\gamma m = \text{constant.} \tag{13.23}$$

(C) Uniform Spherical Shell

We consider a thin spherical shell of radius a and surface density m. Let P be a point at a distance r from the centre O of the sphere (FIG. 13.5).

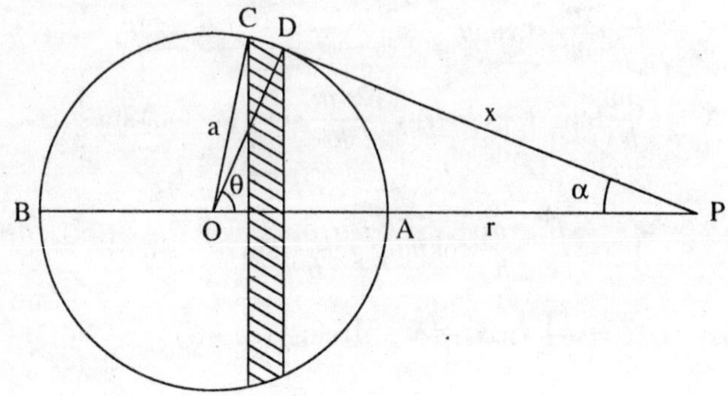

FIG. 13.5

We take an element of arc $CD = a\,d\theta$ and generate a ring of radius $a\sin\theta$.

Then its mass $= (2\pi a\sin\theta)(a\,d\theta)\cdot m = 2\pi a^2 m\sin\theta\,d\theta$.

Hence, attraction at P due to this elementary ring is

$$dF = \frac{2\pi a^2 m\gamma\sin\theta\,d\theta}{PD^2}, \qquad \text{along } PD$$

$$= \frac{2\pi a^2 m\gamma\sin\theta\,d\theta\cdot\cos\alpha}{PD^2}, \qquad \text{along } PO$$

From the figure, we have $PD^2 = x^2 = a^2 + r^2 - 2ar\cos\theta$, which yields

$$\sin\theta\,d\theta = \frac{x\,dx}{ar} \tag{13.24}$$

Also

$$\cos\alpha = \frac{x^2 + r^2 - a^2}{2xr} \tag{13.25}$$

Using these values, we find the total attraction

$$F = 2\pi a^2 m\gamma\int_\xi^\beta \frac{1}{x^2}\cdot\frac{x^2+r^2-a^2}{2xr}\cdot\frac{x}{ar}dx$$

$$= \pi a m\gamma\int_\xi^\beta \frac{1}{r^2}\left(1+\frac{r^2-a^2}{x^2}\right)dx$$

$$= \frac{\pi a m\gamma}{r^2}\left[x - \frac{r^2-a^2}{x}\right]_{x=\xi}^{x=\beta} \tag{13.26}$$

where $\xi = PA$ and $\beta = PB$.

We consider the following cases:

• Case 1 : *P* is an external point

In this case,

$$\left.\begin{aligned}\xi &= PA = r - a\\ \beta &= PB = r + a\end{aligned}\right\} \tag{13.27}$$

Then, from Eqn (13.26), we get

$$F = \frac{\pi a m\gamma}{r^2}\left[(r+a)-(r-a)-(r^2-a^2)\left\{\frac{1}{r+a}-\frac{1}{r-a}\right\}\right]$$

$$\text{i.e., } F = \frac{\pi a m\gamma}{r^2}\left[2a-(r^2-a^2)\left\{\frac{(r-a)-(r+a)}{(r^2-a^2)}\right\}\right]$$

$$= \frac{\pi a m\gamma}{r^2}(2a+2a) = \frac{4\pi a^2 m\gamma}{r^2} = \frac{M\gamma}{r^2} \tag{13.28}$$

where $M = 4\pi a^2 m$ = mass of the shell. Thus, ***the attraction due to a thin spherical shell at an external point is the same as that of a mass of the shell concentrated at its centre.***

• Case 2 : *P* is an internal point

In this case,

$$\left.\begin{aligned}\xi &= PA = a - r\\ \beta &= PB = a + r\end{aligned}\right\} \tag{13.29}$$

Then the total attraction is

$$F = \frac{\pi a m\gamma}{r^2}\left[(a+r)-(a-r)-(r^2-a^2)\left\{\frac{1}{a+r}-\frac{1}{a-r}\right\}\right]$$

$$= \frac{\pi a m\gamma}{r^2}\left[2r+(a^2-r^2)\frac{(a-r)-(a+r)}{a^2-r^2}\right]$$

$$= \frac{\pi a m\gamma}{r^2}(2r-2r) = 0 \tag{13.30}$$

Thus, ***the attraction of a spherical shell at an internal point is zero.***

• Case 3 : *P* is on the surface of the shell

In this case, $r = a, \xi = PA = 0, \beta = PB = 2a$.

Then from Eqn (13.26), the total attraction is

$$F = \frac{\pi a m\gamma}{a^2}\left[x-0\right]_0^{2a} = 2\pi m\gamma = \frac{1}{2}\gamma\cdot\frac{4\pi a^2 m}{a^2} = \frac{1}{2}\frac{M\gamma}{a^2} \tag{13.31}$$

(D) Uniform Solid Sphere

We consider a sphere of radius a and density m. The sphere can be assumed to be composed of an infinite number of concentric shells. We consider one such shell of radius x and thickness dx (FIG. 13.6).

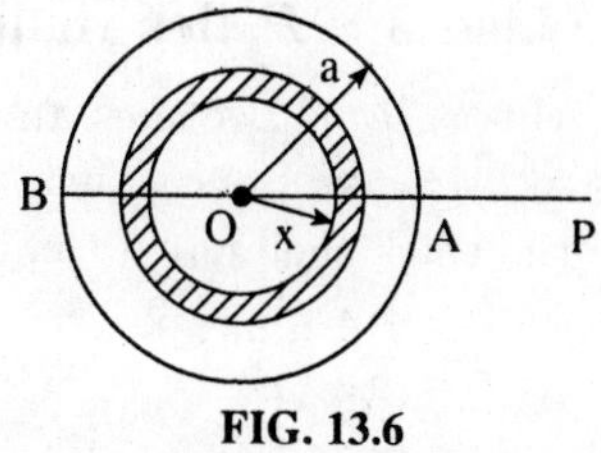

FIG. 13.6

The mass of the elementary shell is

$$dM = \left(4\pi x^2 dx\right)\cdot m \tag{13.32}$$

We consider the following cases:

• Case 1 : *P* is an external point

Let $PO = r$, where O is the centre of the shell. Then, according to Eqn (13.28), attraction at P due to the elementary shell is

$$dF = \gamma\frac{dM}{r^2} = \frac{4\pi m\gamma}{r^2}x^2 dx$$

Hence, the total attraction at P due to the whole sphere is

$$F = \frac{4\pi m\gamma}{r^2}\int_0^a x^2 dx = \frac{4}{3}\pi m a^3 \cdot \frac{\gamma}{r^2} = \frac{\gamma M}{r^2} \tag{13.33}$$

where $M = \frac{4}{3}\pi a^3 m$ = mass of the sphere.

Hence, ***the attraction of a solid sphere at an external point is the same as that of a mass equal to that of the sphere concentrated at its centre.***

• Case 2 : *P* is an internal point

Let $OP = r$. Then, for all the elementary shells with $x > r$, the point P is an internal point. Hence, by Eqn (13.30), attraction at P due to all these shells is zero. The attraction at P will be due to the solid sphere of radius r. Thus,

$$F = \frac{4\pi m\gamma}{r^2}\int_0^r x^2 dx = \frac{4}{3}\pi m\gamma r \tag{13.34}$$

• Case 3 : *P* is on the surface of the sphere

In this case, $r = a$. Then by Eqn (13.34),

$$F = \frac{4}{3}\pi m\gamma a = \frac{4}{3}\pi m a^3 \cdot \frac{\gamma}{a^2} = \gamma\frac{M}{a^2} \tag{13.35}$$

(E) Thick Spherical Shell

We consider a thick spherical shell of external and internal radii a and b respectively.

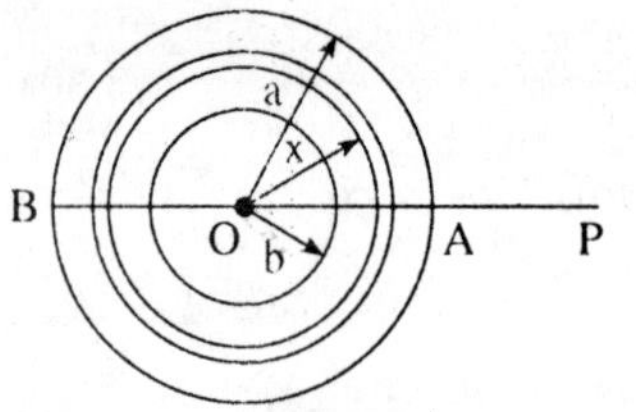

FIG. 13.7

• Case 1 : *P* is an external point

Let $OP = r$. We consider a thin elementary shell of radius x and thickness dx (FIG. 13.7).

By Eqn (13.33), the total attraction at P is

$$F = \frac{4\pi m\gamma}{r^2}\int_b^a x^2 dx = \frac{4}{3}\pi m\gamma\left(\frac{a^3 - b^3}{r^2}\right) \tag{13.36}$$

• Case 2 : *P* is an internal point

In this case, P becomes an internal point to all the elementary shells of radii x, $b \le x \le a$. Hence, by Eqn (13.30), the total attraction at P is zero.

• Case 3 : *P* lies inside the shell

In this case, P becomes an internal point to all the elementary shells whose radii are greater than $OP = r$. Hence, their attraction will be zero. Attraction will be only due to the portion of the thick shell of external radius = r and internal radius = b. Hence, the total attraction is

$$F = \frac{4\pi m\gamma}{r^2}\int_b^r x^2 dx = \frac{4}{3}\pi m\gamma\left(\frac{r^3 - b^3}{r^2}\right) \tag{13.37}$$

• Case 4 : *P* lies on the external surface

In this case, $OP = r = a$. Hence, the total attraction, by Eqn (13.36), is

$$F = \frac{4}{3}\pi m\gamma\left(\frac{a^3 - b^3}{a^2}\right) = \frac{4}{3}\pi m\gamma a(1 - \lambda^3) \tag{13.38}$$

where $\lambda = b/a$.

(F) Shell Bounded by Non-intersecting Eccentric Spheres

We consider a shell of uniform density bounded by two non-intersecting spheres of radii

$$a \text{ and } b(< a).$$

• Case 1 : *P* is external to both spheres

In the figure (FIG. 13.8), let $OP = r$ and $O'P = r'$, where O, O' are centres of the spheres. By Eqn (13.33), attraction at P due to the sphere of radius a is

$$F_1 = \frac{4}{3}\frac{\pi m\gamma a^3}{r^2}, \text{ along } PO \tag{13.39}$$

and that due to the sphere of radius b is

$$F_2 = \frac{4}{3}\frac{\pi m\gamma b^3}{r'^2}, \text{ along } O'P. \tag{13.40}$$

The resultant of these two forces will be the attraction at P due to the shell.

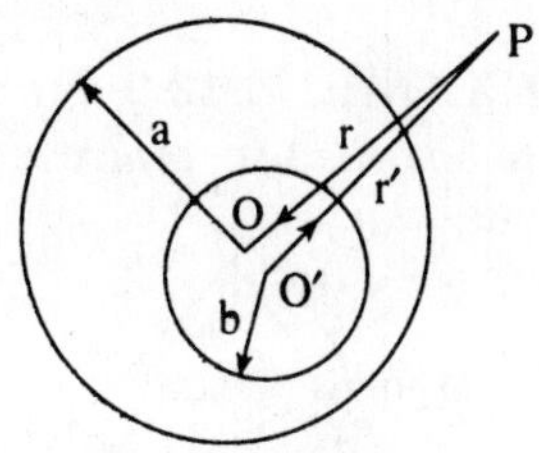

FIG. 13.8

• Case 2 : *P* is inside the shell material

In this case, P is internal to the sphere of radius a, but external to the sphere of radius b.

Hence, attraction due to sphere of radius a, by Eqn (13.34), is

$$F_1 = \frac{4}{3}\pi m\gamma r, \text{ along } PO \tag{13.41}$$

and that due to the sphere of radius b is

$$F_2 = \frac{4}{3}\frac{\pi m\gamma b^3}{r'^2}, \text{ along } O'P. \tag{13.42}$$

The net attraction will be the resultant of these two forces.

• Case 3 : *P* is within the cavity

In this case, P lies within both the spheres (FIG. 13.9).

Hence,

$$\left.\begin{aligned} F_1 &= \frac{4}{3}\pi m\gamma r, \text{ along } PO \\ F_2 &= \frac{4}{3}\pi m\gamma r', \text{ along } O'P \end{aligned}\right\} \tag{13.43}$$

Hence, the resultant attraction is

$$F = \frac{4}{3}\pi m\gamma \cdot OO' \tag{13.44}$$

acting in a direction parallel to OO'. Thus, attraction inside the shell is constant in magnitude and direction. Let us now find some applications of these formulae.

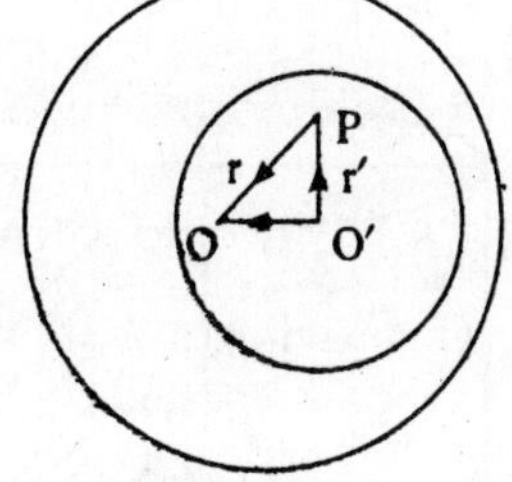

FIG. 13.9

EXAMPLE 13.1 *If a uniform rod be of infinite length, show that the work done in removing a particle of unit mass against its attraction from a perpendicular distance y_1 to y_2 is*

$$2\gamma m \log\left(\frac{y_2}{y_1}\right).$$

Solution. Let at any moment the particle be at a perpendicular distance y from the rod.

Then the attraction on it is $F = 2\gamma m/y$.

If the particle is removed by a distance dy upwards, work done against the attraction

$$= F \cdot dy = \frac{2\gamma m}{y} dy$$

Hence, the total work done in removing the particle from distance y_1 to y_2

$$= \int_{y_1}^{y_2} \frac{2\gamma m}{y} dy = 2\gamma m \log\left(\frac{y_2}{y_1}\right)$$

EXAMPLE 13.2 *Find the attraction of a uniform rod bent into the form of a circular arc of radius R at its centre.*

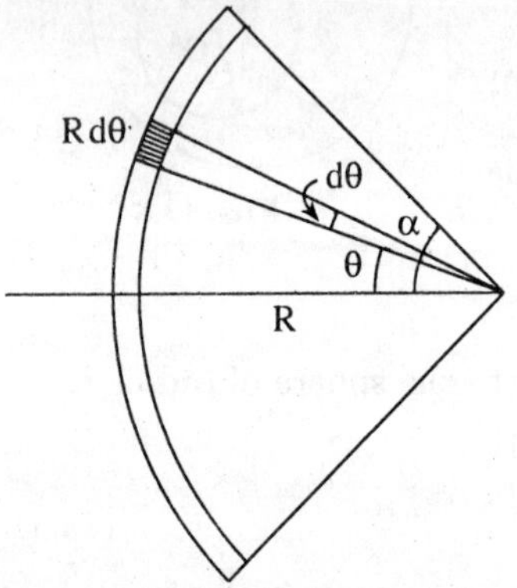

FIG. 13.10

Solution. Considering an elementary length of the arc (FIG. 13.10), attraction at P

$$= \frac{\gamma(Rd\theta)m}{R^2} = \frac{\gamma m}{R} d\theta$$

Its vertical component will be neutralised by the attraction of a similar element below and hence is neglected.

Its horizontal component is $dF = \dfrac{\gamma m}{R} \cos\theta d\theta$.

Hence, the total attraction at P is

$$F = \int_{-\alpha}^{\alpha} \frac{\gamma m}{R} \cos\theta \, d\theta = \frac{2\gamma m}{R} \sin\alpha \tag{1}$$

Here m is the mass per unit length of the rod.

If M = total mass of the rod, then

$$M = 2R\alpha m, \text{ i.e., } m = \frac{M}{2R\alpha} \tag{2}$$

From Eqns (1) and (2), we write, $F = \dfrac{\gamma M}{R^2} \dfrac{\sin\alpha}{\alpha}$.

EXAMPLE 13.3 *Determine the attraction of a uniform cylindrical shell of radius R and length l at a point on its axis.*

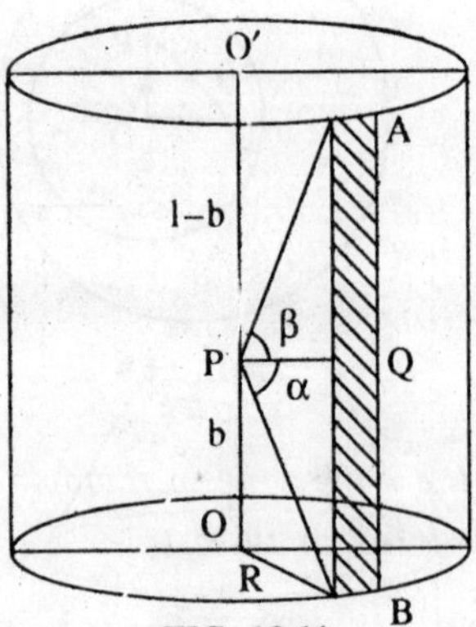

FIG. 13.11

Solution. Let the point P on the axis be at a height b from the centre of the base (FIG. 13.11). Then $OP = b$ and $O'P = l - b$.

We consider an elementary rod AB. Due to symmetry, the component of attraction perpendicular to the axis becomes zero.

By Eqn (13.18), the component of attraction along the rod AB is

$$X = \frac{\gamma m}{PQ}(\cos\alpha - \cos\beta) \tag{1}$$

where m = mass per unit area of the shell.

Now

$$\cos\alpha = \frac{PQ}{PB} = \frac{PQ}{\sqrt{b^2 + R^2}}$$

$$\text{and } \cos\beta = \frac{PQ}{AP} = \frac{PQ}{\sqrt{R^2 + (l-b)^2}} \tag{2}$$

Hence,

$$X = \gamma m \left[\frac{1}{\sqrt{R^2 + b^2}} - \frac{1}{\sqrt{R^2 + (l - b)^2}} \right] \quad (3)$$

The component of attraction due to the whole cylindrical surface is

$$F = (2\pi R) \cdot X = 2\pi R \gamma m \left[\frac{1}{\sqrt{R^2 + b^2}} - \frac{1}{\sqrt{R^2 + (l - b)^2}} \right] \quad (4)$$

If M be the total mass of the cylinder, then

$$M = (2\pi R) \cdot lm, \text{ i.e., } m = \frac{M}{2\pi Rl}$$

Hence, $F = \dfrac{\gamma M}{l} \left[\dfrac{1}{\sqrt{R^2 + b^2}} - \dfrac{1}{\sqrt{R^2 + (l - b)^2}} \right]$.

EXAMPLE 13.4 *Determine the attraction of a uniform circular ring of radius R at a point on its axis.*

Solution. We consider an elementary length of mass dm of the ring at Q (FIG. 13.12).

Let the distance $OP = r$ and $\angle OPQ = \theta$.

Due to symmetry, the component of attraction along horizontal direction will be zero.

Attraction at P due to mass dm at Q is

$$dF = \frac{\gamma dm}{PQ^2}, \text{ along } PQ.$$

Its component along PO

$$= dF \cos\theta = \frac{\gamma dm}{PQ^2} \cos\theta \quad (1)$$

Now

$$\cos\theta = \frac{r}{PQ} \quad (2)$$

Hence, total attraction along PO

$$= \sum \frac{\gamma dm}{PQ^2} \cdot \frac{r}{PQ} = \frac{\gamma M r}{PQ^3} = \frac{\gamma M r}{(a^2 + R^2)^{3/2}}$$

where M = total mass of the ring.

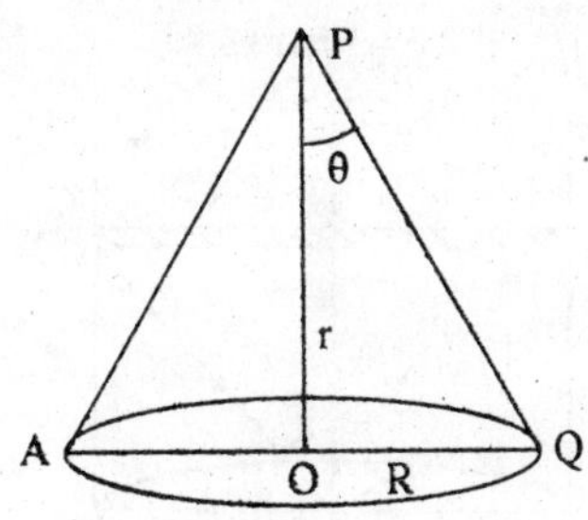

FIG. 13.12

EXAMPLE 13.5 *Find the attraction of a uniform solid right circular cone of height h and vertical angle 2α at its vertex.*

Solution. We consider an elementary circular disc of thickness dx at a depth of x from the vertex A of the cone (FIG. 13.13).

If ρ = density of the material, then mass per unit area of the disc is $m = \rho dx$.

Hence, by Eqn (13.22) the attraction due to elementary disc at the vertex is

$$dF = 2\pi\gamma\rho(1 - \cos\alpha)dx$$

Thus, the total attraction due to the whole cone is

$$F = \int_0^h 2\pi\gamma\rho(1 - \cos\alpha)dx = 2\pi\gamma\rho h(1 - \cos\alpha)$$

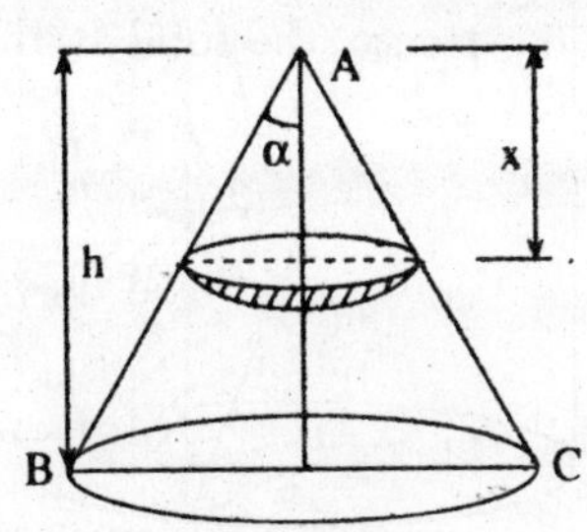

FIG. 13.13

EXAMPLE 13.6 *Show that the attraction of a solid hemisphere at the centre of its plane base is* $\frac{3}{2}\frac{\gamma M}{R^2}$, *where M is the mass and R the radius of the hemisphere.*

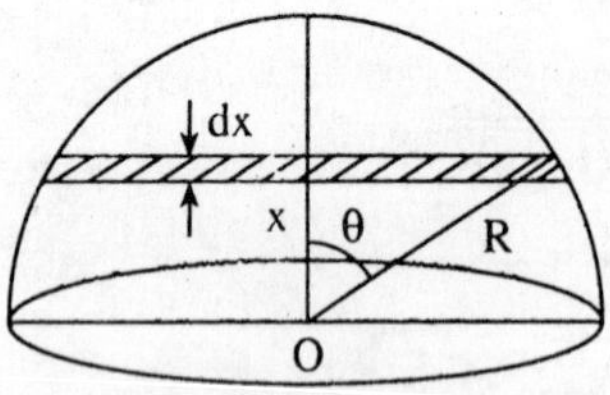

FIG. 13.14

Solution. Let ρ = density of the hemisphere. Then its mass

$$M = \frac{2}{3}\pi R^3 \rho \tag{1}$$

We consider an elementary disc of thickness dx at a distance x from the centre O of the base (FIG. 13.14).

Let the radius of the disc subtend an angle θ at the centre O of the base. Then

$$\cos\theta = \frac{x}{R} \tag{2}$$

By Eqn (13.22), the attraction at O due to the elementary disc is

$$dF = 2\pi\gamma\rho(1 - \cos\theta)dx = 2\pi\gamma\rho\left(1 - \frac{x}{R}\right)dx$$

Hence, the total attraction at O due to the hemisphere is

$$F = \int_0^R 2\pi\gamma\rho\left(1 - \frac{x}{R}\right)dx = 2\pi\gamma\rho\left[x - \frac{x^2}{2R}\right]_0^R = \pi\gamma\rho R \tag{3}$$

Eleminating ρ from Eqns (1) and (3), we get $F = \frac{3}{2}\frac{\gamma M}{R^2}$.

EXAMPLE 13.7 *A homogeneous right circular cylinder is of infinite length in one direction and at the other end, the section is perpendicular to the generators. Prove that the attraction at the centre of this end is* $2\gamma M/R$, *where M is the mass of the cylinder per unit of length and R the radius.*

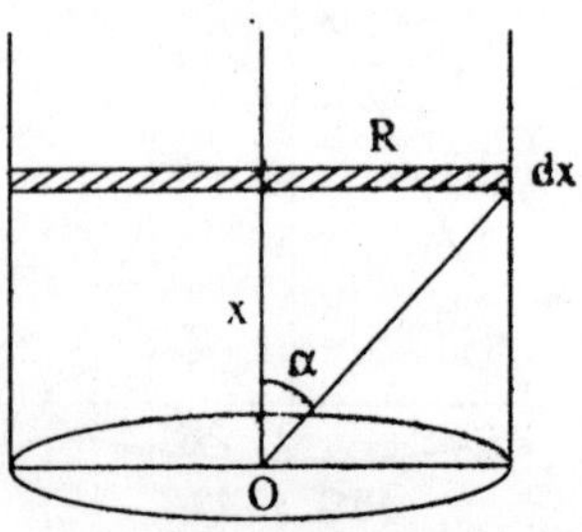

FIG. 13.15

Solution. We consider an elementary disc of thickness dx at a distance of x from the centre of the base. If the radius of the disc subtends an angle α at the centre O of the base, then (FIG. 13.15)

$$\cos\alpha = \frac{x}{\sqrt{x^2 + R^2}} \tag{1}$$

Let ρ be the density of the material. Then mass per unit length of the cylinder is

$$M = \pi R^2 \rho \tag{2}$$

By Eqn (13.22), the attraction at O due to the elementary disc is

$$dF = 2\pi\gamma\rho(1 - \cos\alpha)dx = 2\pi\gamma\rho\left(1 - \frac{x}{\sqrt{x^2 + R^2}}\right)dx$$

Hence, the total attraction due to the cylinder is

$$\begin{aligned}
F &= \int_0^\infty 2\pi\gamma\rho\left(1 - \frac{x}{\sqrt{x^2 + R^2}}\right)dx = 2\pi\gamma\rho\left[x - \sqrt{x^2 + R^2}\right]_0^\infty \\
&= 2\pi\gamma\rho\left[\frac{x^2 - (x^2 + R^2)}{x + \sqrt{x^2 + R^2}}\right]_0^\infty = -2\pi\gamma\rho R^2\left[\frac{1}{x + \sqrt{x^2 + R^2}}\right]_0^\infty \\
&= 2\pi\gamma\rho R = 2\pi\gamma R \cdot \frac{M}{\pi R^2} \quad \text{[using Eqn (2)]} \\
&= \frac{2\gamma M}{R}
\end{aligned}$$

EXAMPLE 13.8 *Show that the attraction of a uniform thin hollow frustum of a cone at its vertex is* $2\pi\gamma\rho \sin\alpha \cos\alpha \log\left(\frac{R}{r}\right)$, *where* R, r *are the radii of circular ends,* α *the semi-vertical angle of the cone and* ρ *the surface density.*

Solution. We consider an elementary ring of breadth dx at a distance x from the vertex so that $PQ = x$ (FIG. 13.16).

Then the radius of the elementary ring is $x \sin\alpha$ so that its mass is $2\pi(x\sin\alpha)\rho\, dx$. Hence, attraction of the ring at P along PO is

$$dF = 2\pi\gamma\rho(x\sin\alpha)\cdot\cos\alpha\cdot\frac{1}{x^2}dx$$

$$= 2\pi\gamma\rho\sin\alpha\cos\alpha\cdot\frac{1}{x}dx$$

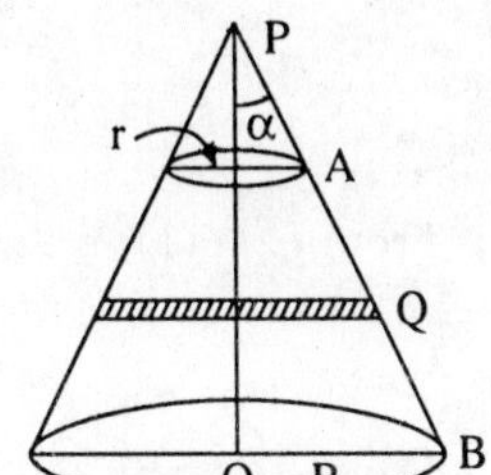

FIG. 13.16

Then the total attraction at P is

$$F = 2\pi\gamma\rho\sin\alpha\cos\alpha\int_{x_1}^{x_2}\frac{dx}{x} \tag{1}$$

The limits of integration in Eqn (1) are $x_1 = PA = r\,\text{cosec}\,\alpha,\ x_2 = PB = R\,\text{cosec}\,\alpha$.

Hence, $F = 2\pi\gamma\rho\sin\alpha\cos\alpha\Big[\log x\Big]_{r\,\text{cosec}\,\alpha}^{R\,\text{cosec}\,\alpha} = 2\pi\gamma\rho\sin\alpha\cos\alpha\log\left(\frac{R}{r}\right)$.

EXAMPLE 13.9 *A uniform solid sphere of mass* M *and density* ρ *is cut into two halves by a diametral plane. Show that the resultant attraction between the halves is* $\frac{3\gamma M^2}{16R^2}$, *where* R *is the radius of the sphere.*

Solution. Mass of the sphere is

$$M = \frac{4}{3}\pi\rho R^3 \tag{1}$$

The sphere is cut into two halves by the plane AOB (FIG. 13.17).

The attraction of portion I on itself is zero, since there will be pairs of equal and opposite forces.

Hence, attraction of portion I on II is the same as the attraction of the whole sphere on portion II.

We consider an element $r\,dr\,d\theta$ at P and generate an elementary ring by revolving it about OC. Its radius is $r\sin\theta$. Hence, its mass is

$$dm = 2\pi(r\sin\theta)r\,dr\,d\theta\,\rho \tag{2}$$

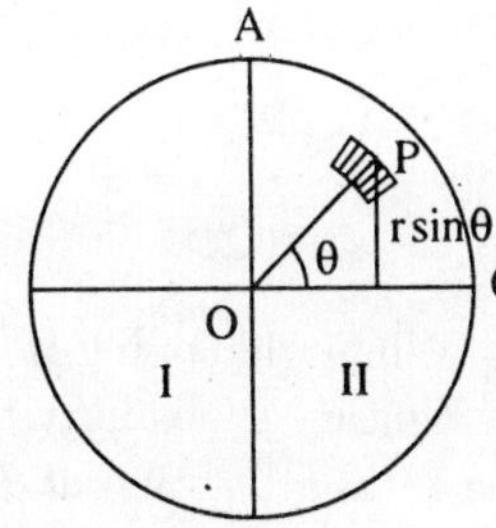

FIG. 13.17

where ρ = density of the sphere.

Now attraction of a solid sphere on a unit mass at any internal point distant r from the centre is, by Eqn (13.34) $= \frac{4}{3}\pi\gamma\rho r$.

Hence, attraction of the whole sphere on the elementary ring along OC is

$$dF = \frac{4}{3}\pi\gamma\rho r(2\pi r\sin\theta)r\,d\theta\,dr\cdot\rho\cos\theta$$

Thus, the attraction of the whole sphere on the half II is

$$F = \frac{8}{3}\pi^2\gamma\rho^2 \int_0^{\pi/2}\int_0^R r^3 \sin\theta \cos\theta dr\, d\theta$$

$$= \frac{8}{3}\pi^2\gamma\rho^2 \left[\frac{r^4}{4}\right]_0^R \int_0^{\pi/2} \frac{1}{2}\sin 2\theta\, d\theta$$

$$= \frac{1}{3}\pi^2\gamma\rho^2 R^4 \left[-\frac{\cos 2\theta}{2}\right]_0^{\pi/2}$$

$$= \frac{1}{3}\pi^2\gamma\rho^2 R^4 \tag{3}$$

From Eqn (1), we get

$$M^2 = \frac{16}{9}\pi^2\rho^2 R^6, \text{ i.e., } \pi^2\rho^2 = \frac{9}{16}\frac{M^2}{R^6} \tag{4}$$

From Eqns (3) and (4), we get

$$F = \frac{1}{3}\gamma R^4 \cdot \frac{9}{16}\frac{M^2}{R^6} = \frac{3}{16}\frac{\gamma M^2}{R^2}$$

EXAMPLE 13.10 *Determine the attractions of a solid sphere at an external and internal point, if its density at a distance x from the centre is $\lambda\left(\frac{x}{a}\right)^n$, where a is its radius and λ = constant.*

Solution. Consider an elementary shell of radius x and thickness dx.

Its surface density being $= \lambda\left(\frac{x}{a}\right)^n$, its mass is

$$dm = 4\pi\lambda x^2 \left(\frac{x}{a}\right)^n dx = \frac{4\pi\lambda}{a^n} x^{n+2} dx \tag{1}$$

If the point is external at a distance r from the centre, total attraction will be equal to

$$F = \int_0^a \frac{\gamma dm}{r^2} \text{ [by Eqn (13.28)]}$$

$$= \frac{4\pi\lambda\gamma}{a^n}\int_0^a \frac{x^{n+2}}{r^2}dx = \frac{4\pi\lambda\gamma}{a^n r^2}\left(\frac{a^{n+3}}{n+3}\right)$$

$$= \frac{4\pi\lambda\gamma}{r^2}\cdot\frac{a^3}{n+3} \tag{2}$$

When the point is internal at a distance r from the centre, the attractions of all the elementary shells for which radius $x > r$ will be zero, since P will be an internal point. Hence, the total attraction at P is

$$F = \int_0^r \frac{\gamma dm}{r^2} = \frac{4\pi\lambda\gamma}{a^n r^2}\int_0^r x^{n+2}dx$$

$$= \frac{4\pi\lambda\gamma}{a^n r^2}\left(\frac{r^{n+3}}{n+3}\right) = \frac{4\pi\lambda\gamma}{a^n}\cdot\frac{r^{n+1}}{n+3}$$

EXAMPLE 13.11 *A sphere is described on the radius of another sphere as diameter. If the latter be a homogeneous sphere of mass M and radius a, show that the resultant attraction on the portion included in the smaller sphere is $\frac{\gamma M^2}{16a^2}$.*

Solution. Equation to the smaller sphere with origin at the centre of the bigger one is (FIG. 13.18)

$$r = R\cos\theta \tag{1}$$

The attraction of the smaller sphere on itself being zero, the attraction of portion II is the same as that of the whole bigger sphere on the smaller one.

We consider a small element $r\,dr\,d\theta$ at P and generate an elementary ring of radius $r\sin\theta$ by revolving about OA. Its mass is

$$dm = \rho(2\pi r\sin\theta)r\,dr\,d\theta \tag{2}$$

ρ being the density of the material. Now, attraction of the whole sphere at any internal point is, by Eqn (13.34),

$$= \frac{4}{3}\pi\gamma\rho r$$

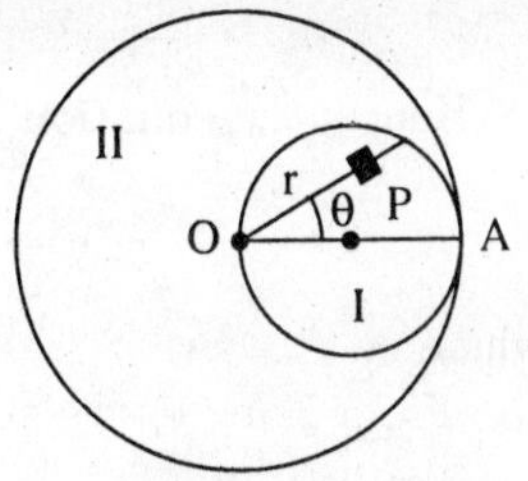

FIG. 13.18

Hence, the attraction on the elementary ring along OA is

$$dF = \frac{4}{3}\pi\gamma\rho r(2\pi r\sin\theta)r\,dr\,d\theta\cdot\rho\cos\theta = \frac{8}{3}\gamma\pi^2\rho^2 r^3\sin\theta\cos\theta\,dr\,d\theta \tag{3}$$

The total attraction is

$$F = \frac{8}{3}\gamma\pi^2\rho^2\int_0^{\pi/2}\int_0^{a\cos\theta} r^3\sin\theta\cos\theta\,dr\,d\theta$$

$$= \frac{8}{3}\gamma\pi^2\rho^2\int_0^{\pi/2}\sin\theta\cos\theta\left[\frac{r^4}{4}\right]_0^{a\cos\theta} = \frac{2}{3}\gamma\pi^2\rho^2a^4\int_0^{\pi/2}\sin\theta\cos^5\theta\,d\theta$$

$$= \frac{2}{3}\gamma\pi^2\rho^2a^4\left(-\frac{\cos^6\theta}{6}\right)_0^{\pi/2} = \frac{1}{9}\gamma\pi^2\rho^2a^4 \tag{4}$$

Now, the mass of the smaller shell is

$$M = \frac{4}{3}\pi a^3\rho,\ \text{i.e.,}\ \pi^2\rho^2 = \frac{9}{16}\frac{M^2}{a^6}. \tag{5}$$

From Eqns (4) and (5),

$$F = \frac{1}{9}\gamma a^4\cdot\frac{9}{16}\frac{M^2}{a^6} = \frac{1}{16}\frac{\gamma M^2}{a^2}$$

EXAMPLE 13.12 *Discuss the motion of a particle in a tunnel bored through the earth considered as a sphere of radius a.*

Solution. Let the mass of the particle $= m$. We consider the earth to be a sphere with centre at C. A tunnel AB is bored (FIG. 13.19) through the sphere. We draw CO perpendicular on AB.

Let the particle be at P at a distance r from the centre and x from O.

Then, by Eqn (13.34), the force of attraction on mass m at an internal point is

$$F = \left(\frac{4}{3}\pi\gamma\rho r\right)\cdot m \tag{1}$$

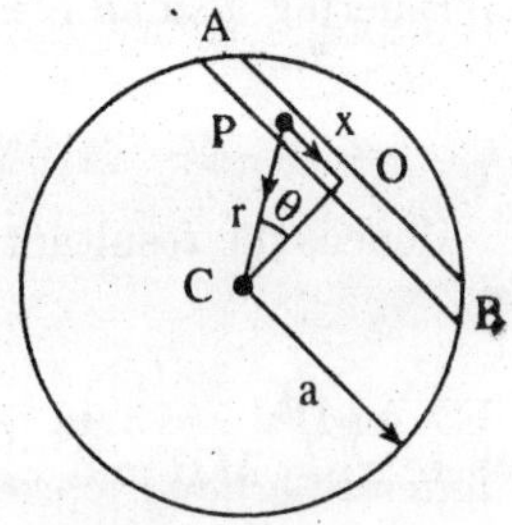

FIG. 13.19

along PC.

The component of this force along PO which causes the motion in the tunnel

$$= F\cos\theta = F\cdot\frac{x}{r} = \left(\frac{4}{3}\pi\gamma\rho m\right)x = \xi x \tag{2}$$

where

$$\xi = \frac{4}{3}\pi\gamma\rho m = \text{constant.} \tag{3}$$

Hence, the equation of motion of the particle is

$$m\frac{d^2x}{dt^2} = -\xi x, \text{ i.e., } \frac{d^2x}{dt^2} = -\omega^2 x \tag{4}$$

where

$$\omega^2 = \frac{\xi}{m} = \frac{4}{3}\pi\gamma\rho \tag{5}$$

Eqn (4) suggests that the motion is simple harmonic. The time period of such a motion is

$$T = \frac{2\pi}{\omega} = \frac{2\pi}{\sqrt{\frac{4}{3}\pi\gamma\rho}} = \sqrt{\frac{3\pi}{\gamma\rho}} \tag{6}$$

At the extreme point A, $r = a$, so that the force of attraction on the particle which is on the surface of the earth is

$$F = \frac{4}{3}\pi\gamma\rho a m \tag{7}$$

Here a is the radius of the earth. But this force is equal to the weight of the particle, so that

$$F = mg \tag{8}$$

where g is the acceleration due to gravity.

From Eqns (7) and (8), we get

$$mg = \frac{4}{3}\pi\gamma\rho a m, \text{ i.e., } \gamma\rho = \frac{3g}{4\pi a} \tag{9}$$

From Eqns (6) and (9), we get

$$T = \sqrt{\frac{3\pi}{3g}\cdot 4\pi a} = 2\pi\sqrt{\frac{a}{g}}$$

13.5 Potential of Some Solids

(A) Thin Uniform Straight Rod

Considering a small element dx as in figure (FIG. 13.3), potential at P due to this element is

$$dV = \frac{\gamma m}{r}dx = \frac{\gamma m}{\sqrt{h^2+x^2}}dx$$

Hence, the resultant potential due to the whole rod is

$$V = \gamma m\int_{x_1}^{x_2}\frac{dx}{\sqrt{h^2+x^2}} \tag{13.45}$$

where $x_1 = MB$ and $x_2 = MA$.

From the figure, we have $x = h\tan\theta$, i.e., $dx = h\sec^2\theta\, d\theta$.

Then

$$V = \gamma m \int_{\beta}^{\alpha} \sec\theta \, d\theta = \gamma m \left[\log\left\{\tan\left(\frac{\pi}{4} + \frac{\theta}{2}\right)\right\}\right]_{\beta}^{\alpha}$$

$$= \gamma m \log\left[\frac{\tan\left(\frac{\pi}{4} + \frac{\alpha}{2}\right)}{\tan\left(\frac{\pi}{4} + \frac{\beta}{2}\right)}\right] = \gamma m \log\left[\cot\left(\frac{\pi}{4} - \frac{\alpha}{2}\right)\cot\left(\frac{\pi}{4} + \frac{\beta}{2}\right)\right]$$

$$= \gamma m \log\left[\cot\frac{\angle PAB}{2} \cot\frac{\angle PBA}{2}\right] \tag{13.46}$$

since, from the figure, we have

$$\angle PBA = \beta + \frac{\pi}{2}, \text{ and } \frac{\pi}{2} + \alpha + \angle PAB = \pi, \text{ i.e., } \angle PAB = \frac{\pi}{2} - \alpha$$

(B) Uniform Circular Disc

Considering an elementary ring of mass (FIG. 13.4),

$$dM = 2\pi m x \, dx$$

we get the potential at P by definition due to this ring as

$$dV = \frac{\gamma dM}{r} = \gamma \cdot \frac{2\pi m x \, dx}{PA} = 2\pi\gamma m \frac{x \, dx}{\sqrt{h^2 + x^2}}$$

Hence, potential at P due to the whole disc is

$$V = 2\pi\gamma m \int_0^R \frac{x \, dx}{\sqrt{h^2 + x^2}} = 2\pi\gamma m \left[\sqrt{h^2 + x^2}\right]_0^R$$

$$= 2\pi\gamma m \left(\sqrt{h^2 + R^2} - h\right) \tag{13.47}$$

where R = radius of the disc.

(C) Uniform Spherical Shell

We consider an elementary ring as in FIG. 13.5 whose mass is

$$dM = 2\pi a^2 m \sin\theta \, d\theta$$

The potential at P due to this element is

$$dV = \gamma \cdot \frac{2\pi a^2 m \sin\theta \, d\theta}{x}$$

Since $x^2 = a^2 + r^2 - 2ar\cos\theta$, we have

$$\sin\theta \, d\theta = \frac{x \, dx}{ar}$$

Hence, potential due to the whole spherical shell is

$$V = \gamma \cdot 2\pi a^2 m \int \frac{dx}{ar} = \gamma \cdot \frac{2\pi a m}{r} \int dx \tag{13.48}$$

If P is an external point, x varies from $PA = r - a$ to $PB = r + a$.

Hence,

$$V = \gamma \frac{2\pi a m}{r} [x]_{r-a}^{r+a} = \gamma \cdot \frac{4\pi a^2 m}{r} = \gamma \cdot \frac{M}{r} \tag{13.49}$$

where M is the mass of the shell. This implies that ***potential at an external point due to a thin uniform spherical shell is the same as if the whole mass were concentrated at its centre.***

If P be an internal point, the limits of x in Eqn (13.48) are from

$$PA = a - r \text{ to } PB = a + r.$$

Hence,

$$V = \gamma \cdot \frac{2\pi am}{r}[x]_{a-r}^{a+r} = \gamma \cdot \frac{2\pi am}{r} \cdot 2r = \gamma \cdot \frac{4\pi a^2 m}{a} = \gamma \cdot \frac{M}{a} \tag{13.50}$$

which is a constant quantity.

(D) Uniform Solid Sphere

We consider an elementary shell of radius x and thickness dx whose mass is (FIG. 13.6)

$$dM = (4\pi x^2 dx)m$$

• Case 1 : *P* is an external point

In this case, P will be an external point to all the elementary shells composing the solid sphere.

Hence, potential at P due to the elementary shell is

$$dV = \gamma \frac{dM}{OP} = \gamma \cdot \frac{(4\pi x^2 dx)m}{r} \tag{13.51}$$

Then the potential at P due to the whole sphere is

$$V = \frac{4\pi\gamma m}{r}\int_0^a x^2 dx = \frac{4}{3}\frac{\pi\gamma a^3 m}{r} = \frac{\gamma M}{r} \tag{13.52}$$

where M = mass of the sphere.

Eqn (13.52) implies that ***the potential of a solid sphere at an external point is the same as if the whole mass were concentrated at its centre***.

• Case 2 : *P* is an internal point

In this case, P is an external point with respect to the elementary shell if $r > x$. Then its potential is

$$dV = \gamma \cdot \frac{(4\pi x^2 dx)m}{r}$$

Hence, potential for all such shells is

$$V_1 = \gamma \cdot \frac{4\pi m}{r}\int_0^r x^2 dx = \frac{4}{3}\gamma\pi m r^2 \tag{13.53}$$

But P will be an internal point to all elementary shells for which $x > r$. In this case, potential by Eqn (13.50) of the elementary shell is

$$dV = \gamma \cdot \frac{(4\pi x^2 dx)m}{x} = \gamma \cdot (4\pi x\, dx)m$$

Hence, potential due to all such shells is

$$V_2 = 4\pi\gamma m \int_r^a x\, dx = 2\pi\gamma m(a^2 - r^2) \tag{13.54}$$

Thus, the total potential at an internal point is

$$V = V_1 + V_2 = \frac{4}{3}\gamma\pi m r^2 + 2\pi\gamma m(a^2 - r^2) = 2\pi\gamma m\left(a^2 - \frac{r^2}{3}\right) \tag{13.55}$$

• Case 3 : *P* is on the surface of the sphere

In this case, $r = a$. Then by Eqn (13.55), we get the potential at P

$$V = 2\pi\gamma m\left(a^2 - \frac{a^2}{3}\right) = \frac{4\pi}{3}\gamma m a^2 = \gamma\frac{M}{a} \tag{13.56}$$

(E) Thick Spherical Shell

We consider an elementary shell of radius x and thickness dx as in figure (FIG. 13.7). Its mass is

$$dM = (4\pi x^2 dx)m$$

• Case 1 : *P* is an external point

Here P is an external point to all the elementary shells composing the thick shell. Thus, the potential at P due to the elementary shell is

$$dV = \gamma\cdot\frac{(4\pi x^2 dx)m}{r}$$

Hence, the potential due to the whole body is

$$V = \frac{4\pi\gamma m}{r}\int_b^a x^2 dx = \frac{4}{3}\pi\gamma m\left(\frac{a^3 - b^3}{r}\right) \tag{13.57}$$

• Case 2 : *P* is an internal point

In this case, P remains within the cavity and hence an internal point to the whole matter. The potential at P due to the elementary shell, by Eqn (13.50), is

$$dV = \gamma\cdot\frac{(4\pi x^2 dx)m}{x} = 4\pi\gamma m x\, dx$$

Hence, potential at P due to the whole body is

$$V = 4\pi\gamma m\int_b^a x\, dx = 2\pi\gamma m(a^2 - b^2) \tag{13.58}$$

• Case 3 : *P* is within the matter

In this case, P will be an external point to shells for which $x < r$. For such shells, potential at P is

$$V_1 = \int_b^r \frac{\gamma(4\pi x^2 dx)m}{r} = \frac{4}{3}\pi\gamma m\left(\frac{r^3 - b^3}{r}\right) \tag{13.59}$$

For those shells to which P is an internal point, $x > r$. The potential of all such shells, by Eqn (13.50), is

$$V_2 = \int_r^a \frac{\gamma(4\pi x^2 dx)m}{x} = 4\pi\gamma m\int_r^a x\, dx = 2\pi\gamma m(a^2 - r^2) \tag{13.60}$$

Hence, the total potential is

$$\begin{aligned} V &= V_1 + V_2 \\ &= \frac{4}{3}\pi\gamma m\left(\frac{r^3 - b^3}{r}\right) + 2\pi\gamma m(a^2 - r^2) \\ &= 2\pi\gamma m\left[a^2 - r^2 - \frac{2}{3}\left(\frac{b^3}{r} - r^2\right)\right] \\ &= 2\pi\gamma m\left(a^2 - \frac{r^2}{3} - \frac{2b^3}{3r}\right) \end{aligned} \tag{13.61}$$

EXAMPLE 13.13 *The density of an elliptic lamina varies as the distance from major axis, the mass of an element of unit area at unit distance being λ. Show that the potential due to the lamina at the focus is $2\gamma\lambda b^2$, where b = semi-minor axis of the ellipse.*

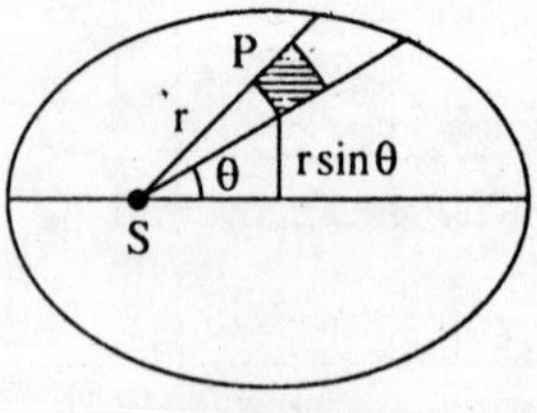

FIG. 13.20

Solution. The polar equation of an ellipse with its focus S as pole is (FIG. 13.20)

$$\frac{l}{r} = 1 - e\cos\theta, \text{ i.e., } r = \frac{l}{1 - e\cos\theta} \tag{1}$$

where

$$l = \frac{b^2}{a} \text{ and } b^2 = a^2(1 - e^2) \tag{2}$$

a, b being the semi-major and semi-minor axes of the ellipse.

We consider an elementary area $r\,dr\,d\theta$ at P. Its distance from major axis is $r\sin\theta$. Hence, by the condition given, mass of the element is

$$dM = \lambda r\sin\theta\, r\,dr\,d\theta \tag{3}$$

Its potential at S is

$$dV = \gamma\frac{dM}{r} = \lambda\gamma r\sin\theta\,dr\,d\theta \tag{4}$$

Hence, potential at S due to the whole lamina is

$$\begin{aligned}
V &= \iint \lambda\gamma r\sin\theta\,dr\,d\theta \\
&= 2\lambda\gamma\int_0^{\pi} \sin\theta\left[\frac{r^2}{2}\right]_0^r d\theta \\
&= 2\lambda\gamma\int_0^{\pi} \frac{\sin\theta}{2}\cdot\frac{l^2 d\theta}{(1 - e\cos\theta)^2} \quad \text{[using Eqn (1)]} \\
\text{i.e., } V &= \frac{-\lambda\gamma l^2}{e}\left[\frac{1}{1 - e\cos\theta}\right]_0^{\pi} \\
&= -\frac{\lambda\gamma l^2}{e}\left[\frac{1}{1+e} - \frac{1}{1-e}\right] \\
&= -\frac{\lambda\gamma l^2}{e}\cdot\frac{-2e}{1-e^2} \\
&= \frac{2\lambda\gamma l^2}{1-e^2} \\
&= 2\lambda\gamma\cdot\left(\frac{b^2}{a}\right)^2\cdot\frac{a^2}{b^2} \quad \text{[using Eqn (2)]} \\
&= 2\lambda\gamma b^2
\end{aligned}$$

EXAMPLE 13.14 *A solid right circular cylinder is of density ρ and length l and the radius of each end is R. Prove that the potential due to it at the centre of each end is*

$$\pi\gamma\rho\left[l\sqrt{R^2 + l^2} - l^2 + R^2\sinh^{-1}\left(\frac{l}{R}\right)\right]$$

Solution. We consider an elementary circular disc having thickness dx at a distance x from the centre of the upper end (FIG. 13.21). Its potential at P, by Eqn (13.47), is

$$dV = 2\pi\gamma\rho dx\left[\sqrt{R^2 + x^2} - x\right]$$

Hence, the total potential at P due to the cylinder is

$$V = 2\pi\gamma\rho \int_0^l \left[\sqrt{R^2 + x^2} - x\right] dx$$

$$= 2\pi\rho\gamma \left[\frac{x}{2}\sqrt{R^2 + x^2} + \frac{R^2}{2}\sinh^{-1}\left(\frac{x}{R}\right) - \frac{x^2}{2}\right]_0^l$$

$$= \pi\gamma\rho \left[l\sqrt{R^2 + l^2} + R^2 \sinh^{-1}\left(\frac{l}{R}\right) - l^2\right]$$

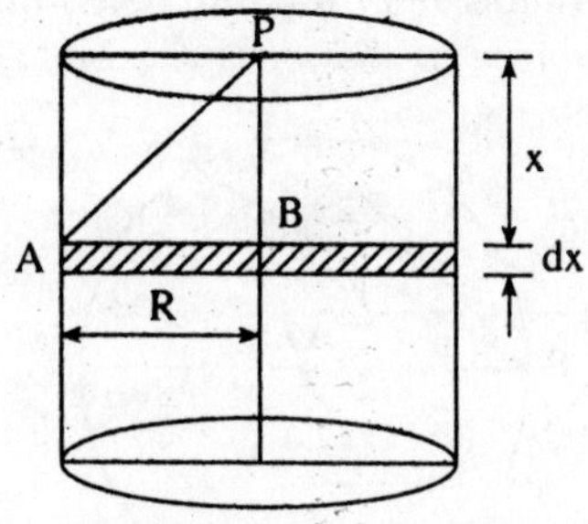

FIG. 13.21

NOTE. To determine the potential at the mid-point of the axis, we find the potential of a cylinder of length $l/2$ at the centre of the circular ends and then multiply the result by 2.

EXAMPLE 13.15 *Show that the potential of a solid hemisphere of radius R and density ρ at an external point P on the convex side situated on the axis at a distance d from the centre is*

$$\frac{2\pi\gamma\rho}{3d}\left[R^3 + \left\{\left(R^2 + d^2\right)^{3/2} - d^3 - \frac{3}{2}a^2 d\right\}\right]$$

Solution. We consider an elementary disc of thickness dx at a distance x from the centre of the base (FIG. 13.22).

Then the potential of this elementary disc at P is

$$dV = 2\pi\gamma\rho(PA - PQ) \qquad (1)$$

Now, since $OP = d$, we have

$$PQ = d - x$$

$$\text{and } AQ = \sqrt{R^2 - x^2}$$

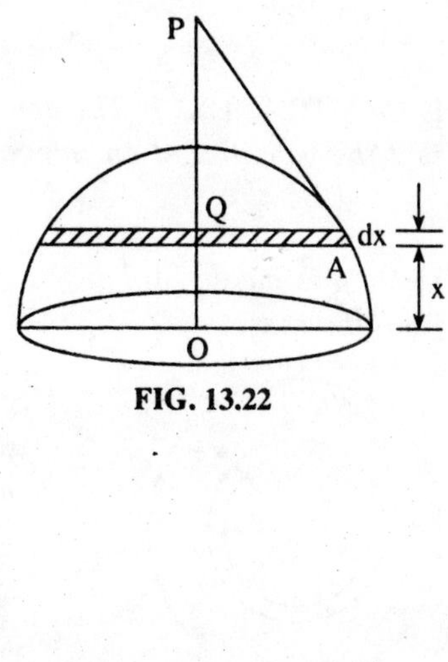

FIG. 13.22

Hence, $PA = \left\{(d-x)^2 + \left(R^2 - x^2\right)\right\}^{1/2}$.

Then $dV = 2\pi\gamma\rho\left[\left\{(d-x)^2 + \left(R^2 - x^2\right)\right\}^{1/2} - (d-x)\right] dx.$

Hence, the potential at P due to the hemisphere is

$$V = 2\pi\gamma\rho \int_0^R \left[\left(d^2 - 2dx + R^2\right)^{1/2} - d + x\right] dx$$

$$= 2\pi\gamma\rho\left[-\frac{1}{2d}\cdot\frac{2}{3}\left(d^2 - 2dx + R^2\right)^{3/2} - dx + \frac{x^2}{2}\right]_0^R$$

$$= 2\pi\gamma\rho\left[-\frac{1}{3d}\left\{\left(d^2 - 2Rd + R^2\right)^{3/2} - \left(d^2 + R^2\right)^{3/2}\right\} - Rd + \frac{R^2}{2}\right]$$

$$= 2\pi\gamma\rho\left[-\frac{1}{3d}(d-R)^3 + \frac{1}{3d}\left(d^2 + R^2\right)^{3/2} - Rd + \frac{R^2}{2}\right]$$

$$= \frac{2\pi\gamma\rho}{3d}\left[-\left(d^3 - 3d^2R + 3dR^2 - R^3\right) + \left(d^2 + R^2\right)^{3/2} - 3Rd^2 + \frac{3}{2}R^2 d\right]$$

$$= \frac{2\pi\gamma\rho}{3d}\left[R^3 - d^3 + \left(d^2 + R^2\right)^{3/2} - \frac{3}{2}R^2 d\right]$$

$$= \frac{2\pi\gamma\rho}{3d}\left[R^3 + \left(d^2 + R^2\right)^{3/2} - d^3 - \frac{3}{2}R^2 d\right]$$

EXAMPLE 13.16 *Show that the potential of a circular disc of uniform surface density ρ and radius R at a point on its circumference is $4\gamma\rho R$.*

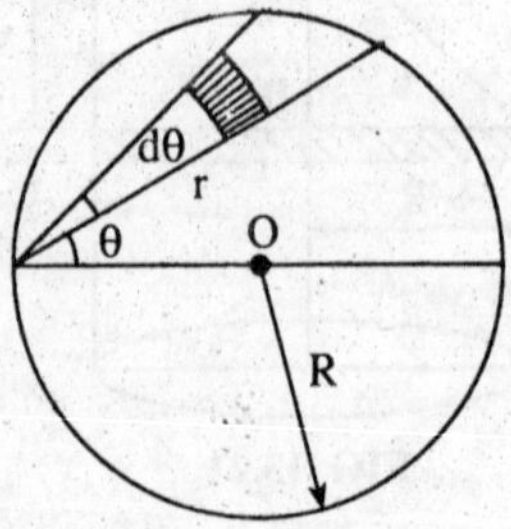

FIG. 13.23

Solution. Equation of the circle with origin at a point on the circumference is (FIG. 13.23)

$$r = 2R\cos\theta \tag{1}$$

We take an element $r\,dr\,d\theta$, whose mass is

$$dM = \rho r\,dr\,d\theta \tag{2}$$

Potential at the point P due to this element is

$$dV = \gamma\frac{\rho r\,dr\,d\theta}{r} = \gamma\rho\,dr\,d\theta \tag{3}$$

Hence, the potential at P due to the whole disc is

$$\begin{aligned} V &= \gamma\rho\int_{-\pi/2}^{\pi/2}\int_0^{2R\cos\theta} dr\,d\theta \\ &= \gamma\rho\int_{-\pi/2}^{\pi/2}(2R\cos\theta)d\theta = 2\gamma\rho R\big[\sin\theta\big]_{-\pi/2}^{\pi/2} \\ &= 4\gamma\rho R \end{aligned}$$

EXAMPLE 13.17 *The density of a solid sphere varies as the distance from the centre. Show that the potential at an internal point is*

$$V = \frac{\pi\gamma\rho}{3R}(4R^3 - r^3)$$

where ρ is the density at the surface, R the radius of the sphere and r the distance of the point from the centre.

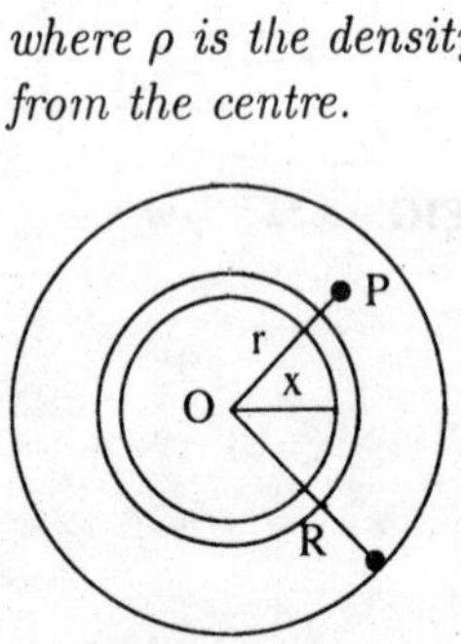

FIG. 13.24

Solution. Let P be an internal point so that $OP = r$. We are to determine potential at P. We consider an elementary shell of radius x and thickness dx (FIG. 13.24).

Density at distance x from the centre is λx, λ = constant. Since density is ρ at the surface, $\rho = \lambda R$, i.e., $\lambda = \rho/R$.

Hence, density at a distance x

$$= \frac{\rho}{R}x \tag{1}$$

Then the mass of the elementary shell is

$$dM = (4\pi x^2 dx)\,\frac{\rho}{R}x = \frac{4\pi\rho}{R}x^3 dx \tag{2}$$

Now, P is an external point for all shells, if $x < r$. Then potential at P is

$$dV = \frac{\gamma dM}{r} = \frac{4\pi\rho\gamma}{rR}x^3 dx$$

Hence, potential at P due to all such shells for which $x < r$ is

$$V_1 = \frac{4\pi\rho\gamma}{rR}\int_0^r x^3 dx = \frac{\pi\rho\gamma r^3}{R} \tag{3}$$

Similarly, P is an internal point for all shells if $x > r$. Then potential at P is

$$dV = \frac{\gamma dM}{x} = \frac{4\pi\rho\gamma}{R}x^2 dx$$

Hence, potential at P due to all such shells for which $x > r$ is

$$V_2 = \frac{4\pi\rho\gamma}{R}\int_r^R x^2 dx = \frac{4\pi\rho\gamma}{3R}\left(R^3 - r^3\right) \tag{4}$$

Thus, the total potential at P is

$$V = V_1 + V_2 = \frac{\pi\rho\gamma r^3}{R} + \frac{4\pi\rho\gamma}{3R}(R^3 - r^3) = \frac{\pi\rho\gamma}{3R}\left(4R^3 - r^3\right)$$

EXAMPLE 13.18 *A solid homogeneous sphere of radius R and density $3\mu/2\pi$ has on its surface a uniform distribution of a repelling matter of surface density $\mu R/2\pi$.*

Find the potential at any point.

Solution. Let P be a point at which the potential is to be determined; we take $OP = r$.

Case 1 : *P* is an external point

Mass of the sphere is

$$M = \frac{4}{3}\pi R^3 \cdot \frac{3\mu}{2\pi} = 2\mu R^3 \tag{1}$$

Potential at P due to the solid sphere is

$$V_1 = \frac{\gamma M}{r} = \frac{2\mu\gamma R^3}{r} \tag{2}$$

Mass of repelling matter is

$$M_1 = 4\pi R^2 \cdot \frac{\mu R}{2\pi} = 2\mu R^3 \tag{3}$$

Potential at P due to this repelling mass

$$V_2 = -\frac{\gamma M_1}{r} = -\frac{2\mu\gamma R^3}{r} \tag{4}$$

Hence, the total potential is

$$V = V_1 + V_2 = \frac{2\mu\gamma R^3}{r} - \frac{2\mu\gamma R^3}{r} = 0 \tag{5}$$

Case 2 : *P* is an internal point

Potential at P due to the solid sphere is

$$V_1 = 2\pi\gamma m\left(R^2 - \frac{r^2}{3}\right) \quad \text{[by Eqn (13.55)]}$$

$$= 2\pi\gamma\left(\frac{3\mu}{2\pi}\right)\left(R^2 - \frac{r^2}{3}\right)$$

$$= 3\mu\gamma\left(R^2 - \frac{r^2}{3}\right) \tag{6}$$

Potential at P due to the repelling matter is

$$V_2 = -\frac{\gamma M_1}{R} = -\frac{\gamma}{R}(2\mu R^3) = -2\mu\gamma R^2 \tag{7}$$

Hence, the total potential at P is

$$V = V_1 + V_2 = 3\mu\gamma\left(R^2 - \frac{r^2}{3}\right) - 2\mu\gamma R^2 = \mu\gamma\left(R^2 - r^2\right)$$

13.6 Relation between Attraction and Potential

We have seen that the space derivative of potential in any direction gives the component of attraction in that direction. Thus, if we know the potential at a point due to a given solid, we can determine the attraction just by finding the derivative.

As an example, we consider the case of a circular disc.

The potential at a point on its axis, by Eqn (13.47) is

$$V = 2\pi\gamma m\left\{\sqrt{x^2+R^2}-x\right\} \tag{13.62}$$

Then the force of attraction towards the centre is

$$F = -\frac{dV}{dx} = 2\pi\gamma m\left\{1-\frac{x}{\sqrt{x^2+R^2}}\right\} \tag{13.63}$$

which agrees with Eqn (13.22).

We can also determine the potential from attraction. For example, we know that attraction due to a solid sphere at an external point distant r from the centre is

$$F = \frac{\gamma M}{r^2} \tag{13.64}$$

where M = mass of the sphere.

If V is the potential at that point, then

$$-\frac{dV}{dr} = F = \frac{\gamma M}{r^2},$$

which yields

$$V = \frac{\gamma M}{r} \tag{13.65}$$

as is expected.

If we consider an internal point, where attraction is $F = \frac{4}{3}\pi\gamma m r$, we have

$$\frac{dV}{dr} = -\frac{4}{3}\pi\gamma m r$$

Hence,

$$V = C - \frac{2}{3}\pi\gamma m r^2 \tag{13.66}$$

where C is a constant of integration. We can determine it from the condition that the potential at the surface must be the same obtained from either Eqn (13.65) or Eqn (13.66).

Thus, for $r = a$,

$$\gamma\frac{M}{a} = C - \frac{2}{3}\pi\gamma\rho a^2, \text{ i.e., } C = \frac{4}{3}\pi\gamma\rho a^2 + \frac{2}{3}\pi\gamma\rho a^2 = 2\pi\gamma\rho a^2$$

Hence, from Eqn (13.66),

$$V = \frac{2}{3}\pi\gamma\rho\left(3a^2 - r^2\right) \tag{13.67}$$

which coincides with Eqn (13.55).

13.7 Self-attracting Systems

We consider a self-attracting system of particles of masses $m_1, m_2, m_3, \ldots$ placed at positions $A_1, A_2, A_3, \ldots$. We intend to determine the work done in bringing these particles one by one from infinity to their respective assigned positions.

We denote the distance between two particles m_p and m_q as r_{pq}, so that distance between m_1 and m_2 is r_{12}, that between m_2 and m_3 is r_{23} and so on.

First, we bring the mass m_1 from infinity to its position at A_1. In doing so, no work has been done, since there is no other attracting particle of the system. Now we bring the mass m_2 from infinity to its position A_2. The work done by the attraction of m_1 is equal to

$$\frac{\gamma m_1 m_2}{r_{12}},$$

which is the potential of m_1 at A_2.

The masses m_1, m_2 being in their assigned positions, we now bring the mass m_3 from infinity to its assigned position at A_3. The work done by the attraction of masses m_1 and m_2

$$= \frac{\gamma m_1 m_3}{r_{13}} + \frac{\gamma m_2 m_3}{r_{23}}$$

Proceeding as above, the total work done in bringing all the masses to their respective positions

$$= \frac{\gamma m_1 m_2}{r_{12}} + \left(\frac{\gamma m_1 m_3}{r_{13}} + \frac{\gamma m_2 m_3}{r_{23}}\right) + \left(\frac{\gamma m_1 m_4}{r_{14}} + \frac{\gamma m_2 m_4}{r_{24}} + \frac{\gamma m_3 m_4}{r_{34}}\right) + \cdots$$

$$= \gamma \sum \frac{m_p m_q}{r_{pq}} \tag{13.68}$$

The summation here is extended to all pairs of masses.

Let V_1 = the potential at A_1 of masses $m_2, m_3, m_4, \ldots,$

V_2 = the potential at A_2 of masses $m_1, m_3, m_4, \ldots$ and so on.

Then

$$V_1 = \frac{\gamma m_2}{r_{12}} + \frac{\gamma m_3}{r_{13}} + \frac{\gamma m_4}{r_{14}} + \cdots$$
$$V_2 = \frac{\gamma m_1}{r_{21}} + \frac{\gamma m_3}{r_{23}} + \frac{\gamma m_4}{r_{24}} + \cdots$$
$$V_3 = \frac{\gamma m_1}{r_{31}} + \frac{\gamma m_2}{r_{32}} + \frac{\gamma m_4}{r_{34}} + \cdots$$

Multiplying these equations respectively by $m_1, m_2, m_3, \ldots$ and adding the results, we get

$$V_1 m_1 + V_2 m_2 + V_3 m_3 + \cdots = 2\gamma \sum \frac{m_p m_q}{r_{pq}}$$

$$\text{or, } \frac{1}{2} \sum V_p m_p = \gamma \sum \frac{m_p m_q}{r_{pq}} \tag{13.69}$$

From Eqns (13.68) and (13.69), we conclude that the total work done in bringing the masses from infinity to their positions in the system is

$$W = \frac{1}{2} \sum V_p m_p \tag{13.70}$$

If the system forms a continuous body, then the total work done is

$$W = \frac{1}{2} \int V\, dm \tag{13.71}$$

where V is the potential of the body at a point within it and dm is an element of the body at that point.

Eqn (13.71) gives the work done by the mutual attraction of the particles.

Similarly, if the particles of a body are scattered from their present position to infinite diffusion, the work done by mutual attraction is

$$W = -\frac{1}{2}\int V\,dm \tag{13.72}$$

If a continuous body is changed from its configuration A to a new configuration B, then the work done is the sum of work done in scattering the particles of A to infinity and the work done in collecting those particles from infinity to form the body B. Thus, in this case,

$$W = -\frac{1}{2}\int_A V\,dm + \frac{1}{2}\int_B V'dm' \tag{13.73}$$

EXAMPLE 13.19 *A self-attracting sphere of uniform density and radius 'a' changes to one of uniform density and radius 'b'. Show that the work done by its mutual attractive forces is*

$$\frac{3}{5}\gamma M^2\left(\frac{1}{b}-\frac{1}{a}\right)$$

where M is the mass of the sphere.

Solution. We consider an internal point at a distance r from the centre of the sphere. Then the potential at that point by Eqn (13.55) is

$$V = 2\pi\gamma\rho\left(a^2 - \frac{r^2}{3}\right) \tag{1}$$

We take a thin spherical shell of radius r and thickness dr. Then elementary mass is

$$dm = (4\pi r^2 dr)\,\rho \tag{2}$$

Hence, work done in scattering the particles of the sphere of radius a to infinity is

$$\begin{aligned} W_1 &= -\frac{1}{2}\int V\,dm = -\frac{1}{2}\int_0^a 2\pi\gamma\rho\left(a^2-\frac{r^2}{3}\right)\cdot 4\pi\rho r^2 dr \\ &= -4\pi^2\gamma\rho^2\int_0^a\left(a^2r^2-\frac{r^4}{3}\right)dr \\ &= -\frac{16\pi^2\rho^2\gamma}{15}a^5 \end{aligned} \tag{3}$$

Mass of the sphere is

$$M = \frac{4}{3}\pi a^3\rho,\ \text{i.e.,}\ \rho = \frac{3M}{4\pi a^3} \tag{4}$$

From Eqns (3) and (4), we get

$$W_1 = -\frac{3}{5}\frac{\gamma M^2}{a} \tag{5}$$

Similarly, the work done in bringing the particles from infinity to form the sphere of radius b is

$$W_2 = \frac{3}{5}\frac{\gamma M^2}{b} \tag{6}$$

Hence, the total work done in changing the sphere of radius a to a sphere of radius b

$$W = W_1 + W_2 = \frac{3}{5}\gamma M^2\left(\frac{1}{b}-\frac{1}{a}\right)$$

13.8 Theorems on Potential

(A) Laplace's Theorem

Let V be the potential of a body of density ρ at a point $P(x, y, z)$ outside the body. Considering a small volume element $d\tau$ at a point (a, b, c) of the body, we have (FIG. 13.25)

$$V = \int \frac{\rho d\tau}{r} \tag{13.74}$$

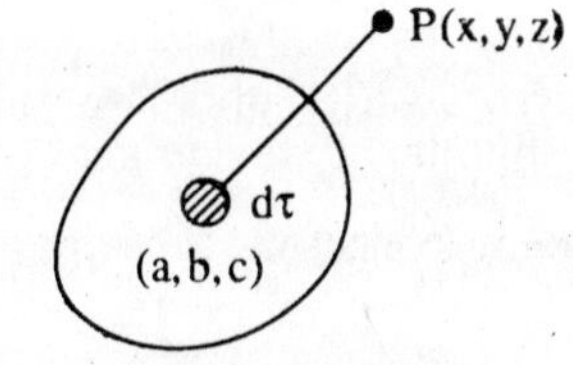

FIG. 13.25

where

$$r^2 = (x-a)^2 + (y-b)^2 + (z-c)^2 \tag{13.75}$$

Here r is measured from an external point, so that the integral in Eqn (13.74) will be always convergent. Moreover, we can differentiate under the integral sign, since the derivatives with respect to x, y, z under integral sign are continuous and yield convergent integrals. Thus, we have

$$\frac{\partial V}{\partial x} = -\int \frac{\rho(x-a)}{r^3} d\tau$$

$$\frac{\partial^2 V}{\partial x^2} = -\int \left[\frac{\rho}{r^3} - \frac{3\rho(x-a)^2}{r^5}\right] d\tau$$

Similarly,

$$\frac{\partial^2 V}{\partial y^2} = -\int \left[\frac{\rho}{r^3} - \frac{3\rho(y-b)^2}{r^5}\right] d\tau$$

$$\frac{\partial^2 V}{\partial z^2} = -\int \left[\frac{\rho}{r^3} - \frac{3\rho(z-c)^2}{r^5}\right] d\tau$$

Adding all these and using Eqn (13.75), we get

$$\frac{\partial^2 V}{\partial x^2} + \frac{\partial^2 V}{\partial y^2} + \frac{\partial^2 V}{\partial z^2} = 0, \text{ i.e., } \nabla^2 V = 0 \tag{13.76}$$

Thus, ***the potential of an attracting system at every point of which there is no matter satisfies Laplace's equation***. This is Laplace's theorem on potential.

(B) Poisson's Theorem

We consider a point $P(x, y, z)$ inside the attracting matter. We describe a small sphere of radius ϵ with centre at (a, b, c) and the point P inside it (FIG. 13.26). We assume that the density ρ of the matter within the sphere is uniform.

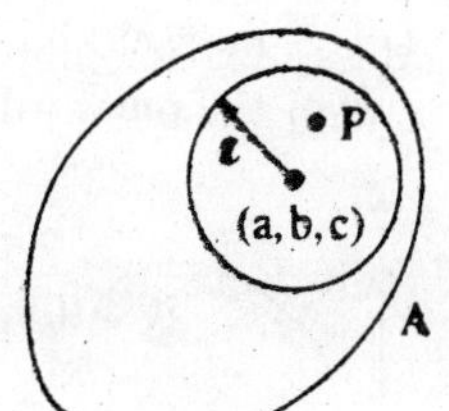

FIG. 13.26

Let V be the potential at P due to the continuous body A.

Then $V = V_1 + V_2$, where

V_1 = potential due to the matter outside the sphere,

and V_2 = potential due to the small sphere.

Since the point P is not in contact with the matter outside the sphere, we have by Laplace's theorem,

$$\nabla^2 V_1 = 0 \tag{13.77}$$

The potential V_2 at a point inside the sphere of radius ϵ is given by

$$V_2 = 2\pi\gamma\rho\left(\epsilon^2 - \frac{1}{3}r^2\right) = 2\pi\gamma\rho\left[\epsilon^2 - \frac{1}{3}\{(x-a)^2 + (y-b)^2 + (z-c)^2)\}\right]$$

Hence,

$$\frac{\partial V_2}{\partial x} = -\frac{4}{3}\pi\gamma\rho(x-a)$$

$$\frac{\partial^2 V_2}{\partial x^2} = -\frac{4}{3}\pi\gamma\rho$$

Similarly, we can get

$$\frac{\partial^2 V_2}{\partial y^2} = -\frac{4}{3}\pi\gamma\rho$$

$$\frac{\partial^2 V_2}{\partial z^2} = -\frac{4}{3}\pi\gamma\rho$$

Hence,

$$\frac{\partial^2 V_2}{\partial x^2} + \frac{\partial^2 V_2}{\partial y^2} + \frac{\partial^2 V_2}{\partial z^2} = -4\pi\gamma\rho$$

$$\text{i.e., } \nabla^2 V_2 = -4\pi\gamma\rho \tag{13.78}$$

From Eqns (13.77) and (13.78), we find

$$\nabla^2 V = \nabla^2(V_1 + V_2) = -4\pi\gamma\rho \tag{13.79}$$

which is a ***Poisson's equation***.

Thus, ***the potential of an attracting system at every point of which there is matter of volume density ρ satisfies the Poisson's equation* (13.79)**. This is ***Poisson's theorem*** on potential.

(C) Gauss's Theorem

The theorem states:

> ***The outward flux of the force of attraction over any closed surface in a gravitational field of force is equal to $-4\pi\gamma$ times the mass enclosed by the surface.***

We will prove it with the use of Green's theorem.

Let S be a closed surface over which an element dS is taken. Let l, m, n be the direction cosines of the outward-drawn normal to dS (FIG. 13.27). If X, Y, Z be the components of force of attraction over dS along the coordinate axes, and V be the corresponding potentials, then

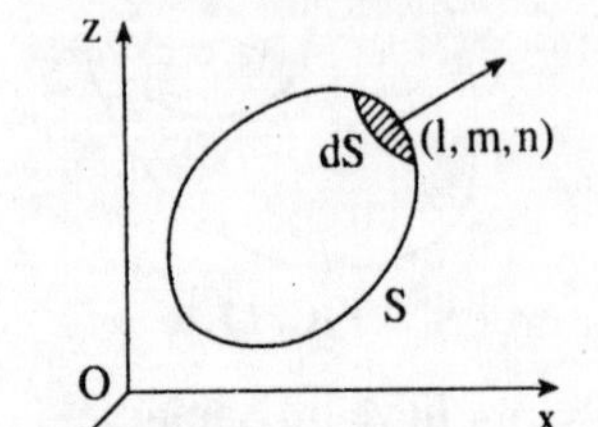

FIG. 13.27

$$\left.\begin{aligned} X &= \frac{\partial V}{\partial x} \\ Y &= \frac{\partial V}{\partial y} \\ Z &= \frac{\partial V}{\partial z} \end{aligned}\right\} \tag{13.80}$$

The outward normal component of force across the whole surface S is

$$\int_S (lX + mY + nZ)dS$$

which, by Green's theorem, assumes the form:

$$\int\left(\frac{\partial X}{\partial x}+\frac{\partial Y}{\partial y}+\frac{\partial Z}{\partial z}\right)d\tau, \text{ [}d\tau\text{ is an element of volume enclosed]}$$

$$=\int\left(\frac{\partial^2 V}{\partial x^2}+\frac{\partial^2 V}{\partial y^2}+\frac{\partial^2 V}{\partial z^2}\right)d\tau, \text{ [using Eqn (13.80)]}$$

$$=-4\pi\gamma\int\rho d\tau, \text{ [by Poisson's theorem]}$$

$$=-4\pi\gamma \text{ (mass enclosed)}.$$

Thus, the theorem is proved.

Discussion

(1) If the potential V of a distribution of matter is given throughout all space and $\nabla^2 V \neq 0$, we can determine the corresponding density ρ of distribution from Poisson's equation. Thus,

$$\rho=-\frac{1}{4\pi\gamma}\nabla^2 V \tag{13.81}$$

(2) If there is a surface distribution of matter, the potential function assumes different values V_1, V_2 on opposite sides of the surface. In this case, we must have

$$\left.\begin{aligned}&\text{(a) } V_1=V_2, \text{ on the surface}\\&\text{(b) } \frac{\partial V_1}{\partial n}-\frac{\partial V_2}{\partial n}=4\pi\gamma\sigma\end{aligned}\right\} \tag{13.82}$$

where δn is an element of normal directed from region 1 to region 2, and σ is the surface density of matter. Hence,

$$\sigma=\frac{1}{4\pi\gamma}\left[\frac{\partial V_1}{\partial n}-\frac{\partial V_2}{\partial n}\right] \tag{13.83}$$

EXAMPLE 13.20 *Find the density of distribution throughout a sphere of radius a and on its surface which gives potential*

$$V_1=\lambda\left(2a^3-r^3\right), \text{ for } r<a$$
$$V_2=\lambda\frac{a^4}{r}, \qquad \text{for } r>a$$

Solution. Since the potential is a function of r alone, Poisson's equation in spherical polar coordinates is

$$\frac{\partial^2 V}{\partial r^2}+\frac{2}{r}\frac{\partial V}{\partial r}=-4\pi\gamma\rho \tag{1}$$

We have, for $r<a$, $V_1=\lambda\left(2a^3-r^3\right)$.

Hence,

$$\frac{\partial V_1}{\partial r}=-3\lambda r^2, \quad \frac{\partial^2 V_1}{\partial r^2}=-6\lambda r \tag{2}$$

Substituting these values into Eqn (1), we get

$$-6\lambda r+\frac{2}{r}\left(-3\lambda r^2\right)=-4\pi\gamma\rho, \text{ i.e., } -12\lambda r=-4\pi\gamma\rho$$

$$\therefore\ \rho=\frac{3\lambda r}{\pi\gamma} \tag{3}$$

For $r > a$, we have $V_2 = \lambda \dfrac{a^4}{r}$.

Hence,

$$\frac{\partial V_2}{\partial r} = -\frac{\lambda a^4}{r^2}, \quad \frac{\partial^2 V_2}{\partial r^2} = \frac{2\lambda a^4}{r^3} \tag{4}$$

Substituting these into Eqn (1), we get

$$\frac{2\lambda a^4}{r^3} + \frac{2}{r}\left(-\frac{\lambda a^4}{r^2}\right) = -4\pi\lambda\rho$$

$$\therefore \ \rho = 0$$

That is, there is no attracting matter outside the sphere.

For $r = a$, it can be verified that $V_1 = V_2$. If σ be the surface density, then

$$\left(\frac{\partial V_1}{\partial r} - \frac{\partial V_2}{\partial r}\right)_{r=a} = 4\pi\gamma\sigma$$

$$\text{i.e.,} \quad \left(-3\lambda r^2 + \frac{\lambda a^4}{r^2}\right)_{r=a} = 4\pi\gamma\sigma, \ \text{i.e.,} \ -3\lambda a^2 + \lambda a^2 = 4\pi\gamma\sigma$$

$$\therefore \sigma = -\frac{\lambda a^2}{2\pi\gamma} \tag{5}$$

This shows that the surface contains a repelling matter of density σ given by Eqn (5).

EXAMPLE 13.21 *The values of potential at any point at a distance r from a fixed point are given as*

$$V_1 = 2\pi\gamma\rho\left(a^2 - b^2\right), \qquad \textit{for } r < b < a,$$

$$V_2 = 2\pi\gamma\rho\left(a^2 - \frac{r^2}{3} - \frac{2b^3}{3r}\right), \ \textit{for } b < r < a$$

$$V_3 = \frac{4}{3}\pi\gamma\rho\left(\frac{a^3 - b^3}{r}\right), \qquad \textit{for } b < a < r$$

Show that the attracting system is a spherical shell of density ρ whose boundaries are spheres of radii a and b with centre at that fixed point.

Solution. The potential in each case being function of r only, the Poisson's equation in spherical polar coordinates is

$$\frac{\partial^2 V}{\partial r^2} + \frac{2}{r}\frac{\partial V}{\partial r} = -4\pi\gamma\rho \tag{1}$$

Case 1 : $r < b < a$

It is clear that outer radius is a and the inner radius is b. Now $V_1 = 2\pi\gamma\rho\left(a^2 - b^2\right)$, which is a constant quantity. Hence, all derivatives are zero, so that Eqn (1) yields $\rho = 0$. That is, there is no attracting matter inside the sphere of radius b.

Case 2 : $b < r < a$

Here $V_2 = 2\pi\gamma\rho\left(a^2 - \dfrac{r^2}{3} - \dfrac{2b^3}{3r}\right)$

$$\left.\begin{aligned} \therefore \quad \frac{\partial V_2}{\partial r} &= 2\pi\gamma\rho\left(-\frac{2r}{3} + \frac{2b^3}{3r^2}\right) \\ \frac{\partial^2 V_2}{\partial r^2} &= 2\pi\gamma\rho\left(-\frac{2}{3} - \frac{4}{3}\frac{b^3}{r^3}\right) \end{aligned}\right\} \tag{2}$$

Substituting Eqn (2) into Eqn (1), if density is ρ_2,

$$2\pi\gamma\rho\left[\left(-\frac{2}{3}-\frac{4}{3}\frac{b^3}{r^3}\right)-\frac{4}{3}+\frac{4}{3}\frac{b^3}{r^3}\right] = -4\pi\gamma\rho_2$$

$$\text{i.e.,} \quad -4\pi\gamma\rho = -4\pi\gamma\rho_2$$

which yields $\rho_2 = \rho$.

That is, attracting matter of density ρ exists within the space enclosed by two spheres of radii a and b.

Case 3 : $b < a < r$

In this case, $V_3 = \frac{4}{3}\pi\gamma\rho\left(\frac{a^3 - b^3}{r}\right)$.

Let the density of matter be ρ_3. We write,

$$V_3 = \frac{\xi}{r} \tag{3}$$

where

$$\xi = \frac{4}{3}\pi\gamma\rho\,(a^3 - b^3) \tag{4}$$

Then

$$\frac{\partial V_3}{\partial r} = -\frac{\xi}{r^2},\ \frac{\partial^2 V_3}{\partial r^2} = \frac{2\xi}{r^3} \tag{5}$$

Substituting Eqn (5) into Eqn (1),

$$2\xi\left(\frac{1}{r^3}-\frac{1}{r^3}\right) = -4\pi\gamma\rho_3$$

which yields $\rho_3 = 0$.

Thus, there is no attracting matter outside the sphere of radius a. Hence, the attracting matter consists of a spherical shell of radii a and b, where $a > b$.

EXAMPLE 13.22 *Applying Gauss's theorem, determine the attraction at internal and external points due to*

(i) *a thin spherical shell and*

(ii) *a solid sphere.*

Solution. (i) **Thin spherical shell.** Let R be the radius of the shell and M its mass. We draw a sphere of radius $r < R$. The outward normal force F, due to symmetry, has the same value at all points on the surface S of this sphere (FIG. 13.28).

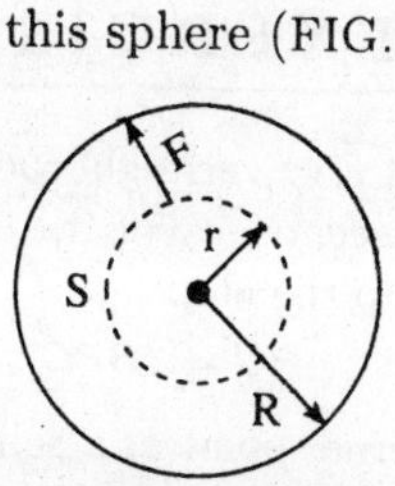

FIG. 13.28

Applying Gauss's theorem, since there is no matter enclosed by S, we have $(4\pi r^2)\cdot F = 0$, which yields $F = 0$ at all points inside the shell.

Now we draw a sphere of radius $r > R$. In this case (FIG. 13.29), the surface S encloses a mass M.

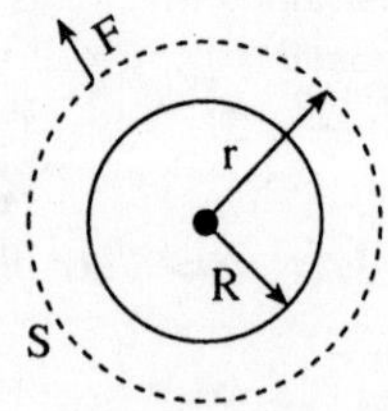

FIG. 13.29

Then, by Gauss's theorem,

$$(4\pi r^2) \cdot F = -4\pi\gamma M$$

$$\text{i.e., } F = -\frac{\gamma M}{r^2} \tag{1}$$

which coincides with the known result.

(ii) **Solid sphere**. Let ρ be the uniform density of the sphere. When the point is inside the sphere, we as before draw a sphere of radius $r < R$. The surface S now encloses a mass

$$= \frac{4}{3}\pi r^3 \rho$$

By Gauss's theorem,

$$(4\pi r^2 \cdot F) = -4\pi\gamma\left(\frac{4}{3}\pi r^3 \rho\right)$$

$$\text{i.e., } F = -\frac{4}{3}\pi\gamma\rho r \tag{2}$$

When the point is outside the sphere, we draw a sphere of radius $r > R$.

In this case, the surface S encloses a mass $= \frac{4}{3}\pi R^3 \rho$

Then, by Gauss's theorem,

$$(4\pi r^2) \cdot F = -4\pi\gamma\left(\frac{4}{3}\pi R^3 \rho\right)$$

$$\text{i.e., } F = -\frac{4}{3}\frac{\pi\gamma R^3 \rho}{r^2} = -\frac{M}{r^2} \tag{3}$$

where M is the mass of the sphere.

EXERCISE

1. Prove that the resultant attraction at P of a straight rod AB of uniform line density m is $2\gamma me/PG$ directed along PG, where e is the eccentricity of the ellipse of foci A, B passing through P, and G is the point at which the normal to the ellipse at P cuts AB.
   [**Hints.** $AG = e \cdot AP, BG = e \cdot BP$ and $\angle APG = \angle BPG$]

2. Prove that the attraction at an internal point of a solid sphere when the mass density obeys the law $\rho = a + br^2$, is

$$4\pi\mu r\left(\frac{a}{3} + \frac{br^2}{5}\right),$$

   where r is the distance of the point from the centre of the sphere.

3. A and B are the centres of two spheres of masses m_1 and m_2 respectively. Find the point O at which the gravitational intensity vanishes. Show that, if a particle at rest at O be given a very small displacement in the direction perpendicular to AB, it will execute simple harmonic motion. In case $m_1 = m_2 = m$, show that the time period will be

$$T = \frac{\pi}{2}\sqrt{\frac{r^3}{Gm}}, \text{ where } r = AB.$$ **[C.U. (HONS)]**

4. A uniform rod of mass M and length L lies on x-axis. Find the gravitational force on a point mass m placed on x-axis at a distance $a (a > L)$ from the left end of the rod.

If the whole mass of the rod were concentrated at a point x_0 such that the force on m remained unchanged, find the location of x_0. **[Ans.** $F = \frac{GMm}{a(a-L)}; x_0 = a\left\{1 - \left(1 - \frac{L}{a}\right)^{1/2}\right\}$]

5. A hemi-sphere of mass M and radius R has a particle of mass m at its centre. Find the force of attraction on it, if the hemi-sphere is (i) a thin shell, (ii) a solid. **[C.U. (HONS.)]**

[Ans. (i) $\frac{GMm}{4R^2}$; (ii) $\frac{3GMm}{4R^2}$]

6. Show that the attraction of a spherical segment of height h on a unit particle at its vertex is

$$2\pi G\rho\left[1 - \frac{1}{2}\sqrt{\frac{2h}{R}}\right],$$

where R = radius and ρ = density of the sphere.

7. The density of a sphere varies as the depth below the surface. Show that the resultant attraction is greatest at a depth equal to one-third of the radius and that its value is 4/3 of the value at the surface.

8. Show that, if M be the mass of a sphere of radius R, the loss of gravitational potential energy in assembling the particles from a state of diffusion at an infinite distance is

$$\frac{3}{5}\frac{GM^2}{R}.$$

9. A solid sphere of mass M and radius R is such that its density at any point is proportional to the n-th power of distance from the centre. Show that potential at an external point distant d from the centre is GM/d.

10. The density of a sphere of mass M and radius R at a distance r from the centre is proportional to $(R - r)$. Prove that the potential energy exhausted in collecting the elements of the sphere is

$$\frac{26}{35}\frac{GM^2}{R}.$$

11. The potential within an attracting medium having spherical symmetry is given by

$$V = \left(c^2 + r^2\right)^{-1/2},$$

where r is the distance from the centre and c is a constant. Prove that the density is given by

$$\frac{3c^2}{4\pi\left(c^2 + r^2\right)^{3/2}}$$

and that the potential energy of the whole medium is $3\pi/32c$.

12. In the interior of a spherical distribution of matter of radius R, the gravitational potential at a distance r from the centre is

$$\frac{A \sin(\pi r/R)}{(\pi r/R)} + \text{constant}.$$

Prove that the density at any point is

$$\frac{\pi A}{4GR^2} \frac{\sin(\pi r/R)}{(\pi r/R)}$$

and that the mass of the whole distribution is AR/G. [Use Poisson's equation in spherical coordinates.]

CHAPTER 14

Mechanics of Ideal Fluids

14.1 Fundamental Concepts

Fluid mechanics is a physical science dealing with the state of rest or of motion of fluids under the action of different forces. It has three subdivisions—(1) fluid statics which is concerned with the fluids at rest; (2) fluid kinematics which is restricted to the discussion of the geometry of motion of fluids, and (3) fluid dynamics which is concerned with the motion of fluids under different forces acting on it.

Liquids and gases are combined to give a common name of fluids. Fluids behave differently from solids under external forces. Solids can resist shearing forces up to their elastic limits, but fluids under shearing forces deform continuously even if the magnitude of the force is negligible, and flows continuously under the action of the shear force.

The property of fluids by virtue of which they resist the continuous deformation under shearing forces is called ***viscosity***. Fluids having this property are called ***viscous fluids***. A fluid is called an ***ideal fluid*** when it has no viscosity. For such fluids, the pressure exerted at any point of a surface in contact with it is in the direction of the normal to the surface at that point.

In practice, every fluid has viscosity to some extent. For convenience in our study, we divide fluids into two groups—(1) ***ideal*** for which viscosity is neglected and (2) ***real*** for which viscosity is taken into account.

Following are some of the important properties of fluids.

(A) Compressibility

Fluids may be ***compressible*** or ***incompressible*** under the action of external forces which cause a change in volume. Compressibility is a measure of this change in volume and ***the coefficient of compressibility η is the reciprocal of bulk modulus of elasticity β of the substance.***

If a substance of original volume V undergoes a change in volume dV under external pressure dP, then the bulk modulus of elasticity is defined as

$$\beta = -\frac{dP}{dV/V} \tag{14.1}$$

If m be the mass of the substance and ρ its density, then

$$V\rho = m = \text{constant}. \tag{14.2}$$

Differentiating this,

$$V d\rho + \rho dV = 0, \text{ i.e., } -\frac{dV}{V} = \frac{d\rho}{\rho} \tag{14.3}$$

Then,

$$\beta = \rho \frac{dP}{d\rho} \tag{14.4}$$

so that the compressibility of the substance is

$$\eta = \frac{1}{\rho}\frac{d\rho}{dP} \tag{14.5}$$

Compressibility of liquids is very small and is neglected for all practical purposes. Thus, ***all liquids are considered incompressible***.

For an ideal gas undergoing isothermal process,

$$PV = RT = \text{constant},$$

which yields

$$PdV + VdP = 0, \text{ i.e., } -\frac{dV}{V} = \frac{dP}{P} \tag{14.6}$$

Then, from Eqn (14.1), the bulk modulus of the gas

$$\beta = P \tag{14.7}$$

That is, the compressibility of the gas undergoing isothermal process is

$$\eta = \frac{1}{P} \tag{14.8}$$

For an ideal gas undergoing adiabatic process,

$$PV^{\gamma} = \text{constant}, \tag{14.9}$$

where γ is the ratio of specific heats of the gas under constant pressure and constant volume, i.e.,

$$\gamma = \frac{C_p}{C_V} \tag{14.10}$$

Differentiating Eqn (14.9),

$$\gamma P V^{\gamma-1} dV + V^{\gamma} dP = 0, \text{ i.e., } -\frac{dV}{V} = \frac{dP}{\gamma P} \tag{14.11}$$

Hence, the bulk modulus

$$\beta = \gamma P \tag{14.12}$$

so that the compressibility of the gas undergoing an adiabatic process is

$$\eta = \frac{1}{\gamma P} \tag{14.13}$$

(B) Thermal Conductivity of Fluids

Thermal conductivity of a substance is a measure of its capacity to transfer heat by molecular conduction. The steady state heat conduction in one-dimensional heat flow is governed by Fourier's equation,

$$H = -kA\frac{dT}{dx} \tag{14.14}$$

where H = rate of heat conduction in x-direction,

A = cross-sectional area normal to heat flow,

k = thermal conductivity of the medium,

and $\frac{dT}{dx}$ = temperature gradient in x-direction.

Thermal conductivities of most liquids are small and decrease with the increase of temperature. However, there are exceptions; for example, the thermal conductivity of water increases with the rise in temperature.

The conduction of heat in gases occurs through the mechanism of diffusion and collision of their molecules. The thermal conductivity of gases in general increases with the rise in temperature. For most gases at atmospheric pressure, the thermal conductivity is independent of pressure. Thermal conductivity of a gas decreases with increase of its molecular weight. Thus, hydrogen being the lightest gas, it has the highest thermal conductivity of 0.149 kcal/mK at 0 °C and 0.192 kcal/mK at 100 °C.

(C) Surface Tension and Capillarity

The force of attraction between the molecules within a liquid is known as ***cohesion***. The force of attraction between the molecules of two different liquids or between the liquid molecules and the molecules of solid boundary containing the liquid is known as ***adhesion***.

On the free liquid surface in contact with air, the upward and downward forces of attraction are not balanced. This causes the liquid surface to act as a membrane under tension. ***The surface tension, γ, is the force in the liquid surface normal to a line of unit length in the surface***. Its value decreases with rise of temperature.

We consider equilibrium of an element $ABCD$ of a curved surface having principal radii of curvature ρ_1 and ρ_2 (FIG. 14.1). Let $AB = x$, $BC = y$ and p = excess of pressure on the concave side. Let the surface be given a small normal displacement dz so that it occupies the position $A'B'C'D'$.

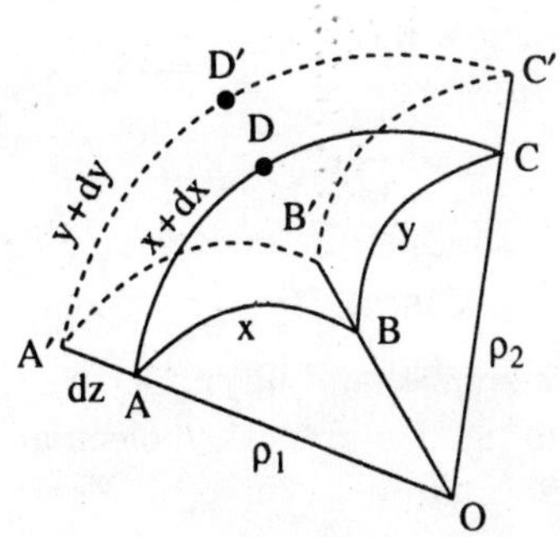

FIG. 14.1

The change in surface area is

$$dS = (x + dx)(y + dy) - xy = x\,dy + y\,dx$$

Also from similar triangles $AOB, A'OB'$, we have

$$\frac{x + dx}{\rho_1 + dz} = \frac{x}{\rho_1} = \frac{dx}{dz}, \text{ i.e., } dx = \frac{x}{\rho_1}dz$$

Similarly, $dy = \frac{y}{\rho_2}dz$. Hence, $dS = \frac{xy}{\rho_1}dz + \frac{xy}{\rho_2}dz$.

For equilibrium of the element, we must have

$$p\,xy \cdot dz = 2\gamma \cdot dS, \text{ i.e., } p\,xy \cdot dz = 2\gamma\left(\frac{xy}{\rho_1}dz + \frac{xy}{\rho_2}dz\right)$$

Hence,

$$p = 2\gamma\left(\frac{1}{\rho_1} + \frac{1}{\rho_2}\right) \tag{14.15}$$

Here we have considered two surfaces. If there is only one surface as in a liquid drop or air-bubble in a liquid,

$$p = \gamma\left(\frac{1}{\rho_1} + \frac{1}{\rho_2}\right) \tag{14.16}$$

If the air-bubble in a liquid is spherical, then $\rho_1 = \rho_2 = R =$ radius of the bubble.

In this case,

$$p = \frac{2\gamma}{R} \tag{14.17}$$

In the case of a spherical soap bubble, there are two surfaces, so that

$$p = \frac{4\gamma}{R} \tag{14.18}$$

For a cylindrical liquid surface, $\rho_1 = R$ and $\rho_2 = \infty$, so that

$$p = \frac{\gamma}{R} \tag{14.19}$$

If a glass tube of small diameter called capillary tube having both ends open is dipped into a liquid, the liquid level in the tube either rises or falls. This phenomenon is called ***capillarity***.

If h be the rise of a liquid within a tube of radius r, then, for equilibrium, we must have (FIG. 14.2), $\rho g h = 2\gamma/R = 2\gamma\cos\theta/r$ since $r = R\cos\theta$.

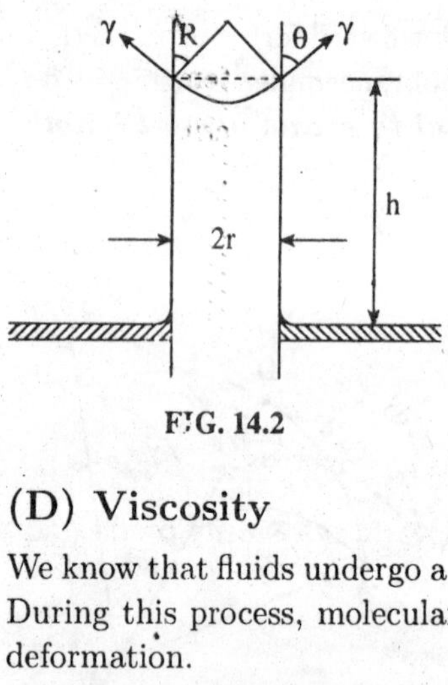

FIG. 14.2

This gives

$$h = \frac{2\gamma\cos\theta}{\rho g r} \tag{14.20}$$

which determines the rise of liquid surface. The angle θ is called the ***angle of contact***.

Since $\gamma =$ force/length, its dimension is

$$[\gamma] = \frac{[M][L]}{[T^2]}\cdot\frac{1}{[L]} = MT^{-2}$$

In S.I. system, its unit is newton/metre and in CGS, the unit is dyne/cm.

(D) Viscosity

We know that fluids undergo a continuous deformation due to externally applied shearing forces. During this process, molecular frictional forces appear within all real fluids which oppose the deformation.

When a liquid flows steadily and slowly over a fixed surface, the layer in contact with the surface remains stationary. The velocity of upper layers increases with the distance from the fixed surface (FIG. 14.3). Thus, between two layers separated by a distance dx, there is a velocity difference of dv. The velocity gradient between these two layers is dv/dx.

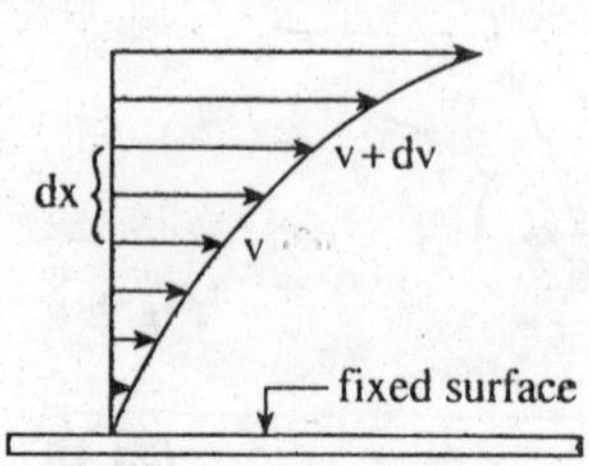

FIG. 14.3

The externally applied shearing force causes tangential stress between two layers. If this is represented by τ, then, according to Newton's law of viscous resistance,

$$\tau \propto \frac{dv}{dx}, \text{ i.e., } \tau = \mu\frac{dv}{dx}, \tag{14.21}$$

where μ is the constant of proportionality known as ***coefficient of viscosity***. It is defined as the tangential stress per unit velocity gradient.

Since $\mu = \dfrac{\text{force/area}}{dv/dx}$, its dimension is

$$[\mu] = \frac{[MLT^{-2}]\,/\,[L^2]}{[LT^{-1}]\,/[L]} = [ML^{-1}T^{-1}]$$

In CGS, the unit of the coefficient of viscosity is known as ***poise*** after the name of Poiseuille.

For ideal fluids, $\mu = 0$; they offer no reistance to shear deformation during flow.

The ***kinetic viscosity*** of fluids is defined as the ratio of coefficient of viscosity to the density. Thus,

$$\nu = \frac{\mu}{\rho} \tag{14.22}$$

where ν = kinetic viscosity.

The viscosity of liquids decreases with increase of temperature at constant pressure. The viscosity of gases, however, increases with the rise of temperature at constant pressure.

There are several empirical formulae relating the viscosity of liquids to the temperature. The relation of kinematic viscosity of water with temperature is given by Poiseuille's formula,

$$\nu = \frac{0.0179}{1 + aT + bT^2} \tag{14.23}$$

where $a = 0.0337$, $b = 0.00022$ and T = temperature in degree centigrade.

The viscosity of gases and of most liquids increases with pressure at constant temperature.

The variation of viscosity is, however, small when the pressure is below the critical. The variation of viscosity for gases with temperature is given by Sutherland's formula,

$$\mu = \mu_0 \frac{A\sqrt{T}}{1 + B/T} \tag{14.24}$$

where

$$\mu_0 = \text{the viscosity at } 0\,°\text{C}$$
$$T = \text{the absolute temperature}$$

and A, B are constants.

This formula fails for high as well as low temperatures. The reciprocal of viscosity coefficient is called ***fluidity*** of the substance. From the view point of viscosity, real fluids are divided into two groups—Newtonian and non-Newtonian. For Newtonian fluids, the viscosity coefficient μ is a constant at a given temperature and pressure. The relation between the shear stress τ and the velocity gradient for such fluids is linear.

For non-Newtonian fluids, the viscosity at a given temperature and pressure is a function of the velocity gradient. Hence, the relation between the shear stress τ and the velocity gradient is non-linear. This is shown in the figure (FIG. 14.4).

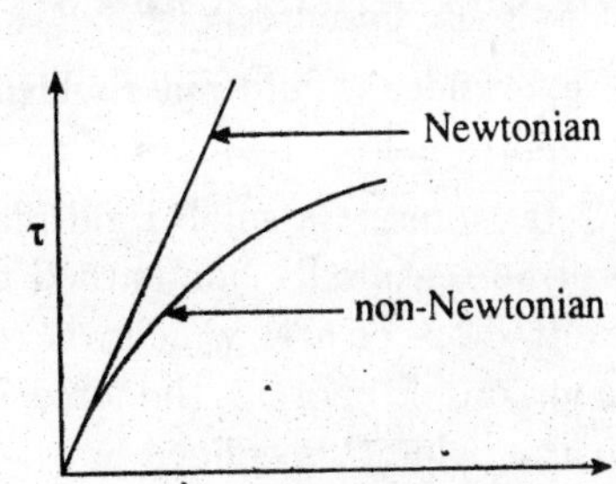

FIG. 14.4

There are several types of non-Newtonian fluids. These are:

(1) **Bingham plastics**, also called ***ideal plastics***, are those which can withstand a certain amount of shear stress. For such fluids, relation between shear stress and velocity gradient is

$$\tau = \tau_0 + \mu \frac{dv}{dx} \tag{14.25}$$

where τ_0 is the yield shear stress after which the fluid begins to deform. Drilling muds, sewage sludge are Bingham plastics.

(2) **Pseudoplastic** liquids are those which deform as soon as a shearing stress is applied, but the viscosity decreases with the velocity gradient. Gels, paper pulp are these types of materials.

(3) **Dilatant** materials are those for which the viscosity increases with the velocity gradient, but the material deforms as soon as a shearing stress is applied. Quick sand and butter are examples of dilatant materials.

For all the non-Newtonian fluids, we may write relation (14.25) in the general form:

$$\tau = \beta + \mu \left(\frac{dv}{dx}\right)^n \tag{14.26}$$

such that, (1) $\beta = \tau_0$, $n = 1$, for Bingham plastics, (2) $\beta = 0$, $n < 1$ for pseudoplastics and (3) $\beta = 0$, $n > 1$, for dilatants.

The shear stress versus the velocity gradient of different types of non-Newtonian liquids are shown in the figure (FIG. 14.5).

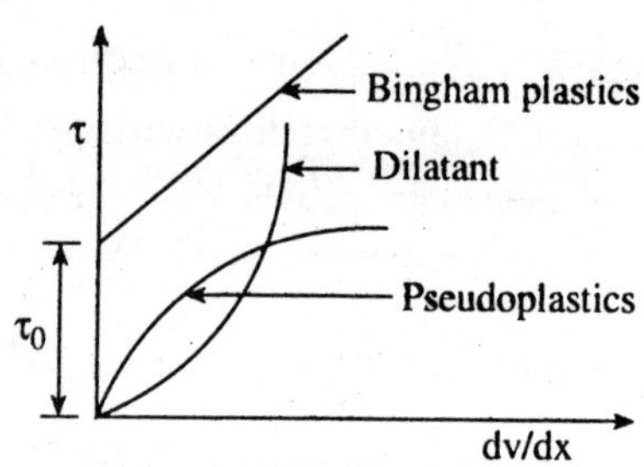

FIG. 14.5

There are some other types of non-Newtonian fluids for which the velocity gradient is a function of time also. Fluids for which the viscosity increases with time are called ***Thixotropic***. Lipstick and some paints are thixotropic fluids. Fluids for which viscosity decreases with time are called ***non-thixotropic***. Gypsum suspension in water and bentonite solutions are non-thixotropic fluids.

14.2 Fluid Statics

We consider a fluid mass which is in equilibrium under some external body forces which are proportional to the mass of an elementary volume of the fluid.

If we insert a small rigid plane area δA within the fluid mass at P, then both sides of δA will be repeatedly bombarded by the moving molecules at random. Thus, a force δF will be exerted on the area which will remain at rest relative to the main fluid mass. Whether the fluid is viscous or inviscid, the force δF will act normally to the area δA. Then, hydrostatic pressure at the point P is defined as

$$p = \lim_{\delta A \to 0} \frac{\delta F}{\delta A}$$

The pressure p at an internal point is same in all direction, so that orientation of δA is immaterial. This we will prove in a later section.

Now we consider the condition of equilibrium of a fluid mass at rest.

For this purpose we take an infinitesimal rectangular parallelepiped of fluid having side lengths dx, dy and dz. Let X, Y, Z be the components of body force at a point P inside the element, p be the pressure and ρ the density there (FIG. 14.6).

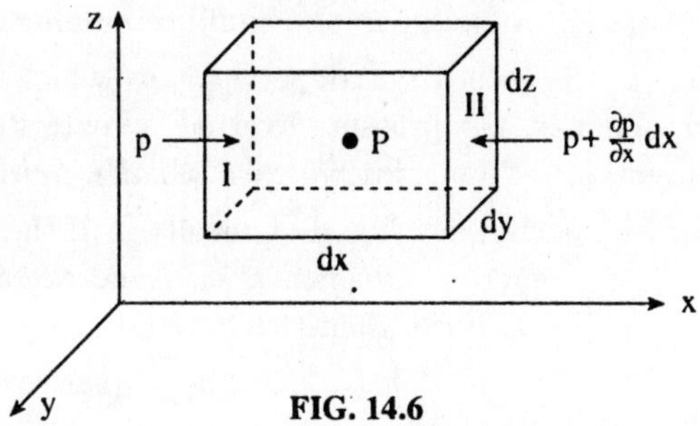

FIG. 14.6

Forces acting on the element along x-direction are:

(1) $p\,dy\,dz$ on face I and $\left(p+\dfrac{\partial p}{\partial x}dx\right)dy\,dz$ on the opposite face II,

(2) Component of body force $= X\rho\,dx\,dy\,dz$.

Thus, for equilibrium in x-direction we must have

$$X\rho\,dx\,dy\,dz + p\,dy\,dz - \left(p+\frac{\partial p}{\partial x}dx\right)dy\,dz = 0, \text{ i.e., } X - \frac{1}{\rho}\frac{\partial p}{\partial x} = 0$$

We can get two other similar equations by considering equilibrium in y- and z-directions.

Thus, we have the conditions of equilibrium

$$\left.\begin{aligned} X &= \frac{1}{\rho}\frac{\partial p}{\partial x} \\ Y &= \frac{1}{\rho}\frac{\partial p}{\partial y} \\ Z &= \frac{1}{\rho}\frac{\partial p}{\partial z} \end{aligned}\right\} \tag{14.27}$$

If the fluid mass be at rest under gravity, then $X = Y = 0$, $Z = -g$. In this case,

$$\frac{\partial p}{\partial x} = 0, \ \frac{\partial p}{\partial y} = 0 \tag{14.28}$$

$$\text{and} \quad -g = \frac{1}{\rho}\frac{\partial p}{\partial z} \tag{14.29}$$

Eqns (14.28) imply that the pressure does not depend on x and y coordinates of the point P. That is, there is no pressure variation on a horizontal plane. Hence, we write Eqn (14.29) as

$$-g = \frac{1}{\rho}\frac{dp}{dz}, \text{ i.e., } dp = -g\rho\,dz \tag{14.30}$$

We consider the following cases:

• Case I : $\rho =$ constant, fluid at rest under gravity

For incompressible fluids at constant temperature, density ρ = constant. **Then integrating** Eqn (14.30) we have $p = -g\rho z + A$, where A is the constant of integration. Let $p = p_0$, **when** $z = 0$. Then $A = p_0$.

Hence,

$$p = p_0 - g\rho z \tag{14.31}$$

This determines the pressure at any point in the liquid remaining at rest under gravity.

Eqn (14.31) implies that $(p - p_0)$ does not depend on p_0 which is the external pressure on the surface of the liquid. That is, the pressure exerted by external forces on the liquid is transmitted equally in all directions. This is known as ***Pascal's principle***.

Let us now consider a body immersed in a liquid of density ρ. If the top and bottom surfaces have cross-sectional area S and are at depths z_1, z_2 from the free surface, then (FIG. 14.7)

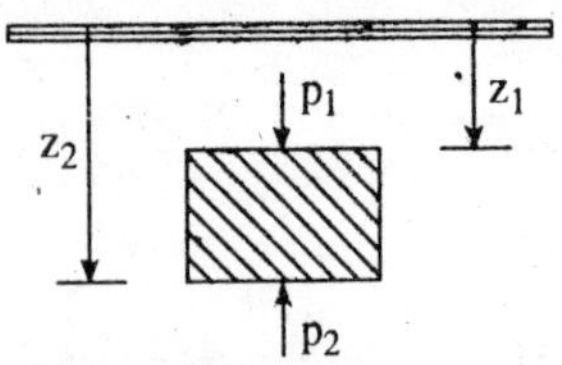

FIG. 14.7

$$p_1 = g\rho z_1 + A$$

$$p_2 = g\rho z_2 + A$$

which yield

$$(p_1 - p_2) = g\rho(z_1 - z_2)$$

$$\text{i.e., } S(p_1 - p_2) = Sg\rho(z_1 - z_2) \tag{14.32}$$

The left-hand side of Eqn (14.32) is the upward thrust by the fluid and the right-hand side gives the weight of the displaced fluid. Thus, ***the upward thrust on a body immersed in a fluid at rest is equal to the weight of the fluid displaced by the body. This is Archimede's principle.***

• Case 2 : $\rho =$ constant, body forces are functions of (x, y, z)

When the components of the body force X, Y and Z are functions of x, y, z, integration of the general Eqn (14.27) is possible, only if we can find a function $\phi(x, y, z)$ such that

$$\left.\begin{aligned} X &= -\frac{\partial \phi}{\partial x} \\ Y &= -\frac{\partial \phi}{\partial y} \\ Z &= -\frac{\partial \phi}{\partial z} \end{aligned}\right\} \tag{14.33}$$

The function $\phi(x, y, z)$ is called ***potential function***.

Multiplying Eqn (14.27) by dx, dy and dz respectively and adding the results, we get

$$X\,dx + Y\,dy + Z\,dz = \frac{1}{\rho}\left(\frac{\partial p}{\partial x}dx + \frac{\partial p}{\partial y}dy + \frac{\partial p}{\partial z}dz\right)$$

$$\text{i.e., } X\,dx + Y\,dy + Z\,dz = \frac{1}{\rho}dp \tag{14.34}$$

$$\text{i.e., } -\left(\frac{\partial \phi}{\partial x}dx + \frac{\partial \phi}{\partial y}dy + \frac{\partial \phi}{\partial z}dz\right) = \frac{1}{\rho}dp$$

$$\text{i.e., } -d\phi = \frac{1}{\rho}dp, \text{ i.e., } d\left(\phi + \frac{p}{\rho}\right) = 0 \tag{14.35}$$

Eqn (14.35) implies that within a fluid mass and on its surface remaining in a potential field, we must have

$$\phi + \frac{p}{\rho} = \text{constant.} \tag{14.36}$$

If γ = weight of unit volume of the fluid so that $\rho = \gamma/g$, we write Eqn (14.36) in the form:

$$\left(\frac{\phi}{g}\right) + \frac{p}{\gamma} = \text{constant} = \Pi \tag{14.37}$$

The quantity Π is called the ***specific potential energy***.

• Case 3 : ρ is a function of pressure

For gases at constant temperature, according to **Boyle's law**,

$$\rho \propto p, \text{ i.e., } \rho = kp \tag{14.38}$$

Then, for the atmosphere under gravity, we have from Eqn (14.30),

$$\frac{dp}{p} = -kg\,dz$$

which on integration yields

$$\log p = -kgz + B$$

If we assume that on the surface of the earth, where $z = 0$, pressure $p = p_0$, then the constant B of integration is $B = \log p_0$.

Hence,

$$\log\frac{p}{p_0} = -kgz, \text{ i.e., } p = p_0\exp(-kgz) \tag{14.39}$$

This gives the atmospheric pressure at different altitudes, provided temperature remains constant.

The temperature in the atmosphere, however, is not same at different altitudes. For a perfect gas, we know

$$\rho = \frac{p}{RT} \tag{14.40}$$

where the absolute temperature T is a prescribed function of altitude z. Then, from Eqn (14.30),

$$\frac{dp}{p} = -\frac{g}{RT(z)}dz \tag{14.41}$$

This can be integrated to get

$$p = p_0\exp\left[-\frac{g}{R}\int\frac{dz}{T(z)}\right] \tag{14.42}$$

where p_0 is the pressure at $z = 0$.

Thus, it is necessary to know the temperature function $T(z)$.

If we assume that air is isothermal at sea level temperature T_0, then Eqn (14.42) yields

$$p = p_0\exp\left(-\frac{gz}{RT_0}\right) \tag{14.43}$$

This is Halley's law which provides fairly accurate results to real problems.

A more realistic static pressure relation with temperature is given by the linear law

$$T(z) = T_0 - \alpha z \tag{14.44}$$

where T_0 is the absolute temperature at $z = 0$ and the constant α is called the ***temperature lapse rate***. A typical value of α for the earth's atmosphere is 6.5 °C/km. This linear variation of temperature is valid from sea level to 45000 ft altitude.

Substituting the function (14.44) into Eqn (14.42) and integrating, we find

$$p = p_0 \left(1 - \frac{\alpha z}{T_0}\right)^{g/R\alpha} \tag{14.45}$$

This is called the ***logarithmic pressure law***.

• Case 4 : ρ = constant, fluid rotating about an axis

We consider a mass of fluid of constant density ρ moving about the z-axis uniformly with an angular velocity ω.

The problem can be changed to a statical one by combining with the existing body forces the reversed effective force $\omega^2 r$ per unit mass of the fluid, where r is the distance from the axis of rotation. This force has components

$$\left.\begin{aligned} \omega^2 r \cos\theta &= \omega^2 x \\ \omega^2 r \sin\theta &= \omega^2 y \end{aligned}\right\} \tag{14.46}$$

along x- and y-directions respectively. Then the general Eqn (14.34) of equilibrium assumes the form:

$$dp = \rho(X\,dx + Y\,dy + Z\,dz + \omega^2 x\,dx + \omega^2 y\,dy)$$

If gravity is the only force acting on the fluid, then $X = 0, Y = 0$ and $Z = -g$.

Hence, $dp = \rho(\omega^2 x dx + \omega^2 y\,dy - g\,dz) = \rho(\omega^2 r\,dr - g\,dz)$.

Integrating, $p = \rho\left(\frac{1}{2}\omega^2 r^2 - gz\right) + C$.

If origin of coordinates be taken at the vertex of the free surface, where pressure is, say, p_0, then $p = p_0$, for $r = 0$ and $z = 0$.

This yields $C = p_0$. Hence,

$$p = p_0 + \rho\left(\frac{1}{2}\omega^2 r^2 - gz\right) \tag{14.47}$$

This shows that the surfaces of constant pressures are given by

$$r^2 = \frac{2g}{\omega^2} z + \text{constant}, \tag{14.48}$$

which are paraboloids of revolution having the same latus rectum $2g/\omega^2$.

EXAMPLE 14.1 *A mass of compressible fluid extends from infinity to the surface of a sphere of radius R. The equation of state is $p = \lambda\rho$, λ = constant. Each element of the fluid is attracted towards the centre of the sphere with a force μ/r^2, μ = constant and r is the radial distance. Show that the pressure exerted on the surface of the sphere is $\lambda\rho_0 \exp\left(\frac{\mu}{\lambda R}\right)$, where ρ_0 is the density at infinity.*

Solution. If p_0 be the pressure and ρ_0 the density at infinity, then, by the equation of state,

$$p = \lambda\rho \tag{1}$$

we have

$$p_0 = \lambda\rho_0 \tag{2}$$

The Eqn (14.34) of equilibrium in this case is

$$\frac{dp}{\rho} = -\frac{\mu}{r^2}dr, \text{ i.e., } \lambda\frac{dp}{p} = -\frac{\mu}{r^2}dr \text{ [using Eqn (1)]} \tag{3}$$

Integrating,

$$\lambda \log p = \frac{\mu}{r} + C \tag{4}$$

Since $p = p_0$ at $r = \infty$, we get

$$C = \lambda \log p_0 \tag{5}$$

Hence, from Eqns (4) and (5),

$$\lambda \log \frac{p}{p_0} = \frac{\mu}{r}, \text{ i.e., } p = p_0 \exp\left(\frac{\mu}{\lambda r}\right), \text{ i.e., } p = \lambda\rho_0 \exp\left(\frac{\mu}{\lambda r}\right) \tag{6}$$

This gives the law of pressure distribution in the fluid mass.

On the surface of the sphere, $r = R$. Hence, pressure on the surface of the sphere is given by

$$p = \lambda\rho_0 \exp\left(\frac{\mu}{\lambda R}\right)$$

EXAMPLE 14.2 *Assuming the equation of state as $p = \lambda\rho^k$ (λ, k are constants) and the value of the acceleration due to gravity at an altitude z from the earth's surface as $\frac{g}{(1+\frac{z}{R})^2}$, $R=$ radius of the earth, determine the pressure $p(z)$ in the atmosphere.*

Solution. We are given the relation

$$p = \lambda\rho^k \tag{1}$$

If p_0 and ρ_0 be the pressure and density at the surface of the earth, then

$$p_0 = \lambda\rho_0^k \tag{2}$$

From Eqns (1) and (2),

$$\frac{p}{p_0} = \left(\frac{\rho}{\rho_0}\right)^k, \text{ i.e., } \rho = \rho_0\left(\frac{p}{p_0}\right)^{1/k} \tag{3}$$

Equation of equilibrium is

$$dp = -g\rho\, dz = -\frac{g\rho_0 p^{1/k}}{p_0^{1/k}\left(1+\frac{z}{R}\right)^2}dz, \text{ i.e., } \frac{dp}{p^{1/k}} = -\frac{g\rho_0}{p_0^{1/k}\left(1+\frac{z}{R}\right)^2}dz$$

Integrating between the limits zero to z,

$$\frac{k}{k-1}\left(p^{\frac{k-1}{k}} - p_0^{\frac{k-1}{k}}\right) = -\frac{g\rho_0}{p_0^{1/k}}\left[R - \frac{R}{1+\frac{z}{R}}\right]$$

$$\text{or, } p^{\frac{k-1}{k}} - p_0^{\frac{k-1}{k}} = -\frac{g\rho_0(k-1)}{kp_0^{1/k}}\left[\frac{z}{1+\frac{z}{R}}\right]$$

$$\text{Hence, } p = \left[p_0^{\frac{k-1}{k}} - \frac{g\rho_0(k-1)z}{kp_0^{1/k}\left(1+\frac{z}{R}\right)}\right]^{\frac{k}{k-1}}.$$

EXAMPLE 14.3 *A closed circular cylinder of radius R is just filled with a liquid of density ρ and rotated with uniorm angular velocity ω about its axis. If the total thrust on the bottom is n times that on the top, then show that $\omega = \frac{2}{R}\sqrt{\frac{gH}{n-1}}$, where H is the height of the cylinder.*

Solution. The equation of equilibrium in the liquid is $dp = \rho(\omega^2 r\, dr - g\, dz)$.

Integrating,

$$p = \rho\left(\frac{\omega^2 r^2}{2} - gz\right) + C \tag{1}$$

At the centre of the top surface $p = 0$, for $r = 0$ and $z = H$.

Hence, $C = \rho g H$. Then from Eqn (1),

$$p = \rho\left(\frac{\omega^2 r^2}{2} - gz + gH\right) \tag{2}$$

This gives the pressure variation in the liquid. Then the pressure variation on the bottom ($z = 0$) is

$$p_1 = \rho\left(\frac{\omega^2 r^2}{2} + gH\right) \tag{3}$$

and that on the top ($z = H$) is

$$p_2 = \frac{1}{2}\rho\omega^2 r^2 \tag{4}$$

Hence, the total thrust on the bottom is

$$T_1 = \int_0^{2\pi}\int_0^R p_1 r\, d\theta\, dr = \int_0^{2\pi}\int_0^R \rho\left(\frac{\omega^2 r^2}{2} + gH\right) r\, d\theta\, dr$$

$$= 2\pi\rho\left(\frac{\omega^2 R^4}{8} + g\frac{R^2 H}{2}\right) \tag{5}$$

and that on the top is

$$T_2 = \int_0^{2\pi}\int_0^R p_2 r\, d\theta\, dr$$

$$= \int_0^{2\pi}\int_0^R \left(\frac{1}{2}\rho\omega^2 r^2\right) r\, d\theta\, dr = 2\pi\rho\left(\frac{\omega^2 R^4}{8}\right) \tag{6}$$

By the given condition, $T_1 = nT_2$. Hence, from Eqns (5) and (6),

$$2\pi\rho\left(\frac{\omega^2 R^4}{8} + g\frac{R^2 H}{2}\right) = n \cdot 2\pi\rho\frac{\omega^2 R^4}{8}, \text{ i.e., } \frac{\omega^2 R^4}{8}(n-1) = g\frac{R^2 H}{2}$$

Hence, $\omega^2 = \dfrac{4gH}{R^2(n-1)}$, i.e., $\omega = \dfrac{2}{R}\sqrt{\dfrac{gH}{n-1}}$.

14.3 Kinematics of Fluids

We consider the fluid to be a continuous medium. The fundamental hydrodynamical parameters of fluid motion are pressure, velocity and density. The general problem of fluid motion is dealt with by two methods: (1) ***Lagrange's method*** and (2) ***Euler's method***.

In Lagrange's method, any particle of fluid is selected in space and its motion is observed throughout its course.

In Euler's method, any point in the space occupied by the fluid is selected and the changes of hydrodynamical parameters occurring at that point during the course of time is investigated.

(A) Velocity of Fluid Flow

If we consider some given points in space of fluid motion, the velocity will be different at these points and also will change with time. Thus, the fluid motion is characterised by the formation of velocity field. The velocity components u, v, w along rectangular coordinate axes will be functions of (x, y, z, t) so that

$$\left.\begin{aligned} u &= u(x, y, z, t) \\ v &= v(x, y, z, t) \\ w &= w(x, y, z, t) \end{aligned}\right\} \tag{14.49}$$

Hence, we can write the velocity vector $\mathbf{q}$ at any point as

$$\mathbf{q} = u\mathbf{i} + v\mathbf{j} + w\mathbf{k} \tag{14.50}$$

(B) Acceleration of Fluid Flow

We consider a fluid at a point $P(x, y, z)$ where the x-component of velocity is $u(x, y, z, t)$. Let after a short interval of time δt, the fluid element moves to a point Q whose coordinates are $x + u\,\delta t,\ y + v\,\delta t,\ z + w\,\delta t$. Then its velocity component in x-direction becomes

$$\begin{aligned} u + \delta u &= u(x + u\delta t, y + v\delta t, z + w\delta t) \\ &= u(x, y, z, t) + \left(u\frac{\partial u}{\partial x} + v\frac{\partial v}{\partial y} + w\frac{\partial w}{\partial z}\right)\delta t \end{aligned}$$

Here we have used Taylor's expansion and neglected terms containing higher powers of δt.

Hence, the x-component of acceleration is

$$a_x = \lim_{\delta t \to 0} \frac{(u + \delta u) - u}{\delta t} = u\frac{\partial u}{\partial x} + v\frac{\partial u}{\partial y} + w\frac{\partial u}{\partial z} + \frac{\partial u}{\partial t}$$

We can get similar expressions for other two components of acceleration. Hence, we write,

$$\left.\begin{aligned} a_x &= u\frac{\partial u}{\partial x} + v\frac{\partial u}{\partial y} + w\frac{\partial u}{\partial z} + \frac{\partial u}{\partial t} \\ a_y &= u\frac{\partial v}{\partial x} + v\frac{\partial v}{\partial y} + w\frac{\partial v}{\partial z} + \frac{\partial v}{\partial t} \\ a_z &= u\frac{\partial w}{\partial x} + v\frac{\partial w}{\partial y} + w\frac{\partial w}{\partial z} + \frac{\partial w}{\partial t} \end{aligned}\right\} \tag{14.51}$$

In these expressions, the quantities $\frac{\partial u}{\partial t}, \frac{\partial v}{\partial t}, \frac{\partial w}{\partial t}$ indicate the acceleration of the fluid element at the fixed point $P(x, y, z)$ as time passes without any variation in (x, y, z). This is known as ***local acceleration***. Other terms in the expressions (14.51) represent the acceleration of the fluid element due to its shifting during the course of time. This is known as ***convective acceleration***.

The acceleration vector at any point in the fluid motion may be written as

$$\mathbf{a} = a_x\mathbf{i} + a_y\mathbf{j} + a_z\mathbf{k} \tag{14.52}$$

If the fluid-flow parameters like density, velocity and pressure at a point do not change with time, the flow is said to be ***steady***. Thus, for steady motion,

$$\left.\begin{aligned} \frac{\partial u}{\partial t} = \frac{\partial v}{\partial t} = \frac{\partial w}{\partial t} = 0 \\ \frac{\partial \rho}{\partial t} = \frac{\partial p}{\partial t} = 0 \end{aligned}\right\}$$

The acceleration components in the steady flow are:

$$\left.\begin{aligned} a_x &= u\frac{\partial u}{\partial x} + v\frac{\partial u}{\partial y} + w\frac{\partial u}{\partial z} \\ a_y &= u\frac{\partial v}{\partial x} + v\frac{\partial v}{\partial y} + w\frac{\partial v}{\partial z} \\ a_z &= u\frac{\partial w}{\partial x} + v\frac{\partial w}{\partial y} + w\frac{\partial w}{\partial z} \end{aligned}\right\} \tag{14.53}$$

(C) Stream Lines and Path Lines

A stream line is a curve traced by a fluid element in steady motion such that, at any instant of time, the tangent at any point of it gives the direction of the resultant velocity. Hence, a fluid element is constrained to move continuously along a stream line and can have no motion perpendicular to it (FIG. 14.8)

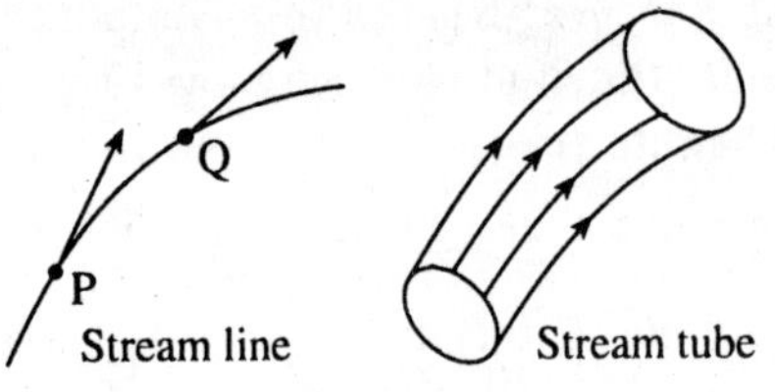

FIG. 14.8

A tubular space in the fluid bounded by stream lines is called a ***stream tube***.

If a point $P(x, y, z)$ on the stream line moves to Q after a time interval of dt, then the coordinates of Q will be $(x + dx, y + dy, z + dz)$, such that $dx = u\,dt, dy = v\,dt, dz = w\,dt$.

These relations yield

$$\frac{dx}{dy} = \frac{u}{v}, \ \frac{dx}{dz} = \frac{u}{w}, \ \text{i.e.,} \ \frac{dx}{u} = \frac{dy}{v} = \frac{dz}{w} \tag{14.54}$$

which are the differential equations of stream lines.

A path line is the curve or trajectory along which a particular fluid particle travels during its motion. The differential equations of the path lines are

$$\frac{dx}{dt} = u, \ \frac{dy}{dt} = v, \ \frac{dz}{dt} = w.$$

The stream lines are not, in general, the same as the path lines. The stream lines show how each particle is moving at a given instant of time, while the path lines show the trajectory of particles at each instant.

(D) Velocity Potentials

If there exists a function $\phi(x, y, z)$ such that the velocity components in a fluid-flow are given by

$$u = -\frac{\partial \phi}{\partial x}, \ v = -\frac{\partial \phi}{\partial y}, \ w = -\frac{\partial \phi}{\partial z} \tag{14.55}$$

then the function ϕ is called a ***velocity potential*** or ***velocity function***.

In this case, the expression $u\,dx + v\,dy + w\,dz$ is an exact differential, since

$$u\,dx + v\,dy + w\,dz = -\left[\frac{\partial \phi}{\partial x}dx + \frac{\partial \phi}{\partial y}dy + \frac{\partial \phi}{\partial z}dz\right] = -d\phi \tag{14.56}$$

Surfaces along which the velocity potential is constant are called ***equipotential surfaces***. Hence, for equipotential surfaces,

$$d\phi = 0, \text{ i.e., } u\,dx + v\,dy + w\,dz = 0 \tag{14.57}$$

Eqns (14.54) and (14.57) show that ***the stream lines cut the equipotential surfaces at right angles.***

(E) Rotation of Fluid Elements

We consider an elementary rectangle $ABCD$ of fluid of sides dx and dy in the xy-plane. The velocity components at A along x- and y-directions are u and v respectively (FIG. 14.9).

Then the velocity components at B are

$$u + \frac{\partial u}{\partial x}dx \text{ along } x\text{-direction}$$

$$\text{and } v + \frac{\partial v}{\partial x}dx \text{ along } y\text{-direction.}$$

Similarly, the velocity components at D are

$$u + \frac{\partial u}{\partial y}dy \text{ along } x\text{-direction}$$

$$\text{and } v + \frac{\partial v}{\partial y}dy \text{ along } y\text{-direction.}$$

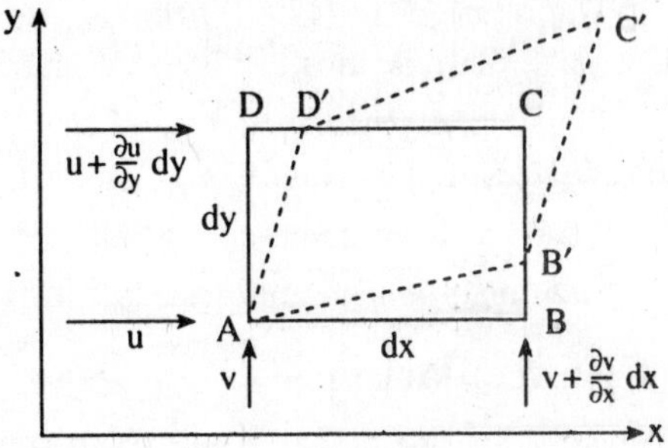

FIG. 14.9

The velocity component at B along y-direction causes the side AB of the fluid element to rotate anti-clockwise about z-axis, so that AB' is the new position. The angular velocity of the linear element AB about z-axis is

$$\frac{\left(v + \frac{\partial v}{\partial x}dx\right) - v}{dx} = \frac{\partial v}{\partial x}$$

Similarly, the angular velocity of the linear element AD due to the velocity component at D along x-direction is

$$\frac{\left(u + \frac{\partial u}{\partial y}dy\right) - u}{dy} = \frac{\partial u}{\partial y}$$

which is in the clockwise direction. Considering the anti-clockwise rotation to be positive, the average angular velocity of the fluid element about z-axis is

$$\omega_z = \frac{1}{2}\left(\frac{\partial v}{\partial x} - \frac{\partial u}{\partial y}\right)$$

We will have similar expressions for angular velocities about x- and y-directions. Thus we write,

$$\left.\begin{aligned} \omega_x &= \frac{1}{2}\left(\frac{\partial w}{\partial y} - \frac{\partial v}{\partial z}\right) \\ \omega_y &= \frac{1}{2}\left(\frac{\partial u}{\partial z} - \frac{\partial w}{\partial x}\right) \\ \omega_z &= \frac{1}{2}\left(\frac{\partial v}{\partial x} - \frac{\partial u}{\partial y}\right) \end{aligned}\right\} \tag{14.58}$$

We define a vector $\boldsymbol{\Omega}$, such that

$$\boldsymbol{\Omega} = 2w_x\mathbf{i} + 2w_y\mathbf{j} + 2w_z\mathbf{k} \tag{14.59}$$

This vector $\boldsymbol{\Omega}$ is known as ***vorticity vector***.

A fluid motion is said to be ***irrotational*** when the vorticity components are zero. That is,

$$\left.\begin{aligned} \frac{\partial w}{\partial y} - \frac{\partial v}{\partial z} &= 0 \\ \frac{\partial u}{\partial z} - \frac{\partial w}{\partial x} &= 0 \\ \frac{\partial v}{\partial x} - \frac{\partial u}{\partial y} &= 0 \end{aligned}\right\} \tag{14.60}$$

The relations in (14.60) are possible when there exists a velocity potential $\phi(x, y, z)$. In fact, if we use the relations (14.55), we will find that the system of equations (14.60) is satisfied.

Thus, ***in a fluid motion, if there exists a velocity potential, the motion should be irrotational***. For this reason, an irrotational flow is often called ***potential flow***.

If the vorticity is non-zero at all points in a flow region, the fluid flow is said to be ***rotational***.

Flow near solid boundaries is rotational.

(F) Circulation

We consider two points A and B in the fluid space. The value of the integral

$$I = \int_A^B (u\,dx + v\,dy + w\,dz)$$

taken along any path from A to B is called ***the flow along that path*** from A to B.

When a velocity potential $\phi(x, y, z)$ exists, the flow from A to B is

$$\therefore I = -\int_A^B \left(\frac{\partial \phi}{\partial x}dx + \frac{\partial \phi}{\partial y}dy + \frac{\partial \phi}{\partial z}dz\right) = -\int_A^B d\phi = \phi_A - \phi_B \tag{14.61}$$

Thus, ***the flow from one point to another in a fluid space is equal to the change in the velocity potential, when it exists.***

Circulation is the fluid-flow round a closed curve. Thus,

$$\Gamma = \oint (u\,dx + v\,dy + w\,dz) \tag{14.62}$$

where Γ represents circulation. It is considered positive in anti-clockwise direction.

Considering an elementary rectangle of sides dx, dy in the fluid space (FIG. 14.10), the differential circulation around the rectangle is

$$d\Gamma = u\,dx + \left(v + \frac{\partial v}{\partial x}dx\right)dy - \left(u + \frac{\partial u}{\partial y}dy\right)dx - v\,dy$$

$$= \left(\frac{\partial v}{\partial x} - \frac{\partial u}{\partial y}\right)dx\,dy$$

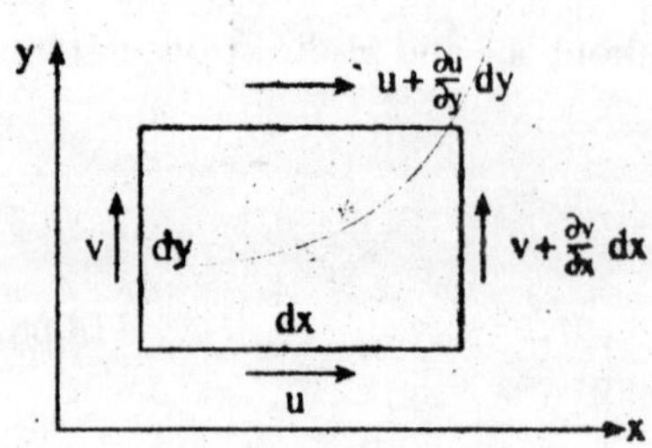

FIG. 14.10

Thus, the circulation per unit area of the rectangle is

$$\frac{d\Gamma}{dx\,dy} = \frac{\partial v}{\partial x} - \frac{\partial u}{\partial y} \tag{14.63}$$

which is the same as the vorticity component along z-axis.

Thus, ***vorticity may also be defined as the circulation per unit area enclosed by the curve.***

(G) Shear Deformation

In an elementary rectangle in the fluid space, we have seen that the angular velocity of side dx is $\partial v/\partial x$, and that of side dy is $-\partial u/\partial y$. Then the net rate of angular velocity

$$= \frac{\partial v}{\partial x} - \left(-\frac{\partial u}{\partial y}\right) = \frac{\partial v}{\partial x} + \frac{\partial u}{\partial y}$$

This gives the rate of shear deformation of the element in xy-plane. Denoting this by S_{xy}, we write,

$$S_{xy} = \frac{\partial v}{\partial x} + \frac{\partial u}{\partial y} \tag{14.64}$$

We will have similar two expressions

$$\left.\begin{aligned} S_{yz} &= \frac{\partial w}{\partial y} + \frac{\partial v}{\partial z} \\ S_{zx} &= \frac{\partial u}{\partial z} + \frac{\partial w}{\partial x} \end{aligned}\right\} \tag{14.65}$$

which give shear rates in corresponding planes.

(H) Conditions at a Rigid Boundary

We consider a fluid-flow with velocity vector **q** past a solid boundary moving with velocity vector **U**. If the fluid is inviscid, then there must be no relative velocity along the normal to the surface at any point. Thus, if **n** represents a unit normal vector at a point on the surface, then the two normal components of the velocities must be equal. That is,

$$\mathbf{q} \cdot \mathbf{n} = \mathbf{U} \cdot \mathbf{n} \tag{14.66}$$

at every point on the surface.

If the solid boundary is at rest, the condition (14.66) reduces to

$$\mathbf{q} \cdot \mathbf{n} = 0 \tag{14.67}$$

at every point of the boundary.

(I) Equation of Continuity

We consider the fluid motion to be continuous. Hence, during the fluid motion, the law of conservation of mass should be retained.

Let $P(x, y, z)$ be a point in the fluid space. We consider an infinitesimal parallelepiped of fluid about this point (FIG. 14.11). Let u, v, w be the velocity components and ρ the density at P.

Then the mass of fluid entering through the face 1 per unit time along x-direction is $-\frac{\partial}{\partial x}(\rho u)dx\,dy\,dz$. Similarly, mass of fluid retained per unit time due to flow in y- and z-directions respectively are

$$-\frac{\partial}{\partial y}(\rho v)dx\,dy\,dz \text{ and } -\frac{\partial}{\partial z}(\rho w)dx\,dy\,dz$$

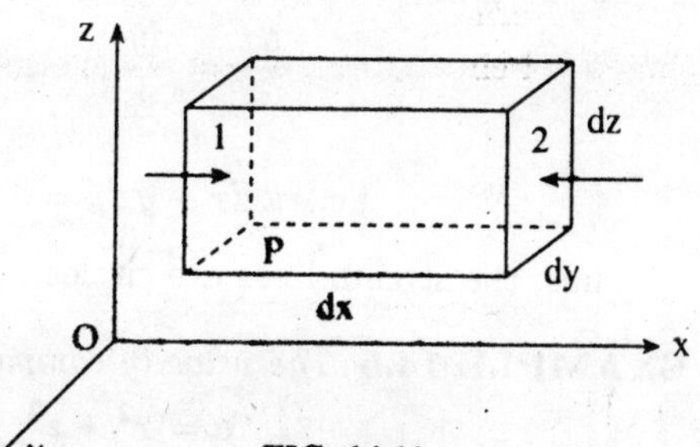

FIG. 14.11

Thus, the total gain of fluid mass within the volume per unit time is

$$-\left[\frac{\partial}{\partial x}(\rho u) + \frac{\partial}{\partial y}(\rho v) + \frac{\partial}{\partial z}(\rho w)\right] dx\,dy\,dz \tag{14.68}$$

Now, the original fluid mass within the element being $\rho\, dx\, dy\, dz$, its increase per unit time is

$$\frac{\partial \rho}{\partial t} dx\, dy\, dz \tag{14.69}$$

But the increase in the mass of fluid within the element per unit time must be equal to the excess of mass that flows in over the mass that flows out per unit time. Hence, equating (14.68) and (14.69) and dividing by $dx\, dy\, dz$, we get

$$\frac{\partial \rho}{\partial t} + \frac{\partial}{\partial x}(\rho u) + \frac{\partial}{\partial y}(\rho v) + \frac{\partial}{\partial z}(\rho w) = 0 \tag{14.70}$$

This is the ***equation of continuity***.

For steady flow, $\frac{\partial \rho}{\partial t} = 0$, and the equation of continuity assumes the form:

$$\frac{\partial}{\partial x}(\rho u) + \frac{\partial}{\partial y}(\rho v) + \frac{\partial}{\partial z}(\rho w) = 0 \tag{14.71}$$

For incompressible fluids, $\rho =$ constant and Eqn (14.71) becomes

$$\frac{\partial u}{\partial x} + \frac{\partial v}{\partial y} + \frac{\partial w}{\partial z} = 0 \tag{14.72}$$

EXAMPLE 14.4 *In a two-dimensional fluid motion, the velocity components are given by*

$$u = -\frac{ay}{x^2 + y^2},\ v = \frac{ax}{x^2 + y^2}, w = 0$$

Show that the continuity equation is satisfied. Also find the equation of stream lines.

Solution. We have

$$u = -\frac{ay}{x^2 + y^2},\ \therefore\ \frac{\partial u}{\partial x} = \frac{2axy}{(x^2 + y^2)^2} \tag{1}$$

$$\text{and } v = \frac{ax}{x^2 + y^2},\ \therefore\ \frac{\partial v}{\partial y} = -\frac{2axy}{(x^2 + y^2)^2} \tag{2}$$

The equation of continuity in two-dimensional flow is

$$\frac{\partial u}{\partial x} + \frac{\partial v}{\partial y} = 0 \tag{3}$$

Using Eqns (1) and (2), it is obvious that Eqn (3) is satisfied.

The stream lines are given by

$$\frac{dx}{u} = \frac{dy}{v}, \text{i.e., } \frac{dx}{-\dfrac{ay}{x^2 + y^2}} = \frac{dy}{\dfrac{ax}{x^2 + y^2}}, \text{ i.e., } \frac{dx}{-y} = \frac{dy}{x}$$

$$\text{i.e., } x\, dx + y\, dy = 0, \text{ i.e., } x^2 + y^2 = \text{constant}.$$

Thus, the stream lines are circles.

EXAMPLE 14.5 *The velocity components in a fluid-flow are given by*

$$u = x^2 + z^2,\ v = y^2 + z^2,\ w = -2z(x + y)$$

Show that the flow is possible. Also examine whether the flow is rotational or not.

Solution. We have $u = x^2 + z^2$.

Hence, $\frac{\partial u}{\partial x} = 2x,\ \frac{\partial u}{\partial y} = 0,\ \frac{\partial u}{\partial z} = 2z.$

Similarly, $v = y^2 + z^2$

$$\therefore \frac{\partial v}{\partial x} = 0, \ \frac{\partial v}{\partial y} = 2y, \ \frac{\partial v}{\partial z} = 2z$$

and $w = -2z(x+y)$

$$\therefore \frac{\partial w}{\partial x} = -2z, \ \frac{\partial w}{\partial y} = -2z, \ \frac{\partial w}{\partial z} = -2(x+y)$$

Hence,

$$\frac{\partial u}{\partial x} + \frac{\partial v}{\partial y} + \frac{\partial w}{\partial z} = 2x + 2y - 2(x+y) = 0.$$

Therefore, the equation of continuity is satisfied so that such a fluid-flow is possible.

Now, the components of vorticity are:

$$\Omega_x = \frac{\partial w}{\partial y} - \frac{\partial v}{\partial z} = -2z - 2z = -4z$$

$$\Omega_y = \frac{\partial u}{\partial z} - \frac{\partial w}{\partial x} = 2z - (-2z) = 4z$$

$$\Omega_z = \frac{\partial v}{\partial x} - \frac{\partial u}{\partial y} = 0 - 0 = 0$$

Since all the vorticity components are not zero, the motion is rotational.

EXAMPLE 14.6 *The velocity components of a fluid-flow are given as in the* Example 14.4. *Determine the potential function, if it exists.*

Solution. We have seen that such a flow is possible.

Since $w = 0$ and velocity components do not depend on z, there is only one vorticity component

$$\Omega_z = \frac{\partial v}{\partial x} - \frac{\partial u}{\partial y} \tag{1}$$

Since $u = -\dfrac{ay}{x^2+y^2}$,

$$\frac{\partial u}{\partial y} = \frac{ay(2y) - (x^2+y^2)\cdot a}{(x^2+y^2)^2} = \frac{a(y^2-x^2)}{(x^2+y^2)^2} \tag{2}$$

Similarly, $v = \dfrac{ax}{x^2+y^2}$,

$$\therefore \frac{\partial v}{\partial x} = \frac{a(x^2+y^2) - ax(2x)}{(x^2+y^2)^2} = \frac{a(y^2-x^2)}{(x^2+y^2)^2} \tag{3}$$

Substituting Eqns (2) and (3) into Eqn (1), we get $\Omega_z = 0$.

Hence, the flow is irrotational and a potential function exists.

Then

$$\frac{\partial \phi}{\partial x} = -u = \frac{ay}{x^2+y^2} \tag{4}$$

$$\frac{\partial \phi}{\partial y} = -v = -\frac{ax}{x^2+y^2} \tag{5}$$

$$\text{and} \quad \frac{\partial \phi}{\partial z} = -w = 0 \tag{6}$$

From Eqn (6), we find that

$$\phi = \phi(x, y) \tag{7}$$

Integrating Eqn (4),

$$\phi(x, y) = a \tan^{-1}\left(\frac{x}{y}\right) + f(y) \tag{8}$$

Hence,

$$\frac{\partial \phi}{\partial y} = f'(y) - \frac{ax}{x^2 + y^2} \tag{9}$$

Comparing Eqns (5) and (9), we find

$$f'(y) = 0, \text{ i.e., } f(y) = \text{constant}.$$

We may take $f(y) = 0$. Hence,

$$\phi(x, y) = a \tan^{-1}\left(\frac{x}{y}\right) \tag{10}$$

EXAMPLE 14.7 *Derive the equation of continuity for two-dimensional flow in polar coordinates.*

Solution. Let the velocity components of a fluid-flow be u in the radial and v in the tangential direction. We consider an element of fluid bounded by sides dr and $r\,d\theta$ (FIG. 14.12).

Then the excess of flow-in over the flow-out through faces ab and cd per unit time is

$$-\frac{\partial}{\partial r}(\rho u r\, d\theta) dr \tag{1}$$

Similar quantity arising from faces bc and ad is

$$-\frac{\partial}{\partial \theta}(\rho v\, dr) d\theta \tag{2}$$

Total gain of fluid mass per unit time within the element is

$$-\left[\frac{\partial}{\partial r}(\rho u r) + \frac{\partial}{\partial \theta}(\rho v)\right] dr\, d\theta \tag{3}$$

Increase in mass per unit time is

$$\frac{\partial \rho}{\partial t} r\, dr\, d\theta \tag{4}$$

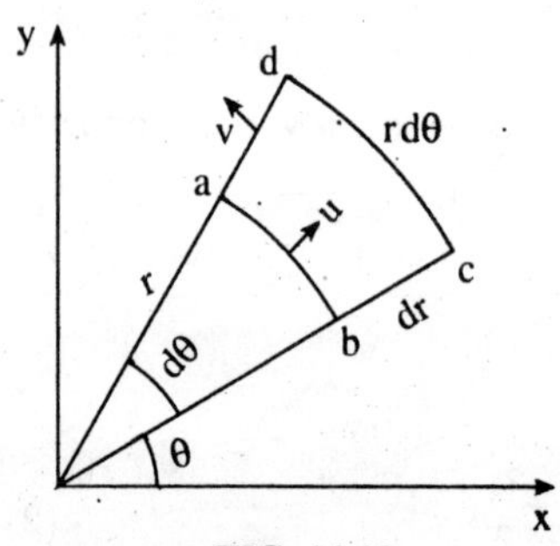

FIG. 14.12

For continuity of flow, we must have

$$\frac{\partial \rho}{\partial t} r\, dr\, d\theta = -\left[\frac{\partial}{\partial r}(\rho u r) + \frac{\partial}{\partial \theta}(\rho v)\right] dr\, d\theta$$

Hence,

$$\frac{\partial \rho}{\partial t} + \frac{1}{r}\frac{\partial}{\partial r}(\rho u r) + \frac{1}{r}\frac{\partial}{\partial \theta}(\rho v) = 0 \tag{5}$$

This is the equation of continuity for two-dimensional flow in polar coordinates. For **steady** flow, this equation becomes

$$\frac{\partial}{\partial r}(\rho u r) + \frac{\partial}{\partial \theta}(\rho v) = 0 \tag{6}$$

For steady flow of incompressible fluid for which $\rho =$ constant, the equation of continuity is

$$\frac{\partial}{\partial r}(ur) + \frac{\partial v}{\partial \theta} = 0 \tag{7}$$

14.4 Dynamics of Ideal Fluids

(A) Pressure at a Point in a Moving Fluid

We consider a point P within a mass of inviscid fluid moving with velocity vector $\mathbf{q}$, and insert a rigid small plane area δA within the fluid at P. The plane area is allowed to move with the same velocity $\mathbf{q}$, i.e., the area δA is at rest relative to the local fluid.

If δF denotes the force exerted normally on one side of the area, the fluid pressure at P is defined as

$$p = \lim_{\delta A \to 0} \frac{\delta F}{\delta A}$$

assuming the limit to exist uniquely.

We now establish that the pressure p is same in all directions. For this purpose, we consider the motion of an elemental tetrahedron of fluid (FIG. 14.13). The edges of the tetrahedron are parallel to coordinate axes.

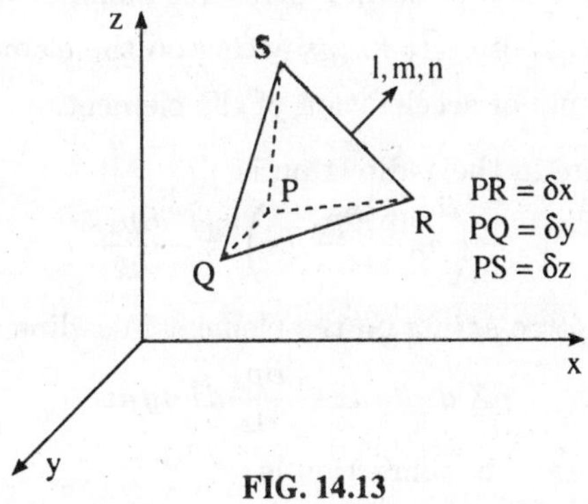

FIG. 14.13

Let area QRS be δS and (l, m, n) be the direction cosines of the normal to this area. Then

$$l\,\delta S = \text{area of face } PQS = \frac{1}{2}\delta y\,\delta z$$

and x-component of force exerted on area QRS is

$$-pl\delta S = -\frac{1}{2}p\,\delta y\,\delta z$$

Let p_x, p_y, p_z be the pressures on the surfaces of the tetrahedron, F_x, F_y, F_z be the components of body forces and a_x, a_y, a_z be the components of acceleration.

The mass of the tetrahedron being $\frac{1}{6}\rho\,\delta x\,\delta y\,\delta z$, the x-component of equation of motion gives

$$\frac{1}{2}(p_x - p)\delta y\,\delta z + \frac{1}{6}\rho F_x\,\delta x\,\delta y\,\delta z = \left(\frac{1}{6}\rho\,\delta x\,\delta y\,\delta z\right) a_x$$

This contains quantities of second and third orders. In the limit when the edges of the tetrahedron tend to zero, we have to the second order

$$\frac{1}{2}(p_x - p)\delta y\,\delta z = 0$$

which yields $p_x = p$.

Similarly, considering the equation of motion along y- and z-directions, we can show that $p = p_y$, $p = p_z$.

Thus, pressures at P are same in all directions. This result also confirms for the static fluids.

(B) Euler's Equation of Motion

We consider an elementary parallelepiped of fluid of sides dx, dy, dz at a point $P(x, y, z)$ as shown in the figure (FIG. 14.14).

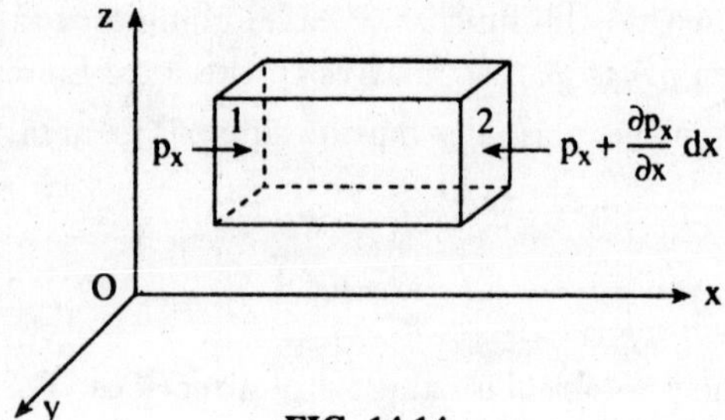

FIG. 14.14

Let p_x, p_y, p_z be the components of normal pressures along x-, y- and z-directions,

X, Y, Z — the components of body forces acting on the element of fluid of density ρ,

a_x, a_y, a_z — the components of acceleration of the element.

The resultant fluid pressure in the x-direction is

$$p_x - \left(p_x + \frac{\partial p_x}{\partial x} dx\right) = -\frac{\partial p_x}{\partial x} dx$$

Hence, the total external force acting on the element in x-direction is

$$\rho X\, dx\, dy\, dz - \frac{\partial p_x}{\partial x} dx\, dy\, dz$$

Then equation of fluid motion in x-direction is

$$(\rho\, dx\, dy\, dz) a_x = \rho X\, dx\, dy\, dz - \frac{\partial p_x}{\partial x} dx\, dy\, dz$$

$$\text{i.e., } a_x = X - \frac{1}{\rho} \frac{\partial p_x}{\partial x}$$

Similarly, considering the motion along y- and z-directions, we can get

$$a_y = Y - \frac{1}{\rho} \frac{\partial p_y}{\partial y}$$

$$a_z = Z - \frac{1}{\rho} \frac{\partial p_z}{\partial z}$$

Recalling that $p_x = p_y = p_z = p$, and substituting the expressions for the acceleration components from Eqn (14.51), we write the equations of motion in the form:

$$\left.\begin{aligned}
\frac{\partial u}{\partial t} + u\frac{\partial u}{\partial x} + v\frac{\partial u}{\partial y} + w\frac{\partial u}{\partial z} &= X - \frac{1}{\rho}\frac{\partial p}{\partial x} \\
\frac{\partial v}{\partial t} + u\frac{\partial v}{\partial x} + v\frac{\partial v}{\partial y} + w\frac{\partial v}{\partial z} &= Y - \frac{1}{\rho}\frac{\partial p}{\partial y} \\
\frac{\partial w}{\partial t} + u\frac{\partial w}{\partial x} + v\frac{\partial w}{\partial y} + w\frac{\partial w}{\partial z} &= Z - \frac{1}{\rho}\frac{\partial p}{\partial z}
\end{aligned}\right\} \tag{14.73}$$

These are known as ***Euler's dynamical equations*** of ideal fluids.

We consider the following cases:

• Case 1 : Irrotational Motion

In this case, a velocity potential $\phi(x, y, z)$ exists, so that

$$\left.\begin{aligned} u &= -\frac{\partial \phi}{\partial x} \\ v &= -\frac{\partial \phi}{\partial y} \\ w &= -\frac{\partial \phi}{\partial z} \end{aligned}\right\} \tag{14.74}$$

We also assume that the body forces are derivable from a potential function $U(x, y, z)$ so that

$$\left.\begin{aligned} X &= -\frac{\partial U}{\partial x} \\ Y &= -\frac{\partial U}{\partial y} \\ Z &= -\frac{\partial U}{\partial z} \end{aligned}\right\} \tag{14.75}$$

Then, the first of Euler's Eqn (14.73) can be written as

$$-\frac{\partial^2 \phi}{\partial x \partial t} + u\frac{\partial u}{\partial x} + v\left(-\frac{\partial^2 \phi}{\partial y \partial x}\right) + w\left(-\frac{\partial^2 \phi}{\partial z \partial x}\right) = -\frac{\partial U}{\partial x} - \frac{1}{\rho}\frac{\partial p}{\partial x} \tag{14.76}$$

Since the function $\phi(x, y, z, t)$ must be continuous, the order of its differentiation can be interchanged.

Hence,

$$\frac{\partial^2 \phi}{\partial y \partial x} = \frac{\partial^2 \phi}{\partial x \partial y} = \frac{\partial}{\partial x}\left(\frac{\partial \phi}{\partial y}\right) = -\frac{\partial v}{\partial x}$$

$$\text{and } \frac{\partial^2 \phi}{\partial z \partial x} = \frac{\partial^2 \phi}{\partial x \partial z} = \frac{\partial}{\partial x}\left(\frac{\partial \phi}{\partial z}\right) = -\frac{\partial w}{\partial x}$$

Using these relations in Eqn (14.76),

$$-\frac{\partial^2 \phi}{\partial x \partial t} + u\frac{\partial u}{\partial x} + v\frac{\partial v}{\partial x} + w\frac{\partial w}{\partial x} + \frac{\partial U}{\partial x} + \frac{1}{\rho}\frac{\partial p}{\partial x} = 0$$

$$\text{i.e., } -\frac{\partial}{\partial x}\left(\frac{\partial \phi}{\partial t}\right) + \frac{1}{2}\frac{\partial}{\partial x}(u^2 + v^2 + w^2) + \frac{\partial U}{\partial x} + \frac{1}{\rho}\frac{\partial p}{\partial x} = 0$$

$$\text{i.e., } -\frac{\partial}{\partial x}\left(\frac{\partial \phi}{\partial t}\right) + \frac{1}{2}\frac{\partial}{\partial x}(q^2) + \frac{\partial U}{\partial x} + \frac{1}{\rho}\frac{\partial p}{\partial x} = 0 \tag{14.77}$$

where $q^2 = u^2 + v^2 + w^2$.

Similarly, we can have other two equations of the forms:

$$\left.\begin{aligned} -\frac{\partial}{\partial y}\left(\frac{\partial \phi}{\partial t}\right) + \frac{1}{2}\frac{\partial}{\partial y}(q^2) + \frac{\partial U}{\partial y} + \frac{1}{\rho}\frac{\partial p}{\partial y} = 0 \\ -\frac{\partial}{\partial z}\left(\frac{\partial \phi}{\partial t}\right) + \frac{1}{2}\frac{\partial}{\partial z}(q^2) + \frac{\partial U}{\partial z} + \frac{1}{\rho}\frac{\partial p}{\partial z} = 0 \end{aligned}\right\} \tag{14.78}$$

Multiplying Eqns (14.77) and (14.78) by dx, dy and dz respectively, and adding the results,

$$-d\left(\frac{\partial \phi}{\partial t}\right) + \frac{1}{2}d(q^2) + dU + \frac{1}{\rho}dp = 0$$

Integrating this, we get

$$-\frac{\partial \phi}{\partial t}+\frac{1}{2}q^2+U+\int\frac{dp}{\rho}=f(t) \tag{14.79}$$

where $f(t)$ is the constant of integration which is a function of time in general.

Eqn (14.79) is the ***Bernoulli's equation for unsteady irrotational flow***. If the fluid is homogeneous and incompressible, then $\rho =$ constant, and Eqn (14.79) assumes the form:

$$-\frac{\partial \phi}{\partial t}+\frac{1}{2}q^2+U+\frac{p}{\rho}=f(t) \tag{14.80}$$

If the motion is steady, $\partial\phi/\partial t = 0$ and $f(t)$ is an absolute constant. In this case,

$$\frac{1}{2}q^2+U+\frac{p}{\rho}=\text{constant}. \tag{14.81}$$

When gravity is the only external force acting on the fluid, $U = gz$, where z is the distance of the fluid particle above some datum line where the potential energy is zero. Hence, in this case,

$$\frac{1}{2}q^2+gz+\frac{p}{\rho}=\text{constant}$$

$$\text{i.e.,}\quad \frac{q^2}{2g}+z+\frac{p}{\gamma}=\text{constant} \tag{14.82}$$

where $\gamma = g\rho =$ weight of unit volume of the fluid, called ***specific weight***.

Each term in Eqn (14.81) expresses energy per unit mass of the fluid. Thus,

$$\frac{q^2}{2}=\text{kinetic energy per unit mass,}$$

$$U=\text{potential energy per unit mass,}$$

$$\text{and}\quad \frac{p}{\rho}=\text{pressure energy per unit mass.}$$

The sum of all these gives the total mechanical specific energy. Thus, Bernoulli's equation physically means that ***the total mechanical specific energy in the steady irrotational flow of an ideal incompressible fluid remains the same at all points in space.***

• Case 2 : Flow is steady, but rotational

In this case, the velocity potential does not exist. We, however, assume that the body forces are conservative so that relations in (14.75) still holds good.

Then the first equation of motion is

$$u\frac{\partial u}{\partial x}+v\frac{\partial u}{\partial y}+w\frac{\partial u}{\partial z}+\frac{\partial U}{\partial x}+\frac{1}{\rho}\frac{\partial p}{\partial x}=0$$

Multiplying throughout by u,

$$u^2\frac{\partial u}{\partial x}+vu\frac{\partial u}{\partial y}+wu\frac{\partial u}{\partial z}+u\frac{\partial U}{\partial x}+\frac{u}{\rho}\frac{\partial p}{\partial x}=0$$

$$\text{or,}\quad u\frac{\partial}{\partial x}\left(\frac{1}{2}u^2\right)+v\frac{\partial}{\partial y}\left(\frac{1}{2}u^2\right)+w\frac{\partial}{\partial z}\left(\frac{1}{2}u^2\right)+u\left(\frac{\partial U}{\partial x}+\frac{1}{\rho}\frac{\partial p}{\partial x}\right)=0$$

$$\text{i.e.,}\quad \left(u\frac{\partial}{\partial x}+v\frac{\partial}{\partial y}+w\frac{\partial}{\partial z}\right)\left(\frac{1}{2}u^2\right)+u\frac{\partial}{\partial x}\left(U+\int\frac{\partial p}{\rho}\right)=0 \tag{14.83}$$

We can get similar other two equations like

$$\left.\begin{aligned}\left(u\frac{\partial}{\partial x}+v\frac{\partial}{\partial y}+w\frac{\partial}{\partial z}\right)\left(\frac{1}{2}v^2\right)+v\frac{\partial}{\partial y}\left(\dot{U}+\int\frac{\partial p}{\rho}\right)=0\\ \left(u\frac{\partial}{\partial x}+v\frac{\partial}{\partial y}+w\frac{\partial}{\partial z}\right)\left(\frac{1}{2}w^2\right)+w\frac{\partial}{\partial z}\left(U+\int\frac{\partial p}{\rho}\right)=0\end{aligned}\right\} \tag{14.84}$$

Adding all these three equations,

$$\left(u\frac{\partial}{\partial x}+v\frac{\partial}{\partial y}+w\frac{\partial}{\partial z}\right)\left(\frac{1}{2}q^2+U+\int\frac{dp}{\rho}\right)=0 \tag{14.85}$$

If we consider an arc length ds in a stream line, then

$$q\frac{\partial}{\partial s}=u\frac{\partial}{\partial x}+v\frac{\partial}{\partial y}+w\frac{\partial}{\partial z}$$

In view of this, Eqn (14.85) becomes

$$q\frac{\partial}{\partial s}\left(\frac{1}{2}q^2+U+\int\frac{dp}{\rho}\right)=0.$$

Integrating,

$$\frac{1}{2}q^2+U+\int\frac{dp}{\rho}=\text{constant}, \tag{14.86}$$

for a particular stream line, but the value is different for different stream lines.

Thus, we get a similar equation of Bernoulli for rotational motion.

14.5 Applications of Bernoulli's Theorem

(A) Venturi Meter

It is an instrument for measuring the flow rate of a fluid through a pipe. In its simplest form, it consists of a short length of pipe tapering to a narrow throat in the middle (FIG. 14.15). Pressures of the fluid at the enlarged end and at the throat are measured by tubes inserted at these sections.

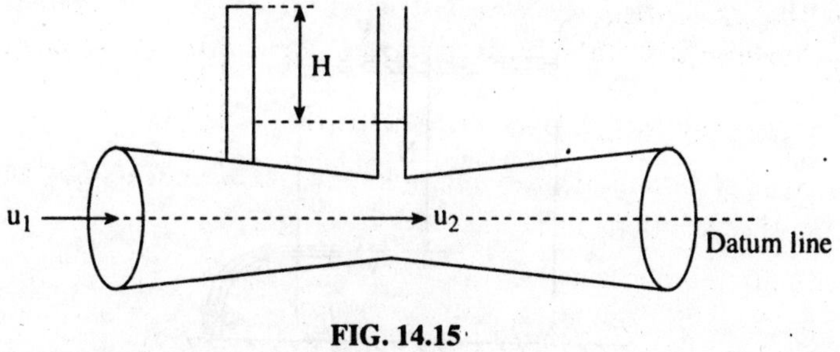

FIG. 14.15

Let A_1 = area of cross section at enlarged end,
A_2 = area of cross section at the throat,
u_1 = velocity of fluid at enlarged end,
u_2 = velocity of fluid at the throat,
ρ = density of fluid which is constant,
H = difference of pressure heads in two tubes,
p_1 = fluid pressure at enlarged end,
p_2 = fluid pressure at the throat.

From the conservation of mass rate flow, we have

$$A_1 u_1 \rho = A_2 u_2 \rho, \text{ i.e., } u_1 = \frac{A_2}{A_1} u_2 \tag{14.87}$$

Applying Bernoulli's equation at two sections,

$$\frac{u_1^2}{2g} + \frac{p_1}{\rho g} = \frac{u_2^2}{2g} + \frac{p_2}{\rho g}$$

$$\text{or, } \frac{p_1 - p_2}{\rho g} = \frac{1}{2g}(u_2^2 - u_1^2)$$

$$\text{or, } H = \frac{u_2^2}{2g}\left(1 - \frac{u_1^2}{u_2^2}\right) = \frac{u_2^2}{2g}\left(1 - \frac{A_2^2}{A_1^2}\right) \text{ [using Eqn (14.87)]}$$

Hence,

$$u_2 = \frac{A_1}{\sqrt{A_1^2 - A_2^2}}\sqrt{2gH} \tag{14.88}$$

Then the flow rate in volume/second

$$Q = A_2 u_2 = \left[\frac{A_1 A_2 \sqrt{2g}}{\sqrt{A_1^2 - A_2^2}}\right]\sqrt{H} \tag{14.89}$$

$$\text{i.e., } Q = C\sqrt{H}, \tag{14.90}$$

where

$$C = \frac{A_1 A_2}{\sqrt{A_1^2 - A_2^2}}\sqrt{2g} \tag{14.91}$$

which is called as a ***Venturi meter constant***.

(B) Torricelli's Theorem

Consider a vessel kept constantly full of a liquid having a horizontal orifice at its vertical side (FIG. 14.16).

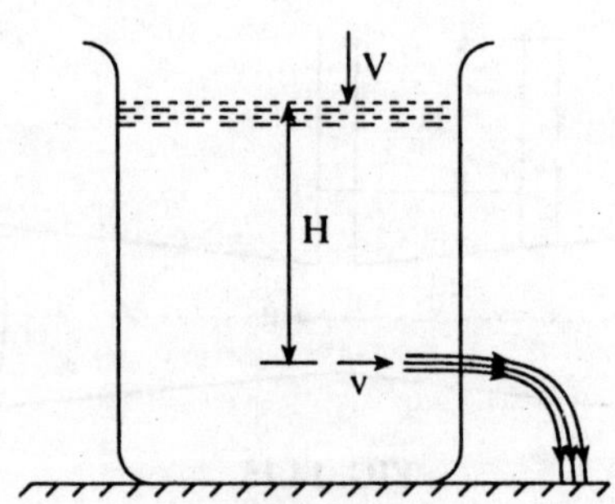

FIG. 14.16

Let A = area of the free surface of the liquid in the vessel,

a = area of the cross section of the orifice,

V = velocity of liquid at the free surface,

v = velocity of liquid at the orifice,

H = depth of centre of orifice from the free surface of liquid,

P_{atm} = atmospheric pressure.

Taking the horizontal line through the centre of the orifice as the datum line and applying Bernoulli's theorem, we have

$$\frac{V^2}{2g} + H + \frac{P_{\text{atm}}}{\rho} = \frac{v^2}{2g} + 0 + \frac{P_{\text{atm}}}{\rho}, \text{ i.e., } v^2 - V^2 = 2gH \tag{14.92}$$

From the condition of continuity, $AV = av$, i.e., $V = (a/A)v$. Substituting this value into Eqn (14.92), we get

$$v^2\left[1 - \left(\frac{a}{A}\right)^2\right] = 2gH \tag{14.93}$$

If the orifice be very small so that we can ignore $(a/A)^2$ in comparison to unity, we get

$$v^2 = 2gH, \text{ i.e., } v = \sqrt{2gH} \tag{14.94}$$

This gives the velocity of efflux of the liquid through the orifice. Eqn (14.94) shows that ***the velocity of the liquid through the orifice is the same as the velocity attained by a body falling freely under gravity through a distance H***. This is Torricelli's theorem.

However, due to the viscosity of the liquid, Eqn (14.94) is never satisfied.

(C) Pitot Tube

It is an instrument for measuring fluid velocity. It consists of an inner tube AB and an outer tube having holes H.

In order to measure the velocity of a stream of water, the tube AB is placed such that the opening is towards the upstream of the flow. An U-tube containing mercury is attached.

Let p be the pressure where the fluid velocity is v. Water entering through A comes to rest at C. Let p_0 be the pressure at C. The mercury level on the other side of the U-tube rises to D, where the pressure is also p (FIG. 14.17).

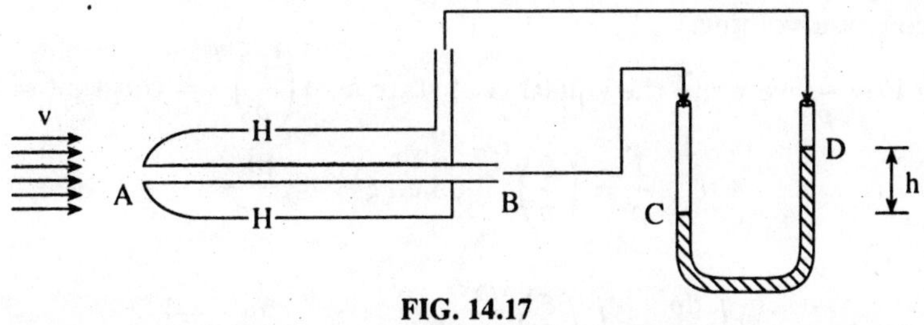

FIG. 14.17

Then applying Bernoulli's equation, we have $\dfrac{p}{\rho} + \dfrac{1}{2}v^2 = \dfrac{p_0}{\rho}$, where ρ is the density of water.

Hence,

$$v = \sqrt{\frac{2(p_0 - p)}{\rho}} \tag{14.95}$$

If h be the difference of mercury levels in the U-tube and δ be the density of mercury, then

$$p_0 - p = \delta gh \tag{14.96}$$

Hence,

$$v = \sqrt{\frac{2\delta gh}{\rho}} \tag{14.97}$$

This measures the velocity of the stream of water.

14.6 Bernoulli's Equation for Gases

In the case of gases, the density is a function of pressure. Thus, gases undergoing steady flows must satisfy Bernoulli's equation:

$$\frac{1}{2}v^2 + U + \int \frac{dp}{\rho} = \text{constant}. \tag{14.98}$$

For gases flowing under gravity, this equation takes the form:

$$\frac{1}{2}v^2 + gz + \int \frac{dp}{\rho} = \text{constant}, \tag{14.99}$$

where z denotes the height from some datum line.

We consider ideal gases and discuss the following two cases.

• Case 1 : Isothermal Process

In this case, the equation of state is given by $p = \lambda\rho$, λ = constant. Then

$$\int \frac{dp}{\rho} = \lambda \int \frac{dp}{p} = \lambda \log p \tag{14.100}$$

Thus, Eqn (14.99) assumes the form:

$$\frac{1}{2}v^2 + gz + \lambda \log p = \text{constant} \tag{14.101}$$

for flow of ideal gases undergoing isothermal process.

• Case 2 : Adiabatic Process

In this case, the equation of state is given by

$$pV^\gamma = \text{constant}, \tag{14.102}$$

where V is the volume of the gas and γ is the ratio of specific heats of the gas at constant pressure and constant volume.

Now, since $V \propto \dfrac{1}{\rho}$, we write the equation of state as $p\left(\dfrac{1}{\rho}\right)^\gamma = \text{constant} = \xi$ (say).

$$\therefore \ \frac{1}{\rho} = \left(\frac{\xi}{p}\right)^{1/\gamma}, \text{ i.e., } \xi^{1/\gamma} = \frac{p^{1/\gamma}}{\rho} \tag{14.103}$$

Then

$$\int \frac{dp}{\rho} = \int \left(\frac{\xi}{p}\right)^{1/\gamma} dp = \xi^{1/\gamma} \cdot \frac{\gamma}{\gamma - 1} p^{\frac{\gamma-1}{\gamma}}$$

$$= \frac{\gamma}{\gamma - 1}\frac{p}{\rho} \ \text{[using Eqn (14.103)]}$$

Thus, Bernoulli's equation in this case becomes

$$\frac{1}{2}v^2 + gz + \frac{\gamma}{\gamma - 1}\frac{p}{\rho} = \text{constant}. \tag{14.104}$$

As an application of this equation, let us consider efflux of gas of density ρ_1 stored in a vessel at pressure p_1 through an orifice with velocity v to atmosphere of density ρ_0 and pressure p_0.

Neglecting the external forces, we have by Eqn (14.104)

$$\frac{\gamma}{\gamma - 1}\frac{p_1}{\rho_1} = \frac{\gamma}{\gamma - 1}\frac{p_0}{\rho_0} + \frac{1}{2}v^2$$

$$\text{or, } v^2 = \frac{2\gamma}{\gamma - 1}\left(\frac{p_1}{\rho_1} - \frac{p_0}{\rho_0}\right) = \frac{2\gamma}{\gamma - 1}\frac{p_1}{\rho_1}\left(1 - \frac{\rho_1}{\rho_0} \cdot \frac{p_0}{p_1}\right) \tag{14.105}$$

Since $p_1/\rho_1^\gamma = p_0/\rho_0^\gamma$, we have

$$\frac{\rho_1}{\rho_0} = \left(\frac{p_1}{p_0}\right)^{1/\gamma}.$$

Hence, Eqn (14.105) becomes

$$v^2 = \frac{2\gamma}{\gamma - 1}\frac{p_1}{\rho_1}\left[1 - \left(\frac{p_0}{p_1}\right)^{\frac{\gamma-1}{\gamma}}\right] \tag{14.106}$$

which determines the velocity of the gas flowing out.

EXAMPLE 14.8 *Show that, for a steady potential flow of an incompressible fluid, the velocity potential satisfies Laplace's equation.*

Solution. The equation of continuity for a steady flow of incompressible fluid is

$$\frac{\partial u}{\partial x} + \frac{\partial v}{\partial y} + \frac{\partial w}{\partial z} = 0 \tag{1}$$

Since the velocity potential exists, we must have

$$u = -\frac{\partial \phi}{\partial x}, \quad v = -\frac{\partial \phi}{\partial y}, \quad w = -\frac{\partial \phi}{\partial z} \tag{2}$$

Inserting these values in Eqn (1), we get

$$-\frac{\partial}{\partial x}\left(\frac{\partial \phi}{\partial x}\right) - \frac{\partial}{\partial y}\left(\frac{\partial \phi}{\partial y}\right) - \frac{\partial}{\partial z}\left(\frac{\partial \phi}{\partial z}\right) = 0$$

$$\therefore \frac{\partial^2 \phi}{\partial x^2} + \frac{\partial^2 \phi}{\partial y^2} + \frac{\partial^2 \phi}{\partial z^2} = 0$$

Thus, the velocity potential function $\phi(x, y, z)$ satisfies Laplace's equation.

EXAMPLE 14.9 *Steam is rushing from a boiler through a conical pipe, the diameters at the ends of which are D and d. If V and v be the corresponding velocities of the steam, and if the motion be supposed to be that of divergence from the vertex of the cone, prove that*

$$\frac{v}{V} = \frac{D^2}{d^2}\exp\left[\frac{v^2 - V^2}{2k}\right],$$

where $k = p/\rho$ = constant.

Solution. Bernoulli's equation in this case is (FIG. 14.18)

$$\frac{1}{2}q^2 + \int \frac{dp}{\rho} = \text{constant}$$

$$\text{i.e., } \frac{1}{2}q^2 + \int k\frac{dp}{p} = \text{constant} \quad \left[\because k = \frac{p}{\rho}\right]$$

$$\text{i.e., } \frac{1}{2}q^2 + k\log p = \text{constant}. \tag{1}$$

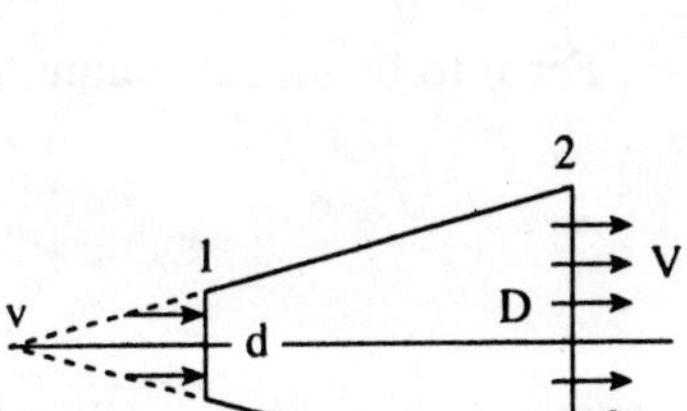

FIG. 14.18

Applying this at the two ends of the pipe,

$$\frac{1}{2}v^2 + k\log p_1 = \frac{1}{2}V^2 + k\log p_2, \text{ i.e., } \frac{v^2 - V^2}{2k} = \log\left(\frac{p_2}{p_1}\right) = \log\left(\frac{\rho_2}{\rho_1}\right) \tag{2}$$

where ρ_1, ρ_2 are the densities of steam at sections 1 and 2 respectively.

From the conservation of mass,

$$\rho A q = \text{constant}, \tag{3}$$

where A is the cross-sectional area of the pipe.

Hence, at two ends

$$\rho_1 \left(\frac{\pi}{4} d^2\right) v = \rho_2 \left(\frac{\pi}{4} D^2\right) V, \text{ i.e., } \frac{\rho_2}{\rho_1} = \frac{d^2}{D^2}\frac{v}{V} \tag{4}$$

Hence, from Eqns (2) and (4),

$$\frac{\rho_2}{\rho_1} = \exp\left[\frac{v^2 - V^2}{2k}\right], \text{ i.e., } \frac{d^2}{D^2}\frac{v}{V} = \exp\left[\frac{v^2 - V^2}{2k}\right]$$

$$\therefore \frac{v}{V} = \frac{D^2}{d^2} \exp\left[\frac{v^2 - V^2}{2k}\right]$$

EXAMPLE 14.10 *Water stands at a depth H in a large open tank whose side-walls are vertical. A hole is made in one of the walls at a depth of z below the water surface. Calculate the horizontal distance y from the wall at which the emerging stream of water strikes the ground. What should be the value of z which makes y maximum?*

Solution. The horizontal velocity of emerging water is

$$v = \sqrt{2gz} \tag{1}$$

This jet of water falls under gravity at a horizontal distance y and a vertical distance $(h - z)$.

If t be the time of fall, then

$$y = vt, \text{ i.e., } t = \frac{y}{v} \tag{2}$$

$$\text{and } \frac{1}{2} g t^2 = (h - z), \text{ i.e., } t^2 = \frac{2(h - z)}{g} \tag{3}$$

Eliminating t from Eqns (2) and (3),

$$\left(\frac{y}{v}\right)^2 = \frac{2(h - z)}{g}, \text{ i.e., } y^2 = 2gz \cdot \frac{2(h - z)}{g} \text{ [using Eqn (1)]}$$

$$= 4z(h - z)$$

$$\therefore y = 2\sqrt{z(h - z)} \tag{4}$$

For y to be an extremum, $dy/dz = 0$

$$\text{i.e., } 2\left[\frac{1}{2}\sqrt{h - z} \cdot \frac{1}{\sqrt{z}} - \frac{1}{2}\frac{\sqrt{z}}{\sqrt{h - z}}\right] = 0 \tag{5}$$

$$\text{or, } \frac{\sqrt{h - z}}{\sqrt{z}} = \frac{\sqrt{z}}{\sqrt{h - z}}, \text{ or, } h - z = z$$

$$\therefore z = \frac{h}{2} \tag{6}$$

Thus, y becomes an extremum for $z = h/2$.

Again, from Eqn (5), we have

$$\frac{dy}{dz} = \frac{\sqrt{h - z}}{\sqrt{z}} - \frac{\sqrt{z}}{\sqrt{h - z}} = \frac{h - 2z}{\sqrt{z(h - z)}} = \frac{2(h - 2z)}{y} \text{ [using Eqn (4)]} \tag{7}$$

Hence,

$$\frac{d^2y}{dz^2} = \frac{-4y - 2(h-2z)\frac{dy}{dz}}{y^2} \tag{8}$$

For $z = h/2$,

$$\frac{d^2y}{dz^2} = -\frac{4y}{y^2} = -\frac{4}{y} < 0 \tag{9}$$

Thus, y is maximum for $z = h/2$.

EXAMPLE 14.11 *A long pipe is of length l and has slowly tapering cross section. It is inclined at an angle α to the horizontal. Water flows steadily through it from upper to the lower end. The section at the upper end has twice the radius of the lower end, where water is delivered at atmospheric pressure. If the pressure at upper end is twice atmospheric, find the velocity of delivery.*

Solution. Let $2R, R$ be the radii of cross sections at upper and lower ends respectively,

u, v — the velocities at entry and exit respectively,

$2P, P$ — the pressures at entry and exit respectively, where P is the atmospheric pressure.

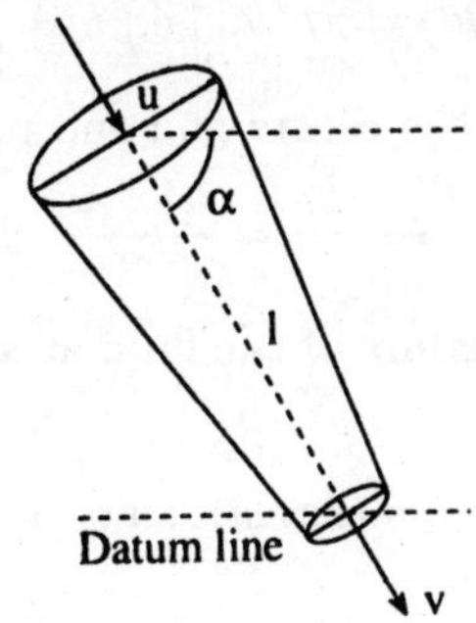

FIG. 14.19

Taking the datum line through the centre of the lower end (FIG. 14.19), we write Bernoulli's equation at the two ends

$$\frac{2P}{\rho} + \frac{1}{2}u^2 + gl\sin\alpha = \frac{P}{\rho} + \frac{1}{2}v^2 \tag{1}$$

The equation of continuity for incompressible fluid is

$$u(4\pi R^2) = v(\pi R^2), \text{ i.e., } u = \frac{v}{4} \tag{2}$$

Using this in Eqn (1),

$$\frac{P}{\rho} + \frac{1}{2}\left(\frac{v^2}{16}\right) + gl\sin\alpha = \frac{1}{2}v^2, \text{ i.e., } \frac{P}{\rho} + gl\sin\alpha = v^2\left(\frac{1}{2} - \frac{1}{32}\right) = \frac{15}{32}v^2$$

Hence,

$$v = \left[\frac{32}{15}\left(\frac{P}{\rho} + g\sin\alpha\right)\right]^{1/2}.$$

EXAMPLE 14.12 *Two circular flat plates of radius R are placed parallel with a small distance d between them. A homogeneous liquid of density ρ enters through a central pipe of radius r upwards and then flows radially between the plates discharging at atmospheric pressure. Neglecting the dynamic force of entering fluid, determine the net fluid thrust on the upper plate.*

Solution. Let (FIG. 14.20)

u = velocity of the fluid at entry, v = velocity of the fluid at exit to atmosphere,

P_0 = absolute pressure of fluid in the pipe, P_a = pressure of the atmosphere.

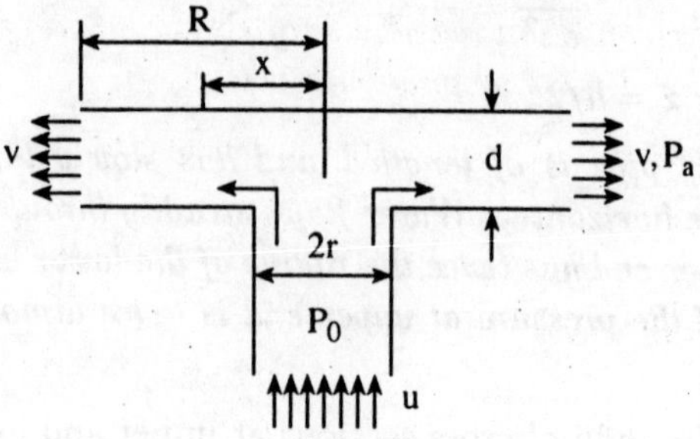

FIG. 14.20

Equating volumetric flow rate through the pipe and plates,

$$v(2\pi Rd) = \pi r^2 u, \text{ i.e., } v = \frac{r^2}{2Rd}u \tag{1}$$

Applying Bernoulli's equation at the centre and end point of plates, we have

$$\frac{P_a}{\rho} + \frac{v^2}{2} = \frac{P_0}{\rho} + \frac{u^2}{2} \text{ or, } \frac{P_a - P_0}{\rho} = \frac{u^2 - v^2}{2} \text{ or, } v^2 = u^2 - \frac{2}{\rho}(P_a - P_0) \tag{2}$$

Let v_x, P_x be the velocity and pressure of the fluid at any distance x from the centre. Then from continuity condition,

$$(2\pi xd)v_x = (2\pi Rd)v, \text{ i.e., } v_x = \frac{R}{x}v \tag{3}$$

By Bernoulli's equation,

$$\frac{P_x}{\rho g} + \frac{v_x^2}{2g} = \frac{P_0}{\rho g} + \frac{u^2}{2g} = \text{constant} = C \text{ (say)}.$$

$$\therefore \frac{P_x}{\rho g} = C - \frac{v_x^2}{2g} = C - \frac{R^2v^2}{2x^2g} \text{ [using Eqn (3)]} \tag{4}$$

Now, considering a thin ring of thickness dx and radius x from the centre, fluid thrust on this ring is

$$dF = (2\pi x dx)P_x = 2\pi\rho g\left(C - \frac{R^2v^2}{2gx^2}\right)x\,dx \text{ [using Eqn (4)]} \tag{5}$$

Hence, the fluid thrust on the top plate due to the flow within circular plates is

$$F_1 = 2\pi\rho g\int_r^R \left(C - \frac{R^2v^2}{2gx^2}\right)x\,dx = 2\pi\rho g\left[C\frac{x^2}{2} - \frac{R^2v^2}{2g}\log x\right]_r^R$$

$$= 2\pi\rho g\left[\frac{C}{2}(R^2 - r^2) - \frac{R^2v^2}{2g}\log\left(\frac{R}{r}\right)\right] \tag{6}$$

The thrust on the upper plate due to the fluid-flowing through the pipe is

$$F_2 = \pi r^2 P_0 \tag{7}$$

The thrust on the top of the upper plate due to the atmosphere in downward direction is

$$F_3 = \pi R^2 P_a \tag{8}$$

Hence, net fluid force in the upper plate

$$F = F_1 + F_2 - F_3$$
$$= 2\pi\rho g\left[\frac{C}{2}(R^2 - r^2) - \frac{R^2v^2}{2g}\log\left(\frac{R}{r}\right)\right] + \pi r^2 P_0 - \pi R^2 P_a$$

EXAMPLE 14.13 *Find the condition that the surface $f(x, y, z, t) = 0$ moving with velocity* $\mathbf{U}$ *may be a boundary surface for a fluid moving with velocity* $\mathbf{q}$.

Solution. If $\mathbf{n}$ be the unit normal vector on the surface at any point, then the boundary condition is

$$(\mathbf{q} - \mathbf{U}) \cdot \mathbf{n} = 0 \tag{1}$$

Now, the direction ratios of $\mathbf{n}$ are

$$\frac{\partial f}{\partial x}, \frac{\partial f}{\partial y} \text{ and } \frac{\partial f}{\partial z}.$$

Again, we know

$$\nabla f = \mathbf{i}\frac{\partial f}{\partial x} + \mathbf{j}\frac{\partial f}{\partial y} + \mathbf{k}\frac{\partial f}{\partial z} \tag{2}$$

Hence, ∇f and $\mathbf{n}$ are parallel vectors, so that we can write:

$$\mathbf{n} = \lambda\nabla f \tag{3}$$

Then condition (1) becomes

$$(\mathbf{q} - \mathbf{U}) \cdot \lambda\nabla f = 0, \text{ i.e., } \mathbf{q} \cdot \nabla f = \mathbf{U} \cdot \nabla f \tag{4}$$

Let a point $P(\mathbf{r}, t)$ move to a point $Q(\mathbf{r} + \delta\mathbf{r}, t + \delta t)$ in time δt. Since the point Q lies on the given surface, we must have $f(\mathbf{r} + \delta\mathbf{r}, t + \delta t) = 0$.

Expanding by Taylor's theorem, we get

$$f(\mathbf{r}, t) + \delta\mathbf{r} \cdot \nabla f + \delta t\left(\frac{\partial f}{\partial t}\right) = 0 \text{ or, } \frac{\partial f}{\partial t} + \frac{\delta\mathbf{r}}{\delta t} \cdot \nabla f = 0 \tag{5}$$

since we have the equation of given surface as $f(x, y, z, t) = 0$, i.e., $f(\mathbf{r}, t) = 0$.

Proceeding to the limit as $\delta\mathbf{r} \to 0$, $\delta t \to 0$, and noting that $\lim\limits_{\delta t \to 0} \delta\mathbf{r}/\delta t = \mathbf{U}$, we get from Eqn (5)

$$\frac{\partial f}{\partial t} + \mathbf{U} \cdot \nabla f = 0, \text{ i.e., } \frac{\partial f}{\partial t} + \mathbf{q} \cdot \nabla f = 0 \tag{6}$$

where we have used equation (4).

This is the condition that the boundary surface should satisfy.

Taking the dot product of vectors $\mathbf{q} = \mathbf{i}u + \mathbf{j}v + \mathbf{k}w$ and ∇f in Eqn (2), we get the condition (6) as

$$\frac{\partial f}{\partial t} + u\frac{\partial f}{\partial x} + v\frac{\partial f}{\partial y} + w\frac{\partial f}{\partial z} = 0$$

EXAMPLE 14.14 *A stream in a horizontal pipe, after passing a contraction in the pipe at which its sectional area is A, is delivered at atmospheric pressure at a place where the sectional area is B. Show that if a side tube is connected with the pipe at the former place, water will be sucked up through it into the pipe from a reservoir at a depth*

$$\frac{Q^2}{2g}\left(\frac{1}{A^2} - \frac{1}{B^2}\right)$$

below the pipe, Q being the delivery per second.

Solution. Let

v, p be the velocity of fluid and the pressure respectively in the smaller tube and V, P be respectively the corresponding quantities in the bigger tube (FIG. 14.21).

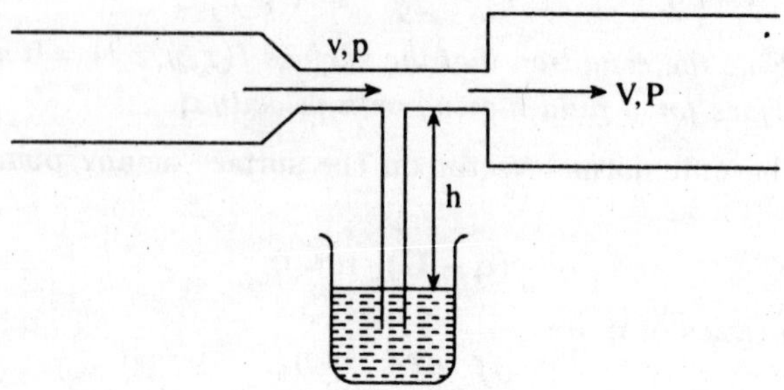

FIG. 14.21

Applying Bernoulli's equation, we get

$$\frac{p}{\rho} + \frac{1}{2}v^2 = \frac{P}{\rho} + \frac{1}{2}V^2, \text{ i.e., } \frac{1}{\rho}(P - p) = \frac{1}{2}\left(v^2 - V^2\right) \tag{1}$$

If h be the height through which water of density ρ is sucked up, then

$$h\rho g = P - p \tag{2}$$

The equation of continuity is

$$Q = Av = BV, \text{ i.e., } v = \frac{Q}{A} \text{ and } V = \frac{Q}{B} \tag{3}$$

Using Eqns (2) and (3) in Eqn (1), we get

$$\frac{1}{\rho}(h\rho g) = \frac{Q^2}{2}\left(\frac{1}{A^2} - \frac{1}{B^2}\right), \quad \therefore\ h = \frac{Q^2}{2g}\left(\frac{1}{A^2} - \frac{1}{B^2}\right)$$

EXAMPLE 14.15 *Derive the equation of continuity in the vector form.*

Solution. We consider a mass of fluid of density ρ within a small closed surface S enclosing a volume V (FIG. 14.22).

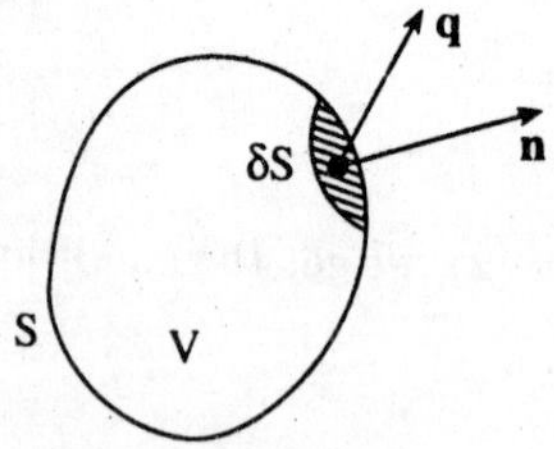

FIG. 14.22

Let P be any point on the surface where $\mathbf{q}$ is the fluid velocity and $\mathbf{n}$ is the unit outward-drawn normal to an element δS surrounding P.

The normal component of $\mathbf{q}$ in the outward direction is $\mathbf{n} \cdot \mathbf{q}$.

Then the rate of mass flow across δS is $\rho(\mathbf{n} \cdot \mathbf{q})\delta S$.

Hence, total rate of mass flow through S

$$= \int_S \rho(\mathbf{n} \cdot \mathbf{q})\delta S \tag{1}$$

Using Gauss's divergence theorem, we write Eqn (1)

$$= \int_V \nabla \cdot (\rho\mathbf{q})dV \tag{2}$$

Therefore, the total rate of mass flow into the volume V

$$= -\int_V \nabla \cdot (\rho\mathbf{q})dV \tag{3}$$

Again, mass of the fluid within S at time t

$$= \int_V \rho dV \tag{4}$$

Then total rate of mass increase within S

$$= \frac{\partial}{\partial t}\left(\int_V \rho dV\right) \tag{5}$$

We assume that there are no inlets or outlets through which fluid can enter or leave the region within S. Then, by the law of conservation of mass, the increase of mass rate of fluid within V must be equal to the total mass rate of flow into V. Thus, from Eqns (3) and (5),

$$\int_V \frac{\partial \rho}{\partial t} dV = -\int_V \nabla \cdot (\rho \mathbf{q}) dV \text{ or, } \int_V \left[\frac{\partial \rho}{\partial t} + \nabla(\rho \mathbf{q})\right] dV = 0 \tag{6}$$

The volume V being arbitrary, Eqn (6) holds good, if

$$\frac{\partial \rho}{\partial t} + \nabla(\rho \mathbf{q}) = 0 \tag{7}$$

This is the equation of continuity in vector form.

If the fluid is incompressible, $\partial \rho / \partial t = 0$. Then Eqn (7) becomes

$$\nabla \cdot (\rho \mathbf{q}) = 0 \tag{8}$$

If the fluid is incompressible and homogeneous, $\rho =$ constant. Then Eqn (8) of continuity becomes

$$\nabla \cdot \mathbf{q} = 0 \tag{9}$$

In the case of potential flow, $\mathbf{q} = \nabla \cdot \phi$. Then, from Eqn (9), we get

$$\nabla^2 \phi = 0 \tag{10}$$

Thus, the velocity potential function satisfies Laplace's equation.

14.7 Potential Flow in Spherical Coordinates

We consider the potential flow of incompressible fluids in spherical polar coordinates (r, θ, ψ). If $\phi(r, \theta, \psi)$ be the velocity potential,then we know that

$$\nabla^2 \phi = 0 \tag{14.107}$$

where the Laplace's operator in spherical coordinates is

$$\nabla^2 = \sin\theta \frac{\partial}{\partial r}\left(r^2 \frac{\partial}{\partial r}\right) + \frac{\partial}{\partial \theta}\left(\sin\theta \frac{\partial}{\partial \theta}\right) + \frac{1}{\sin\theta}\left(\frac{\partial^2}{\partial \psi^2}\right) \tag{14.108}$$

If q_r, q_θ, q_ψ be the velocity components of the fluid along the coordinate lines, then

$$\left.\begin{aligned} q_r &= -\frac{\partial \phi}{\partial r} \\ q_\theta &= -\frac{1}{r}\frac{\partial \phi}{\partial \theta} \\ q_\psi &= -\frac{1}{r\sin\theta}\frac{\partial \phi}{\partial \psi} \end{aligned}\right\} \tag{14.109}$$

If we assume the symmetry of flow about the line $\theta = 0$, then $\phi = \phi(r, \theta)$. Eqn (14.107) then assumes the form:

$$\sin\theta \frac{\partial}{\partial r}\left(r^2 \frac{\partial \phi}{\partial r}\right) + \frac{\partial}{\partial \theta}\left(\sin\theta \frac{\partial \phi}{\partial \theta}\right) = 0 \tag{14.110}$$

Solution of this equation can be obtained by the method of separation of variables. However, by inspection, we find the solutions:

$$\phi_1 = r\cos\theta, \phi_2 = \frac{1}{r^2}\cos\theta .$$

Thus, we accept a more general solution of Eqn (14.110) as

$$\phi(r,\theta) = Ar\cos\theta + \frac{B}{r^2}\cos\theta \tag{14.111}$$

where A, B are unknown constants to be determined from initial and boundary conditions.

We can solve many problems with the use of the velocity potential function in (14.111).

As an example, consider a sphere of radius R remaining stationary in a uniform stream of incompressible fluid for which the undisturbed velocity is U in the direction of negative x-axis (FIG. 14.23).

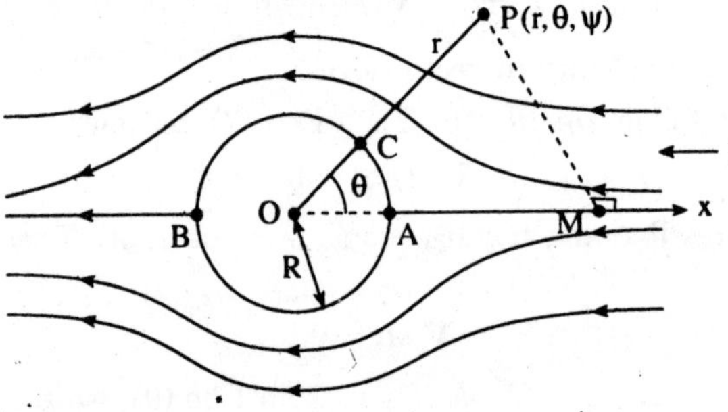

FIG. 14.23

If the sphere were not there, the fluid velocity would have been uniform and parallel to x-axis. Hence, the velocity potential would have been

$$\phi(r,\theta) = Ux = Ur\cos\theta \tag{14.112}$$

Due to the presence of the sphere, there will be local perturbation in its vicinity. As a result, the velocity potential as given by Eqn (14.111) needs to be modified to account for the perturbation. The additional term must satisfy two conditions:

(i) It must satisfy Laplace's equation and

(ii) It must tend to zero as $r \to \infty$.

A function satisfying these conditions is $\cos\theta/r^2$. Thus, we take the velocity potential function in the present case as

$$\phi(r,\theta) = Ur\cos\theta + \frac{B\cos\theta}{r^2} \tag{14.113}$$

We determine the constant B from the condition that the normal velocity is zero for $r = R$, i.e.,

$$\left(\frac{\partial\phi}{\partial r}\right)_{r=R} = 0 \tag{14.114}$$

Using the function (14.113), we get

$$U\cos\theta - \frac{2B}{R^3}\cos\theta = 0, \text{ i.e., } B = \frac{1}{2}UR^3 \tag{14.115}$$

Thus, we finally get

$$\phi(r,\theta) = Ur\cos\theta + \frac{UR^3}{2r^2}\cos\theta,\ (r \geq R) \tag{14.116}$$

Then the velocity components are:

$$\left.\begin{aligned} q_r &= -\frac{\partial\phi}{\partial r} = -U\cos\theta\left(1-\frac{R^3}{r^3}\right)\\ q_\theta &= -\frac{1}{r}\frac{\partial\phi}{\partial\theta} = U\sin\theta\left(1+\frac{R^3}{2r^3}\right)\\ q_\psi &= -\frac{1}{r\sin\theta}\frac{\partial\phi}{\partial\psi} = 0 \end{aligned}\right\} \tag{14.117}$$

The ***stagnation points*** in the flow field are those where all the velocity components are zero. In the present case, the stagnation points are given by

$$\left.\begin{aligned} U\cos\theta\left(1-\frac{R^3}{r^3}\right) &= 0\\ U\sin\theta\left(1+\frac{R^3}{2r^3}\right) &= 0 \end{aligned}\right\} \tag{14.118}$$

Since $r \geq R$, these are consistently satisfied, when

$$\left.\begin{aligned} r = R \text{ and } \sin\theta &= 0\\ \text{i.e., } r = R \text{ and } \theta &= 0\\ \theta &= \pi \end{aligned}\right\}$$

These points are marked A and B in the figure.

Stagnation pressure is the fluid pressure at stagnation points. If p_0 be the stagnation pressure at A and p_∞ the pressure in the fluid at infinity where velocity is $-U$, then applying Bernoulli's equation,

$$\frac{p_0}{\rho} + 0 = \frac{p_\infty}{\rho} + \frac{1}{2}U^2, \text{ i.e., } p_0 = p_\infty + \frac{1}{2}\rho U^2 \tag{14.119}$$

Let us consider any point $C(r,\theta,\psi)$ on the surface of the sphere. The velocity components for $r = R$ from Eqn (14.117) are:

$$q_r = 0,\ q_\theta = \frac{3}{2}U\sin\theta,\ q_\psi = 0$$

so that square of the resultant velocity

$$q^2 = \frac{9}{4}U^2\sin^2\theta \tag{14.120}$$

If p denotes the pressure at the point C, then

$$\frac{p}{\rho} + \frac{1}{2}\left(\frac{9}{4}U^2\sin^2\theta\right) = \frac{p_\infty}{\rho} + \frac{1}{2}U^2$$

$$\text{i.e., } p = p_\infty + \frac{1}{2}\rho U^2\left(1 - \frac{9}{4}\sin^2\theta\right) \tag{14.121}$$

The pressure on the surface will be minimum when $\sin^2\theta = 1$, i.e., $\theta = \pi/2$. Thus,

$$p_{\min} = p_\infty - \frac{5}{8}\rho U^2 \tag{14.122}$$

We can determine the fluid thrust on the hemisphere $r = R$, $0 \leq \theta \leq \pi/2$. For this purpose, we consider an arc $R\,d\theta$ and generate an elementary ring of radius $R\sin\theta$ about x-axis. Area of the ring is (FIG. 14.24), $dS = (2\pi R\sin\theta)R\,d\theta$.

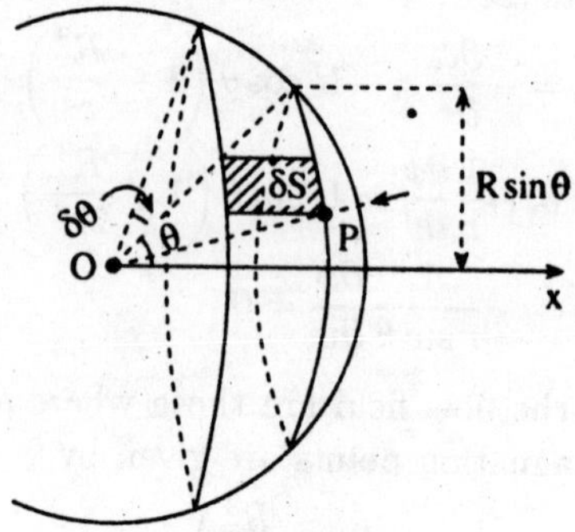

FIG. 14.24

Fluid pressure at any point on this ring is p as given by Eqn (14.121) directed towards the centre. Its component along x-axis is $p\cos\theta$. Hence, thrust on elementary ring is

$$(p\cos\theta)dS = 2\pi R^2 p\cos\theta\sin\theta d\theta$$

Hence, the total thrust on the hemisphere is

$$F = 2\pi R^2 \int_0^{\pi/2} p\sin\theta\cos\theta d\theta$$

$$\text{i.e., } F = 2\pi R^2 \int_0^{\pi/2} \sin\theta\cos\theta \left[p_\infty + \frac{1}{2}\rho U^2\left(1 - \frac{9}{4}\sin^2\theta\right)\right] d\theta$$

$$= \pi R^2\left(p_\infty - \frac{1}{16}\rho U^2\right) \tag{14.123}$$

NOTE. If we consider the whole sphere, the fluid thrust obtained will be zero. This is true in general, i.e., ***the total thrust on a solid body of any shape placed in a uniform stream is zero***. This is known as ***D'Alembert's Paradox***.

14.8 Potential Flow in Cylindrical Coordinates

In cylindrical coordinates (r, θ, z), the velocity potential function $\phi(r, \theta, z)$ satisfies the Laplace's equation

$$\frac{1}{r}\frac{\partial}{\partial r}\left(r\frac{\partial\phi}{\partial r}\right) + \frac{1}{r^2}\frac{\partial^2\phi}{\partial\theta^2} + \frac{\partial^2\phi}{\partial z^2} = 0 \tag{14.124}$$

The velocity components along coordinate lines are given by

$$\left.\begin{aligned} q_r &= -\frac{\partial\phi}{\partial r} \\ q_\theta &= -\frac{1}{r}\frac{\partial\phi}{\partial\theta} \\ q_z &= -\frac{\partial\phi}{\partial z} \end{aligned}\right\} \tag{14.125}$$

Let us consider two-dimensional flow so that all the physical quantities are independent of z.

Then Eqn (14.124) assumes the form:

$$\frac{1}{r}\frac{\partial}{\partial r}\left(r\frac{\partial \phi}{\partial r}\right)+\frac{1}{r^2}\frac{\partial^2 \phi}{\partial \theta^2}=0 \tag{14.126}$$

which can be used in solving many practical problems.

As an example, we consider the uniform flow past a fixed infinite circular cylinder of radius R (FIG. 14.25). If the uniform velocity of the fluid-flow be U in the negative x-direction, the velocity potential due to this uniform stream is

$$\phi_1(r,\theta)=Ux=Ur\cos\theta \tag{14.127}$$

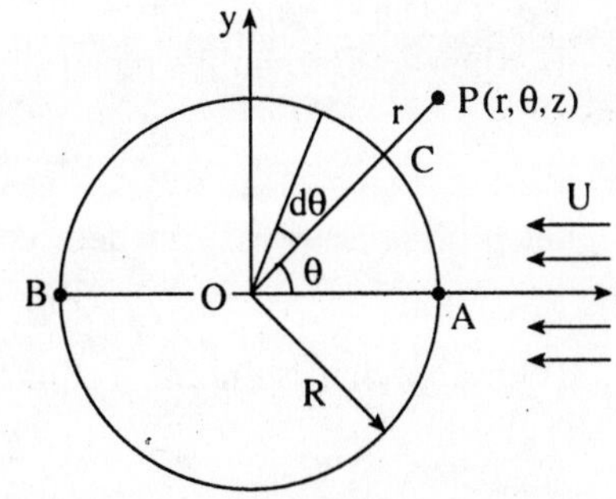

FIG. 14.25

Due to the presence of the cylinder, there will be some perturbation in the flow. Hence, we have to add another function $\phi_2(r,\theta)$ which must satisfy Laplaces Eqn (14.126) as well as the condition that it should tend to zero as $r\to\infty$. The simplest function satisfying these conditions is

$$\phi_2(r,\theta)=\frac{B\cos\theta}{r},\quad B=\text{constant}. \tag{14.128}$$

Hence, the total potential function is

$$\phi(r,\theta)=\phi_1(r,\theta)+\phi_2(r,\theta)=Ur\cos\theta+\frac{B\cos\theta}{r} \tag{14.129}$$

The constant B is determined from the condition that normal velocity of flow on the surface of the cylinder is zero. That is,

$$\left(\frac{\partial \phi}{\partial r}\right)_{r=R}=0,\ \text{for all } \theta$$

This yields

$$\left(U-\frac{B}{R^2}\right)\cos\theta=0,\ \text{i.e., } B=UR^2$$

Hence,

$$\phi(r,\theta)=U\cos\theta\left(r+\frac{R^2}{r}\right) \tag{14.130}$$

Then the velocity components at any point in the fluid is $(r\geq R)$

$$\left.\begin{aligned} q_r&=-\frac{\partial \phi}{\partial r}=-U\cos\theta\left(1-\frac{R^2}{r^2}\right)\\ q_\theta&=-\frac{1}{r}\frac{\partial \phi}{\partial \theta}=U\sin\theta\left(1+\frac{R^2}{r^2}\right)\\ q_z&=0\end{aligned}\right\} \tag{14.131}$$

Stagnation points are given by

$$\left.\begin{aligned} r=R \text{ and } \sin\theta=0\\ \text{i.e., } r=R \text{ and } \theta=0\\ \theta=\pi\end{aligned}\right\}$$

These are the points A and B as shown in the figure.

If p_0 be the stagnation pressure at A and p_∞ the pressure at infinity where velocity of flow is $-U$, then by Bernoulli's equation

$$\frac{p_0}{\rho}+0=\frac{p_\infty}{\rho}+\frac{1}{2}U^2, \text{ i.e., } p_0=p_\infty+\frac{1}{2}\rho U^2 \tag{14.132}$$

Let p be pressure at C where velocity components are

$$q_r=0,\ q_\theta=2U\sin\theta,\ q_\psi=0$$

Then applying Bernoulli's equation along the stream line $x-A-C$,

$$\frac{p}{\rho}+\frac{1}{2}(2U\sin\theta)^2=\frac{p_\infty}{\rho}+\frac{1}{2}U^2$$

$$\text{i.e., } p=p_\infty+\frac{1}{2}\rho U^2(1-4\sin^2\theta) \tag{14.133}$$

Minimum pressure occurs at $r=R$, $\theta=\pi/2$. Its value is

$$p_{\min}=p_\infty-\frac{3}{2}\rho U^2 \tag{14.134}$$

The minimum pressure becomes zero, when

$$U=\sqrt{\frac{2p_\infty}{3\rho}} \tag{14.135}$$

If $U>\sqrt{2p_\infty/3\rho}$, the fluid will tend to break away from the surface, and ***cavitation*** occurs.

Let us now find the fluid thrust on the semi-circular cylinder.

Considering an arc $R\,d\theta$ on the surface, total thrust along x-direction per unit length is

$$F=\int_{-\pi/2}^{\pi/2}(p\cos\theta)R\,d\theta$$

$$=R\int_{-\pi/2}^{\pi/2}\left[p_\infty+\frac{1}{2}\rho U^2(1-4\sin^2\theta)\right]\cos\theta d\theta=R\left(2p_\infty-\frac{1}{3}\rho U^2\right) \tag{14.136}$$

It can be verified that the total thrust on the whole surface, when θ varies from $\theta=0$ to $\theta=2\pi$, will be zero.

EXERCISE

1. The density of a heterogenous liquid at a depth z is $\rho(z)=\lambda z$, $\lambda=$ constant. If π be the atmospheric pressure, then show that the pressure at the point is

$$p=\pi+\frac{1}{2}\lambda g z^2.$$

2. In the lower half of a uniform circular tube, one quadrant is occupied by a liquid of density 2ρ and the other by two liquids, A of density 2ρ and B of density ρ. Prove that volume of B is twice that of A.

3. In a vertical column of a perfect gas, the pressure and absolute temperature at any height z are p and T. Prove that

$$z=\frac{p_0}{g\rho_0 T_0}\int_p^{p_0}\frac{T}{p}dp,$$

where p_0, ρ_0, T_0 are pressure, density and temperature at the bottom.

4. Absolute temperature at height z from the surface of the earth is

$$T = T_0\left(1 - \frac{z}{nH}\right),$$

where H is the height of homogeneous atmosphere. Show that pressure at that height is

$$p = p_0\left(1 - \frac{z}{nH}\right)^n$$

where p_0 is the pressure at $z = 0$.

5. Given the velocity field

$$\mathbf{q} = \mathbf{i}x^2y + \mathbf{j}y^2zt + \mathbf{k}zt^2,$$

determine the components of acceleration.

6. The velocity components in a two-dimensional steady flow are given as

$$u = \frac{x^2 - y^2}{(x^2 + y^2)^2}, \quad v = \frac{2xy}{(x^2 + y^2)^2}.$$

Show that the equation of continuity is satisfied.

7. Air, obeying Boyle's law, is in motion in a uniform tube of small section. If ρ be the density and v the velocity at a distance x from a fixed point at time t, then prove that

$$\frac{\partial^2 \rho}{\partial t^2} = \frac{\partial^2}{\partial x^2}\left[\rho\left(v^2 + k\right)\right],$$

where $p = k\rho$.

8. If ω be the area of cross section of a stream filament, prove that the equation of continuity is

$$\frac{\partial}{\partial t}(\rho\omega) + \frac{\partial}{\partial s}(\rho\omega q) = 0,$$

where ρ = density, q = speed of flow and δs is an element of the filament in the direction of flow.

9. Show that the variable ellipsoid

$$\frac{x^2}{a^2t^4} + t^2\left(\frac{y^2}{b^2} + \frac{z^2}{c^2}\right) = 1$$

is a possible form of a boundary surface of a liquid at time t if $u = 2x/t$, $v = -y/t$ and $w = -z/t$.

[**Hints.** Use the condition in Example 14.13]

10. Find the stream lines and paths of the particles when

$$u = \frac{x}{1+t}, \quad v = \frac{y}{1+t}, \quad w = \frac{z}{1+t}.$$

11. Determine the vorticity components when the velocity distribution is

$$\mathbf{q} = \mathbf{i}(x^2yt) + \mathbf{j}(y^2zt) + \mathbf{k}(zt^2).$$

Is the flow rotational?

12. Test whether the motion specified by

$$\mathbf{q} = \frac{k^2(\mathbf{j}x - \mathbf{i}y)}{x^2 + y^2} \quad (k = \text{constant})$$

is possible for an incompressible fluid. Also determine the velocity potential, if it exists.

13. Show that

$$u = -\frac{2xyz}{(x^2+y^2)^2}, \quad v = \frac{(x^2-y^2)z}{(x^2+y^2)^2}, \quad w = \frac{y}{x^2+y^2}$$

are the velocity components of a possible liquid motion. Is this motion irrotational?

14. Show that $\phi = (x-t)(y-t)$ represents the velocity potential of an incompressible two-dimensional fluid. Show that the stream lines at time t are the curves

$$(x-t)^2 - (y-t)^2 = \text{constant}.$$

15. Water is maintained at a height of 15 cm in a vessel which has a small circular hole in its horizontal base. The radius of the base is 1 mm and the area of the free surface is 30 sq. cm. Find the rate at which the liquid leaves the vessel if the coefficient of contraction is 0.624.

[**C.U. (HONS.)**]

[**Ans.** 3.36 cc/sec]

16. A horizontal tube has radii 0.5 cm and 0.3 cm at two places. For flowing water, the pressure difference at these two places is 1 cm of water. What is the rate of flow? [**C.U. (HONS.)**]

[**Ans.** 13.42 cc/sec]

17. The velocity field of the motion of an incompressible fluid is

$$\mathbf{q} = \mathbf{i}(x^2y + y^2) - \mathbf{j}xy^2$$

Determine the vorticity vector and pressure gradient in x- and y-directions on the horizontal plane. [**Hints.** Use Euler equation for $X = Y = 0$ to find the pressure gradients $\partial p/\partial x$ and $\partial p/\partial y$.]

18. A cylinder of cross-sectional area A is full of liquid between radius r_1 and r_2. It revolves about a vertical axis with constant angular velocity w. Show that the total centrifugal force impressed is

$$\frac{\rho A}{2g}\left(v_2^2 - v_1^2\right),$$

where v_1, v_2 are tangential velocities at $r = r_1$ and $r = r_2$ respectively.

19. A tank of cross-sectional area A and height H is full of water. There arises a small hole at the base of cross-sectional area a through which water flows out. Show that time taken to empty the tank is

$$\frac{2A}{a}\sqrt{\frac{H}{g}}.$$

CHAPTER 15

Special Theory of Relativity

15.1 Introduction

The study of physics begins with the measurement of certain physical quantities. In mechanics, these quantities are primarily displacement and velocity. In order to measure these quantities, we have to define some frame of reference which is considered to be at rest.

In Newtonian mechanics, translatory motion is recognised as a motion relative to some other material bodies. This motion will cause some passage of time. According to Newton, space and time are completely separable. Time flows on uniformly regardless of anything external. This is Newtonian Relativity. Thus, if c be the velocity of light taken in some direction and u be the velocity of an observer in the same direction, then the velocity of light to the observer will appear to be $(c - u)$. But experiments have confirmed that speed of light in any direction in vacuum is a constant of nature.

This led Einstein to formulate the special theory of relativity in the year 1905 which revolutionised the physical concepts. According to this theory, space and time are not independent of each other. This idea of relativity incorporated into mechanics gives rise to relativistic mechanics which is indispensable when the particles forming a system are moving with high velocities comparable to that of light.

The special theory of relativity deals with the problems of uniform rectilinear motion where one frame of reference moves with uniform velocity relative to the other. But there is another part of Einstein's theory of relativity which deals with more general cases of non-uniform velocity where one frame of reference moves with an accelerated velocity relative to the other. This is known as ***General Theory of Relativity***. The frames of refeence in the former case are called ***inertial*** while those in the latter case are called ***non-inertial***.

15.2 Galilean Transformation

Consider a Cartesian frame of reference S which is at rest. Let there be another Cartesian frame of reference S' whose axes x', y', z' are parallel to the axes x, y, z of S. The frame of reference S' is assumed to move with uniform velocity u along x' (or x) axis. We further assume that at $t = 0$, the origins O and O' were coincident and all the axes overlapped (FIG. 15.1).

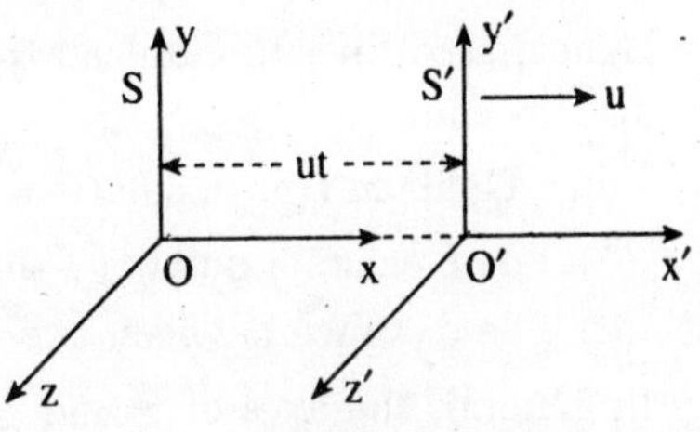

FIG. 15.1

Then, the coordinates of a particle w.r.t. the S' frame at any time t will be

$$x' = x - ut \tag{15.1}$$

$$y' = y \tag{15.2}$$

$$\text{and} \quad z' = z \tag{15.3}$$

Since time remains same in both the S and S' frames, above equations are supplemented by the equation of transformation of time

$$t' = t \tag{15.4}$$

These transformations are known as ***Galilean transformations***.

The inverse Galilean transformations are:

$$\left.\begin{aligned} x &= x' + ut \\ y &= y' \\ z &= z' \\ t &= t' \end{aligned}\right\} \tag{15.5}$$

Suppose, there is a rod of length L in S frame with end coordinates x_1 and x_2. Then

$$L = x_2 - x_1 \tag{15.6}$$

If at the same time, the end coordinates of the rod in S' frame are x'_1 and x'_2, then

$$L' = x'_2 - x'_1 \tag{15.7}$$

From Eqn (15.1), for any time t,

$$x'_2 - x'_1 = x_2 - x_1$$

Hence,

$$L' = L \tag{15.8}$$

Therefore, the length or distance between two points is invariant under Galilean transformation. The relations between velocity components in the two frames are obtained by differentiating Eqns (15.1)-(15.3) w.r.t. time:

$$\left.\begin{aligned} v_{x'} &= v_x - u \\ v_{y'} &= v_y \\ v_{z'} &= v_z \end{aligned}\right\} \tag{15.9}$$

A further differentiation w.r.t. time yields the acceleration components in the two frames:

$$\left.\begin{aligned} a_{x'} &= a_x \\ a_{y'} &= a_y \\ a_{z'} &= a_z \end{aligned}\right\} \tag{15.10}$$

Hence, according to Galilean transformations, the accelerations of a particle relative to the frames are equal.

Thus, Galilean transformations are based on two assumptions:

(1) There exists a universal time t which is the same in all reference frames.

(2) The distance between two points in all inertial systems is the same.

In general, the axes of S and S' are not parallel. Further, the relative velocity between S and S' also may not be parallel to x-axis. In this case, we assume that the origins O and O' were coincident at $t = 0$ and the relative velocity is along OO' (FIG. 15.2).

If the position vectors of a point be $\mathbf{r}$ and $\mathbf{r}'$ relative to the frames S and S' respectively, then

$$\mathbf{r}' = \mathbf{r} - \mathbf{u}t \tag{15.11}$$

$$\text{and } t' = t \tag{15.12}$$

The velocity components are related as

$$\mathbf{v}' = \mathbf{v} - \mathbf{u} \tag{15.13}$$

and the acceleration components are related as

$$\mathbf{a}' = \mathbf{a} \tag{15.14}$$

The inverse transformations of (15.11) and (15.12) are

$$\mathbf{r} = \mathbf{r}' + \mathbf{u}t, \quad t = t' \tag{15.15}$$

From Eqn (15.13),

$$\mathbf{v} = \mathbf{v}' + \mathbf{u} \tag{15.16}$$

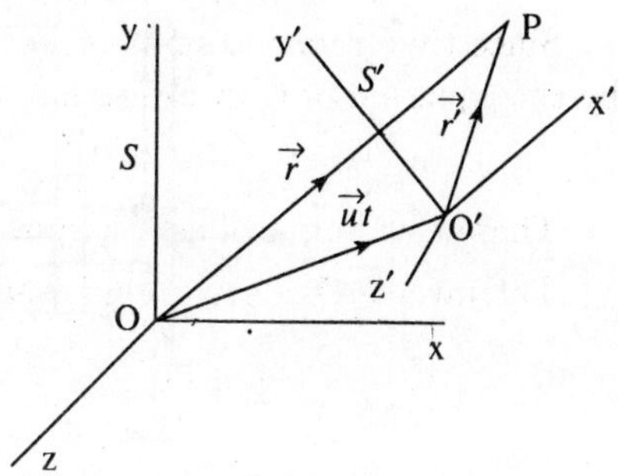

FIG. 15.2

Eqn (15.16) transforms the velocity of a particle from one frame to the other and is known as ***Galilean law of addition*** of velocities.

Suppose that, on a particle of mass m in the frame S, a force $\mathbf{F}$ acts. Then, by Newton's second law,

$$\mathbf{F} = m\mathbf{a} \tag{15.17}$$

The force acting on the particle in S' frame is

$$\mathbf{F}' = m'\mathbf{a}' \tag{15.18}$$

But, in Newtonian mechanics, the mass is independent of velocity. Hence,

$$m = m' \tag{15.19}$$

In view of Eqns (15.14) and (15.19), we get

$$\mathbf{F} = m\mathbf{a} = m'\mathbf{a}' = \mathbf{F}' \tag{15.20}$$

Thus, in all inertial systems, the force $\mathbf{F}$ will have same magnitude and direction independent of relative velocity of the reference frames. That is, Newton's second law and hence all the laws of motion are invariant under Galilean transformations.

It is to be noted that, if there be a source of light at the origin O of the S frame, and the velocity of light in this system be c in all directions, then the velocity of light in the S' frame by using Galilean transformation will be

$$c' = c - u \tag{15.21}$$

Hence, the speed of light is not constant in all inertial systems. This contradicts the principle of relativity.

15.3 Fizeau's Experiment

Fizeau in 1851 made an experiment to establish the effect of the velocity of water on the velocity of light passing through the flowing water. The experimental setup is shown in the FIG. 15.3.

$ABCD$ is a U-shaped tube, ends of the two limbs AB and CD being closed with glass plates. Water enters at upper end and exits at the lower end of the tube. Monochromatic light from a source S is incident on a half-silvered glass plate P. Three mirrors M_1, M_2, M_3 are set as shown in the figure. A telescope T is there for observations.

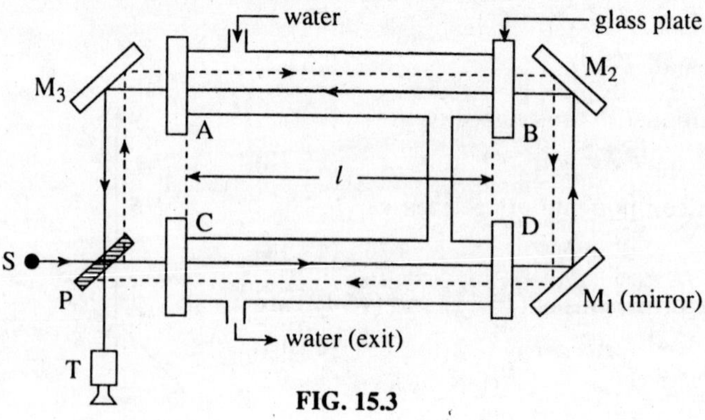

FIG. 15.3

A part of the incident light transmits through the glass plate P along the path $PM_1M_2M_3T$ and the other part gets reflected and moves along the path $PM_3M_2M_1PT$.

Let n = refractive index of water

v = velocity of water

and l = distance as shown in the figure.

Then both the paths traversed by light are of equal lengths $2l$. Hence,

$$t_1 = \text{time to traverse the path } PM_1M_2M_3T = \frac{2l}{\dfrac{c}{n} - v} \tag{15.22}$$

$$\text{and } t_2 = \text{time to traverse the path } PM_3M_2M_1PT = \frac{2l}{\dfrac{c}{n} + v} \tag{15.23}$$

Then

$$\Delta t = t_1 - t_2 = 2l\left[\frac{1}{\dfrac{c}{n} - v} - \frac{1}{\dfrac{c}{n} + v}\right] = \frac{4lv}{\dfrac{c^2}{n^2} - v^2} = \frac{4lvn^2}{c^2}\left(1 - \frac{v^2n^2}{c^2}\right)^{-1} \tag{15.24}$$

Since $vn \ll c$, we can write:

$$\Delta t \simeq \frac{4lvn^2}{c^2} \tag{15.25}$$

Due to this difference in time, there will be a shift in the interference fringes before and after the start of water flow. This shift is

$$\Delta s = \nu \Delta t = \frac{4lvn^2\nu}{c^2} = \frac{4lvn^2}{c\lambda} \tag{15.26}$$

where ν is the frequency of light and λ is its wavelength.

In the experiment, the shift was found to be

$$\Delta s = f\frac{4lvn^2}{c\lambda} \tag{15.27}$$

where the factor $f < 1$.

The factor called the ***dragging coefficient*** was found to be

$$f = 1 - \frac{1}{n^2}. \tag{15.28}$$

15.4 Michelson-Morley Experiment

Michelson-Morley experiment performed in 1881 contradicted the assumption that the velocity of light changes due to the motion of the observer. The basic idea of the experimental setup is shown in the FIG. 15.4.

Here S is a source of monochromatic light; P is a half-silvered glass plate inclined at 45° to the direction of light; M_1, M_2 are two mirrors placed at equal distances from P and T, the telescope for observations.

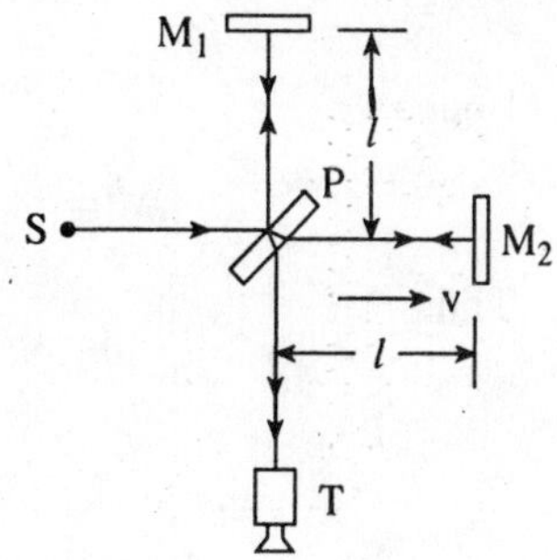

FIG. 15.4

A part of the ray incident on glass plate P gets reflected towards the mirror M_1, retraces back the same path and enters the telescope. The other part of the ray incident on the glass plate P gets transmitted, then being reflected by mirror M_2 retraces back the same path and enters the telescope after reflection through P. Thus, the two rays unite at P after some time and the interference fringes are obtained.

In the beginning of the experiment, the arm PM_2 is taken along the direction of the motion of the earth in space. The whole apparatus is floated on mercury in a large vessel so that it can move freely. Hence, by the time the ray reflected from P reaches the mirror M_1, both the plate P and the mirror M_1 shift to new positions P' and M_1' due to the earth's motion (FIG. 15.5).

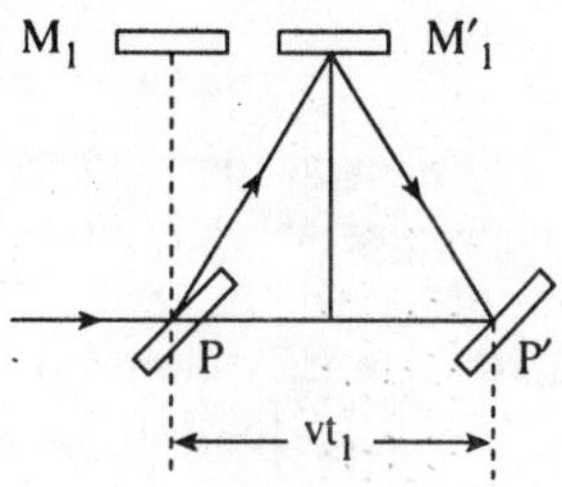

FIG. 15.5

Let v = speed of the earth, t_1 = time taken by the ray to traverse the path $PM_1'P'$ and $PM_1 = PM_2 = l$.

Then

$$PP' = vt_1 \text{ and } PM_1'P' = ct_1 \tag{15.29}$$

But

$$PM_1'P' = 2PM_1' = 2\sqrt{l^2 + v^2t_1^2/4} \tag{15.30}$$

From Eqns (15.29) and (15.30), we get

$$ct_1 = 2\sqrt{l^2 + v^2t_1^2/4}$$

which yields

$$t_1 = \frac{2l}{\sqrt{c^2 - v^2}} = \frac{2l}{c}\left(1 - \frac{v^2}{c^2}\right)^{-1/2} \quad \text{or, } t_1 = \frac{2l}{c}\left(1 + \frac{v^2}{2c^2}\right) \tag{15.31}$$

Here we have neglected in the Binomial expansion higher order terms due to smallness.

Now, the time taken by the transmitted ray through P to travel the distance PM_2 and back is

$$t_2 = \frac{l}{c - v} + \frac{l}{c + v} = \frac{2lc}{c^2 - v^2} = \frac{2l}{c}\left(1 - \frac{v^2}{c^2}\right)^{-1} \quad \text{or, } t_2 = \frac{2l}{c}\left(1 + \frac{v^2}{c^2}\right) \tag{15.32}$$

Hence,

$$\Delta t = t_2 - t_1 = \frac{2l}{c}\left(1 + \frac{v^2}{c^2}\right) - \frac{2l}{c}\left(1 + \frac{v^2}{2c^2}\right) \quad \text{or, } \Delta t = \frac{lv^2}{c^3} \tag{15.33}$$

Now the whole apparatus was turned by 90° so that the two arms PM_1 and PM_2 were interchanged. In this case,

$$\bar{t}_1 = \frac{2l}{c}\left(1+\frac{v^2}{c^2}\right) \tag{15.34}$$

$$\text{and } \bar{t}_2 = \frac{2l}{c}\left(1+\frac{v^2}{2c^2}\right) \tag{15.35}$$

Hence,

$$\Delta\bar{t} = \bar{t}_2 - \bar{t}_1 = \frac{2l}{c}\left(1+\frac{v^2}{2c^2}\right) - \frac{2l}{c}\left(1+\frac{v^2}{c^2}\right) = -\frac{lv^2}{c^3} \tag{15.36}$$

Thus,

$$\Delta\bar{t} - \Delta t = -\frac{lv^2}{c^3} - \frac{lv^2}{c^3} = -\frac{2lv^2}{c^3} \tag{15.37}$$

Hence, the shift in the number of fringes is

$$\frac{\Delta s}{s} = \nu\,(\Delta\bar{t} - \Delta t) = -\frac{\nu}{c}\frac{2lv^2}{c^2} \tag{15.38}$$

$$\text{or, } \frac{\Delta s}{s} = -\frac{2l}{\lambda}\left(\frac{v}{c}\right)^2 \tag{15.39}$$

The orbital velocity of the earth is $v = 3 \times 10^4$ m/sec.
Michelson took $\lambda = 6000$ Å and $l = 1.2$ m. Hence,

$$\frac{\Delta s}{s} = -\frac{2 \times 1.2}{6000 \times 10^{-10}}\left(\frac{3 \times 10^4}{3 \times 10^8}\right)^2 = -0.04 \tag{15.40}$$

This expected shift could be measured easily in the experiment. However, no shift of fringes was actually observed. The experiment was repeated many times, but no such shift was observed. In 1887, Michelson in collaboration with Morley repeated the experiment with more accurately designed instrument taking $l = 11$ m and $\lambda = 5500$ Å. This theoretically yields $\Delta s/s = 0.4$. However, no shift was observed.

This experiment suggests that the value v of the speed of the earth relative to absolute frame should be zero. That is:

The speed of light in vacuum must be the same in all inertial frames.

15.5 Postulates of Einstein's Relativity

Einstein propounded the special theory of relativity in which both time and space are variables at high speeds for various observers. The two fundamental postulates of special theory of relativity are:

(1) ***All the laws of physics remain invariant in all Galilean frames.***
Here the laws of physics include both the laws of mechanics and electrodynamics. According to this postulate, no scientific experiment will be able to find a difference between two reference frames.

(2) ***The velocity of light in vacuum is a constant in every inertial system.***
This postulate is an experimental fact. It has been proved that the velocity of light is independent of the direction of propagation in vacuum and the relative velocity of the source and the observer.

15.6 Lorentz Transformations

Consider two frames of reference S and S' where S' is moving along positive direction of x-axis with uniform velocity u relative to the frame S. We assume that origins O and O' of the two systems coincide at $t = t' = 0$.

We recall the Galilean transformations connecting the measurements in the two frames

$$x' = x - ut \tag{15.41}$$

$$y' = y \tag{15.42}$$

$$z' = z \tag{15.43}$$

$$t' = t \tag{15.44}$$

We have seen that Galilean transformations violate the postulates of the special theory of relativity. It is, therefore, necessary to obtain different transformation equations.

We assume that Eqn (15.41) in the new transformation has the form:

$$x' = \alpha(x - ut) \tag{15.45}$$

where α is a constant which does not depend on x and t; it may, however, be a function of u.

Eqn (15.45) has been assumed on the basis of following considerations:

(1) The relation between x and x' should be linear, so that a single event in S frame corresponds to a single event in S' frame.

(2) At very low values of $u \ll c$, the transformation should be reduced to the Galilean which is known to be correct. Here we take $\alpha = 1$.

The inverse transformation can be written by interchanging the primed and unprimed coordinates and by putting $u = -u$. Thus,

$$x = \alpha'(x' + ut') \tag{15.46}$$

Eqns (15.42) and (15.43) will remain the same, while Eqn (15.44) will be inapplicable since times t and t' are not equal. This can be seen if we substitute the value of x' from Eqn (15.45) into (15.46). Thus,

$$x = \alpha\alpha'(x - ut) + \alpha' ut'$$

which gives

$$t' = \alpha t + \left(\frac{1 - \alpha\alpha'}{\alpha' u}\right) x \tag{15.47}$$

Now suppose that at time $t = t' = 0$, a signal of light is given out from the common origin of S and S'. Then the signal will propagate in the two systems such that

$$x = ct \tag{15.48}$$

$$\text{and } x' = ct' \tag{15.49}$$

Substituting the values of x' and t' from Eqns (15.45) and (15.47) into Eqn (15.49), we get

$$\alpha(x - ut) = c\alpha t + \left(\frac{1 - \alpha\alpha'}{\alpha' u}\right) cx$$

This gives

$$x = ct \left[\frac{1 + \frac{u}{c}}{1 - \left\{\frac{1}{\alpha\alpha'} - 1\right\}\frac{c}{u}}\right] \tag{15.50}$$

Comparing Eqn (15.48) with Eqn (15.50), we obtain

$$1 + \frac{u}{c} = 1 - \left(\frac{1}{\alpha\alpha'} - 1\right)\frac{c}{u} \quad \text{or,} \ \sqrt{\alpha\alpha'} = \left(1 - \frac{u^2}{c^2}\right)^{-1/2} \tag{15.51}$$

Eqn (15.51) shows that $\alpha\alpha'$ depends on u^2 and hence, is not affected by the sign of u. So, we choose

$$\alpha = \alpha' = \frac{1}{\sqrt{1 - u^2/c^2}} \tag{15.52}$$

With these values of α and α', Eqns (15.45) and (15.47) assume the forms:

$$x' = \frac{x - ut}{\sqrt{1 - u^2/c^2}} \tag{15.53}$$

$$\text{and} \ t' = \frac{t - ux/c^2}{\sqrt{1 - u^2/c^2}} \tag{15.54}$$

Other two transformation Eqns (15.42) and (15.43) remain the same. Hence, we write the transformation equations of the two systems as

$$\left.\begin{aligned} x' &= \frac{x - ut}{\sqrt{1 - u^2/c^2}} \\ y' &= y \\ z' &= z \\ t' &= \frac{t - ux/c^2}{\sqrt{1 - u^2/c^2}} \end{aligned}\right\} \tag{15.55}$$

Eqns (15.55) are called ***Lorentz transformations***. The inverse Lorentz transformations are:

$$\left.\begin{aligned} x &= \frac{x' + ut'}{\sqrt{1 - u^2/c^2}} \\ y &= y' \\ z &= z' \\ t &= \frac{t' + ux'/c^2}{\sqrt{1 - u^2/c^2}} \end{aligned}\right\} \tag{15.56}$$

These transformations are reduced to those of Galilean when the relative velocity u is very small compared to the speed of light. For convenience, we sometimes put

$$\beta = \frac{u}{c} \ \text{and} \ \frac{1}{\sqrt{1 - u^2/c^2}} = \frac{1}{\sqrt{1 - \beta^2}} = \gamma \tag{15.57}$$

Hence, the Lorentz transformations are written as

$$\left.\begin{aligned} x' &= \gamma(x - ut) \\ y' &= y \\ z' &= z \\ t' &= \gamma(t - ux/c^2) \end{aligned}\right\} \tag{15.58}$$

and the inverse Lorentz transformations as

$$\left.\begin{aligned} x &= \gamma(x' + ut') \\ y &= y' \\ z &= z' \\ t &= \gamma(t' + ux'/c^2) \end{aligned}\right\} \tag{15.59}$$

These transformations are effective only when the relative velocity u is practically comparable with the speed of light.

15.7 Consequences of Lorentz Transformations

Following are some of the most important consequences of the Lorentz transformations:

A. Relativity of Simultaneity

If two events occur at the same time in a particular frame of reference, they are said to be ***simultaneous***.

Let two events occur simultaneously in S frame at two points $P_1(x_1, y_1, z_1, t_1)$ and $P_2(x_2, y_2, z_2, t_2)$ respectively as measured by an observer at O. Since the events are simultaneous

$$t_2 = t_1, \text{ i.e., } t_2 - t_1 = 0 \tag{15.60}$$

If t_1' and t_2' are the corresponding times of the same two events in S' frame as measured by an observer at O', then, by Lorentz transformation equations, we have

$$t_1' = \gamma(t_1 - ux_1/c^2) \text{ and } t_2' = \gamma(t_2 - ux_2/c^2)$$

so that

$$t_2' - t_1' = \gamma(t_2 - t_1) - \frac{\gamma u}{c^2}(x_2 - x_1) = -\frac{\gamma u}{c^2}(x_2 - x_1) \text{ [by Eqn (15.60)]}$$

Hence,

$$t_2' - t_1' = \frac{\gamma u}{c^2}(x_1 - x_2) \tag{15.61}$$

which is not zero. This means that two events which are simultaneous in S frame are not simultaneous in S' frame which is moving with velocity u along x-axis relative to S. Thus, ***simultaneity is not absolute, but relative***.

B. Length Contraction

Consider a rod placed on the x-axis of S frame remaining at rest. If the coordinates of the end points of the rod measured simultaneously at time t are $(x_1, 0, 0)$ and $(x_2, 0, 0)$, then length of the rod as measured by an observer in S frame will be

$$L = x_2 - x_1 \tag{15.62}$$

Let the coordinates of the end points of the rod in S' frame (measured simultaneously at time t') which is moving with uniform velocity u parallel to x' (or x) axis be $(x_2', 0, 0)$ and $(x_1', 0, 0)$. Then the length of the rod in S' frame will be

$$L' = x_2' - x_1' \tag{15.63}$$

By the transformation Eqns (15.59),

$$x_1 = \gamma(x_1' + ut') \text{ and } x_2 = \gamma(x_2' + ut')$$

so that

$$x_2 - x_1 = \gamma(x_2' - x_1'), \text{ or, } L' = L/\gamma = L\sqrt{1 - u^2/c^2} \tag{15.64}$$

Thus, the observer in S' finds the length of the rod in S frame contracted. Therefore, the length of the rod which is in motion with respect to an observer appears to the observer to be shorter than the length when it is at rest with respect to him. This phenomenon is known as ***Lorentz-FitzGerald contraction***.

The rod will appear to have its maximum length in the coordinate system in which it is at rest. This maximum length is called ***proper length***. The effect of length contraction becomes significant only when the value of the relative velocity u is comparable to the speed of light.

Thus,

$$\text{for } \frac{u}{c} = 0.5, \quad \frac{L'}{L} = 0.87$$

whereas

$$\text{for } \frac{u}{c} = 0.9, \quad \frac{L'}{L} = 0.44.$$

C. Time Dilation

Consider a clock placed at point $(x', 0, 0)$ in the S' frame which is moving with uniform velocity u parallel to x'-axis w.r.t. the frame S which is at rest. Let t'_1 and t'_2 be two instants of time recorded by an observer in S' frame. Then the time interval measured by him will be

$$t' = t'_2 - t'_1 \tag{15.65}$$

Let t_1 and t_2 be the corresponding instants of time measured by an observer in S frame. Then, by Lorentz transformation

$$t_1 = \gamma(t'_1 + ux'/c^2) \tag{15.66}$$

$$\text{and} \quad t_2 = \gamma(t'_2 + ux'/c^2) \tag{15.67}$$

Therefore, the time interval measured by the observer in frame S is

$$t = t_2 - t_1 = \gamma(t'_2 - t'_1) \tag{15.68}$$

Combining Eqns (15.65) and (15.68),

$$t = \gamma t' = \frac{t'}{\sqrt{1 - u^2/c^2}} \tag{15.69}$$

Thus, a clock measures a longer time interval placed in a frame at rest than the time interval when it is placed in a frame moving relative to it. This phenomenon is called ***time dilation***.

15.8 Addition of Velocities

Let the coordinates of a particle in S frame be (x, y, z, t) and those in S' frame be (x', y', z', t'). Then, according to inverse Lorentz transformation,

$$\left.\begin{aligned} x &= \gamma(x' + ut') \\ y &= y' \\ z &= z' \\ t &= \gamma\left(t' + \frac{ux'}{c^2}\right) \end{aligned}\right\} \tag{15.70}$$

The differentials of Eqns (15.70) are

$$\left.\begin{aligned} dx &= \gamma(dx' + udt') \\ dy &= dy' \\ dz &= dz' \\ dt &= \gamma\left(dt' + \frac{udx'}{c^2}\right) \end{aligned}\right\} \tag{15.71}$$

The velocity components of the particle in S frame are

$$V_x = \frac{dx}{dt}, \quad V_y = \frac{dy}{dt}, \quad V_z = \frac{dz}{dt} \tag{15.72}$$

and those in S' frame are

$$V_{x'} = \frac{dx'}{dt'},\ V_{y'} = \frac{dy'}{dt'},\ V_{z'} = \frac{dz'}{dt'} \tag{15.73}$$

From Eqns (15.71)-(15.73), we get

$$V_x = \frac{dx}{dt} = \frac{\gamma(dx' + udt')}{\gamma(dt' + udx'/c^2)} = \frac{(dx'/dt') + u}{1 + \frac{u}{c^2}(dx'/dt')} = \frac{V_{x'} + u}{1 + (uV_{x'})/c^2} \tag{15.74}$$

Similarly,

$$V_y = \frac{dy}{dt} = \frac{V_{y'}}{\gamma\{1 + (uV_{x'})/c^2\}} \tag{15.75}$$

$$V_z = \frac{dz}{dt} = \frac{V_{z'}}{\gamma\{1 + (uV_{x'})/c^2\}} \tag{15.76}$$

Eqns (15.74)-(15.76) give the ***relativistic law of addition of velocities***. In classical mechanics, these relations are

$$V_x = V_{x'} + u,\ V_y = V_{y'},\ V_z = V_{z'} \tag{15.77}$$

Taking the Lorentz transformation, it can be proved in the same way that

$$V_{x'} = \frac{V_x - u}{1 - (uV_x)/c^2} \tag{15.78}$$

$$V_{y'} = \frac{V_y}{\gamma\{1 - (uV_x)/c^2\}} \tag{15.79}$$

$$V_{z'} = \frac{V_z}{\gamma\{1 - (uV_x)/c^2\}} \tag{15.80}$$

• Particular Cases

(1) Let $V_{x'}$ = speed of light = c. Then, from Eqn (15.74)

$$V_x = \frac{c + u}{1 + cu/c^2} = c \tag{15.81}$$

Therefore, the speed of light in a given direction is constant and equal to c for all observers moving in that direction at any speed u.

(2) Let $V_{x'} = 0$, $V_{y'} = 0$ and $V_{z'} = c$. Then, from Eqns (15.74)-(15.76), we find

$$V_x = u,\ V_y = 0 \text{ and } V_z = \frac{c}{\gamma} = c\sqrt{1 - u^2/c^2}$$

Then the resultant velocity is

$$V = \sqrt{V_x^2 + V_y^2 + V_z^2} = \sqrt{u^2 + c^2 - u^2} = c \tag{15.82}$$

Therefore, the speed of light to an observer moving at right angles to the beam of light is also c.

We, therefore, find that Lorentz transformation equations are consistent with the postulates of relativity theory.

EXAMPLE 15.1 *Fizeau's equation for the speed of light parallel to the stream velocity v of a water flow is*

$$w = \frac{c}{n} + v\left(1 - \frac{1}{n^2}\right)$$

where n is the refractive index of water. Prove this considering only first order in v/c.

Solution. Using the relativistic formula for addition of velocities

$$w = \frac{\frac{c}{n} + v}{1 + \frac{c}{n}\frac{v}{c^2}} = \left(\frac{c}{n} + v\right)\left(1 + \frac{v}{nc}\right)^{-1}$$

$$= \left(\frac{c}{n} + v\right)\left(1 - \frac{v}{nc}\right), \text{ neglecting higher powers of } v/c.$$

$$\therefore\ w = \frac{c}{n} + v - \frac{v}{n^2} - \frac{v^2}{n^2c^2}$$

Neglecting the last term,

$$w = \frac{c}{n} + v\left(1 - \frac{1}{n^2}\right)$$

The factor $(1 - 1/n^2)$ is called ***drag coefficient***. Taking the value of $n = 1.333$, the drag coefficient is $1 - (1.333)^{-2} = 0.437$. Fizeau found this value experimentally to be 0.434.

EXAMPLE 15.2 *An observer moving along x-axis of a reference frame S with a velocity v sees a body of proper volume V_0 moving in the same direction with a velocity u. Show that the apparent volume of the body measured by the observer is*

$$V = V_0 \frac{\sqrt{c^2 - u^2}\sqrt{c^2 - v^2}}{c^2 - uv}.$$

Solution. Let the observer be in S' frame so that the relative velocity of S' w.r.t. S is v along positive x-axis. The body moves with velocity u relative to S frame along x-axis.

Then

$$u = \frac{dx}{dt} \tag{1}$$

Similarly, the velocity of the body relative to S' is

$$u' = \frac{dx'}{dt'} \tag{2}$$

Differentiating equations of Lorentz transformation, we get

$$dx' = \gamma(dx - vdt) \text{ and } dt' = \gamma(dt - vdx/c^2) \tag{3}$$

Hence, from Eqns (2) and (3),

$$u' = \frac{dx'}{dt'} = \frac{dx - vdt}{dt - vdx/c^2} = \frac{dx/dt - v}{1 - \frac{v}{c^2}\frac{dx}{dt}} \quad \text{or,} \quad u' = \frac{u - v}{1 - uv/c^2} \tag{4}$$

Hence, the volume of the body relative to the observer will be

$$V = V_0\sqrt{1 - (u'/c)^2} = V_0\left[1 - \left(\frac{u - v}{c - uv/c}\right)^2\right]^{1/2}$$

$$= V_0\left[1 - \frac{c^2(u - v)^2}{(c^2 - uv)^2}\right]^{1/2}$$

$$= V_0\frac{\sqrt{c^2 - u^2}\sqrt{c^2 - v^2}}{c^2 - uv}.$$

EXAMPLE 15.3 *Show that the D'Alembertian operator*

$$\Box^2 = \frac{\partial^2}{\partial x^2} + \frac{\partial^2}{\partial y^2} + \frac{\partial^2}{\partial z^2} - \frac{1}{c^2}\frac{\partial^2}{\partial t^2}$$

is invariant under Lorentz transformation.

Solution. The equations of Lorentz transformation are:

$$x' = \gamma(x - ut),\ y' = y,\ z' = z \text{ and } t' = \gamma(t - ux/c^2) \tag{1}$$

Then

$$\frac{\partial}{\partial x} = \frac{\partial}{\partial x'}\frac{\partial x'}{\partial x} + \frac{\partial}{\partial t'}\frac{\partial t'}{\partial x} = \gamma\frac{\partial}{\partial x'} - \frac{u\gamma}{c^2}\frac{\partial}{\partial t'}$$

$$\therefore\ \frac{\partial^2}{\partial x^2} = \left(\gamma\frac{\partial}{\partial x'} - \frac{u\gamma}{c^2}\cdot\frac{\partial}{\partial t'}\right)^2 = \gamma^2\frac{\partial^2}{\partial x'^2} - \frac{2u\gamma^2}{c^2}\frac{\partial^2}{\partial x'\partial t'} + \frac{u^2\gamma^2}{c^4}\frac{\partial^2}{\partial t'^2} \tag{2}$$

$$\frac{\partial}{\partial y} = \frac{\partial}{\partial y'}\frac{\partial y'}{\partial y} = \frac{\partial}{\partial y'}$$

$$\therefore\ \frac{\partial^2}{\partial y^2} = \frac{\partial^2}{\partial y'^2} \tag{3}$$

Similarly,

$$\frac{\partial^2}{\partial z^2} = \frac{\partial^2}{\partial z'^2} \tag{4}$$

$$\frac{\partial}{\partial t} = \frac{\partial}{\partial t'}\frac{\partial t'}{\partial t} + \frac{\partial}{\partial x'}\frac{\partial x'}{\partial t'} = \gamma\frac{\partial}{\partial t'} - u\gamma\frac{\partial}{\partial x'}$$

$$\therefore\ \frac{\partial^2}{\partial t^2} = \left(\gamma\frac{\partial}{\partial t'} - u\gamma\frac{\partial}{\partial x'}\right)^2 = \gamma^2\frac{\partial^2}{\partial t'^2} - 2u\gamma^2\frac{\partial^2}{\partial x'\partial t'} + u^2\gamma^2\frac{\partial^2}{\partial x'^2} \tag{5}$$

Using Eqns (2) – (5)

$$\frac{\partial^2}{\partial x^2} + \frac{\partial^2}{\partial y^2} + \frac{\partial^2}{\partial z^2} - \frac{1}{c^2}\frac{\partial^2}{\partial t^2}$$

$$= \gamma^2\frac{\partial^2}{\partial x'^2} - \frac{2u\gamma^2}{c^2}\frac{\partial^2}{\partial x'\partial t'} + \frac{u^2\gamma^2}{c^4}\frac{\partial^2}{\partial t'^2} + \frac{\partial^2}{\partial y'^2} + \frac{\partial^2}{\partial z'^2} - \frac{\gamma^2}{c^2}\frac{\partial^2}{\partial t'^2} + \frac{2u\gamma^2}{c^2}\frac{\partial^2}{\partial x'\partial t'} - \frac{u^2\gamma^2}{c^2}\frac{\partial^2}{\partial x'^2}$$

$$= \gamma^2\left(1 - \frac{u^2}{c^2}\right)\frac{\partial^2}{\partial x'^2} + \frac{\partial^2}{\partial y'^2} + \frac{\partial^2}{\partial z'^2} - \frac{\gamma^2}{c^2}\left(1 - \frac{u^2}{c^2}\right)\frac{\partial^2}{\partial t'^2}$$

$$= \frac{\partial^2}{\partial x'^2} + \frac{\partial^2}{\partial y'^2} + \frac{\partial^2}{\partial z'^2} - \frac{1}{c^2}\frac{\partial^2}{\partial t'^2}$$

since $\gamma = \left(1 - u^2/c^2\right)^{-1/2}$.

Hence, the given operator is invariant under Lorentz transformation.

EXAMPLE 15.4 *A frame S' moves with uniform velocity u along x'-axis relative to a frame S. At time $t = t' = 0$, the origins of S and S' coincided and axes overlapped. In S frame, a projectile thrown with velocity V describes a parabola given by*

$$x = Vt,\ y = \frac{1}{2}ft^2$$

Find its trajectory in S' frame.

Solution. We are given

$$x = Vt \text{ and } y = \frac{1}{2}ft^2 \tag{1}$$

in S frame. If x', t' are coordinates of the point in S' frame, then

$$x = \gamma(x' + ut') \text{ and } t = \gamma(t' + ux'/c^2) \tag{2}$$

Combining Eqns (1) and (2),

$$\gamma(x' + ut') = V\gamma(t' + ux'/c^2)$$

$$\text{or, } x' + ut' = V(t' + ux'/c^2) \text{ or, } x'(1 - uV/c^2) = (V - u)t'$$

$$\therefore\ x' = \frac{V-u}{1 - uV/c^2}t' \tag{3}$$

Again,

$$y' = y = \frac{1}{2}ft^2$$

$$\text{or, } y' = \frac{1}{2}f\gamma^2\left(t' + \frac{ux'}{c^2}\right)^2 = \frac{1}{2}f\gamma^2\left[t' + \frac{u}{c^2}\cdot\frac{V-u}{1-uV/c^2}t'\right]^2$$

$$= \frac{1}{2}f\gamma^2\left[\frac{c^2 - u^2}{c^2 - uV}\right]^2 t'^2 = \frac{1}{2}f\frac{(1 - u^2/c^2)}{(1 - uV/c^2)^2}t'^2 \tag{4}$$

Eleminating t' from Eqns (3) and)4), we get

$$y' = \frac{1}{2}f\frac{1 - u^2/c^2}{(V-u)^2}x'^2 \text{ or, } y' = Ax'^2 \tag{5}$$

where

$$A = \frac{1}{2}f\frac{1 - u^2/c^2}{(V-u)^2} \tag{6}$$

Thus, the trajectory is a parabola.

EXAMPLE 15.5 *In the two frames S and S' used in Lorentz transformation, there is only one plane in S on which the clocks agree with those of S' after time t. Prove that this plane moves with a velocity*

$$V = \frac{c^2}{u}\left[1 - \left(1 - \frac{u^2}{c^2}\right)^{1/2}\right] \text{ in } S \text{ frame.}$$

Solution. Lorentz transformation of time in two frames is given by

$$t' = \gamma\left(t - \frac{ux}{c^2}\right) = \frac{1}{\sqrt{1 - \dfrac{u^2}{c^2}}}\left(t - \frac{ux}{c^2}\right) \tag{1}$$

Since the clocks agree after time t, $t = t'$. Hence, from (1) we get

$$\left(1 - \sqrt{1 - \frac{u^2}{c^2}}\right)t = \frac{u}{c^2}x \text{ or, } x = \frac{c^2}{u}\left[1 - \sqrt{1 - \frac{u^2}{c^2}}\right]t$$

Hence, the plane in S moves with a velocity

$$V = \frac{dx}{dt} = \frac{c^2}{u}\left[1 - \sqrt{1 - \frac{u^2}{c^2}}\right].$$

EXAMPLE 15.6 *A light signal moves in the xy-plane of S frame at an angle θ with x-axis. If θ' be the angle between x'-axis and direction of motion of the light signal in S' frame, show that the aberration angle*

$$\Delta\theta = \theta' - \theta = \frac{u}{c}\sin\theta$$

Solution. The velocity components of the light signal in S frame are

$$V_x = c\cos\theta,\ V_y = c\sin\theta \tag{1}$$

In S' frame, these components are

$$V_{x'} = c\cos\theta',\ V_{y'} = c\sin\theta' \tag{2}$$

By Eqns (15.78) and (15.79),

$$c\cos\theta' = \frac{c\cos\theta - u}{1 - \dfrac{u\cos\theta}{c}} \tag{3}$$

$$\text{and}\quad c\sin\theta' = \frac{c\sin\theta\sqrt{1-u^2/c^2}}{1 - \dfrac{u\cos\theta}{c}} \tag{4}$$

From Eqns (3) and (4),

$$\tan\theta' = \frac{\sin\theta\sqrt{1-u^2/c^2}}{\cos\theta - u/c} \tag{5}$$

Eqn (5) shows that the direction of light rays depends on the relative velocity u of the source and the observer. If $u = 0$, $\tan\theta = \tan\theta'$.

Now from Eqn (5), we have

$$\tan\theta' = \frac{\tan\theta\sqrt{1-u^2/c^2}}{1 - \dfrac{u}{c}\sec\theta} \tag{6}$$

For $u/c \ll 1$, Eqn (6) gives

$$\tan\theta' = \tan\theta\left(1 - \frac{u}{c}\sec\theta\right)^{-1} = \tan\theta + \frac{u}{c}\frac{\sin\theta}{\cos^2\theta} \tag{7}$$

The aberration angle is $\Delta\theta = \theta' - \theta$, i.e., $\theta' = \Delta\theta + \theta$. Hence, from Eqn (7)

$$\tan(\Delta\theta + \theta) - \tan\theta = \frac{u}{c}\frac{\sin\theta}{\cos^2\theta} \tag{8}$$

since $\Delta\theta$ is small, $\sin\Delta\theta = \Delta\theta$ and $\cos\Delta\theta = 1$. Then, from Eqn (8), we get

$$\frac{\sin\Delta\theta}{\cos\theta\cos(\theta + \Delta\theta)} = \frac{u}{c}\frac{\sin\theta}{\cos^2\theta}\quad \text{or,}\ \Delta\theta = \frac{u}{c}\sin\theta.$$

Hence proved.

15.9 Lorentz Group

Consider three frames S, S' and S'' where the frame S' is moving parallel to x-axis (or x'-axis) with velocity u_1 relative to the frame S and the frame S'' in turn is moving parallel to x'-axis (or x''-axis) with a velocity u_2 relative to S'. Let the spatial and time coordinates of a point in the frames S, S' and S'' are respectively $(x, y, z, t), (x', y', z', t')$ and (x'', y'', z'', t'').

Then, by Lorentz transformation from $S \to S'$, we have

$$\left.\begin{aligned} x' &= \gamma_1(x - u_1 t),\ y' = y,\ z' = z \\ t' &= \gamma_1(t - u_1 x/c^2) \end{aligned}\right\} \tag{15.83}$$

where

$$\gamma_1 = \left(1 - u_1^2/c^2\right)^{-1/2}$$

Similarly, for the transformation from $S' \to S''$

$$\left.\begin{aligned} x'' &= \gamma_2(x' - u_2 t'),\ y'' = y',\ z'' = z' \\ t'' &= \gamma_2(t' - u_2 x'/c^2) \end{aligned}\right\} \tag{15.84}$$

where

$$\gamma_2 = \left(1 - u_2^2/c^2\right)^{-1/2} \tag{15.85}$$

For the composite transformation from $S \to S''$, we combine Eqns (15.83) and (15.84), and get, after some algebraic manipulations

$$\left.\begin{aligned} x'' &= \frac{x - Vt}{\sqrt{1 - V^2/c^2}},\ y'' = y;\ z'' = z \\ t'' &= \frac{t - Vx/c^2}{\sqrt{1 - V^2/c^2}} \end{aligned}\right\} \tag{15.86}$$

where

$$V = \frac{u_1 + u_2}{1 + u_1 u_2/c^2} \tag{15.87}$$

Hence, the composite transformation from $S \to S''$ is also a Lorentz transformation. The parameter V is symmetric w.r.t. u_1 and u_2, so that the process of composite transformation is commutative. Therefore, Lorentz transformation for all values of the parameter V constitute a group called ***Lorentz group***. This group contains only one parameter V since we have not considered the most general transformation taking into account three translations and three rotations of axes. Therefore, the group discussed above would be called a ***subgroup of the general group***.

In matrix notations, the above Lorentz transformations can be written as

$$\begin{bmatrix} x' \\ y' \\ z' \\ t' \end{bmatrix} = \begin{bmatrix} \gamma_1 & 0 & 0 & -u_1\gamma_1 \\ 0 & 1 & 0 & 0 \\ 0 & 0 & 1 & 0 \\ -\gamma_1 u_1/c^2 & 0 & 0 & \gamma_1 \end{bmatrix} \begin{bmatrix} x \\ y \\ z \\ t \end{bmatrix} \tag{15.88}$$

Similarly,

$$\begin{bmatrix} x'' \\ y'' \\ z'' \\ t'' \end{bmatrix} = \begin{bmatrix} \gamma_2 & 0 & 0 & -u_2\gamma_2 \\ 0 & 1 & 0 & 0 \\ 0 & 0 & 1 & 0 \\ -\gamma_2 u_2/c^2 & 0 & 0 & \gamma_2 \end{bmatrix} \begin{bmatrix} x' \\ y' \\ z' \\ t' \end{bmatrix} \tag{15.89}$$

Denoting the matrices by $L(u_1)$ and $L(u_2)$ respectively, it can be verified that

$$L(u_1)\, L(u_2) = L(u_2)\, L(u_1). \tag{15.90}$$

15.10 Relativistic Mechanics

(a) Variation of Mass

In Newtonian mechanics, mass of a particle remains constant during its motion. In that case, the velocity of the particle will go on increasing indefinitely by a continuous application of a force and ultimately may reach the speed of light. But no material particle may have the speed of light.

In relativistic mechanics, therefore, the mass of a particle is assumed to depend on its velocity u and may be taken as

$$m(u) = m_0 f(u) \tag{15.91}$$

where m_0 is the mass when the particle is at rest and is called ***rest mass*** and the function $f(u)$ is such that

$$f(u) \to 1 \text{ as } u \to 0 \text{ and } f(u) \to \infty \text{ as } u \to c \tag{15.92}$$

Consider two frames S and S'. We take two particles each of mass m' moving in S' frame along x' axis, opposite to each other, with velocity v'. Then, velocities of two particles as observed from S frame, by the relativistic law for addition of velocities, are given by

$$v_1 = \frac{v' + u}{1 + uv'/c^2} \tag{15.93}$$

$$\text{and } v_2 = \frac{-v' + u}{1 - uv'/c^2} \tag{15.94}$$

Let the masses of the particles in S frame be m_1 and m_2 respectively. Let the particles in S' frame collide and coalesce into one. Assuming that the law of conservation of momentum as well as the law of conservation of relativistic masses hold good, we write

$$m_1 v_1 + m_2 v_2 = (m_1 + m_2)u \tag{15.95}$$

Substituting the values of v_1 and v_2 from Eqns (15.93) and (15.94) into (15.95), and making some algebraic manipulations, we get

$$\frac{m_1}{m_2} = \frac{1 + uv'/c^2}{1 - uv'/c^2} \tag{15.96}$$

Now, using Eqns (15.93) and (15.94), we can prove the following two identities:

$$1 - \frac{v_1^2}{c^2} = \frac{(1 - u^2/c^2)(1 - v'^2/c^2)}{(1 + uv'/c^2)^2} \tag{15.97}$$

$$\text{and } 1 - \frac{v_2^2}{c^2} = \frac{(1 - u^2/c^2)(1 - v'^2/c^2)}{(1 - uv'/c^2)^2} \tag{15.98}$$

With the use of Eqns (15.97) and (15.98), we write Eqn (15.96) as

$$\frac{m_1}{m_2} = \frac{\sqrt{1 - v_2^2/c^2}}{\sqrt{1 - v_1^2/c^2}} \tag{15.99}$$

Let the velocity of the second particle v_2 in S frame be zero, i.e., $v_2 = 0$. Then its mass m_2 becomes the rest mass and can be denoted by m_0. Further, let the velocity of the first particle be u relative to S frame, so that $v_1 = u$. We denote its mass m_1 as m. Then, from Eqn (15.99), we get

$$m = \frac{m_0}{\sqrt{1 - u^2/c^2}} \tag{15.100}$$

so that the function $f(u)$ in Eqn (15.91) is

$$f(u) = \frac{1}{\sqrt{1 - u^2/c^2}} \tag{15.101}$$

which satisfies the conditions in Eqn (15.92).

Eqn (15.100) determines the variation of mass with its velocity. This is quite significant at high velocities comparable to that of light.

Thus, the relativistic momentum is given by

$$\mathbf{p} = m\mathbf{u} = \frac{m_0 u}{\sqrt{1 - u^2/c^2}} \tag{15.102}$$

The relativistic force is given by

$$\mathbf{F} = \frac{d\mathbf{p}}{dt} = \frac{d}{dt}(m\mathbf{u}), \text{ i.e., } \mathbf{F} = m\frac{du}{dt} + u\frac{dm}{dt} \tag{15.103}$$

where the term $u(dm/dt)$ does not vanish since m depends on velocity, which in turn varies with time.

(b) Relativistic Energy

Let a force F be acting on a particle of mass m so that its kinetic energy increases. If at any instant its velocity is u, then

$$F = \frac{d}{dt}(mu) \tag{15.104}$$

Let dr = infinitesimal displacement of the particle along the line of action of the force. Then work done on the particle, which is equal to the gain in kinetic energy is

$$dT = F dr = \frac{d}{dt}(mu)dr = u\, d(mu) \tag{15.105}$$

since $u = dr/dt$.

If initially, the particle were at rest, then the total gain in kinetic energy by the particle will be

$$\begin{aligned} T &= \int_0^u u\, d(mu) = u(mu)\Big|_0^u - \int_0^u (mu)du \text{ (integrating by parts)} \\ &= mu^2 - \int_0^u \frac{m_0 u}{\sqrt{1 - u^2/c^2}} du \\ &= \frac{m_0 u^2}{\sqrt{1 - u^2/c^2}} + m_0 c^2 \sqrt{1 - u^2/c^2}\Big|_0^u \quad \left[\text{put } z^2 = 1 - \frac{u^2}{c^2}\right] \\ &= \frac{m_0 u^2}{\sqrt{1 - u^2/c^2}} + m_0 c^2 \sqrt{1 - u^2/c^2} - m_0 c^2 \\ &= \frac{m_0 c^2}{\sqrt{1 - u^2/c^2}} - m_0 c^2 = mc^2 - m_0 c^2 \end{aligned}$$

Therefore,

$$T = (m - m_0)c^2 \tag{15.106}$$

Eqn (15.106) gives the ***relativistic kinetic energy***. The quantity $m_0 c^2$ is the kinetic energy due to rest mass and is called ***rest energy*** or ***proper energy*** and denoted by E_0.

Thus,

$$E_0 = m_0 c^2 \tag{15.107}$$

The total energy of the particle moving with velocity u is given by

$$E = T + E_0 = (m - m_0)c^2 + m_0 c^2 = mc^2$$

Therefore, the total energy is

$$E = mc^2 = \frac{m_0 c^2}{\sqrt{1 - u^2/c^2}} \tag{15.108}$$

This energy E is called ***relativistic total energy*** of the particle moving with velocity u. Eqn (15.108) shows a close relation between mass and energy, and is the well-known ***Einstein's mass-energy relation***.

The relativistic kinetic energy can be expressed as

$$T = (m - m_0)c^2 = m_0 c^2 \left[\left(1 - \frac{u^2}{c^2}\right)^{-1/2} - 1\right]$$

$$= m_0 c^2 \left[1 + \frac{1}{2}\frac{u^2}{c^2} + \cdots - 1\right] = \frac{1}{2} m_0 u^2 \tag{15.109}$$

Here we have neglected higher powers of u/c considering $u/c \ll 1$. Eqn (15.109) is the kinetic energy in classical mechanics.

(c) Relation between Energy and Momentum

A particle of rest mass m_0 moving with velocity u has the momentum

$$p = \frac{m_0 u}{\sqrt{1 - u^2/c^2}} = \gamma m_0 u, \text{ or, } p^2 = \gamma^2 m_0^2 u^2 \tag{15.110}$$

Here

$$\gamma^2 = \frac{1}{1 - u^2/c^2} = \frac{c^2}{c^2 - u^2}$$

which gives

$$\gamma^2 u^2 = \gamma^2 c^2 - c^2 \tag{15.111}$$

Combining Eqns (15.110) and (15.111), we get

$$p^2 = m_0^2 \gamma^2 c^2 - m_0^2 c^2, \text{ or, } p^2 c^2 = m_0^2 \gamma^2 c^4 - m_0^2 c^4 \tag{15.112}$$

But, by Eqn (15.108),

$$E = mc^2 = m_0 \gamma c^2$$

$$\therefore\ E^2 = m_0^2 \gamma^2 c^4 \tag{15.113}$$

From Eqns (15.112) and (15.113),

$$E^2 = p^2 c^2 + m_0^2 c^4 \tag{15.114}$$

Eqn (15.114) gives the relation between total energy of a particle to its momentum. It is to be noted that the quantity $m_0 c^4$ in Eqn (15.114) is constant and equal to the square of rest energy. Hence,

$$E^2 = p^2 c^2 + E_0^2 \tag{15.115}$$

For a particle with zero rest mass, $E^2 = p^2c^2$, i.e., $E = pc$. But

$$p = mu = \frac{E}{c^2}u, \text{ since } E = mc^2$$

Hence,

$$E = \left(\frac{E}{c^2}u\right)c \text{ or, } u = c$$

Thus, a particle of zero rest mass travels with the speed of light.

(d) Transformation of Momentum and Energy

Consider a particle of rest mass m_0 moving in S frame with velocity V. Then its momentum can be written as

$$p = \frac{m_0 V}{\sqrt{1 - V^2/c^2}} = \Gamma m_0 V \tag{15.116}$$

where we have denoted

$$\Gamma = \frac{1}{\sqrt{1 - V^2/c^2}} \tag{15.117}$$

The total energy of the particle is

$$E = mc^2 = \frac{m_0 c^2}{\sqrt{1 - V^2/c^2}} = \Gamma m_0 c^2 \tag{15.118}$$

Let the momentum and energy of the particle in S' frame be expressed as

$$p' = \frac{m_0 V'}{\sqrt{1 - V'^2/c^2}} = \Gamma' m_0 V' \tag{15.119}$$

$$\text{and } E' = \frac{m_0 c^2}{\sqrt{1 - V'^2/c^2}} = \Gamma' m_0 c^2 \tag{15.120}$$

where we have denoted

$$\Gamma' = \frac{1}{\sqrt{1 - V'^2/c^2}} \tag{15.121}$$

If the components of the velocity in S frame are V_x, V_y, V_z and those in S' frame are $V_{x'}, V_{y'}$ and $V_{z'}$, then, using Eqns (15.74) to (15.76), we have

$$V^2 = V_x^2 + V_y^2 + V_z^2 = \frac{\gamma^2 \left(V_{x'} + u\right)^2 + V_{y'}^2 + V_{z'}^2}{\gamma^2 \left[1 + (uV_{x'})/c^2\right]^2} \tag{15.122}$$

It can be verified that

$$1 - \frac{V^2}{c^2} = \frac{(1 - u^2/c^2)(1 - V'^2/c^2)}{\left[1 + (uV_{x'})/c^2\right]^2} \tag{15.123}$$

where $V'^2 = V_{x'}^2 + V_{y'}^2 + V_{z'}^2$.

From Eqn (15.123), we find

$$\Gamma = \left(1 - \frac{V^2}{c^2}\right)^{-1/2} = \frac{1 + (uV_{x'})/c^2}{\sqrt{1 - u^2/c^2}\sqrt{1 - V'^2/c^2}}, \text{ i.e., } \Gamma = \gamma\left(1 + \frac{uV_{x'}}{c^2}\right)\Gamma' \tag{15.124}$$

Changing the sign on u and the primes,

$$\Gamma' = \gamma\left(1 - \frac{uV_x}{c^2}\right)\Gamma \tag{15.125}$$

From Eqns (15.119) and (15.125), we write

$$p' = m_0\gamma\left(1 - \frac{uV_x}{c^2}\right)V'\Gamma \tag{15.126}$$

In component forms:

$$p'_x = m_0\gamma\left(1 - \frac{uV_x}{c^2}\right)V_{x'}\Gamma = m_0\gamma\left(1 - \frac{uV_x}{c^2}\right)\frac{V_x - u}{1-(uV_x)/c^2}\Gamma$$

$$\text{or, } p'_x = m_0\gamma(V_x - u)\Gamma \tag{15.127}$$

Similarly,

$$p'_y = m_0 V_{y'}\Gamma' = m_0\gamma\left(1 - \frac{uV_x}{c^2}\right)\frac{V_y}{\gamma\,(1 - uV_x/c^2)} = m_0 V_y\Gamma \tag{15.128}$$

$$\text{and } p'_z = m_0 V_z\Gamma \tag{15.129}$$

Also

$$E' = m_0c^2\Gamma' = m_0c^2\gamma\left(1 - \frac{uV_x}{c^2}\right)\Gamma \tag{15.130}$$

But

$$p_x = \Gamma m_0 V_x,\ p_y = \Gamma m_0 V_y \text{ and } p_z = \Gamma m_0 V_z \tag{15.131}$$

Eqns (15.127) – (15.130) can be written in view of Eqns (15.118) and (15.131) in the form:

$$\left.\begin{aligned} p'_x &= \gamma\left(p_x - \frac{uE}{c^2}\right)\\ p'_y &= p_y\\ p'_z &= p_z\\ E' &= \gamma\,(E - up_x)\end{aligned}\right\} \tag{15.132}$$

The inverse transformations are:

$$\left.\begin{aligned} p_x &= \gamma\left(p'_x + \frac{uE'}{c^2}\right)\\ p_y &= p'_y\\ p_z &= p'_z\\ E &= \gamma\,(E' + up'_x)\end{aligned}\right\} \tag{15.133}$$

Again, since $E = mc^2$ and $E' = m'c^2$, we get from Eqn (15.130)

$$m'c^2 = (\Gamma m_0c^2)\gamma\left(1 - \frac{uV_x}{c^2}\right) = mc^2\gamma\left(1 - \frac{uV_x}{c^2}\right)$$

which gives the transformation equation of mass in the form:

$$m' = m\gamma\left(1 - \frac{uV_x}{c^2}\right) \tag{15.134}$$

If the particle be at rest in the S' frame, then its mass is m_0 and

$$p'_x = p'_y = p'_z = 0,\ E' = m_0c^2 \tag{15.135}$$

Eqns (15.132) assume the form:

$$p_x = \gamma m_0 u,\ p_y = p_z = 0,\ E = \gamma m_0c^2 \tag{15.136}$$

(e) Lagrangian and Hamiltonian for a Particle

Consider a particle of rest mass m_0 moving in an inertial frame with velocity v. Then

$$v^2 = \dot{x}^2 + \dot{y}^2 + \dot{z}^2 \tag{15.137}$$

Following the non-relativistic mechanics, we assume that the momentum components are given by

$$\frac{\partial L}{\partial \dot{x}} = p_x = \frac{m_0 \dot{x}}{\sqrt{1 - v^2/c^2}} \tag{15.138}$$

$$\frac{\partial L}{\partial \dot{y}} = p_y = \frac{m_0 \dot{y}}{\sqrt{1 - v^2/c^2}} \tag{15.139}$$

$$\frac{\partial L}{\partial \dot{z}} = p_z = \frac{m_0 \dot{z}}{\sqrt{1 - v^2/c^2}} \tag{15.140}$$

where L is the Lagrangian of the particle. Integrating Eqn (15.138), we get

$$L = \int \frac{m_0 \dot{x} d\dot{x}}{\sqrt{1 - v^2/c^2}} + \text{const.} = -m_0 c^2 \sqrt{1 - v^2/c^2} - V \tag{15.141}$$

We may take the constant of integration V as the potential energy of the particle which is a function of x, y, z only. Then the relativistic force components will be given by

$$F_x = -\frac{\partial V}{\partial x}, \quad F_y = -\frac{\partial V}{\partial y}, \quad F_z = -\frac{\partial V}{\partial z} \tag{15.142}$$

The correctness of the Eqn (15.141) will be shown by deducing the correct relativistic equations of motion .

$$F_x = \frac{dp_x}{dt}, \quad F_y = \frac{dp_y}{dt}, \quad F_z = \frac{dp_z}{dt} \tag{15.143}$$

from Lagrange's equations of motion.

Lagrange's equation of motion in x-direction is

$$\frac{d}{dt}\left(\frac{\partial L}{\partial \dot{x}}\right) - \frac{\partial L}{\partial x} = 0 \tag{15.144}$$

Now, from Eqn (15.141),

$$\frac{\partial L}{\partial \dot{x}} = p_x \text{ and } \frac{\partial L}{\partial x} = -\frac{\partial V}{\partial x} \tag{15.145}$$

which, when substituted in Eqn (15.144), yields

$$\frac{dp_x}{dt} + \frac{\partial V}{\partial x} = 0 \tag{15.146}$$

Thus,

$$F_x = -\frac{\partial V}{\partial x} = \frac{dp_x}{dt} \tag{15.147}$$

which is the relativistic equation of motion of the particle. We, therefore, conclude that Eqn (15.141) is the correct Lagrangian of the particle.

We, however, note that, in relativistic mechanics, the Lagrangian is not equal to $(T - V)$, since the first part of the Eqn (15.141) does not represent the kinetic energy of the particle.

The Hamiltonian of the particle moving under conservative force is given by

$$H = p_x \dot{x} + p_y \dot{y} + p_z \dot{z} - L \tag{15.148}$$

Substituting the expressions for p_x, p_y, p_z and L, we get

$$H = \frac{m_0}{\sqrt{1 - v^2/c^2}}\left[\dot{x}^2 + \dot{y}^2 + \dot{z}^2\right] + m_0c^2\sqrt{1 - v^2/c^2} + V$$

$$= \frac{m_0v^2}{\sqrt{1 - v^2/c^2}} + m_0c^2\sqrt{1 - v^2/c^2} + V$$

$$= mc^2 + V = E + V = \text{total energy.} \tag{15.149}$$

Since $E^2 = p^2c^2 + m_0^2c^4$, the Hamiltonian of the particle may be written as

$$H = E + V = \sqrt{p^2c^2 + m_0^2c^4} + V \tag{15.150}$$

EXAMPLE 15.7

(a) *Express the rest mass m_0 of a particle in terms of its momentum p and kinetic energy T.*

(b) *Prove that, $E^2 = p^2c^2 + m_0^2c^4$ is invariant under Lorentz transformations.*

Solution.

(a) We know that

$$E = T + m_0c^2, \quad \text{by Eqn (15.106)} \tag{1}$$

$$\text{and} \quad E^2 = p^2c^2 + m_0^2c^4 \tag{2}$$

Combining Eqns (1) and (2),

$$\left(T + m_0c^2\right)^2 = p^2c^2 + m_0^2c^4 \quad \text{or,} \quad T^2 + 2Tm_0c^2 = p^2c^2$$

$$\therefore \; m_0 = \frac{p^2c^2 - T^2}{2Tc^2}$$

(b) The given expression is

$$E^2 = p^2c^2 + m_0^2c^4, \text{ i.e., } E^2 - p^2c^2 = m_0^2c^4 = \text{const.}$$

In S' frame,

$$E'^2 - p'^2c^2 = E'^2 - \left(p_x'^2 + p_y'^2 + p_z'^2\right)c^2 \tag{3}$$

Using the transformation relations in Eqn (15.132),

$$E'^2 - p'^2c^2 = \gamma^2(E - up_x)^2 - \left[\gamma^2\left(p_x - \frac{uE}{c^2}\right)^2 + p_y^2 + p_z^2\right]c^2$$

$$= \gamma^2\left(E^2 - 2Eup_x + u^2p_x^2\right) - c^2\left(p_y^2 + p_z^2\right) - c^2\gamma^2\left(p_x^2 - 2Eup_x/c^2 + u^2E^2/c^4\right)$$

$$= \gamma^2E^2\left(1 - \frac{u^2}{c^2}\right) - c^2\left[\gamma^2\left(1 - \frac{u^2}{c^2}\right)p_x^2 + p_y^2 + p_z^2\right]$$

$$= E^2 - c^2\left(p_x^2 + p_y^2 + p_z^2\right)$$

$$= E^2 - p^2c^2$$

Hence, $E'^2 - p'^2c^2 = E^2 - p^2c^2 = m_0^2c^4$.

EXAMPLE 15.8 *A particle of rest mass m_0 moves with a velocity v and collides with a stationary particle of mass m. If the two particles coalesce, determine the mass and velocity of the composite body.*

Solution. Let M = mass of the composite body
and V = velocity of the composite body.

Then, by the theorem of conservation of momentum,

$$\frac{m_0 v}{\sqrt{1-v^2/c^2}} = \frac{MV}{\sqrt{1-V^2/c^2}} \tag{1}$$

By the theorem of conservation of energy,

$$\frac{m_0 c^2}{\sqrt{1-v^2/c^2}} + mc^2 = \frac{Mc^2}{\sqrt{1-V^2/c^2}} \tag{2}$$

From Eqn (2), we get

$$\frac{M}{\sqrt{1-V^2/c^2}} = \frac{m_0}{\sqrt{1-v^2/c^2}} + m \tag{3}$$

From Eqns (1) and (3),

$$\frac{m_0 v}{\sqrt{1-v^2/c^2}} = \left[\frac{m_0}{\sqrt{1-v^2/c^2}} + m\right] V \quad \text{or,} \quad m_0 v = \left[m_0 + m\sqrt{1-v^2/c^2}\right] V$$

Hence,

$$V = \frac{m_0 v}{m_0 + m\sqrt{1-v^2/c^2}} \tag{4}$$

Squaring both sides of Eqn (1),

$$\frac{m_0^2 v^2}{c^2 - v^2} = \frac{M^2 V^2}{c^2 - V^2}$$

$$\therefore \; M^2 = \frac{m_0^2 v^2}{c^2 - v^2} \frac{c^2 - V^2}{V^2} \tag{5}$$

Substituting the value of V from Eqn (4) into Eqn (5), we find

$$M = \left[m_0 + m + \frac{2m_0 m}{\sqrt{1-v^2/c^2}}\right]^{1/2}.$$

EXAMPLE 15.9 *A meson of rest mass π comes to rest and disintegrates into a muon of rest mass μ and a neutrino of zero mass. Show that the kinetic energy of the motion of muon is*

$$T = \frac{(\pi - \mu)^2 c^2}{2\pi}$$

Solution. Let v be the velocity of muon, and p be the momentum of neutrino. Then, from the principle of conservation of momentum,

$$p = \frac{\mu v}{\sqrt{1-v^2/c^2}} \tag{1}$$

Energy of neutrino of zero rest mass is [by Eqn (15.115)] pc. Hence, from the principle of conservation of energy

$$\pi c^2 = pc + \frac{\mu c^2}{\sqrt{1-v^2/c^2}} = \frac{\mu v c}{\sqrt{1-v^2/c^2}} + \frac{\mu c^2}{\sqrt{1-v^2/c^2}} \quad \text{[by using Eqn (1)]}$$

$$= \mu c^2 \frac{1 + v/c}{\sqrt{1-v^2/c^2}} = \mu c^2 \sqrt{\frac{1+v/c}{1-v/c}} \tag{2}$$

Kinetic energy of muon is

$$T = E - E_0 = \mu c^2 \left[\frac{1}{\sqrt{1 - v^2/c^2}} - 1 \right] \tag{3}$$

Now, from Eqn (2), we have

$$\frac{\pi}{\mu} = \sqrt{\frac{1 + v/c}{1 - v/c.}} \quad \text{or,} \quad \frac{\pi^2}{\mu^2} = \frac{1 + v/c}{1 - v/c}$$

Hence,

$$\frac{v}{c} = \frac{\pi^2 - \mu^2}{\pi^2 + \mu^2}$$

Therefore,

$$1 - \frac{v^2}{c^2} = 1 - \left(\frac{\pi^2 - \mu^2}{\pi^2 + \mu^2} \right)^2 = \frac{4\pi^2\mu^2}{(\pi^2 + \mu^2)^2} \quad \text{or,} \quad \sqrt{1 - v^2/c^2} = \frac{2\pi\mu}{\pi^2 + \mu^2} \tag{4}$$

Using Eqn (4) in Eqn (3), we get

$$T = \mu c^2 \left[\frac{\pi^2 + \mu^2}{2\pi\mu} - 1 \right] = \frac{c^2(\pi - \mu)^2}{2\pi}.$$

EXAMPLE 15.10 *A particle of rest mass m_0 carrying a charge q moves from rest through a potential difference of V. Show that, the relativistic de Broglie wavelength of the particle is*

$$\lambda = h \left[qV \left(2m_0 + \frac{qV}{c^2} \right) \right]^{-1/2}$$

where h is the Planck's constant.

Solution. Total energy of the particle is given by

$$E = T + m_0c^2 = qV + m_0c^2 \tag{1}$$

If p be the momentum of the particle, then

$$E^2 = p^2c^2 + (m_0c^2)^2 \tag{2}$$

Combining Eqns (1) and (2),

$$(qV + m_0c^2)^2 = p^2c^2 + m_0^2c^4$$

$$\text{or,} \quad p^2c^2 = q^2V^2 + 2m_0c^2qV \quad \text{or,} \quad p^2 = qV\,(2m_0 + qV/c^2)$$

Hence,

$$p = \sqrt{qV \left(2m_0 + \frac{qV}{c^2} \right)}. \tag{3}$$

The wavelength λ of de Broglie is $\lambda = h/p$. Hence,

$$\lambda = \frac{h}{p} = h \left[qV \left(2m_0 + \frac{qV}{c^2} \right) \right]^{-1/2}.$$

15.11 Minkowski Space

Minkowski conceived a four-dimensional continuum in which the space coordinates x, y, z specify the position and ict $(i = \sqrt{-1})$ refers to time. The coordinate axes form an orthogonal system. The time coordinate axis is, of course, imaginary.

This four-dimensional space is called ***Minkowski or world space***. It is also called ***space-time continuum*** and sometimes briefly as ***four-space***.

A physical event in Minkowski space is described by a point with coordinates (x, y, z, ict) which is called ***world point***. The path of a particle in this space corresponds to a line called ***world line***. The relative position of one event with respect to another is represented by line element joining the two events.

The sum of the squares of coordinate components has the form:

$$s^2 = x^2 + y^2 + z^2 - c^2t^2 \tag{15.151}$$

and is invariant under Lorentz transformations. In other words, if x', y', z' and ict' are the four coordinates in frame S', while x, y, z, ict are those in S frame, then

$$s^2 = x^2 + y^2 + z^2 - c^2t^2 = x'^2 + y'^2 + z'^2 - c^2t'^2 \tag{15.152}$$

This invariance of s^2 suggests that Lorentz transformation can be simply regarded as a rotation of coordinate axes x, y, z and ict in space-time. We intend to prove this statement.

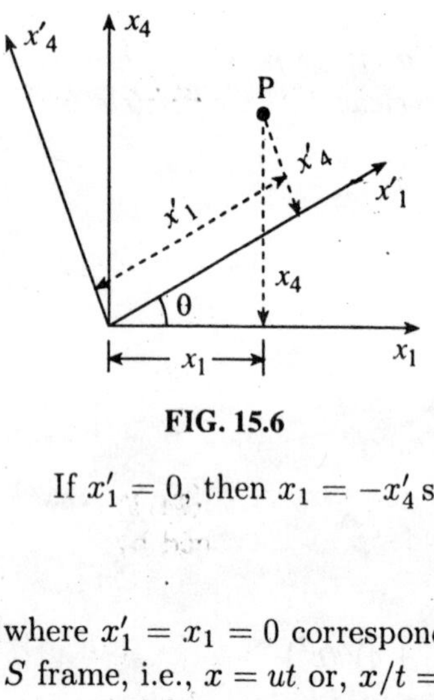

FIG. 15.6

Using the symbols

$$x_1 = x,\ x_2 = y, x_3 = z,\ x_4 = ict \tag{15.153}$$

we consider a rotation of x_1x_4-plane through an angle θ as shown in the FIG. 15.6. The coordinates (x_1, x_4) of a point P are related to the coordinates (x_1', x_4') by the relation

$$\begin{bmatrix} x_1' \\ x_4' \end{bmatrix} = \begin{bmatrix} \cos\theta & \sin\theta \\ -\sin\theta & \cos\theta \end{bmatrix} \begin{bmatrix} x_1 \\ x_4 \end{bmatrix} \tag{15.154}$$

The inverse relation is

$$\begin{bmatrix} x_1 \\ x_4 \end{bmatrix} = \begin{bmatrix} \cos\theta & -\sin\theta \\ \sin\theta & \cos\theta \end{bmatrix} \begin{bmatrix} x_1' \\ x_4' \end{bmatrix} \tag{15.155}$$

If $x_1' = 0$, then $x_1 = -x_4' \sin\theta$ and $x_4 = x_4' \cos\theta$, so that

$$\tan\theta = -\frac{x_1}{x_4} = -\frac{x}{ict} = \frac{iu}{c} \tag{15.156}$$

where $x_1' = x_1 = 0$ corresponds to the coordinate of the point O' in S' frame relative to O in S frame, i.e., $x = ut$ or, $x/t = u$.

From Eqn (15.156), we have

$$\left.\begin{aligned} \sin\theta &= \frac{iu/c}{\sqrt{1 - u^2/c^2}} = \frac{i\gamma u}{c} \\ \text{and}\quad \cos\theta &= \frac{1}{\sqrt{1 - u^2/c^2}} = \gamma \end{aligned}\right\} \tag{15.157}$$

Hence, Eqn (15.154) can be expressed as

$$\left.\begin{aligned} x_1' &= \gamma x_1 + i\gamma\frac{u}{c}x_4 = \gamma\left(x_1 + i\frac{u}{c}x_4\right) \\ x_4' &= -i\gamma\frac{u}{c}x_1 + \gamma x_4 = \gamma\left(-i\frac{u}{c}x_1 + x_4\right) \end{aligned}\right\} \tag{15.158}$$

Thus, the above transformation equations become:

$$\left.\begin{aligned} x_1' &= \gamma\left(x_1 + i\frac{u}{c}x_4\right) = \gamma(x_1 - ut) \\ x_2' &= x_2 \\ x_3' &= x_3 \\ x_4' &= \gamma\left(-i\frac{u}{c}x_1 + x_4\right) = ic\gamma\left(t - \frac{ux_1}{c^2}\right) \end{aligned}\right\} \tag{15.159}$$

These equations are precisely the Lorentz transformations. In matrix notation, this transformation can be expressed as

$$\begin{bmatrix} x_1' \\ x_2' \\ x_3' \\ x_4' \end{bmatrix} = \begin{bmatrix} \gamma & 0 & 0 & i\gamma u/c \\ 0 & 1 & 0 & 0 \\ 0 & 0 & 1 & 0 \\ -i\gamma u/c & 0 & 0 & \gamma \end{bmatrix} \begin{bmatrix} x_1 \\ x_2 \\ x_3 \\ x_4 \end{bmatrix} \tag{15.160}$$

The inverse Lorentz transformations are

$$\begin{bmatrix} x_1 \\ x_2 \\ x_3 \\ x_4 \end{bmatrix} = \begin{bmatrix} \gamma & 0 & 0 & -i\gamma u/c \\ 0 & 1 & 0 & 0 \\ 0 & 0 & 1 & 0 \\ i\gamma u/c & 0 & 0 & \gamma \end{bmatrix} \begin{bmatrix} x_1' \\ x_2' \\ x_3' \\ x_4' \end{bmatrix} \tag{15.161}$$

If $A_{\mu\nu}$ are the elements of the square matrix in Eqn (15.160), then the above transformations are:

$$\left.\begin{aligned} x_\mu' &= \sum_{\nu=1}^{4} A_{\mu\nu} x_\nu \\ \text{and } x_\mu &= \sum_{\nu=1}^{4} A_{\nu\mu} x_\nu' \end{aligned}\right\} \tag{15.162}$$

15.12 Space-Time Intervals

Consider an inertial frame S where two events are defined by the coordinates $P_1(x_1, y_1, z_1, t_1)$ and $P_2(x_2, y_2, z_2, t_2)$. The square of the interval between these two events is defined by

$$\begin{aligned} S_{12}^2 &= (x_2 - x_1)^2 + (y_2 - y_1)^2 + (z_2 - z_1)^2 - c^2(t_2 - t_1)^2 \\ &= r_{12}^2 - c^2(t_2 - t_1)^2 \end{aligned} \tag{15.163}$$

In Newtonian mechanics, the space interval r_{12} and time interval $(t_2 - t_1)$ between two events are separately invariant. However, in special theory of relativity, neither of the above two intervals are separately invariant. But the interval S_{12} defined by Eqn (15.163) is invariant under Lorentz transformation. This interval S_{12} may be considered as a measure of separation between two events in the Minkowski four-dimensional space, and is called ***space-time interval***.

The square of the space-time interval may be positive, zero and negative. When it is positive, the interval is called ***space-like***. From Eqn (15.163), for space-like intervals,

$$S_{12}^2 > 0, \text{ i.e., } \frac{r_{12}}{c} > (t_2 - t_1) \tag{15.164}$$

Eqn (15.164) shows that, the time-interval between two events is less than the time taken by light to cover the distance between them. Therefore, for space-like intervals, the two events cannot be connected by any real physical process.

If the square of the space-time interval is negative,

$$c^2(t^2 - t_1)^2 > r_{12}^2, \text{ i.e., } t_2 - t_1 > \frac{r_{12}}{c} \tag{15.165}$$

In this case, the interval is imaginary and is called ***time-like***.

Eqn (15.165) shows that the time-interval between two events is greater than the time taken by light to cover the distance between them. Therefore, for time-like intervals, the two events can be connected by a real physical process.

If the square of the space-time interval is zero, then

$$r_{12}^2 = c^2(t_2 - t_1)^2 \tag{15.166}$$

The interval in this case is called ***light-like*** or ***singular***.

Thus, the space-time interval S_{12} has been classified as space-like, light-like and time-like according as $S_{12}^2 >, =, < 0$. This clearly demarcates the Minkowski space into three distinct regions, called ***world regions***. Let us call these regions as A, B and C, such that $S_{12}^2 > 0$ in region A, $S_{12}^2 < 0$ in region B and $S_{12}^2 = 0$ on the surface C.

Considering the two events occurring at origin and at (x, y, z, ct)

$$S_{12}^2 = S^2 = x^2 + y^2 + z^2 - c^2t^2 \tag{15.167}$$

Hence, the surface C is defined by

$$S^2 = 0, \text{ i.e., } x^2 + y^2 + z^2 = c^2t^2 \tag{15.168}$$

In four-dimensions, Eqn (15.168) represents the surface of a cone, called ***null cone***, with apex at the origin and time axis *ct* with semi-vertical angle equal to $\pi/4$ (FIG. 15.7). In the upper cone, the time coordinates of all events are positive in all frames of reference and, therefore, we have events which occur after the event at the origin. That is, the upper cone represents ***absolute future***. The lower cone similarly represents ***absolute past***.

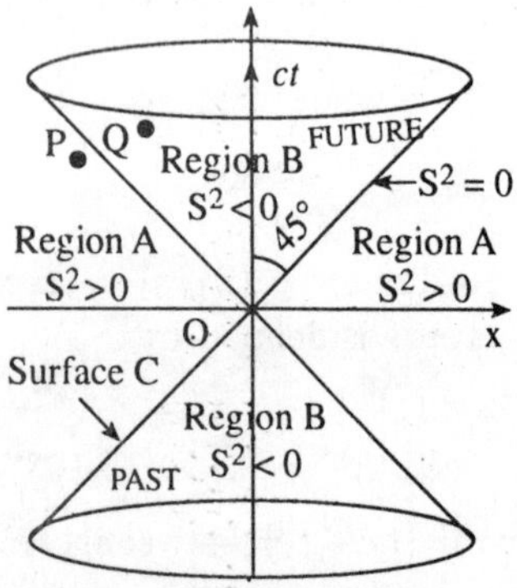

FIG. 15.7

The null cone is the space-time representation of the propagation of light. Hence, it is also called ***light cone***.

The region A corresponds to space-like interval. For any point P in this region, we can always find a frame of reference such that the events at O and P are simultaneous. The region B corresponds to time-like interval. For any point Q in this region, we can always find a frame of reference such that the events at O and Q occur at the same place.

15.13 Four-Vector in Mechanics

A vector in four-dimensional Minkowski space is called a ***four-vector***. The components of a four-vector transform from S frame to S' frame are similar to Lorentz transformation according to Eqns (15.159) and (15.160). The position four-vector of a particle is

$$x_\mu = (x, y, z, ict), \ \mu = 1, 2, 3, 4 \tag{15.169}$$

The differential change of this vector is

$$dx_\mu = (dx, dy, dz, ic\,dt) \tag{15.170}$$

Norm of this vector is

$$ds^2 = dx^2 + dy^2 + dz^2 - c^2dt^2 \tag{15.171}$$

which is invariant.

We define the quantity

$$d\tau^2 = -\frac{1}{c^2}ds^2 = \left\{dt^2 - \frac{1}{c^2}(dx^2 + dy^2 + dz^2)\right\} \tag{15.172}$$

and call it ***proper time differential***. This is also invariant. From Eqn (15.172), we get

$$d\tau = dt\sqrt{1 - V^2/c^2} \tag{15.173}$$

where

$$V^2 = \left(\frac{dx}{dt}\right)^2 + \left(\frac{dy}{dt}\right)^2 + \left(\frac{dz}{dt}\right)^2$$

Since neither time (t) nor the norm of the space coordinates is invariant under Lorentz transformation, we have to discard t in defining the velocity vectors. We instead use the proper time τ which is an invariant parameter. Thus, the velocity four-vector is

$$U_\mu = \frac{d}{d\tau}(x_\mu), \ \mu = 1, 2, 3, 4 \tag{15.174}$$

The components of this vector are

$$U_\mu = \frac{dx_\mu}{d\tau} = \frac{dx_\mu}{dt}\frac{dt}{d\tau} = \frac{dx_\mu}{dt} \cdot \frac{1}{\sqrt{1 - V^2/c^2}} \tag{15.175}$$

Defining, as in Eqn (15.117),

$$\Gamma = \left(1 - \frac{V^2}{c^2}\right)^{-1/2}$$

we write the components of this vector as

$$\left.\begin{aligned} U_1 &= \Gamma V_x \\ U_2 &= \Gamma V_y \\ U_3 &= \Gamma V_z \\ U_4 &= ic\Gamma \end{aligned}\right\} \tag{15.176}$$

Norm of this four-vector is

$$U_\mu U_\mu = \Gamma^2 V^2 - \Gamma^2 c^2 = -c^2 \tag{15.177}$$

• Momentum Four-vector

We define the four-momentum vector as

$$P_\mu = m_0 U_\mu \ (m_0 = \text{rest mass}) \tag{15.178}$$

Hence, its components are:

$$\left.\begin{aligned} P_1 &= m_0 U_1 = m_0 \Gamma V_x = p_x \\ P_2 &= m_0 \Gamma V_y = p_y \\ P_3 &= m_0 \Gamma V_z = p_z \\ P_4 &= imc = iE/c \end{aligned}\right\} \tag{15.179}$$

where p_x, p_y, p_z are components of relativistic momentum.

Hence, the space-part of four-momentum is the relativistic momentum and its time part is the relativistic energy.

We find that the fourth component of the four-momentum vectors is proportional to the total energy E of the particle.

The norm of this vector is

$$P_\mu P_\mu = m^2V^2 - E^2/c^2 = -\left(E^2 - p^2c^2\right)/c^2 = -m_0^2c^2 \tag{15.180}$$

• Acceleration Four-vector

We define the acceleration four-vector as

$$a_\mu = \frac{d}{d\tau}U_\mu = \frac{d}{dt}(U_\mu)\frac{dt}{d\tau} = \frac{d}{dt}(U_\mu)\frac{1}{\sqrt{1-V^2/c^2}} \tag{15.181}$$

Using expressions for velocity components from Eqn (15.176), we write:

$$a_1 = \Gamma\frac{d}{dt}(U_1) = \Gamma\frac{d}{dt}(\Gamma V_x) = \Gamma\left(\Gamma\dot{V}_x + V_x\frac{d\Gamma}{dt}\right) \tag{15.182}$$

$$a_2 = \Gamma\left(\Gamma\dot{V}_y + V_y\frac{d\Gamma}{dt}\right) \tag{15.183}$$

$$a_3 = \Gamma\left(\Gamma\dot{V}_z + V_z\frac{d\Gamma}{dt}\right) \tag{15.184}$$

$$a_4 = \Gamma\frac{d}{dt}(U_4) = \Gamma\frac{d}{dt}(ic\Gamma) = ic\Gamma\frac{d\Gamma}{dt} \tag{15.185}$$

The four-force vectors may be expressed as

$$F_\mu = \frac{d}{d\tau}(m_0U_\mu) = m_0\frac{d}{d\tau}(U_\mu) = m_0a_\mu \tag{15.186}$$

• Minkowski Force

The Minkowski force equation presented in a form similar to Newton's equation is

$$F_\mu = \frac{dP_\mu}{d\tau} = \frac{d}{d\tau}(m_0U_\mu) = m_0\frac{d^2x_\mu}{d\tau^2} \tag{15.187}$$

The four-force vector is

$$F_\mu = \frac{dP_\mu}{d\tau} = \frac{dP_\mu}{dt}\frac{dt}{d\tau} = \frac{dP_\mu}{dt}\cdot\frac{1}{\sqrt{1-V^2/c^2}} \tag{15.188}$$

Using Eqns (15.179), the components of force vector are:

$$F_1 = \Gamma\frac{d}{dt}(p_x) = \Gamma F_x \tag{15.189}$$

$$F_2 = \Gamma\frac{d}{dt}(p_y) = \Gamma F_y \tag{15.190}$$

$$F_3 = \Gamma\frac{d}{dt}(p_z) = \Gamma F_z \tag{15.191}$$

$$F_4 = \Gamma\frac{d}{dt}\left(\frac{iE}{c}\right) = \frac{i\Gamma}{c}\frac{dE}{dt} \tag{15.192}$$

where F_x, F_y, F_z are relativistic components of forces.

EXAMPLE 15.11 *Show that the velocity and force four-vectors are orthogonal.*

Solution. We have

$$U_\mu F_\mu = U_\mu \frac{d}{d\tau}(m_0 U_\mu) = m_0 U_\mu \frac{dU_\mu}{d\tau} \tag{1}$$

Eqn (1) can be written as

$$\begin{aligned} U_\mu F_\mu &= \frac{1}{2} m_0 \frac{d}{d\tau}(U_\mu U_\mu) = \frac{1}{2} m_0 \frac{d}{d\tau}(-c^2) \text{ [by Eqn (15.177)]} \\ &= 0 \end{aligned}$$

Hence, the four-vectors U_μ and F_μ are orthogonal.

EXAMPLE 15.12 *A particle of rest mass m_0 moves with velocity v in an inertial frame S. There is another inertial frame S' which is moving with uniform speed u along x-axis. Determine the transformation rule for the momentum four-vectors.*

Solution. The transformation rule is given by Eqn (15.160). Hence, we write:

$$\begin{bmatrix} P_1' \\ P_2' \\ P_3' \\ P_4' \end{bmatrix} = \begin{bmatrix} \gamma & 0 & 0 & i\gamma u/c \\ 0 & 1 & 0 & 0 \\ 0 & 0 & 1 & 0 \\ -i\gamma u/c & 0 & 0 & \gamma \end{bmatrix} \begin{bmatrix} P_1 \\ P_2 \\ P_3 \\ P_4 \end{bmatrix} \tag{1}$$

where

$$\gamma = \left(1 - u^2/c^2\right)^{-1/2} \tag{2}$$

This yields

$$P_1' = \gamma\left(P_1 + i\frac{u}{c}P_4\right) = \gamma\left(mv_x + i\frac{u}{c}\cdot i\frac{E}{c}\right)$$

$$\text{or,}\quad P_1' = \gamma\left(mv_x - uE/c^2\right) \tag{3}$$

where

$$m = m_0\left(1 - v^2/c^2\right)^{-1/2} \tag{4}$$

and $v_x = x$-component of velocity of the particle in S-frame.

Similarly,

$$P_2' = P_2 = mv_y, \tag{5}$$

$$P_3' = P_3 = mv_z \tag{6}$$

$$\text{and } P_4' = \gamma\left(P_4 - i\frac{u}{c}P_1\right) = \gamma\left(i\frac{E}{c} - i\frac{u}{c}\cdot mv_x\right)$$

$$\text{or,}\quad P_4' = \frac{i\gamma}{c}(E - muv_x) \tag{7}$$

Since $P_4' = i(E'/c)$, Eqn (7) can be written as

$$E' = \gamma(E - muv_x) \tag{8}$$

The inverse transformations are obtained by replacing u by $-u$. Thus,

$$\left.\begin{aligned} P_1 &= \gamma\left(mv_x' + \frac{uE'}{c^2}\right) \\ P_2 &= P_2' \\ P_3 &= P_3' \\ E &= \gamma(E' + muv_x') \end{aligned}\right\} \tag{9}$$

15.14 Conservation of Four-Momentum

Let P^1_μ and P^2_μ be the four-momentum vectors of two particles. If they collide, according to the conservation of four-momentum,

$$P^1_\mu + P^2_\mu = \text{constant} \tag{15.193}$$

before and after the collision.

This means that the sum of relativistic momenta and the sum of total energy of the two particles will remain conserved separately in their collision.

EXAMPLE 15.13 *Show that if l_0^3 is the rest volume of a cube, then*

$$l_0^3\left(1-\beta^2\right)^{1/2}$$

is the volume as viewed from a reference frame moving with a uniform velocity v parallel to an edge of the cube.

Solution. The length of the cube at rest $= l_0$.

Hence, the volume at rest

$$V_0 = l_0^3 \tag{1}$$

Along the direction of motion parallel to an edge, the length l_0 is viewed as

$$= l_0\left(1-\beta^2\right)^{1/2} \tag{2}$$

In the other two directions, there is no Lorentz contraction. Hence, the volume of the moving cube

$$= l_0 \times l_0 \times l_0\left(1-\beta^2\right)^{1/2} = l_0^3\left(1-\beta^2\right)^{1/2} \tag{3}$$

where $\beta = v/c$.

EXAMPLE 15.14 *Two rods of length l_0 move lengthwise towards each other parallel to the common axis with the same velocity v relative to the laboratory frame. Show that the length of each rod in the reference frame fixed to the other end is*

$$l = l_0\frac{1-\beta^2}{1+\beta^2},\ \beta = \frac{v}{c}$$ **[C.U. (HONS.)]**

Solution. Let v_r = velocity of S' frame relative to the laboratory S-frame. Then the length l_0 of the rod contracts to

$$l = l_0\left(1-\frac{v_r^2}{c^2}\right)^{1/2} \tag{1}$$

But

$$v_r = \frac{-v-v}{1-(-v)v/c^2} = \frac{-2v}{1+v^2/c^2} = \frac{-2v}{1+\beta^2} \tag{2}$$

where $\beta = v/c$.

Inserting (2) in (1),

$$l = l_0\left[1-\left(\frac{-2v}{1+\beta^2}\right)^2\frac{1}{c^2}\right]^{1/2} = l_0\left[1-\frac{4\beta^2}{(1+\beta^2)^2}\right]^{1/2} = l_0\frac{1-\beta^2}{1+\beta^2} \quad \textbf{(Proved)}$$

EXAMPLE 15.15 *If* $\gamma = \dfrac{1}{\sqrt{1 - v^2/c^2}} = \dfrac{1}{\epsilon}$, $\epsilon \ll 1$, *show that* (1) $\beta = 1 - \dfrac{\epsilon^2}{2}$, $\beta = \dfrac{v}{c}$ *and* (2) $\dfrac{T}{E} = 1 - \epsilon$, *where* $\beta = \dfrac{v}{c}$, T *and* E *are kinetic energy and total energy of a particle respectively.*

Solution. (1) We are given,

$$\frac{1}{\epsilon} = \frac{1}{\sqrt{1 - v^2/c^2}} = \frac{1}{\sqrt{1 - \beta^2}}$$

Hence,

$$\epsilon^2 = 1 - \beta^2, \text{ i.e., } \beta^2 = 1 - \epsilon^2 \tag{1}$$

$$\therefore \beta = \left(1 - \epsilon^2\right)^{1/2} = 1 - \frac{\epsilon^2}{2}, \text{ since } \epsilon \ll 1.$$

(2) If m_0 and m be the rest mass and moving mass of the particle, then the kinetic energy

$$T = (m - m_0)c^2 \tag{2}$$

and total energy

$$E = mc^2 \tag{3}$$

Hence,

$$\frac{T}{E} = \frac{m - m_0}{m} = 1 - \frac{m_0}{m} \tag{4}$$

But

$$m = \frac{m_0}{\sqrt{1 - v^2/c^2}} \tag{5}$$

From (4) and (5), $\dfrac{T}{E} = 1 - \sqrt{1 - \dfrac{v^2}{c^2}} = 1 - \epsilon$ **(Proved)**

EXAMPLE 15.16 *A photon rocket uses pure radiation as its propellant. If* m_i *and* m_f *be the initial and final rest masses of the rocket, show that the speed* v *of the rocket relative to its initial rest frame is given by*

$$\frac{m_i}{m_f} = \left(\frac{c + v}{c - v}\right)^{1/2} = \left(\frac{1 + \beta}{1 - \beta}\right)^{1/2}$$ [C.U. (HONS.)]

Solution. We are given, m_i = initial rest mass and m_f = final rest mass.

Let E = energy of radiation. From the conservation of energy, we have

$$m_i c^2 = m_f \gamma c^2 + E \tag{1}$$

and from the conservation of momentum, we have

$$0 = m_f \gamma v - \frac{E}{c} \tag{2}$$

where

$$\gamma = \frac{1}{\sqrt{1 - v^2/c^2}} = \frac{1}{\sqrt{1 - \beta^2}} \tag{3}$$

From (2),

$$E = m_f \gamma v c \tag{4}$$

From (1) and (4),

$$m_i c^2 = m_f \gamma c^2 + m_f \gamma v c = m_f \gamma c^2 \left(1 + \frac{v}{c}\right) \tag{5}$$

$$\therefore m_i = m_f \gamma (1+\beta)$$

$$\text{i.e., } \frac{m_i}{m_f} = \gamma(1+\beta) = \frac{1+\beta}{\sqrt{1-\beta^2}} \text{ [using (3)]}$$

$$= \left(\frac{1+\beta}{1-\beta}\right)^{1/2} = \left(\frac{c+v}{c-v}\right)^{1/2} \quad \textbf{(Proved)}$$

EXAMPLE 15.17 *A body of mass m disintegrates into two parts of masses m_1 and m_2 while at rest. Show that the energies E_1 and E_2 of the two pieces satisfy the relation*

$$\frac{E_1}{E_2} = \frac{m^2 + m_1^2 - m_2^2}{m^2 + m_2^2 - m_1^2}$$

Solution. The total energy of the body

$$E = E_1 + E_2 = mc^2 \tag{1}$$

Due to the disintegration, the masses m_1 and m_2 move in the opposite directions having equal momentum p.

Now, we have by Eqn (15.114),

$$E_1^2 = p^2c^2 + m_1^2c^4 \tag{2}$$

$$\text{and } E_2^2 = p^2c^2 + m_2^2c^4 \tag{3}$$

Hence,

$$E_1^2 - E_2^2 = c^4(m_1^2 - m_2^2)$$

$$\text{i.e., } (E_1+E_2)(E_1-E_2) = c^4\left(m_1^2 - m_2^2\right)$$

$$\text{i.e., } E(E_1-E_2) = c^4\left(m_1^2 - m_2^2\right) \tag{4}$$

From (1) and (3),

$$E_1 - E_2 = \frac{c^2}{m}\left(m_1^2 - m_2^2\right) \tag{5}$$

Adding (1) and (4),

$$2E_1 = mc^2 + \frac{c^2}{m}\left(m_1^2 - m_2^2\right) = \frac{c^2}{m}\left(m^2 + m_1^2 - m_2^2\right) \tag{6}$$

Similarly, subtracting (4) from (1),

$$2E_2 = mc^2 - \frac{c^2}{m}\left(m_1^2 - m_2^2\right) = \frac{c^2}{m}\left(m^2 + m_2^2 - m_1^2\right) \tag{7}$$

From (5) and (6),

$$\frac{E_1}{E_2} = \frac{m^2 + m_1^2 - m_2^2}{m^2 + m_2^2 - m_1^2} \quad \textbf{(Proved)}$$

EXAMPLE 15.18 *An excited atom of mass m_0 is at rest in a certain frame of reference. It emits a photon and thereby loses an internal energy ΔE. Taking care of the recoil of the atom, show that the exact frequency of the emitted photon is*

$$\nu = \frac{\Delta E}{h}\left(1 - \frac{\Delta E}{2m_0c^2}\right)$$

Solution. Since the atom is at rest, emission of the photon will cause a recoil of the atom.

Let V = recoil velocity. Then, from momentum conservation,

$$0 = m_0 V + \left(\frac{\Delta E}{c^2}\right) c, \text{ i.e., } V = -\frac{\Delta E}{m_0 c} \tag{1}$$

From energy balance, we have

$$\Delta E = h\nu + \frac{1}{2} m_0 V^2 = h\nu + \frac{1}{2} m_0 \left(-\frac{\Delta E}{m_0 c}\right)^2 \quad \text{[using (1)]}$$

$$= h\nu + \frac{(\Delta E)^2}{2m_0 c^2} \tag{2}$$

Here h = Planck's constant.

From (2), we get

$$h\nu = \Delta E - \frac{(\Delta E)^2}{2m_0 c^2} = \Delta E \left(1 - \frac{\Delta E}{2m_0 c^2}\right)$$

Hence,

$$\nu = \frac{\Delta E}{h}\left(1 - \frac{\Delta E}{2m_0 c^2}\right) \quad \textbf{(Proved)}$$

EXAMPLE 15.19 *A nucleus at rest with rest mass m_0 decays spontaneously into two components of rest masses m_1 and m_2 with speeds v_1 and v_2 respectively. Show that*

$$m_0 = \sqrt{m_1^2 + p^2/c^2} + \sqrt{m_2^2 + p^2/c^2}$$

where p is the momentum of each particle.

Solution. Let E_1 = energy of the particle of mass m_1
E_2 = energy of the particle of mass m_2
and E = total energy.

Then,

$$E = E_1 + E_2 \tag{1}$$

Now

$$E = m_0 c^2 \tag{2}$$

and

$$E_1^2 = p^2c^2 + m_1^2 c^4, \text{ i.e., } E_1 = c^2\sqrt{m_1^2 + p^2/c^2} \tag{3}$$

Similarly,

$$E_2 = c^2\sqrt{m_2^2 + p^2/c^2} \tag{4}$$

Hence,

$$m_0 c^2 = c^2\sqrt{m_1 + p^2/c^2} + c^2\sqrt{m_2 + p^2/c^2}$$

$$\therefore m_0 = \sqrt{m_1 + p^2/c^2} + \sqrt{m_2 + p^2/c^2} \quad \textbf{(Proved)} \tag{5}$$

It is obvious from (5) that

$$m_0 > m_1 + m_2. \tag{6}$$

EXERCISE

1. A particle is moving with velocity V in an inertial S frame. There is another inertial S' frame which is moving with uniform velocity u along x-axis relative to S frame.
   Derive the Lorentz transformation for the components of acceleration.
2. The relativistic mass of a particle moving with a velocity u exceeds its rest mass by a fraction η. Show that
$$u = \frac{[(2+\eta)\eta]^{1/2}}{1+\eta}$$
3. A particle moves with velocity V at an angle ϕ with x-axis in the S frame. Determine its velocity in S' frame.
4. The direction cosines of a particle's velocity w.r.t. the axes are l, m, n in S frame and l', m', n' in S' frame. Derive the relations between them.
5. A rocket moves with uniform velocity V relative to the earth to reach a star L light years away in one year, as measured by clocks at rest on the rocket. Show that
$$V = \left(\frac{L}{L+1}\right)c$$
6. An excited nucleus of rest mass m_0 is at rest in an inertial frame. It goes over to a lower state losing energy ΔE. It emits a γ-ray photon and undergoes a recoil.
   Assuming that the mass of the nucleus is not changed due to emission of photon, show that emitted energy of the photon is
$$\Delta E\left(1 - \frac{\Delta E}{2m_0c^2}\right)$$
7. A particle moves with velocity V in an inertial S frame. There is another inertial frame S' moving with uniform velocity u relative to S. Find the transformation rule for the four velocity-vectors.
8. A particle of rest mass m_0 describes the trajectory
$$x = \alpha t,\ y = \frac{1}{2}\beta t^2,\ z = 0$$
   in an inertial frame S. Find the components of force four-vector.
9. In an inertial frame S, a particle moves with velocity V which is very near to speed of light c. Show that
$$\frac{c-V}{c} = \frac{1}{2}\left(\frac{m_0c^2}{E}\right)^2$$
   where E is the relativistic energy.
10. A nucleus at rest with rest mass m_0 decays spontaneously into two components of rest masses m_1 and m_2. Show that, $m_0 > m_1 + m_2$. Also show that
$$m_0 = \sqrt{m_1^2 + \frac{q^2}{c}} + \sqrt{m_2^2 + \frac{q^2}{c}}$$
   where q is the momentum of each particles.

APPENDIX A

Vector Analysis

A.1 Vectors and Scalars

Physical quantities are of two types. A quantity which can be specified completely by its magnitude with corresponding unit is called a ***scalar***. For examples, mass, volume, density, energy and temperature are scalar quantities.

A physical quantity which can be completely and meaningfully specified by its magnitude and direction is called a ***vector***. For examples, velocity, acceleration, force, displacement and torque are vector quantities. A vector quantity is graphically represented by a straight line with an arrow at the tip; the direction of the quantity is indicated by the head of the arrow and the magnitude of the quantity is given by the length of the straight line. Unit vector has a length of unity.

A.2 Cartesian System

In Cartesian coordinate system, we take unit vectors $\mathbf{i}, \mathbf{j}, \mathbf{k}$ along the coordinate axes. Consider a point $P(x, y, z)$ (FIG. A-1). The ***position*** or ***radius vector*** $\mathbf{r}$ of the point is the vector drawn from the origin of coordinate system to the point P. Thus,

$$\mathbf{r} = \mathbf{i}x + \mathbf{j}y + \mathbf{k}z. \tag{A.1}$$

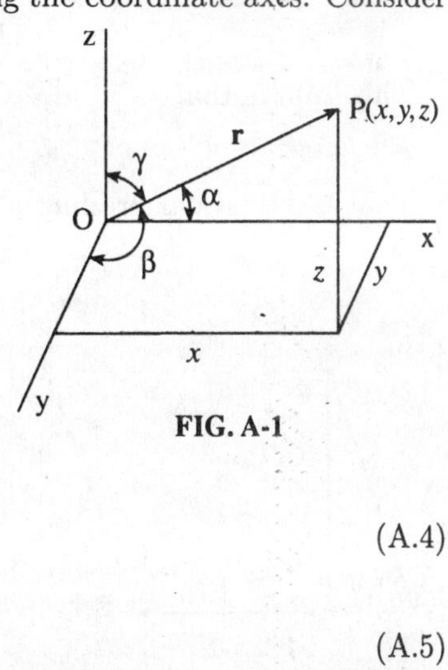

FIG. A-1

The magnitude of the vector $\mathbf{r}$ is given by

$$r^2 = |\mathbf{r}|^2 = x^2 + y^2 + z^2 \tag{A.2}$$

If α, β, γ are the angles that the vector $\mathbf{r}$ makes with coordinate axes, then the direction cosines are defined as

$$l = \cos\alpha,\ m = \cos\beta,\ n = \cos\gamma \tag{A.3}$$

The coordinates of the point P are related as

$$x = rl,\ y = rm,\ z = rn \tag{A.4}$$

Eqns (A.2) and (A.4) yield

$$l^2 + m^2 + n^2 = 1 \tag{A.5}$$

Let us consider an arbitrary vector $\mathbf{F}$ having projections F_x, F_y, F_z along the coordinate axes. Then

$$\mathbf{F} = \mathbf{i}F_x + \mathbf{j}F_y + \mathbf{k}F_z \tag{A.6}$$

Its magnitude is given by

$$|\mathbf{F}| = \sqrt{F_x^2 + F_y^2 + F_z^2} \tag{A.7}$$

The unit vector along **F** is

$$\frac{\mathbf{F}}{|\mathbf{F}|} = \mathbf{i}\cos\alpha + \mathbf{j}\cos\beta + \mathbf{k}\cos\gamma \tag{A.8}$$

A.3 Algebraic Operations with Vectors

We consider two vectors **E** and **F** given by

$$\left.\begin{aligned} \mathbf{E} &= \mathbf{i}E_x + \mathbf{j}E_y + \mathbf{k}E_z \\ \mathbf{F} &= \mathbf{i}F_x + \mathbf{j}F_y + \mathbf{k}F_z \end{aligned}\right\} \tag{A.9}$$

(a) Addition : Addition (or subtraction) of two vectors **E** and **F** is obtained simply by adding (or subtracting) similar components. Thus,

$$\begin{aligned} \mathbf{E} \pm \mathbf{F} &= (\mathbf{i}E_x + \mathbf{j}E_y + \mathbf{k}E_z) \pm (\mathbf{i}F_x + \mathbf{j}F_y + \mathbf{k}F_z) \\ &= \mathbf{i}(E_x \pm F_x) + \mathbf{j}(E_y \pm F_y) + \mathbf{k}(E_z \pm F_z) \end{aligned} \tag{A.10}$$

(b) Scalar (or Dot) Product : The ***scalar product*** of two vectors **E** and **F** is a scalar and is defined as

$$\mathbf{E}\cdot\mathbf{F} = |\mathbf{E}|\,|\mathbf{F}|\cos\theta \tag{A.11}$$

where θ is the angle between the directions of the vectors. Then

$$\left.\begin{aligned} \mathbf{i}\cdot\mathbf{i} = \mathbf{j}\cdot\mathbf{j} = \mathbf{k}\cdot\mathbf{k} = 1 \\ \mathbf{i}\cdot\mathbf{j} = \mathbf{j}\cdot\mathbf{k} = \mathbf{k}\cdot\mathbf{i} = 0 \end{aligned}\right\} \tag{A.12}$$

With these relations, Eqn (A.11) can be written as

$$\mathbf{E}\cdot\mathbf{F} = E_xF_x + E_yF_y + E_zF_z \tag{A.13}$$

This implies that

$$\mathbf{E}\cdot\mathbf{F} = \mathbf{F}\cdot\mathbf{E} \tag{A.14}$$

That is, the scalar product is commutative.

(c) Vector (or Cross) Product : The ***vector or cross product*** of the two vectors is a vector and is defined by

$$\mathbf{E}\times\mathbf{F} = |\mathbf{E}|\,|\mathbf{F}|\sin\theta\mathbf{n} = \mathbf{n}EF\sin\theta \tag{A.15}$$

where **n** is a unit vector along the normal to the plane containing the vectors **E** and **F**. The positive direction of **n** is determined by the right-hand screw rule when it is turned from **E** to **F** (FIG. A-2).

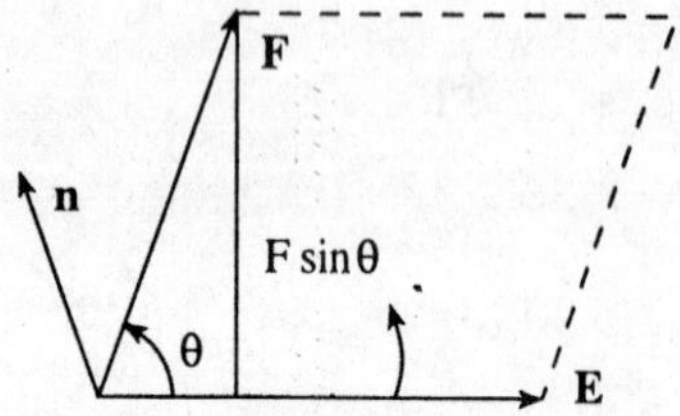

FIG. A-2

Hence,

$$\mathbf{E}\times\mathbf{F} = -\mathbf{F}\times\mathbf{E} \tag{A.16}$$

Moreover, by Eqn (A.15), we have

$$\left.\begin{aligned} \mathbf{i}\times\mathbf{i} = \mathbf{j}\times\mathbf{j} = \mathbf{k}\times\mathbf{k} = 0 \\ \mathbf{i}\times\mathbf{j} = \mathbf{k},\ \mathbf{j}\times\mathbf{k} = \mathbf{i},\ \mathbf{k}\times\mathbf{i} = \mathbf{j} \end{aligned}\right\} \tag{A.17}$$

With the use of Eqn (A.9), Eqn (A.15) can be written as

$$\begin{aligned}\mathbf{E}\times\mathbf{F} &= (\mathbf{i}E_x+\mathbf{j}E_y+\mathbf{k}E_z)\times(\mathbf{i}F_x+\mathbf{j}F_y+\mathbf{k}F_z)\\ &= \mathbf{i}\,(E_yF_z-E_zF_y)+\mathbf{j}\,(E_zF_x-E_xF_z)+\mathbf{k}\,(E_xF_y-E_yF_x)\\ &= \begin{vmatrix}\mathbf{i} & \mathbf{j} & \mathbf{k}\\ E_x & E_y & E_z\\ F_x & F_y & F_z\end{vmatrix}\end{aligned} \tag{A.18}$$

NOTE. The area of the parallelogram in (FIG. A-2) is $EF\sin\theta$. Hence, Eqn (A.15) implies that we can represent the area of a parallelogram by a vector. In this situation, the area of the parallelogram is shown by means of a vector drawn normally to the plane of the parallelogram. The direction of the vector is determined by the right-hand screw rule.

In fact, any plane area can be represented by a vector whose length is proportional to the magnitude of the area and whose direction is normal to the plane of the area. The tip of the arrow-head is determined by rotating the right-hand screw in the direction in which the boundary curve of the plane area is traversed.

A.4 Differentiation of Vector Functions

Consider a vector **F** which depends on a scalar variable t such that for each value of t, there corresponds a value of **F**. Then **F** is a function of t and we write:

$$\mathbf{F}=\mathbf{F}(t) \tag{A.19}$$

Let the scalar variable t change to $t+\Delta t$. Then (FIG. A-3)

$$\mathbf{F}+\Delta\mathbf{F}=\mathbf{F}(t+\Delta t) \tag{A.20}$$

Hence,

$$\frac{\Delta\mathbf{F}}{\Delta t}=\frac{\mathbf{F}(t+\Delta t)-\mathbf{F}(t)}{\Delta t} \tag{A.21}$$

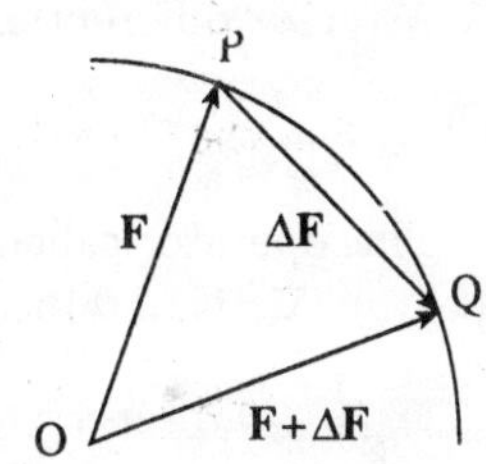

FIG. A-3

The derivative of **F** with respect to t is defined as

$$\lim_{\Delta t\to 0}\frac{\Delta\mathbf{F}}{\Delta t} \tag{A.22}$$

if the limit exists, and is written as $d\mathbf{F}/dt$. Thus,

$$\frac{d\mathbf{F}}{dt}=\lim_{\Delta t\to 0}\frac{\mathbf{F}(t+\Delta t)-\mathbf{F}(t)}{\Delta t} \tag{A.23}$$

In terms of components of **F**, we get

$$\frac{d\mathbf{F}}{dt}=\mathbf{i}\frac{dF_x}{dt}+\mathbf{j}\frac{dF_y}{dt}+\mathbf{k}\frac{dF_z}{dt} \tag{A.24}$$

Thus, derivative of a vector is a vector along the direction of $\Delta\mathbf{F}$.

We can easily prove the following relations:

$$\left.\begin{aligned}&\text{(i)}\quad \frac{d}{dt}(\mathbf{E}+\mathbf{F})=\frac{d\mathbf{E}}{dt}+\frac{d\mathbf{F}}{dt}\\ &\text{(ii)}\quad \frac{d}{dt}(\mathbf{E}\cdot\mathbf{F})=\mathbf{E}\cdot\frac{d\mathbf{F}}{dt}+\frac{d\mathbf{E}}{dt}\cdot\mathbf{F}\\ &\text{(iii)}\quad \frac{d}{dt}(\mathbf{E}\times\mathbf{F})=\mathbf{E}\times\frac{d\mathbf{F}}{dt}+\frac{d\mathbf{E}}{dt}\times\mathbf{F}\\ &\text{(iv)}\quad \frac{d}{dt}(\lambda\mathbf{F})=\lambda\frac{d\mathbf{F}}{dt},\ \lambda=\text{a constant scalar}\end{aligned}\right\} \tag{A.25}$$

In a special case, if $\mathbf{F}$ is a vector of constant magnitude, we have

$$\frac{d\mathbf{F}}{dt} = 0, \text{ i.e., } \mathbf{F} \cdot \frac{d\mathbf{F}}{dt} = 0 \tag{A.26}$$

This shows that the vectors $\mathbf{F}$ and $\frac{d\mathbf{F}}{dt}$ are perpendicular to each other.

A.5 Partial Differentiation

Consider a vector $\mathbf{F}$ which is a function of the rectangular coordinates x, y, z of a point in space. That is

$$\mathbf{F} = \mathbf{F}(x, y, z) \tag{A.27}$$

Keeping the variables y and z constant, we increase x to $x + \Delta x$. Then, by definition,

$$\frac{\partial \mathbf{F}}{\partial x} = \lim_{\Delta x \to 0} \frac{\mathbf{F}(x + \Delta x, y, z) - \mathbf{F}(x, y, z)}{\Delta x} \tag{A.28}$$

when the limit exists.

Similarly, we define the partial derivatives $\frac{\partial \mathbf{F}}{\partial y}$ and $\frac{\partial \mathbf{F}}{\partial z}$.

The partial derivatives represent the rate of increase of the vector with respect to one variable when the other two remain constant.

A vector differential operator $\boldsymbol{\nabla}$ is defined by

$$\boldsymbol{\nabla} = \mathbf{i}\frac{\partial}{\partial x} + \mathbf{j}\frac{\partial}{\partial y} + \mathbf{k}\frac{\partial}{\partial z} \tag{A.29}$$

This operator called ***del*** (***nabla***) ***operator*** is of immense importance in physics. The scalar operator ∇^2 is obtained from Eqn (A.29) as

$$\nabla^2 = \boldsymbol{\nabla} \cdot \boldsymbol{\nabla} = \frac{\partial^2}{\partial x^2} + \frac{\partial^2}{\partial y^2} + \frac{\partial^2}{\partial z^2} \tag{A.30}$$

This is called the ***Laplacian operator***.

The ***divergence*** of a vector $\mathbf{F}$ is defined as

$$\operatorname{div} \mathbf{F} = \boldsymbol{\nabla} \cdot \mathbf{F} \tag{A.31}$$

In terms of the components of $\mathbf{F}$,

$$\begin{aligned} \operatorname{div} \mathbf{F} &= \left(\mathbf{i}\frac{\partial}{\partial x} + \mathbf{j}\frac{\partial}{\partial y} + \mathbf{k}\frac{\partial}{\partial z}\right) \cdot (\mathbf{i}F_x + \mathbf{j}F_y + \mathbf{k}F_z) \\ &= \frac{\partial F_x}{\partial x} + \frac{\partial F_y}{\partial y} + \frac{\partial F_z}{\partial z} \end{aligned} \tag{A.32}$$

Thus, divergence of a vector is a scalar quantity.

The ***curl*** of a vector $\mathbf{F}$ is defined as

$$\operatorname{curl} \mathbf{F} = \boldsymbol{\nabla} \times \mathbf{F} \tag{A.33}$$

In terms of the components of $\mathbf{F}$,

$$\operatorname{curl} \mathbf{F} = \begin{vmatrix} \mathbf{i} & \mathbf{j} & \mathbf{k} \\ \frac{\partial}{\partial x} & \frac{\partial}{\partial y} & \frac{\partial}{\partial z} \\ F_x & F_y & F_z \end{vmatrix} \tag{A.34}$$

If curl $\mathbf{F} = 0$, the vector $\mathbf{F}$ is called an ***irrotational vector***. The gravitational and electrostatic fields are irrotational.

If the vector $\mathbf{F}$ represents a conservative force field, i.e., when the work done by the force on an object to move it from position A to position B is independent of the path, then $\mathbf{F} = \nabla\phi$, where ϕ is a scalar function. In this case also curl $\mathbf{F} = 0$.

A.6 Physical Meaning of Divergence

Consider an elementary parallelepiped of sides dx, dy and dz (FIG. A-4). Then the elementary volume is

$$d\tau = dx\,dy\,dz \tag{A.35}$$

Let $\mathbf{V}$ be a vector point function at the origin of coordinates O. We take this vector to be associated with fluid flow where

$$\mathbf{V} = \rho\mathbf{v} \tag{A.36}$$

Here ρ = density of the fluid and $\mathbf{v}$ = velocity vector of the fluid. Then Eqn (A.36) represents the rate of mass flow of the fluid through unit area normal to the velocity vector $\mathbf{v}$. This is called the ***flux of vector*** $\mathbf{V}$.

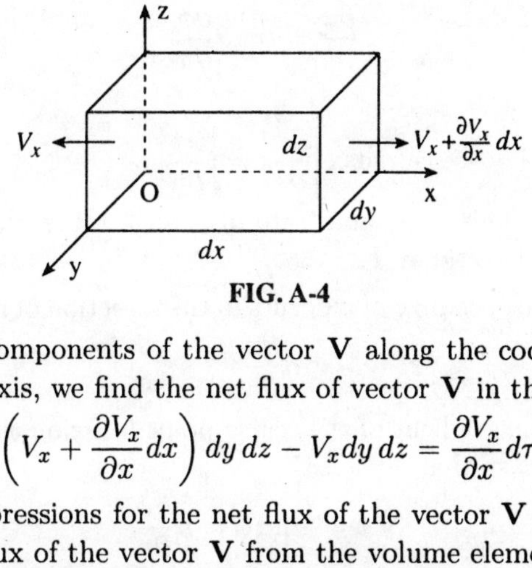

FIG. A-4

Let V_x, V_y, V_z be the components of the vector $\mathbf{V}$ along the coordinate axes. Considering the surfaces normal to x-axis, we find the net flux of vector $\mathbf{V}$ in the direction of x-axis is

$$\left(V_x + \frac{\partial V_x}{\partial x}dx\right)dy\,dz - V_x dy\,dz = \frac{\partial V_x}{\partial x}d\tau \tag{A.37}$$

We will get similar expressions for the net flux of the vector $\mathbf{V}$ in the directions of y- and z-axes. Hence, the total flux of the vector $\mathbf{V}$ from the volume element is

$$\left(\frac{\partial V_x}{\partial x} + \frac{\partial V_y}{\partial y} + \frac{\partial V_z}{\partial z}\right)d\tau \tag{A.38}$$

Hence, the amount of outward flux of the vector $\mathbf{V}$ per unit volume is

$$\frac{\partial V_x}{\partial x} + \frac{\partial V_y}{\partial y} + \frac{\partial V_z}{\partial z} \tag{A.39}$$

By Eqn (A.32), this is the same as div $\mathbf{V}$. Thus, ***divergence of a vector* V *represents the amount of outward flux of that vector per unit volume***. This is the physical meaning of divergence of a vector.

If div $\mathbf{V} > 0$, then occurs outflow of the flux.

If div $\mathbf{V} < 0$, then occurs inflow of the flux.

If div $\mathbf{V} = 0$, then there is no inflow or outflow of the flux. A vector for which divergence is zero is called a ***solenoidal*** vector.

A.7 Gradient of a Scalar Function

We consider a scalar point function $\phi = \phi(x, y, z)$ and draw two neighbouring surfaces S and S' over which the magnitude of the scalar point function remains constant (FIG. A-5).

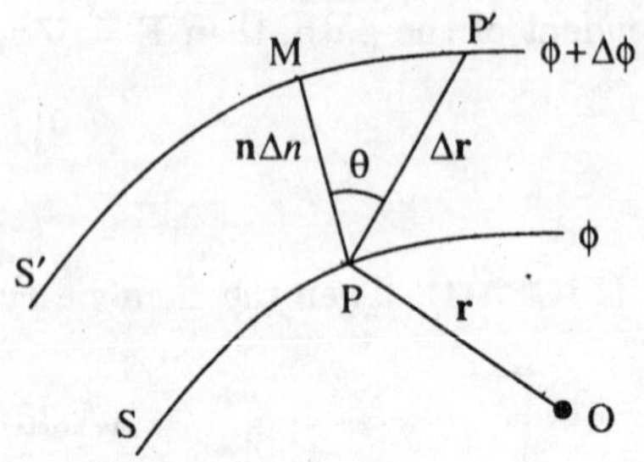

FIG. A-5

Let the function assume the values ϕ and $\phi + \Delta\phi$ on the surfaces S, S' respectively. We take points P, P' on the neighbouring surfaces S, S' respectively and draw PM perpendicular to S'. Then

$$\overline{OP} = \mathbf{r};\ \overline{OP'} = \mathbf{r} + \Delta\mathbf{r} \tag{A.40}$$

Let $PM = \Delta n$, $PP' = \Delta s$ and $\angle MPP' = \theta$.

Then

$$\Delta n = \Delta s \cos\theta. \tag{A.41}$$

Hence, the rate of change of ϕ w.r.t. s as we go from the point P on the surface S to the point P' on the surface S' is

$$\frac{\Delta\phi}{\Delta s} = \frac{\Delta\phi}{\Delta n}\frac{\Delta n}{\Delta s} = \cos\theta\frac{\Delta\phi}{\Delta n} \tag{A.42}$$

In the limit as $P' \to P$ along $\overline{P'P}$, we get

$$\frac{\partial\phi}{\partial s} = \cos\theta\frac{\partial\phi}{\partial n} \tag{A.43}$$

Hence,

$$\max\left|\frac{\partial\phi}{\partial s}\right| = \left|\frac{\partial\phi}{\partial n}\right| \tag{A.44}$$

at a chosen point P on S. This maximum rate of change takes place when $\theta = 0$, i.e., along the direction of unit normal vector $\mathbf{n}$.

Thus, the maximum rate of change of ϕ occurs in the direction of $\mathbf{n}$ and is a vector denoted by

$$\mathbf{n}\frac{\partial\phi}{\partial n}. \tag{A.45}$$

This vector is termed the gradient of the scalar point function $\phi(x, y, z)$ at the point P and is denoted by grad ϕ or $\nabla\phi$ so that

$$\nabla\phi = \text{grad}\,\phi = \mathbf{n}\frac{\partial\phi}{\partial n} \tag{A.46}$$

Hence, ***the gradient of a scalar function ϕ at any point is the vector having a magnitude equal to the maximum rate of change of ϕ at that point in the direction of the normal to the surface $\phi(x, y, z) =$ constant at that point.***

Now, $\phi = \phi(x, y, z)$, and so

$$d\phi = \frac{\partial\phi}{\partial x}dx + \frac{\partial\phi}{\partial y}dy + \frac{\partial\phi}{\partial z}dz \tag{A.47}$$

Since

$$d\mathbf{r} = \mathbf{i}dx + \mathbf{j}dy + \mathbf{k}dz \tag{A.48}$$

Eqn (A.47) can be written as

$$d\phi = \left(\mathbf{i}\frac{\partial\phi}{\partial x} + \mathbf{j}\frac{\partial\phi}{\partial y} + \mathbf{k}\frac{\partial\phi}{\partial z}\right)\cdot d\mathbf{r} \tag{A.49}$$

But

$$d\phi = \frac{\partial \phi}{\partial n} dn = \frac{\partial \phi}{\partial n} \mathbf{n} \cdot d\mathbf{r} = \boldsymbol{\nabla} \phi \cdot d\mathbf{r} \tag{A.50}$$

Eqns (A.49) and (A.50) yield

$$\left[\boldsymbol{\nabla} \phi - \left(\mathbf{i} \frac{\partial \phi}{\partial x} + \mathbf{j} \frac{\partial \phi}{\partial y} + \mathbf{k} \frac{\partial \phi}{\partial z} \right) \right] \cdot d\mathbf{r} = 0 \tag{A.51}$$

Since $d\mathbf{r}$ is a small non-zero vector chosen arbitrarily, we get

$$\text{grad}\, \phi = \boldsymbol{\nabla} \phi = \mathbf{i} \frac{\partial \phi}{\partial x} + \mathbf{j} \frac{\partial \phi}{\partial y} + \mathbf{k} \frac{\partial \phi}{\partial z} \tag{A.52}$$

This equation is generally used to evaluate the gradient of a scalar function.

As an example, let

$$\phi(x, y, z) = 3x^2 + 6xy + y^2 - 9zx + z^2$$

Then

$$\frac{\partial \phi}{\partial x} = 6x + 6y - 9z, \quad \frac{\partial \phi}{\partial y} = 6x + 2y, \quad \frac{\partial \phi}{\partial z} = -9x + 2z$$

Hence,

$$\text{grad}\, \phi = \mathbf{i}(6x + 6y - 9z) + \mathbf{j}(6x + 2y) + \mathbf{k}(-9x + 2z)$$

At the point $(1, 0, 0)$,

$$\text{grad}\, \phi = 6\mathbf{i} + 6\mathbf{j} - 9\mathbf{k}.$$

A.8 Integration of Vectors

Integration of vectors are performed by the usual procedures of integral calculus. Let $\phi(x, y, z)$ be a scalar point function and $\mathbf{F}$ be a vector in space.

(a) Line Integral : Line integrals are performed along some defined contour γ.

Thus,

(1) $$\int_\gamma \phi d\mathbf{r} = \int_\gamma \phi(\mathbf{i}dx + \mathbf{j}dy + \mathbf{k}dz)$$

$$= \mathbf{i} \int_\gamma \phi dx + \mathbf{j} \int_\gamma \phi dy + \mathbf{k} \int_\gamma \phi dz \tag{A.53}$$

(2) $$\int_\gamma \mathbf{F} \cdot d\mathbf{r} = \int_\gamma (\mathbf{i}F_x + \mathbf{j}F_y + \mathbf{k}F_z) \cdot (\mathbf{i}dx + \mathbf{j}dy + \mathbf{k}dz)$$

$$= \int_\gamma F_x dx + \int_\gamma F_y dy + \int_\gamma F_z dz \tag{A.54}$$

(3) $$\int_\gamma \mathbf{F} \times d\mathbf{r} = \int_\gamma (\mathbf{i}F_x + \mathbf{j}F_y + \mathbf{k}F_z) \times (\mathbf{i}dx + \mathbf{j}dy + \mathbf{k}dz)$$

$$= \mathbf{i} \int_\gamma (F_y dz - F_z dy) + \mathbf{j} \int_\gamma (F_z dx - F_x dz) + \mathbf{k} \int_\gamma (F_x dy - F_y dx) \tag{A.55}$$

Here F_x, F_y, F_z are the components of the vector $\mathbf{F}$ along the coordinate axes. The above integrals can be evaluated by the usual methods of integation.

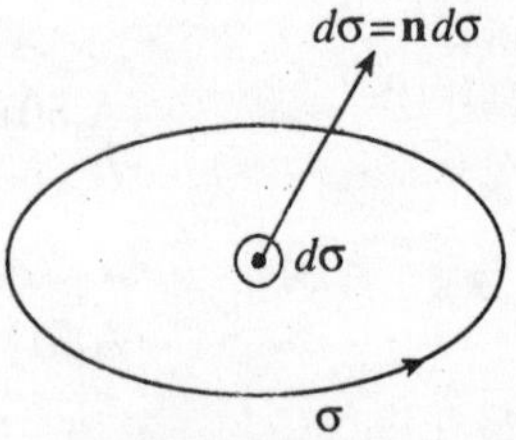

FIG. A-6

(b) Surface Integral : We can represent a plane surface area by a vector, called ***area vector*** (FIG. A-6). Let $d\boldsymbol{\sigma}$ be an elementary surface taken from the surface $\boldsymbol{\sigma}$. Then

$$\int_\sigma \mathbf{F}\cdot d\boldsymbol{\sigma} = \int \mathbf{F}\cdot \mathbf{n}d\sigma \qquad \text{(A.56)}$$

where $\mathbf{n}$ is the unit normal vector on the surface.

The integral in the L.H.S. of Eqn (A.56) is defined as the flux of the vector $\mathbf{F}$ through the surface $\boldsymbol{\sigma}$.

(c) Volume Integral : We consider a volume element $d\tau$ which is a scalar quantity. Hence, the volume integral of a vector $\mathbf{F}$ over the volume τ is

$$\int_\tau \mathbf{F}d\tau = \mathbf{i}\int F_x d\tau + \mathbf{j}\int F_y d\tau + \mathbf{k}\int F_z d\tau \qquad \text{(A.57)}$$

This is simple to evaluate by usual methods.

A.9 Gauss's Divergence Theorem

Gauss's divergence theorem relates a volume integral with the surface integral.

It states that: ***The volume integral of the divergence of the vector point function* F *taken over any volume* τ *is equal to the surface integral of the vector* F *taken over the closed surface surrounding the volume* τ.**

Thus,

$$\int_\tau \operatorname{div}\mathbf{F}d\tau = \int_\sigma \mathbf{F}\cdot\mathbf{n}d\sigma \qquad \text{(A.58)}$$

where $\mathbf{n}$ is the unit outward-drawn normal vector (proof is not given).

As an example of its application, let us evaluate the integral

$$I = \int_\sigma (ax\mathbf{i} + by\mathbf{j} + cz\mathbf{k})\cdot \mathbf{n}d\sigma$$

where $\boldsymbol{\sigma}$ is the surface of the sphere $x^2+y^2+z^2=1$.

Here,

$$\mathbf{F} = ax\mathbf{i} + by\mathbf{j} + cz\mathbf{k}.$$

Hence,

$$\operatorname{div}\mathbf{F} = \frac{\partial}{\partial x}(ax) + \frac{\partial}{\partial y}(by) + \frac{\partial}{\partial z}(cz) = a+b+c.$$

By Gauss's theorem,

$$\int_\sigma \mathbf{F}\cdot\mathbf{n}d\sigma = \int_\tau \operatorname{div}\mathbf{F}\cdot d\tau.$$

Hence, the given integral is

$$\int_\tau (a+b+c)d\tau = (a+b+c)\tau.$$

Volume of the sphere

$$\tau = \frac{4}{3}\pi r^3 = \frac{4}{3}\pi \ (\because\ r=1).$$

Thus, the value of the given integral is $\frac{4}{3}\pi(a+b+c)$.

A.10 Green's Theorem

Green's theorem is an important corollary of divergence theorem.

It states that: ***If ϕ and ψ are two scalar functions having continuous derivatives within a region τ bounded by the closed surface σ, then***

$$\int_\tau (\phi\nabla^2\psi - \psi\nabla^2\phi)\, d\tau = \int_\sigma (\phi\boldsymbol{\nabla}\psi - \psi\boldsymbol{\nabla}\phi)\cdot \mathbf{n}d\sigma \tag{A.59}$$

where $\mathbf{n}$ is the unit outward-drawn normal vector to the surface.

A.11 Stokes' Theorem

Stokes' theorem relates a surface integral with a line integral.

It states that: ***The surface integral of the curl of a vector point function F taken over any surface σ is equal to the line integral of the vector point function F taken around the periphery of the surface.***

Thus,

$$\int_\sigma (\boldsymbol{\nabla}\times\mathbf{F})\cdot\mathbf{n}d\sigma = \int \mathbf{F}\cdot d\mathbf{r} \tag{A.60}$$

where $\mathbf{n}$ is the outward-drawn unit normal vector to the surface (proof is not given).

A.12 Green's Theorem in Plane

It states that: ***If M and N are continuous functions of x and y having continuous derivatives in a region σ bounded by a closed curve C, then***

$$\int_C (Mdx + Ndy) = \int_\sigma \left(\frac{\partial N}{\partial x} - \frac{\partial M}{\partial y}\right) dxdy. \tag{A.61}$$

As an application of this theorem, let us evaluate the integral

$$I = \int_C (e^{-x}\sin y\, dx + e^{-x}\cos y\, dy)$$

where C is a rectangle with vertices $(0,0), (\pi,0), (\pi,\pi/2), (0,\pi/2)$.

$$\left.\begin{array}{l}\text{Here, } M = e^{-x}\sin y;\ N = e^{-x}\cos y. \\ \text{Hence, } \dfrac{\partial M}{\partial y} = e^{-x}\cos y;\ \dfrac{\partial N}{\partial x} = -e^{-x}\cos y.\end{array}\right\}$$

Then

$$\begin{aligned} I = \int_C (Mdx + Ndy) &= \int_{x=0}^{x=\pi}\int_{y=0}^{y=\pi/2} \left(\frac{\partial N}{\partial x} - \frac{\partial M}{\partial y}\right) \\ &= \int_{x=0}^{x=\pi}\int_{y=0}^{y=\pi/2} -2e^{-x}\cos y\, dxdy \\ &= \int_{x=0}^{x=\pi} (-2e^{-x}dx) \int_{y=0}^{y=\pi/2} \cos y\, dy = 2\left(e^{-\pi} - 1\right). \end{aligned}$$

APPENDIX B

Rudiments of Matrices

B.1 Definition

A ***matrix*** is a rectangular array of quantities like

$$A = \begin{bmatrix} a_{11} & a_{12} & \cdots & a_{1n} \\ a_{21} & a_{22} & \cdots & a_{2n} \\ \vdots & & & \\ a_{m1} & a_{m2} & \cdots & a_{mn} \end{bmatrix} \tag{B.1}$$

where a_{ij} are called ***elements*** of the matrix. Here the matrix A has n rows and m columns, and is called a matrix of order $n \times m$. The element at i-th row and j-th column is denoted by a_{ij}. If $m = n$, the matrix is called ***square***. The elements of a matrix may be real or complex numbers and also functions.

The matrix having only one row like

$$A = (a_{11}, a_{12}, \cdots, a_{1n}) \tag{B.2}$$

is called a ***row matrix***. The matrix having only one column like

$$A = \begin{bmatrix} a_{11} \\ a_{21} \\ \vdots \\ a_{n1} \end{bmatrix} \tag{B.3}$$

is called a ***column matrix***.

Two matrices A and B of the same order are said to be equal if and only if

$$a_{ij} = b_{ij} \tag{B.4}$$

for all i and j. For example, if

$$A = \begin{bmatrix} 2 & 4 \\ 3i & 0 \end{bmatrix} \text{ and } B = \begin{bmatrix} 2 & 4 \\ 3i & 0 \end{bmatrix} \tag{B.5}$$

then we write $A = B$.

A ***null matrix*** has got all its elements equal to zero. Thus, for a null matrix,

$$a_{ij} = 0, \text{for all } i \text{ and } j. \tag{B.6}$$

B.2 Matrix Operations

The operation of ***addition*** or ***subtraction*** of two matrices A and B of the same order is defined as

$$C = A \pm B \tag{B.7}$$

where the elements c_{ij} of the resulting matrix C are given by

$$c_{ij} = a_{ij} \pm b_{ij} \tag{B.8}$$

for all i and j.

For example, if

$$A = \begin{bmatrix} 2 & 1 & 0 \\ 4i & 5 & 3 \end{bmatrix} \text{ and } B = \begin{bmatrix} 3 & 0 & 6 \\ -7i & 4 & 9 \end{bmatrix}$$

then

$$A + B = \begin{bmatrix} 5 & 1 & 6 \\ -3i & 9 & 12 \end{bmatrix}$$

The ***product*** of two matrices A and B is obtained by following the definitions

$$C = AB \tag{B.9}$$

where the elements c_{ij} of the matrix C are given by

$$c_{ij} = \sum_{k=1}^{p} a_{ik} b_{kj} \tag{B.10}$$

Here the orders of matrices A, B and C are $n \times p, p \times m$ and $n \times m$ respectively.

Let the matrices A and B are given by

$$A = \begin{bmatrix} a_{11} & a_{12} \\ a_{21} & a_{22} \end{bmatrix} \text{ and } B = \begin{bmatrix} b_{11} & b_{12} \\ b_{21} & b_{22} \end{bmatrix} \tag{B.11}$$

Then the product

$$C = AB = \begin{bmatrix} a_{11} & a_{12} \\ a_{21} & a_{22} \end{bmatrix} \begin{bmatrix} b_{11} & b_{12} \\ b_{21} & b_{22} \end{bmatrix} \tag{B.12}$$

The elements of matrix C are

$$\left.\begin{aligned} c_{11} &= a_{11}b_{11} + a_{12}b_{21} \\ c_{12} &= a_{11}b_{12} + a_{12}b_{21} \\ c_{21} &= a_{21}b_{11} + a_{22}b_{22} \\ c_{22} &= a_{21}b_{12} + a_{22}b_{22} \end{aligned}\right\} \tag{B.13}$$

Consider a system of equations:

$$\left.\begin{aligned} a_{11}x + a_{12}y + a_{13}z &= p \\ a_{21}x + a_{22}y + a_{23}z &= q \\ a_{31}x + a_{32}y + a_{33}z &= r \end{aligned}\right\} \tag{B.14}$$

Following the definition of matrix product, we can write the above system of equations in matrix notation in the form:

$$[A]\{X\} = \{P\} \tag{B.15}$$

where

$$[A] = \begin{bmatrix} a_{11} & a_{12} & a_{13} \\ a_{21} & a_{22} & a_{23} \\ a_{31} & a_{32} & a_{33} \end{bmatrix} \tag{B.16}$$

$$\text{and } \{X\} = \begin{bmatrix} x \\ y \\ z \end{bmatrix} ; \ \{P\} = \begin{bmatrix} p \\ q \\ r \end{bmatrix} \tag{B.17}$$

It can easily be verified that:

(a) $AB \neq BA$. That is, the commutative law of multiplication is not, in general, valid.

(b) $A(BC) = (AB)C$. That is, the associative law of multiplication is valid for matrix product.

The division by a general matrix is not uniquely defined.

B.3 Matrix Calculus

When the elements of a matrix are ***functions*** of a variable x, we can find the ***derivative*** and ***integral*** of that matrix.

The ***derivative*** of a matrix with respect to a variable x is obtained by differentiating each of the elements of the matrix separately with respect to x. Thus, if we have a matrix

$$A(x) = \begin{bmatrix} \dfrac{x^2}{2} & 3x & 2 \\ e^x & \sin x & x^3 \end{bmatrix} \tag{B.18}$$

then its derivative is given by

$$\frac{d}{dx}[A(x)] = \begin{bmatrix} x & 3 & 0 \\ e^x & \cos x & 3x^2 \end{bmatrix} \tag{B.19}$$

The ***integral*** of a matrix with respect to a variable x is obtained by integrating each of the elements of the matrix separately with respect to x. Thus, the integral of the matrix $A(x)$ in Eqn (B.18) is

$$\int [A(x)]dx = \begin{bmatrix} \dfrac{x^3}{6} & \dfrac{3x^2}{2} & 2x \\ e^x & -\cos x & \dfrac{x^4}{4} \end{bmatrix} + \begin{bmatrix} a_1 & a_2 & a_3 \\ a_4 & a_5 & a_6 \end{bmatrix} \tag{B.20}$$

B.4 Properties of Matrices

(a) Unit Matrix. The ***unit matrix*** has all the diagonal elements equal to unity and all the other elements equal to zero. Thus, a 3×3 unit matrix is given by

$$I = \begin{bmatrix} 1 & 0 & 0 \\ 0 & 1 & 0 \\ 0 & 0 & 1 \end{bmatrix} \tag{B.21}$$

(b) Transpose Matrix. The ***transpose*** of an arbitrary matrix A is obtained by interchanging the corresponding rows and columns of A and is denoted by A^T. Thus, if $A = a_{ij}$, then $A^T = a_{ji}$.

For example, if

$$A = \begin{bmatrix} 2 & 4 & 3i \\ 0 & 6 & e^2 \end{bmatrix}, \text{ then } A^T = \begin{bmatrix} 2 & 0 \\ 4 & 6 \\ 3i & e^2 \end{bmatrix}.$$

(c) Symmetric Matrix. The matrix A is said to be ***symmetric***, if

$$A^T = A \tag{B.22}$$

Thus, if

$$A = \begin{bmatrix} 0 & 1 \\ 1 & 0 \end{bmatrix}, \text{ then } A^T = \begin{bmatrix} 0 & 1 \\ 1 & 0 \end{bmatrix} = A$$

Hence, the matrix is symmetric.

(d) Antisymmetric or Skew-Symmetric Matrix. The matrix A is said to be antisymmetric or skew-symmetric matrix if

$$A^T = -A \tag{B.23}$$

Thus, the matrix

$$A = \begin{bmatrix} 0 & -i \\ i & 0 \end{bmatrix}$$

is antisymmetric, since

$$A^T = \begin{bmatrix} 0 & i \\ -i & 0 \end{bmatrix} = -A$$

(e) Orthogonal Matrix. The matrix A is said to be ***orthogonal***, if

$$AA^T = I \tag{B.24}$$

Thus, the matrix

$$A = \begin{bmatrix} \cos\theta & \sin\theta \\ -\sin\theta & \cos\theta \end{bmatrix}$$

is orthogonal, since

$$AA^T = \begin{bmatrix} \cos\theta & \sin\theta \\ -\sin\theta & \cos\theta \end{bmatrix} \begin{bmatrix} \cos\theta & -\sin\theta \\ \sin\theta & \cos\theta \end{bmatrix} = \begin{bmatrix} 1 & 0 \\ 0 & 1 \end{bmatrix} = I$$

(f) Singular Matrix. The determinant with the elements of a matrix A is denoted by $\det|A|$.

The matrix A is said to be a ***singular*** matrix if

$$\det|A| = 0 \tag{B.25}$$

For example, if

$$A = \begin{bmatrix} 2 & 5 & 19 \\ 1 & -2 & -4 \\ -3 & 2 & 0 \end{bmatrix}, \text{ then } \det|A| = \begin{vmatrix} 2 & 5 & 19 \\ 1 & -2 & -4 \\ -3 & 2 & 0 \end{vmatrix} = 0$$

Hence, the matrix A is singular.

(g) Complex-Conjugate Matrix. Let the matrix A contain complex quantities. The ***complex-conjugate*** of the matrix A is formed by taking complex-conjugate of each element and is denoted by A^*. Thus, if

$$A = \begin{bmatrix} a+ib & c+id \\ -2ia & a-ib \end{bmatrix}, \text{ then } A^* = \begin{bmatrix} a-ib & c-id \\ 2ai & a+ib \end{bmatrix}$$

(h) Hermitian Conjugate. The ***Hermitian conjugate*** of a matrix A containing complex quantities is obtained by transposing its complex-conjugate matrix and is denoted by $[A^*]^T$. Thus, if

$$A = \begin{bmatrix} 2+5i & 3-2i \\ 7i & 4-6i \end{bmatrix}$$

then its complex-conjugate is

$$A^* = \begin{bmatrix} 2-5i & 3+2i \\ -7i & 4+6i \end{bmatrix}$$

and its Hermitian conjugate is

$$[A^*]^T = \begin{bmatrix} 2-5i & -7i \\ 3+2i & 4+6i \end{bmatrix}$$

(i) Trace of a Matrix. The ***trace*** of a matrix is the algebraic sum of its diagonal elements. Thus, if the elements of a matrix A are a_{ij}, then

$$\operatorname{Tr} A = \sum_k a_{kk} \tag{B.26}$$

For example, if $A = \begin{bmatrix} 2 & 5 & 9 \\ 1 & 6 & 4 \\ -3 & 2 & 0 \end{bmatrix}$, then $\operatorname{Tr} A = 2 + 6 + 0 = 8$

(j) Inverse Matrix. The ***inverse*** of a matrix A is denoted by A^{-1} and is such that

$$AA^{-1} = I \tag{B.27}$$

Thus, if

$$A = \begin{bmatrix} 2 & 3 & 1 \\ 1 & 2 & 3 \\ 3 & 1 & 2 \end{bmatrix}$$

it can be determined that

$$A^{-1} = \frac{1}{18}\begin{bmatrix} 1 & -5 & 7 \\ 7 & 1 & -5 \\ -5 & 7 & 1 \end{bmatrix}$$

and can also be verified that $AA^{-1} = I$.

(k) Square Matrix. If the number of rows and number of columns in a matrix are equal, then the matrix is called a ***square matrix***. Thus, the matrix

$$A = \begin{bmatrix} 2 & 1 & 1 \\ 4 & -7 & 6 \\ 9 & 3 & 11 \end{bmatrix}$$

is a square matrix.

Every square matrix A can be expressed as the sum of a symmetric and a skew-symmetric matrix. Thus,

$$A = \frac{1}{2}(A + A^T) + \frac{1}{2}(A - A^T) \tag{B.28}$$

where the first part in RHS is a symmetric matrix and the second part is skew-symmetric.

(l) Cofactor Matrix. The ***cofactor*** corresponding to element a_{ij} of a matrix A is the determinant containing elements obtained by suppressing all elements in the i-th row and j-th column with proper sign. It is denoted by A^{ij} and is equal to

$$A^{ij} = (-1)^{i+j} \text{ cofactor.} \tag{B.29}$$

Thus, when

$$A = \begin{bmatrix} a_{11} & a_{12} & a_{13} \\ a_{21} & a_{22} & a_{23} \\ a_{31} & a_{32} & a_{33} \end{bmatrix}, \quad A^{12} = (-1)^{1+2} \begin{bmatrix} a_{21} & a_{23} \\ a_{31} & a_{33} \end{bmatrix}$$

The ***cofactor matrix*** denoted by A^c is the matrix whose elements are the cofactors. Thus,

$$A^c = \begin{bmatrix} A^{11} & A^{12} & A^{13} \\ A^{21} & A^{22} & A^{23} \\ A^{31} & A^{32} & A^{33} \end{bmatrix}$$

(m) Adjoint of a Matrix. The ***adjoint*** of a matrix A is defined as the transpose of the cofactor matrix and is written as adj A. Thus,

$$\text{adj } A = [A^c]^T \tag{B.30}$$

For example, if

$$A = \begin{bmatrix} 1 & 3 \\ 2 & 1 \end{bmatrix}, \text{ then } A^c = \begin{bmatrix} 1 & -2 \\ -3 & 1 \end{bmatrix} \text{ and adj } A = \begin{bmatrix} 1 & -3 \\ -2 & 1 \end{bmatrix}$$

(n) Self-Adjoint Matrix. The matrix A is said to be ***self-adjoint***, if

$$\text{adj } A = A \tag{B.31}$$

For example, if

$$A = \begin{bmatrix} -1 & 0 \\ 0 & -1 \end{bmatrix}, \text{ then } A^c = \begin{bmatrix} -1 & 0 \\ 0 & -1 \end{bmatrix}$$

$$\text{and adj } A = [A^c]^T = \begin{bmatrix} -1 & 0 \\ 0 & -1 \end{bmatrix} = A$$

B.5 Determining Inverse of a Matrix

Every matrix does not have an inverse. The necessary and sufficient condition for the matrix A to have an inverse is that it should be non-singular. That is

$$|A| \neq 0 \tag{B.32}$$

Moreover, the inverse of a matrix, if it exists, is unique; a matrix cannot have two inverse matrices.

The inverse of a matrix A may be computed by using the formula:

$$A^{-1} = \frac{\text{adj}\, A}{|A|} \tag{B.33}$$

For example, let

$$A = \begin{bmatrix} 1 & 3 \\ 2 & 1 \end{bmatrix}$$

Then

$$A_{11} = (-1)^{1+1} \cdot 1 = 1$$
$$A_{12} = (-1)^{1+2} \cdot 2 = -2$$
$$A_{21} = (-1)^{2+1} \cdot 3 = -3$$
$$A_{22} = (-1)^{2+2} \cdot 1 = 1$$

Thus,

$$A^c = \begin{bmatrix} 1 & -2 \\ -3 & 1 \end{bmatrix} \quad \text{and} \quad \text{adj}\, A = \begin{bmatrix} 1 & -3 \\ -2 & 1 \end{bmatrix}.$$

The determinant of matrix A is

$$|A| = \begin{bmatrix} 1 & 3 \\ 2 & 1 \end{bmatrix} = -5$$

Hence, the inverse of A is

$$A^{-1} = \frac{\text{adj}\, A}{|A|} = -\frac{1}{5}\begin{bmatrix} 1 & -3 \\ -2 & 1 \end{bmatrix}$$

It can be verified that $AA^{-1} = I$.

B.6 Transpose and Inverse of Product of Two Matrices

The transpose of the product of two matrices is the product of their transposes taken in the reverse order. Thus, if A and B are two matrices conformable to multiplication, then

$$[AB]^T = B^T A^T \tag{B.34}$$

Similarly, the inverse of the product of two matrices of same order is the product of their inverses taken in reverse order. Thus,

$$[AB]^{-1} = B^{-1} A^{-1}. \tag{B.35}$$

APPENDIX C

Vectors and Tensors

C.1 Introduction

The concept of a tensor was introduced by Gauss, Reimann and Christoffel while developing the subject of differential geometry. Ricci and his student Levi-Civita introduced Tensor analysis as a systematic branch of mathematics in a publication in 1901. Einstein popularised the subject by making an extensive use of tensors in his General Relativity theory. Today, the tensor analysis has emerged as an invaluable tool in many branches of theoretical physics.

The fundamental postulate of physics is that the laws of nature should be valid for all reference frames. The subject matter of tensor analysis has emerged as the study of the behaviour of a physical law from one system of coordinates to any other.

A ***tensor*** is a set of quantities called the coordinates of ***components*** whose ***properties*** are independent of the coordinate system. For two different coordinate systems, the components of a tensor obey specific transformation laws.

C.2 Notations and Symbols

Consider N quantities denoted by $x^1, x^2, \dots, x^N$ which will be called ***coordinates*** of a point. The upper indices $1, 2, \dots,$ which we shall call ***superscripts*** are simply identification labels and not indices of powers. The expression

$$\sum_{i=1}^{N} a_i x^i = a_1 x^1 + a_2 x^2 + \cdots + a_N x^N \tag{C.1}$$

representing the summation of N terms will be written simply $a_i x^i$; the letter i attached with a is called ***subscript***. If the subscript and superscript are numerical, then it will represent a single term, but not summation.

Repeat index in any expression is called a ***dummy index***, and it can be replaced by any other letter. Thus, we can write, $a_i x^i = a_k x^k$.

If, in Eqn (C.1), $N = 4$, then

$$a_k x^k = a_1 x^1 + a_2 x^2 + a_3 x^3 + a_4 x^4 \tag{C.2}$$

Again, if ϕ is a function of $x^1, x^2, \dots, x^N$, then

$$d\phi = \frac{\partial \phi}{\partial x^1} dx^1 + \frac{\partial \phi}{\partial x^2} dx^2 + \cdots + \frac{\partial \phi}{\partial x^N} dx^N \tag{C.3}$$

which is written as

$$d\phi = \frac{\partial \phi}{\partial x^i} dx^i \tag{C.4}$$

The Kronecker Delta is defined and denoted by

$$\delta^i_j = \begin{cases} 1, & \text{if } i = j \\ 0, & \text{if } i \neq j \end{cases} \tag{C.5}$$

Thus, $\delta^1_1 = 1, \delta^2_2 = 1, \delta^2_3 = 0$, etc.

Also

$$\delta^i_k x^k = x^i \tag{C.6}$$

since all other terms will vanish. That is why δ^i_k is called a ***substitution factor***.

The superscripts are used to denote the components of a ***contravariant tensor***. The subscripts are used to denote the components of a ***covariant tensor***. The components of a ***mixed tensor*** is denoted by both superscripts and subscripts. Thus,

(1) a^i are components of a contravariant tensor.

(2) B_{ij} are components of a covariant tensor.

(3) x^k_{ij} are components of a mixed tensor.

The ***rank*** of a tensor is the number of indices appearing in the symbol representing the components of a tensor (or a tensor itself). Thus, the ranks of above three tensors are respectively one, two and three. However, if an index appears as a subscript and again as a superscript, then it will be ignored. Thus, C^{ik}_{jkl} is a mixed tensor of rank three (not five).

C.3 *N*-Dimensional Space

An ordered set of N variables $x^1, x^2, \ldots, x^N$ is called the coordinates of a ***point***.

The ***N-Dimensional space***, denoted by V_N, is the set of all these points.

A ***subspace*** V_M of the space V_N is defined for $M < N$, if the N variables $x^1, x^2, \ldots, x^N$ are given functions of M parameters. Considering a space V_N, the N equations

$$\bar{x}^i = \phi^i \left(x^1, x^2, \ldots, x^N\right) \; (i = 1, 2, \ldots, N) \tag{C.7}$$

where ϕ^i are single-valued continuous and differentiable functions of coordinates and define a new coordinate system $\bar{x}^1, \bar{x}^2, \ldots, \bar{x}^N$. It is essential that the N functions ϕ^i are independent. The Eqn (C.7) are said to define a ***transformation of coordinates***.

Differentiating Eqn (C.7) with respect to x^j, we get

$$d\bar{x}^i = \sum_{j=1}^{N} \frac{\partial \phi^i}{\partial x^j} dx^j = \sum_{j=1}^{N} \frac{\partial \bar{x}^i}{\partial x^j} dx^j = \frac{\partial \bar{x}^i}{\partial x^j} dx^j \text{ (by summation convention).} \tag{C.8}$$

If the Jacobian determinant formed with the partial derivatives $\partial \bar{x}^i / \partial x^j$ does not vanish, then the inverse transformation exists, so that we can obtain

$$x^i = \psi^i \left(\bar{x}^1, \bar{x}^2, \ldots, \bar{x}^N\right) \; (i = 1, 2, \ldots, N) \tag{C.9}$$

C.4 Contravariant and Covariant Vector

A set of N functions A^i of N coordinates x^i are said to form the components of a ***contravariant vector***, if they transform according to the equation

$$\bar{A}^i = \frac{\partial \bar{x}^i}{\partial x^j} A^j \tag{C.10}$$

where $\bar{x}^i$ refers to the new coordinate system.

Multiplying equations (C.10) by $\partial x^k / \partial \bar{x}^i$ and summing over index i from 1 to N, we have

$$\frac{\partial x^k}{\partial \bar{x}^i} \bar{A}^i = \frac{\partial x^k}{\partial \bar{x}^i} \frac{\partial \bar{x}^i}{\partial x^j} A^j = \frac{\partial x^k}{\partial x^j} A^j = \delta^k_j A^j = A^k$$

Thus, the solution of equations (C.10) is

$$A^j = \frac{\partial x^j}{\partial \bar{x}^i} \bar{A}^i \tag{C.11}$$

Comparing equations (C.8) and (C.10), we find that the differentials dx^j form the components of a contravariant vector.

A set of N functions A_i of the coordinates x^i are said to form the components of a ***covariant vector***, if they transform according to the equation

$$\bar{A}_i = \frac{\partial x^j}{\partial \bar{x}^i} A_j \tag{C.12}$$

where $\bar{x}^i$ refers to the new coordinate system.

Multiplying Eqn (C.12) by $\partial \bar{x}^i / \partial x^k$ and summing over the index i from 1 to N, we have

$$\frac{\partial \bar{x}^i}{\partial x^k} \bar{A}_i = \frac{\partial \bar{x}^i}{\partial x^k} \frac{\partial x^j}{\partial \bar{x}^i} A_j = \frac{\partial x^j}{\partial x^k} A_j = \delta^j_k A_j = A_k \tag{C.13}$$

Thus, the solution of Eqn (C.12) is

$$A_j = \frac{\partial \bar{x}^i}{\partial x^j} \bar{A}_i \tag{C.14}$$

From equations (C.11) and (C.14), we find that

$$A^j B_j = \left(\frac{\partial x^j}{\partial \bar{x}^i} \bar{A}^i\right)\left(\frac{\partial \bar{x}^t}{\partial x^j} \bar{B}_t\right) = \delta^t_i \bar{A}^i \bar{B}_t = \bar{A}^t \bar{B}_t$$

Hence,

$$A^j B_j = \bar{A}^t \bar{B}_t = \text{invariant}. \tag{C.15}$$

Thus, if a function ϕ of coordinates x^i is transformed to a function ψ of new coordinates $\bar{x}^i$ such that $\phi = \psi$, then the function ϕ is called an ***invariant*** or a ***scalar***.

EXAMPLE C.1 *In rectangular Cartesian coordinates, the components of acceleration vectors are* $(\ddot{x}, \ddot{y})$. *Find these components in polar coordinates.*

Solution. In polar coordinates, $x = r\cos\theta, y = r\sin\theta$. This gives

$$\left.\begin{aligned} dx &= dr\cos\theta - (r\sin\theta)d\theta \\ dy &= dr\sin\theta + (r\cos\theta)d\theta \end{aligned}\right\} \tag{1}$$

On solving these, we have

$$\left.\begin{aligned} dr &= (\cos\theta)dx + (\sin\theta)dy \\ d\theta &= \left(-\frac{1}{r}\sin\theta\right)dx + \left(\frac{1}{r}\cos\theta\right)dy \end{aligned}\right\} \tag{2}$$

We put

$$\left.\begin{aligned} (x, y) = (x^1, x^2), (r, \theta) = (\bar{x}^1, \bar{x}^2) \\ (\ddot{x}, \ddot{y}) = (A^1, A^2), (a_r, a_\theta) = (\bar{A}^1, \bar{A}^2) \end{aligned}\right\} \tag{3}$$

where a_r, a_θ are components of acceleration vectors in polar coordinates.

Now, from Eqn (2) using Eqn (3), we have

$$d\bar{x}^1 = (\cos\theta)dx^1 + (\sin\theta)dx^2$$

$$d\bar{x}^2 = \left(-\frac{1}{r}\sin\theta\right)dx^1 + \left(\frac{1}{r}\cos\theta\right)dx^2$$

Hence,

$$\left.\begin{aligned} \frac{\partial\bar{x}^1}{\partial x^1} = \cos\theta, \quad \frac{\partial\bar{x}^1}{\partial x^2} = \sin\theta \\ \frac{\partial\bar{x}^2}{\partial x^1} = -\frac{1}{r}\sin\theta, \quad \frac{\partial\bar{x}^2}{\partial x^2} = \frac{1}{r}\cos\theta \end{aligned}\right\} \tag{4}$$

Thus,

$$a_r = \bar{A}^1 = \frac{\partial\bar{x}^1}{\partial x^1}A^1 + \frac{\partial\bar{x}^1}{\partial x^2}A^2 = (\cos\theta)\ddot{x} + (\sin\theta)\ddot{y} \tag{5}$$

$$\text{and } a_\theta = \bar{A}^2 = \frac{\partial\bar{x}^2}{\partial x^1}A^1 + \frac{\partial\bar{x}^2}{\partial x^2}A^2 = \left(-\frac{1}{r}\sin\theta\right)\ddot{x} + \left(\frac{1}{r}\cos\theta\right)\ddot{y} \tag{6}$$

From Eqn (1),

$$\dot{x} = \frac{dx}{dt} = (\cos\theta)\dot{r} - r(\sin\theta)\dot{\theta}$$

$$\ddot{x} = (\cos\theta)\ddot{r} - 2(\sin\theta)\dot{\theta}\dot{r} - r(\cos\theta)\dot{\theta}^2 - r(\sin\theta)\ddot{\theta} \tag{7}$$

$$\dot{y} = (\sin\theta)\dot{r} + r(\cos\theta)\dot{\theta}$$

$$\ddot{y} = (\sin\theta)\ddot{r} + 2(\cos\theta)\dot{\theta}\dot{r} - r(\sin\theta)\dot{\theta}^2 + r(\cos\theta)\ddot{\theta} \tag{8}$$

Using Eqns (7) and (8), we get from Eqns (5) and (6),

$$a_r = \ddot{r} - r\dot{\theta}^2, \; a_\theta = \ddot{\theta} + \frac{2}{r}\dot{r}\dot{\theta}$$

EXAMPLE C.2 *The components of a covariant vector in rectangular Cartesian systems are*

$$A_1 = \frac{y}{x}, \; A_2 = \frac{x}{y}$$

Find these in polar coordinates.

Solution. Let the components in polar coordinates be $\bar{A}_1, \bar{A}_2$. Then, we have

$$\bar{A}_1 = \frac{\partial x^1}{\partial\bar{x}^1}A_1 + \frac{\partial x^2}{\partial\bar{x}^1}A_2 \tag{1}$$

$$\bar{A}_2 = \frac{\partial x^1}{\partial\bar{x}^2}A_1 + \frac{\partial x^2}{\partial\bar{x}^2}A_2 \tag{2}$$

From Eqn (1) of previous example, we get

$$\left.\begin{aligned} dx^1 &= d\bar{x}^1 \cos\theta - (r\sin\theta)d\bar{x}^2 \\ dx^2 &= d\bar{x}^1 \sin\theta + (r\cos\theta)d\bar{x}^2 \end{aligned}\right\}$$

Hence,

$$\left.\begin{aligned} \frac{\partial x^1}{\partial \bar{x}^1} &= \cos\theta, \quad \frac{\partial x^1}{\partial \bar{x}^2} = -r\sin\theta \\ \frac{\partial x^2}{\partial \bar{x}^1} &= \sin\theta, \quad \frac{\partial x^2}{\partial \bar{x}^2} = +r\cos\theta \end{aligned}\right\}$$

Using these relations, we get from Eqns (1) and (2),

$$\bar{A}_1 = \cos\theta\left(\frac{y}{x}\right) + \sin\theta\left(\frac{x}{y}\right) = \sin\theta + \cos\theta$$

$$\begin{aligned} \bar{A}_2 &= -r\sin\theta\left(\frac{y}{x}\right) + r\cos\theta\left(\frac{x}{y}\right) \\ &= -r\frac{\sin^2\theta}{\cos\theta} + r\frac{\cos^2\theta}{\sin\theta} \end{aligned}$$

C.5 Higher Rank Tensors

We have seen the transformation laws of contravariant and covariant vectors which are tensors of rank one. Now we shall discuss the same for higher rank tensors.

Consider N^2 functions A^{ij} which are defined in $x^1, x^2, \ldots, x^N$ coordinate system. If due to a change in coordinate system to $\bar{x}^1, \bar{x}^2, \ldots, \bar{x}^N$, the quantities A^{ij} transform according to the equation

$$\bar{A}^{ij} = \frac{\partial \bar{x}^i}{\partial x^k}\frac{\partial \bar{x}^j}{\partial x^m}A^{km} \tag{C.16}$$

then we call A^{ij} the components of a ***contravariant tensor*** of ***second order***.

Similarly, if the N^2 functions A_{ij} transform according to the law,

$$\bar{A}_{ij} = \frac{\partial x^k}{\partial \bar{x}^i}\frac{\partial x^m}{\partial \bar{x}^j}A_{km} \tag{C.17}$$

we call A_{ij} the components of a ***covariant tensor*** of ***second order***.

Again, if we have N^2 functions A^i_j, which transform according to the equation

$$\bar{A}^i_j = \frac{\partial \bar{x}^i}{\partial x^k}\frac{\partial x^m}{\partial \bar{x}^j}A^k_m \tag{C.18}$$

we call A^i_j the components of a ***mixed tensor*** of second order.

A set of N^{s+p} quantities

$$A^{k_1k_2\cdots k_s}_{m_1m_2\cdots m_p} \tag{C.19}$$

is said to be the components of a ***mixed*** tensor of $(s+p)$th order, if they transform according to the equation

$$\bar{A}^{k_1k_2\cdots k_s}_{m_1m_2\cdots m_p} = \frac{\partial \bar{x}^{k_1}}{\partial x^{t_1}}\cdots\frac{\partial \bar{x}^{k_s}}{\partial x^{t_s}}\frac{\partial x^{q_1}}{\partial \bar{x}^{m_1}}\cdots\frac{\partial x^{q_p}}{\partial \bar{x}^{m_p}}A^{t_1t_2\cdots t_s}_{q_1q_2\cdots q_p} \tag{C.20}$$

C.6 Symmetric and Skew-symmetric Tensors

The order of indices in a tensor is important. In general,

$$A^{ij} \neq A^{ji} \tag{C.21}$$

If, however,

$$\left.\begin{aligned} B_k^{ij} &= -B_k^{ji} \\ \text{or,} \quad B_{ij}^k &= -B_{ji}^k \end{aligned}\right\} \tag{C.22}$$

then we say that the tensors B_k^{ij} or B_{ij}^k are ***skew-symmetric.***

Thus, if two contravariant or two covariant indices are interchanged, each component of a skew-symmetric tensor alters in sign but not in magnitude.

When all the indices of a tensor can be interchanged without altering it, the tensor is called ***symmetric***.

If a tensor is symmetric with respect to two indices, it will remain so in any other coordinate system with respect to those two indices. If we assume that A^{ij} is symmetric, then

$$A^{ij} = A^{ji} \tag{C.23}$$

Now, by Eqn (C.16), we have

$$\bar{A}^{ij} = \frac{\partial \bar{x}^i}{\partial x^k}\frac{\partial \bar{x}^j}{\partial x^m}A^{km} = \frac{\partial \bar{x}^j}{\partial x^m}\frac{\partial \bar{x}^i}{\partial x^k}A^{mk} = \bar{A}^{ji}$$

Thus,

$$\bar{A}^{ij} = \bar{A}^{ji} \tag{C.24}$$

Therefore, we have proved the property of symmetry for a contravariant tensor of second order in another coordinate system.

Similarly, it can be shown that the property of skew symmetry is independent of the coordinate system.

A symmetric tensor of second order has at most $\frac{1}{2}(N+1)N$ different components. If A^{ij} is symmetric, then

$$A^{12} = A^{21},\ A^{13} = A^{31},\ \ldots,\ \text{etc.}$$

The total number of elements $= N^2$.

There are quantities A^{ii} (no summation) of N numbers.

Thus, the total number of symmetric set $= N^2 - N$.

Hence, maximum number of different component is

$$\frac{1}{2}(N^2 - N) + N = \frac{1}{2}N(N+1)$$

Similarly, a skew-symmetric tensor A^{ij} of second order has at most $\frac{1}{2}N(N-1)$ different components. Because, in this case,

$$A^{11} = A^{22} = A^{33} = \cdots = 0$$

$$\text{and } A^{12} = -A^{21},\ A^{13} = -A^{31},\ \cdots \text{etc.}$$

TENSOR ALGEBRA

C.7 Addition (Subtraction) of Tensors

Two tensors are said to be of ***same type*** if they have same number of contravariant and covariant indices. Thus, A^i_{jk} and B^i_{jk} are same type of tensors. These tensors can be added and subtracted to produce a single tensor. Thus, the ***sum*** and ***difference*** of tensors A^i_{jk} and B^i_{jk} are respectively the tensors

$$C^i_{jk} = A^i_{jk} + B^i_{jk};\ D^i_{jk} = A^i_{jk} - B^i_{jk} \tag{C.25}$$

Let us consider the covariant tensor A_{ij} of second order. We can write it as

$$A_{ij} = \frac{1}{2}(A_{ij} + A_{ji}) + \frac{1}{2}(A_{ij} - A_{ji}) \tag{C.26}$$

Now

$$\frac{1}{2}(A_{ij} + A_{ji}) \quad \text{is symmetric}$$
$$\text{and } \frac{1}{2}(A_{ij} - A_{ji}) \quad \text{is skew-symmetric.}$$

Thus, any covariant tensor of second order is the sum of a symmetric and a skew-symmetric tensors. We can show that this is also true for a contravariant tensor of second order.

C.8 Multiplication of Tensors

The ***outer product*** of two tensors denoted by B^{ij}_{klm} and C^{qrs}_{uvwx} is defined as

$$A^{ijqrs}_{klmuvwx} = B^{ij}_{klm} C^{qrs}_{uvwx} \tag{C.27}$$

The division of one tensor of rank greater than zero by another tensor of rank greater than zero is not defined.

C.9 Contraction of Tensors

The process by which we can reduce the number of covariant and contravariant indices of a mixed tensor by one is known as ***contraction***. As an example, we consider a mixed tensor A^{ij}_{klm}. By the Eqn (C.20), we write:

$$\bar{A}^{st}_{pqr} = \frac{\partial \bar{x}^s}{\partial x^i}\frac{\partial \bar{x}^t}{\partial x^j}\frac{\partial x^k}{\partial \bar{x}^p}\frac{\partial x^l}{\partial \bar{x}^q}\frac{\partial x^m}{\partial \bar{x}^r} A^{ij}_{klm} \tag{C.28}$$

Hence,

$$\bar{A}^{sr}_{pqr} = \frac{\partial \bar{x}^s}{\partial x^i}\frac{\partial \bar{x}^r}{\partial x^j}\frac{\partial x^k}{\partial \bar{x}^p}\frac{\partial x^l}{\partial \bar{x}^q}\frac{\partial x^m}{\partial \bar{x}^r} A^{ij}_{klm}$$
$$= \frac{\partial \bar{x}^s}{\partial x^i}\frac{\partial x^k}{\partial \bar{x}^p}\frac{\partial x^l}{\partial \bar{x}^q}\delta^m_j A^{ij}_{klm} = \frac{\partial \bar{x}^s}{\partial x^i}\frac{\partial x^k}{\partial \bar{x}^p}\frac{\partial x^l}{\partial \bar{x}^q} A^{im}_{klm} \tag{C.29}$$

Thus, we have obtained a mixed tensor whose covariant and contravariant indices have been reduced by one. Hence, the process of contraction enables us to obtain a tensor of order $(r-2)$ from a mixed tensor of order r.

C.10 Inner Product of Tensors

The process of combining outer multiplication and contraction to produce new tensors is called ***inner multiplication*** of two tensors. The resulting tensor is called the ***inner product*** of the two tensors.

For example, we consider two tensors A^{ij}_k and B^p_{qrs}. The inner product of these two tensors can be written as

$$C^{ij}_{qrs} = A^{ij}_k B^k_{qrs} \tag{C.30}$$

C.11 Quotient Law

We know that a set of functions forming the components of a tensor should satisfy the transformation equations of the type Eqn (C.20). This is the direct method, although practically not always feasible, to ascertain whether a set of functions forms the components of a tensor.

The ***quotient law*** is a simple indirect method to ascertain whether a set of functions forms the components of a tensor. The quotient law states that—

If there are N^r functions of coordinates x^i forming the components of a tensor of order r, then the inner product of these functions with an ***arbitrary*** tensor is itself a tensor. As an example, consider the set of N^3 functions A^{ijk}. These will form the components of a tensor of the type indicated by the indices, if

$$A^{ijk} B^r_{ij} = C^{rk} \tag{C.31}$$

provided that (i) B^r_{ij} is an ***arbitrary*** tensor and (ii) C^{rk} is a tensor.

In a new $\bar{x}^i$ coordinate system, Eqn (C.31) assumes the form:

$$\bar{C}^{rk} = \bar{A}^{ijk} \bar{B}^r_{ij} \tag{C.32}$$

Using Eqn (C.20), we get

$$\frac{\partial \bar{x}^r}{\partial x^l} \frac{\partial \bar{x}^k}{\partial x^m} C^{lm} = \bar{A}^{ijk} \frac{\partial \bar{x}^r}{\partial x^u} \frac{\partial x^s}{\partial \bar{x}^i} \frac{\partial x^t}{\partial \bar{x}^j} B^u_{st} \tag{C.33}$$

Substituting Eqn (C.31) into Eqn (C.33) and making appropriate change of indices, that is,

$$C^{lm} = A^{stm} B^l_{st} \tag{C.34}$$

we get

$$\frac{\partial \bar{x}^r}{\partial x^l} \left[\bar{A}^{ijk} \frac{\partial x^s}{\partial \bar{x}^i} \frac{\partial x^t}{\partial \bar{x}^j} - \frac{\partial \bar{x}^k}{\partial x^m} A^{stm} \right] B^l_{st} = 0 \tag{C.35}$$

We multiply the Eqn (C.35) by $\partial x^\alpha / \partial \bar{x}^r$, and sum over r, we get

$$\left[\bar{A}^{ijk} \frac{\partial x^s}{\partial \bar{x}^i} \frac{\partial x^t}{\partial \bar{x}^j} - \frac{\partial \bar{x}^k}{\partial x^m} A^{stm} \right] B^\alpha_{st} = 0 \tag{C.36}$$

Since, B^α_{st} is arbitrary, we must have

$$\bar{A}^{ijk} \frac{\partial x^s}{\partial \bar{x}^i} \frac{\partial x^t}{\partial \bar{x}^j} = \frac{\partial \bar{x}^k}{\partial x^m} A^{stm} \tag{C.37}$$

Inner multiplication of this equation by $\dfrac{\partial \bar{x}^r}{\partial x^s} \dfrac{\partial \bar{x}^g}{\partial x^t}$ yields the result

$$\bar{A}^{rgk} = \frac{\partial \bar{x}^r}{\partial x^s} \frac{\partial \bar{x}^g}{\partial x^t} \frac{\partial \bar{x}^k}{\partial x^m} A^{stm} \tag{C.38}$$

Thus, A^{stm} is a tensor of third order having all contravariant indices. It is important to note that the tensor B^l_{st} should be arbitrary without having symmetric and skew-symmetric properties. Otherwise, we cannot get the Eqn (C.37).

C.12 Conjugate Tensors

The covariant tensor A_{ij} and the contravariant tensor B^{ij} are said to be ***conjugate*** tensors, if they satisfy the equation

$$A_{ij}B^{ij} = \delta^k_j \tag{C.39}$$

A tensor of second order has a conjugate, only if the determinant formed by its components is not zero.

LINE ELEMENT

C.13 The Metric Tensor

Let us consider two neighbouring points x^i and $x^i + dx^i$ in the space V_N. The distance between these two points is given by the quadratic differential form:

$$ds^2 = g_{ij}dx^i dx^j \tag{C.40}$$

where it is assumed that (i) the g_{ij} are functions of x^i, (ii) g_{ij} is symmetric covariant tensor, i.e., $g_{ij} = g_{ji}$; (iii) $g = \det|g_{ij}| \neq 0$.

In this case, the space is said to be ***Riemannian space***, and g_{ij}, the symmetric covariant tensor of second order is called the ***fundamental tensor*** or the ***metric tensor*** of the Riemannian space. The quadratic form in Eqn (C.40) is called the ***metric***.

In Euclidean space V_3, referred to rectangular Cartesian coordinates, we write:

$$ds^2 = \left(dx^1\right)^2 + \left(dx^2\right)^2 + \left(dx^3\right)^2 \tag{C.41}$$

so that

$$g_{11} = g_{22} = g_{33} = 1, \; g_{ij} = 0 \tag{C.42}$$

In cylindrical coordinates (r, θ, z), we have $ds^2 = dr^2 + r^2 d\theta^2 + dz^2$ which, upon writing $r = x^1$, $\theta = x^2$, $z = x^3$, becomes

$$ds^2 = \left(dx^1\right)^2 + \left(x^1\right)^2 \left(dx^2\right)^2 + \left(dx^3\right)^2 \tag{C.43}$$

Thus, here

$$g_{11} = 1, g_{22} = \left(x^1\right)^2, g_{33} = 1, g_{ij} = 0 \tag{C.44}$$

Similarly, in spherical coordinates (r, θ, ϕ),

$$\begin{aligned} ds^2 &= dr^2 + r^2 d\theta^2 + (r \sin\theta)^2 d\phi^2 \\ &= \left(dx^1\right)^2 + \left(x^1\right)^2 \left(dx^2\right)^2 + \left(x^1 \sin x^2\right)^2 \left(dx^3\right)^2 \end{aligned} \tag{C.45}$$

so that

$$g_{11} = 1, \; g_{22} = \left(x^1\right)^2, \; g_{33} = \left(x^1 \sin x^2\right)^2, \; g_{ij} = 0 \tag{C.46}$$

We say that ds^2 is ***positive-definite***, if it is positive for all the values of the differentials dx^i, not all equal to zero. Thus, from the equations (C.41)–(C.46), we find that, the metric of a Euclidean space is positive-definite in all the coordinate systems.

In the discussion of special relativity theory, we consider a V_4 space, where the line element ds is given by

$$ds^2 = -\left(dx^1\right)^2 - \left(dx^2\right)^2 - \left(dx^3\right)^2 + c^2\left(dx^4\right)^2 \tag{C.47}$$

Thus, the metric is positive for curves along which x^1, x^2, x^3 are constants and negative for curves along which x^4 is constant. In order that the distance ds between two points is real, we write the Eqn (C.40) as

$$ds^2 = e g_{ij} dx^i dx^j \tag{C.48}$$

where the factor e takes the values $+1$ or -1 so that ds^2 is always positive. This factor is called ***indicator***.

C.14 Conjugate of Metric Tensors

Let us consider det $|g_{ij}|$ whose value is g.

Let G^{ji} be the cofactor of g_{ij}.

Since g_{ij} is symmetric, $g_{ij} = g_{ji}$, hence $G^{ji} = G^{ij}$.

Then the normalised cofactor, denoted by g^{ij} is $g^{ij} = \dfrac{1}{g}G^{ij}$.

Thus, we have

$$g^{ij} g_{jk} = \delta^i_k \tag{C.49}$$

so that by Eqn (C.39), g^{ij} and g_{ij} are conjugate tensors. But, we have to prove that g^{ij} is a contravariant tensor of second order. To do so, we introduce an arbitrary contravariant vector A^i, and put

$$B_i = g_{ij} A^j \tag{C.50}$$

We multiply Eqn (C.50) by g^{ik}, then

$$g^{ik} B_i = g^{ik} g_{ij} A^j = \delta^k_j A^j = A^k \tag{C.51}$$

Since A^i is arbitrary, so must be B_i. Hence, from quotient law, g^{ik} is a contravariant tensor of second order.

The tensors g_{ij} and g^{ij} are called ***conjugate tensors***.

EXAMPLE C.3 *Find g^{ij} in spherical coordinate system.*

Solution. In spherical coordinate, we have

$$g_{11} = 1, g_{22} = \left(x^1\right)^2, g_{33} = \left(x^1 \sin x^2\right)^2, g_{ij} = 0$$

Hence,

$$g = \det|g_{ij}| = g_{11} g_{22} g_{33} = \left(x^1\right)^2 \left(x^1 \sin x^2\right)^2.$$

Cofactor of $g_{11} = g_{22} g_{33}$.

Hence,

$$g^{11} = \frac{1}{g} g_{22} g_{33} = \frac{1}{g_{11}} = 1.$$

Similarly,

$$g^{22} = \frac{1}{g_{22}} = \frac{1}{\left(x^1\right)^2};\ g^{33} = \frac{1}{g_{33}} = \frac{1}{\left(x^1 \sin x^2\right)^2}.$$

EXAMPLE C.4 *Find g^{ij}, if $ds^2 = (dx^1)^2 + 2\cos\xi dx^1 dx^2 + (dx^2)^2$.*

Solution. Here $g_{11} = 1,\ g_{22} = 1,\ g_{12} = g_{21} = \cos\xi$.

Hence,

$$g = \begin{vmatrix} g_{11} & g_{12} \\ g_{21} & g_{22} \end{vmatrix} = \begin{vmatrix} 1 & \cos\xi \\ \cos\xi & 1 \end{vmatrix} = \sin^2\xi.$$

Thus,

$$g^{11} = \frac{g_{22}}{g} = \frac{1}{\sin^2\xi},\ g^{21} = g^{12} = \frac{-\cos\xi}{\sin^2\xi},\ g^{22} = \frac{1}{\sin^2\xi}.$$

C.15 Length of a Curve

Let a curve be given by the equation

$$x^i = x^i(t) \tag{C.52}$$

where t is the parameter.

Then the length of the curve between the points corresponding to $t = t_0$ and $t = t_1$ is defined by

$$s = \int_{t_0}^{t_1} \sqrt{e g_{ij} \frac{dx^i}{dt}\frac{dx^j}{dt}}\, dt \tag{C.53}$$

If at all points of the curve,

$$g_{ij}\frac{dx^i}{dt}\frac{dx^j}{dt} = 0 \tag{C.54}$$

the curve is called a ***null line***.

In Riemannian space whose metric is positive-definite, no real null line exists. But in V_4, there exists some real curves, which are null lines.

These are given by

$$\left.\begin{aligned} x^1 &= c\int r\cos\theta\cos\phi dt, & x^2 &= c\int r\cos\theta\sin\phi dt \\ x^3 &= c\int r\sin\theta dt, & x^4 &= \int rdt \end{aligned}\right\} \tag{C.55}$$

where r, θ, ϕ are functions of t.

C.16 Magnitude of a Vector

The ***magnitude*** A of a contravariant vector A^i is defined by

$$(A)^2 = e g_{ij} A^i A^j \tag{C.56}$$

The magnitude A is invariant.

The magnitude B of a covariant vector B_i is defined by

$$(B)^2 = e g^{ij} B_i B_j \tag{C.57}$$

If A or B equals to one, then the corresponding vector is called a ***unit vector***.

In the Euclidean space under Cartesian coordinates, Eqn (C.56) gives the magnitude of a vector (a, b, c) as $\sqrt{a^2 + b^2 + c^2}$, which is expected.

C.17 Associate Tensors

Let us consider a contravariant vector A^j. We define its ***associate***, the covariant vector A_i given by

$$A_i = g_{ij} A^j \tag{C.58}$$

Similarly, we say that the vector B^i is an associate to the vector B_j, if

$$B^i = g^{ij} B_j \tag{C.59}$$

The process in Eqn (C.58) is called ***lowering*** the superscript, and that in Eqn (C.59) is called ***raising*** the subscripts. These processes may be performed on higher rank tensor also.

C.18 Angle between Two Vectors

We consider unit vectors A^j and B^i. Then the angle θ between them is defined by

$$\cos\theta = g_{ij} A^i B^j = A_j B^j = g^{jk} A_j B_k = A^k B_k \tag{C.60}$$

In Euclidean V_3 space under Cartesian coordinate system, this equation corresponds to

$$\cos\theta = ll' + mm' + nn' \tag{C.61}$$

where (l, m, n) and (l', m', n') are unit vectors.

If the metric of the Riemannian space is positive-definite, Eqn (C.60) always defines a real angle. In fact, for all real values of λ and μ, the magnitude of the vector $\lambda A^i + \mu B^i$ is ≥ 0.

That is, $g_{ij}(\lambda A^i + \mu B^i)(\lambda A^j + \mu B^j) \geq 0$, which reduces to

$$\lambda^2 + 2\lambda\mu\cos\theta + \mu^2 \geq 0 \tag{C.62}$$

The discriminant of this quadratic should be negative or zero. Hence,

$$\cos^2\theta - 1 \leq 0 \text{ or, } \cos^2\theta \leq 1 \tag{C.63}$$

Hence, θ is real.

If A^i and B^i are not unit vectors, then the angle between them is given by

$$\cos\theta = \left(g_{ij} A^i B^j\right) / (AB) \tag{C.64}$$

where A and B are magnitudes of the vectors A^i and B^j taken to be real and positive.

Two vectors are said to be ***orthogonal*** to one another if the angle between them is a right angle. In this case, Eqn (C.64) yields $g_{ij} A^i B^j = 0$.

Bibliography

[1] Goldstein, H : *Classical Mechanics.* Addison-Wesley, Reading, Massachusetts, 1950.

[2] Kibble, T.W. : *Classical Mechanics.* McGraw-Hill, New York, 1966.

[3] Rana, N.C. and Joag, P.S. : *Classical Mechanics.* Tata McGraw-Hill Publishing Co. Ltd., New Delhi, 1994.

[4] Knudsen, J.M. and Hjorth, P.G. : *Elements of Newtonian Mechanics.* Springer-Verlag, Berlin, 1995.

[5] Bohm, D. : *Special Theory of Relativity*, W.A. Benjamin, Inc., New York, 1965.

[6] Kacser, C. : *Introduction to Special Theory of Relativity.* Englewood Cliffs, Prentice-Hall, Inc., N.J., 1967.

[7] Bhattacharyya, B. : *Mathematical Physics.* New Central Book Agency (P) Ltd., Kolkata, 2009.

[8] Rev. Father Goreux, F. : *Introduction to Tensor Calculus*, New Central Book Agency (P) Ltd., Kolkata, 1992.

[9] Besant, W.H. and Ramsey, A.S. : *A Treatise on Hydro-mechanics*, G. Bell and Sons Ltd., London, 1962.

[10] Bhattacharyya, B. : *Fluid Dynamics*, New Central Book Agency (P) Ltd., Kolkata, 2009.

[11] Krylov, N.M. and Bogolubov, N.N. : *Introduction to Non-linear Mechanics* (in Russian), AN USSR, Kieve, 1937.

[12] Ramsey, A.S. : *An Introduction to the Theory of Newtonian Attraction*, University Prss, Cambridge, 1956.

Index

Some exciting gifts are up for grabs with this book!!

AN INTRODUCTION TO CLASSICAL MECHANICS

Dr Binoy Bhattacharyya

FREE Gifts Rs. 1000/-

Dear Readers

By now you must have gone through this book in the entirety. So we invite your assessment on this title with absolute impartiality and candid observations. After all, we publishers faithfully depend on your benchmarking. Friends, please fill up the open spaces below with CAPITAL LETTERS. You may add some extra sheets of paper if the space provided is not sufficient for you.

i. What are the important topics of your syllabus which are not adequately covered in this book?

ii. In which chapters and on what topics the treatment of the subject-matter by the author is not satisfactory and why?

iii. Have you come across the typos, factual inaccuracies, inconsistencies, and other forms of mistakes in the book? If yes, then please specify the mistakes and page numbers.

iv. Name top three books on the same subject in your consideration. Could you provide some reasons?

Your 1st Preference: ....

Your 2nd Preference: ....

Your 3rd Preference: ....

v. Any suggestion to improve the quality of illustrations, diagrams, etc.?

vi. How did you like the cover design? Any suggestion on colours?

vii. Suggestions for further improvement of production-standard:

viii. Who recommended you this book:

☐ Teacher ☐ Friend ☐ Parents ☐ Bookseller ☐ Librarian

ix. Name of the person who has recommended:

Designation: ....

Address: ....

Phone No.: ....

e-mail: ....

x. Name and address of the bookseller from whom you have purchased this book:

....

....

....

xi. Name and address of the library where you would like to find the book:

....

....

xii. Name and address of your institution (Please mention the University or Board):

....

....

xiii. Your name, e-mail, and complete postal address:

....

....

....

xiv. A few other titles published by us that you would like to read:

....

....

The best assessment will be rewarded half-yearly. The award will be in the form of our publications, as decided by the Editorial and Marketing Operations Division, amounting to Rs. 1000/- (Rupees one thousand) only.

....

Please mail the filled up coupon at your earliest to :

Editorial & Marketing Operations Division

NEW CENTRAL BOOK AGENCY (P) LTD

8/1 Chintamoni Das Lane, Kolkata 700 009, West Bengal, INDIA

Ph. No.: 033 + 22418411, 22418573, 22419094

Fax: 033 + 22419573, 22413852

e-mail: ncbapvtltd@eth.net ◆. ***Website:*** www.ncbabooks.com